Topical Problems in Solid Mechanics

Editors

Prof. N.K. Gupta
Henry Ford Chair Emeritus Professor
Department of Applied Mechanics
Indian Institute of Technology Delhi, New Delhi, India

Prof. A.V. Manzhirov
Ishlinsky Institute for Problems in Mechanics of
the Russian Academy of Sciences, Moscow, Russia

Elite Publishing House Pvt. Ltd.

302, JMD House, 4-B, Ansari Road, Daryaganj, New Delhi-110002 Phone: 41004299 E-mail: rg_elite@yahoo.com

ISBN: 81-88901-36-9

Published by R. Gangadharan for Elite Publishing House Pvt. Ltd.
302, JMD House, 4-B, Ansari Road, Daryaganj, New Delhi-110002
Phone : 91-11-41004299, 91-11-26194323
E-mail: rg_elite@yahoo.com

Graphics: Elite Studio, New Delhi-110002
Printed at Elite Offset, New Delhi-110002

PRINTED IN INDIA

PREFACE

Topical problems in solid mechanics contains the papers, which were presented in the Indo–Russian workshop held on 11–14 November 2008 at the BIT Pilani, Goa Campus. This workshop was sponsored by the Department of Science and Technology, Delhi, and the Russian Foundation for Basic Research, Moscow.

The papers presented in this workshop deal with topics which include damage and fracture mechanics, nano- and micromechanics, contact mechanics, dynamics, elasto–viscoplasticity, mechanics of accreted solids, biomechanics, mechanics of composites, and thermo–mechanics.

A good number of Russian scientists from Russian Academy of Sciences Institutes and leading Russian Universities and Indian scientists from various Indian Institutes of Technology and other academic institutes as well as from various research laboratories participated and delivered lectures during this workshop.

The main aim of the workshop was to provide a forum to the scientists for getting familiar with the latest advances made in various aspects of the subject in both the countries, and to facilitate personal interaction between Russian and Indian scientists working in the area.

We hope that coming together of the scientists from Russia and India during this meeting and the discussions that they had during the four days of the workshop would have provided impetus to the participating scientists for initiating joint research activities in the frontiers of experimental, analytical and numerical researches in various topical problems in solid mechanics and in formulating several long term collaborative research projects between the scientists from institutions in both the countries.

We thank Dr. K.E. Kazakov, Dr. D.A. Parshin, and Mr. A.L. Levitin for their help in preparing the manuscripts for publication.

N.K. Gupta,
A.V. Manzhirov

Coordinators:

Prof. N.K. Gupta
Indian Institute of Technology Delhi, New Delhi, India

Prof. A.V. Manzhirov
Ishlinsky Institute for Problems in Mechanics
of the Russian Academy of Sciences, Moscow, Russia

Prof. K.E. Raman
BITS Pilani — Goa Campus, India

Organizing Committee

Hosting Institute

BITS Pilani, Goa Campus,
Zuarinagar — 403726, GOA, INDIA

CONTENTS

WORKSHOP ON TOPICAL PROBLEMS IN SOLID MECHANICS. INAUGURAL ADDRESS

D.V. Singh

Prof. N.K. Gupta, Prof. A.V. Manzhirov, distinguished scientists from Russia and India, Ladies and Gentlemen. I consider it an honour and a privilege to be invited to this Indo-Russian Workshop on Topical Problems in Solid Mechanics.

The world around us is built and held together by solid materials. Understanding their behaviour is in the domain of solid mechanics, which includes many diverse areas. The variety of materials (metals, rocks, glasses, sand, flesh and bone) and their properties (porosity, viscosity, elasticity, plasticity) are reflected by the concepts and techniques needed to understand them through mathematics, physics and experiments. Solid mechanics has a very wide sweep. It is generally understood to encompass composites, biomechanics, crash worthiness, experimental mechanics, computational mechanics, dynamics, behavior of materials, micro- and nano-mechanics, wave propagation, fracture and fatigue, and continuum mechanics. Industrial product development involves modeling and simulation of manufacturing processes as well as analysis of product functionality, for example, analysis of the material behaviour when hardened under very rapid cooling, or material response under high strain rates. Solid Mechanics is a classic engineering subject that historically has played a major role in the development of industry and infrastructure. It is also a modern science that together with related scientific areas provides the basis for development of advanced materials and mechanical components of the future through mathematical description of the mechanisms that determine deformation and fracture of materials and components subjected to load, using analytical, experimental and numerical methods, and also inverse methods to cover material properties on the micro-scale level up to the function of systems and components in industrial applications.

Solid Mechanics research addresses issues related to impact resistance and notch sensitivity for critical applications, such as gas turbine materials, large deformation, residual stresses and residual deformation in solids after fabrication, and the mechanics of thin bonded films and coatings. Mathematical models for accurate prediction of fracture formation can help in prevention of catastrophic failure of materials, some of which under cyclic loading experience formation of dislocation ladders and banded patterns. The mathematical simulation and reliable prediction of material failure is particularly important in applications where safety is critical.

Development of theoretical models for various emerging materials, such as composites, smart materials, nano-materials and biomaterials, is the main focus in the current research of solid mechanics. Applying the basic mechanics principles and material models, together with modern computational tools, in the design and analysis of mechanical components and systems has gained increasing importance in many industries, such as automobile, aerospace, consumer product, electronics and many others.

Solid Mechanics is concerned with deformation, stress, stability, fatigue and fracture of materials and structures under various load conditions. Some important issues addressed by solid mechanics include characterization of new materials, safety analysis of machine components, optimal design for stronger and light-weight mechanical systems, and mechanics of biological materials such as cells, tissues and bones. Analytical, experimental and computational methods are often employed in the research of solid mechanics at the nano, micro, and macro length scales.

Microstructure morphology gives insight into the properties of alloys and the microstructure that arise when the crystal lattice structures of these materials change their geometric shapes at certain temperatures. This will help in understanding microscopic and macroscopic changes in the alloys and for designing new materials and nano-devices. Properties of nano-size objects involve surface contribution to the thermodynamic potential for bulk and nano-dimensional particles of ion crystals, which require estimation with the help of the electron statistical theory of ion crystal lattice. Several size effects are associated with the excess surface energy of ultra disperse particles. In particular, there is a stability loss in the crystal lattice upon the transition of the surface energy to the range of negative values under high pressure.

Estimation of strength of nano-dimensional objects (nanotubes) contained in the form of reinforcing elements of special composites is important for engineering applications. Suitable experiment methodology is required and experiment parameters are to be selected considering the nano-tube deformation processes and its interaction with the matrix, which reflect the specific properties of materials in the scale of nanometer dimensions. The strength of reinforcing elements relate to the fracture mechanism in the transition when the tubes get pulled out of the matrix. Variation in the viscous matrix strain rate in the reinforced filament and variation in the energy of the chemical interaction between the nanotube and the rigid matrix of nano-objects need to be considered to estimate fracture stresses in the glue film reinforced by nano-tubes. The critical conditions of the fracture mechanism change in the end separation region correspond to the effective specific work of structural failure.

The behavior of new and advanced materials, which include: ceramics, polymers, metallic and intermetallic composites, and metallic alloys presents unique challenges, which range from a fundamental characterization of the defect behavior controlling basic flow and fracture processes, to understanding the interactions between the constitutive behavior of various components in composite systems, and to describing the evolution of damage induced by micro-structural instabilities, complex loading and environmental attack. Success requires combining fundamental knowledge of material behavior with novel analysis methods of mechanics.

Engineering issues relate to the fabrication and performance of materials for advanced thermo-structural systems, which require research involving experiment and modeling to tackle problems related to the use of advanced materials in load bearing applications. The materials include advanced ceramics, composites, films/coatings and cellular materials, as well as interfaces and their adhesion. The technologies are diverse: (a) high temperature materials, (b) materials for thermal management in hypersonic vehicles, ships and power electronics, (c) materials for ultra light components used in aerospace systems, (d) materials having especially low weight and compactness through multifunctionality, (e) materials and systems for actuation and shape morphing, (f) materials and structures that impart blast and fragment protection.

Methods using advanced fracture mechanics and statistical reliability assessment procedures for predicting safe lifetime limits in structural components for a wide range of engineering systems are essential for design and operation of new engineering systems as well as lifetime prediction of existing aging systems. The research integrates modeling and experiments to develop a rigorous understanding of microstructural, dimensional and mechanical property changes experienced by materials used in hostile environments. The research emphasis is not only on traditional materials but also on the application of advanced materials where the effects of environmental degradation are much less understood.

Micro-defects such as dislocations and macro-cracking should be controlled during the crystal growth process to obtain high-quality bulk single crystals. Solid mechanics and material strength studies on the single crystals are of importance to solve the problems related to the generation and multiplication of dislocations and the cracking of single crystals. Research will give greater understanding into thermal stresses and dislocation densities during crystal growth process and the cracking of single crystal due to thermal stress.

Experimental techniques in solid mechanics are becoming more and more sophisticated. Three-dimensional images generated by X-ray micro tomography give a view of the inside of the material and make it possible to construct 3D representations of the microstructure without inducing any damage. A miniature compression machine in conjunction allows imaging samples while being loaded. Measurement of the displacements inside the material is possible by digital volume correlation digitally to obtain the mechanical properties of the matrix inside a composite material. New experimental techniques are being developed that will enable quantification of the residual deformation and corresponding stresses in the close vicinity of impact induced edge notches on aero engine components. The methodology relies upon the high speed photography, image processing and numerical simulation of the conducted experiments based upon optimal material properties obtained from material tests at high rates of strain. The results can be used for fatigue life predictions.

For many advanced mechanical products, processes and production systems, e.g., in precision technology, MEMS, NEMS and vehicle systems, it is important to predict their dynamic behavior with very high accuracy. On the required scale of accuracy in many of these systems, friction and contact nonlinearities have a substantial effect on their dynamic behavior. In such cases prediction of dynamic behavior requires consideration of these nonlinearities. Some of the challenging topics in this area are impact systems, synchronization in nonlinear systems, nonlinear stochastic systems and control of nonlinear dynamic systems.

Understanding of soft tissue deformation and growth in the human body, their detection and imaging of tumors, especially in the breast and colon can be invaluable in the practice of medicine. It is often difficult to interpret and to match images taken of the breast with ultrasound, MRI and X-ray mammography because the breast is distorted differently during each type of test. Mathematical models of breast deformation and tissue elasticity can provide the basis to overcome this obstacle. Bio-materials are more difficult to understand and to predict than alloys because they frequently contain residual stress, are inhomogeneous, may be actively moving or growing, and can have extensive deformation.

Fluid-structure interaction has become important in various fields, both on small scales, e.g., in the medical context, or on large scales e.g., oil platforms, for which a two-way coupling is essential, to deal with back reaction of the flexible structure on the flow. The coupling manifests itself in many other areas, such as porous media, rheology, flow of granular materials, biomedical flow, or acoustic radiation. It is therefore clear that the borders between fluid and solid mechanics are fading away. In addition, the interactions with other areas of engineering sciences, such as materials technology, thermodynamics and systems and control have become increasingly important.

Scientific computing is central to both solid and fluid mechanics. Serious progress in these areas requires significant advances in the computational efficiency and robustness of the underlying numerical algorithms. More efficient numerical algorithms frequently have a much larger impact than simple increases in hardware speed, which is becoming increasingly more difficult to achieve. Important areas of research are more accurate and efficient discretization techniques for partial differential equations. In particular, solution adaptive methods with advanced error control are very promising for problems with a wide range of scales. In addition, further advances in multigrid and sparse grid techniques, including efficient preconditioning and operator splitting, are essential for the efficient solution of the very large algebraic systems encountered in these fields. Large-scale computational techniques for the solution of problems in solid mechanics and optimization are often needed. The problems have a wide range from development of numerical tools for optimal design to computer simulation of manufacturing processes.

Solid mechanics integrates knowledge in material behavior, which now include micro-mechanics, continuum mechanics and nano-scale modeling; component performance and reliability that includes dynamics, stochastic and probabilistic analysis; and structural design. The unifying theme that connects these areas is the study of material and structural behaviors that encompass multiple length-scales. Research into behavior at the molecular scale provides unique insight to behavior at the micro scale, which in turn affects macroscopic response. Such multi-scale interactions occur in all sub-disciplines of solid mechanics and require multi-scale models and experiments to provide the physical and mathematical foundations to develop predictive capabilities to design complex structural systems

Advancing the mathematics and computation of solid mechanics is the tool which can open new frontiers in the studies of solids and enhance the understanding of alloys, composites, nano-materials and bio-materials. I am sure that this Indo-Russian Workshop on Topical Problems in Solid Mechanics will stimulate the study of mathematical solid mechanics, particularly in the key areas such as microstructure morphology; fracture; nano-materials; bio-materials; and applications in medicine. These areas share more or less the same governing equations as those of nonlinear elasticity theory and lessons learnt in one area will prove fruitful in the others.

This Workshop is expected to open avenues for joint research activities of Russian and Indian scientists and will foster through cooperative effort, the development of new analytical, mathematical and computational approaches and tools, which will contribute towards the training of a new generation of young scientists in this key field of engineering, physical and life sciences.

With great pleasure I inaugurate the Workshop with my best wishes for its success.

INDO–RUSSIAN COOPERATION IN SCIENCE AND TECHNOLOGY

Y.P. Kumar[1]

Science and Technology system in India with its robust and well defined structure is marching ahead to attain newer heights drawing advantages from front ranking research in Chemistry, Physics, biology, engineering and other disciplines. Research and Development budgets have seen significant increase during recent years. We are focusing on capacity building and developing cutting edge technologies. The culture of patenting is being inculcated and number of patents filed by the Indian scientists is steadily growing. Our scientists are equipped to compete globally even though they have large responsibility of evolving scientific solutions to address local problems of common man.

International cooperation is one of the important components of development of science. Until not very long ago it was confined to international workshops, seminars, meetings and few isolated one to one interactions. It took new dimensions in last two decades due to availability of information in public domain and revolution in information technology triggered by internet. Access to advanced technologies which was limited to a club of few is available online thereby hastening the process of growth of science. Pace of development of science and innovation to meet the aspirations of society is increasing day by day.

Involvement of like minded scientific groups of different regions/countries and availability of sophisticated precision instrumentation took the science to new heights with greater speed, though at much higher costs. Cost of scientific developments further increased due to its lower shelf life resulting from quick advancements. Today s research requires much higher inputs and gives short lived outputs. The solution lies in pooling of inputs and sharing the results.

Scientists share their efforts and resources online. Such network approach turns out proven result in a short span of time like structures of composites which are unimaginably stronger than its constituents. Of course the choice of constituents

[1]Advisor and Head International Cooperation Department of Science and Technology Govt. of India, New Delhi

and the methods of processing play an important role in determining the quality of the end product.

Indian and Russian Scientists have been interacting with each other for several decades. A major milestone of S&T cooperation between our two countries was conclusion of Integrated Long Term Programme of Cooperation (ILTP) on July 3, 1087. ILTP is one of the most exhaustive bilateral collaborative R&D programme covering all aspects of Science and Technology. While there is an emphasis on applied research and technology transfer, basic research is also equally important component of the programme.

ILTP provides for scientist to scientist and institute to institute interaction in frontier areas of Science and Technology. Main instrument of cooperation is joint R&D projects which are identified by the scientists and are supported by the two sides based on merit. Other modes of cooperation include holding of thematic workshops, organizing exploratory visits of senior scientists, fellowships for young Russian Scientists and participation in major scientific events.

Intense interaction between Indian and Russian scientists has led to creation of ten Centres of Excellence which are today engaged on scientific research as well as realization of joint accomplishments in industry. These Centres are:

- Advanced Research Centre For Power Metallurgy and New Materials, Hyderabad
- Oral Polio Vaccine Production Facility, Bulandshahar
- Advanced Computing Research Centre, Moscow
- Earth Quake Research Centre, New Delhi
- Gas Hydrates Science and Technology Centre, Chennai
- Ayurveda Research Centre, Moscow
- Biotechnology Centre, Allahabad
- Non-Ferrous Metallurgy Centre, Jamshedpur
- Biomedical Instrumentation S&T Centre, Trivandrum
- Laser and Accelerator Centre, Indore

With a view to supporting collaborative activities in basic research, India and Russia have embarked upon a new programme of cooperation which is supported by DST and Russian Foundation for Basic Research from Indian and Russian sides respectively. Through joint call for proposal, we have supported 33 projects under this Programme. In addition, we have agreed to support 7 joint workshops. Today's Workshop in the area of Topical problems in Solid Mechanics is one of the most important Workshop under this Programme. We have large participation both from Indian and Russian sides. Scientists would be deliberating on identified topics of mutual interest. The Workshop will not only lead to the greater understanding of the subject but also would bring Indian and Russian scientists together to jointly work on the problems of mutual interest. Joint projects selected through the process of next call for proposals, which is presently open, would be supported by DST and RFBR.

Today's Workshop is a brain child of two eminent scientists, Prof. N.K.Gupta from the Indian Institute of Technology, Delhi and Prof. A.V. Manzhirov from the Ishlinsky Institute for Problems in Mechanics, Moscow, coordinators from Indian and Russian sides. Their untiring efforts have brought such a large team of senior and young scientists in the field of applied mechanics together. On behalf of Department of Science and Technology, I would like to thank all the participants and specifically the coordinators from Indian and Russian sides. I will also like to thank our counterparts, Russian Foundation for Basic Research for cosponsoring this event. Organisation of an event of such magnitude involving participants from India as well as foreign country requires very meticulous planning, large managerial as well as implementation capability. I would like to take this opportunity to thank Prof. K.E. Raman, Director, Pilani and organizing team led by Prof. Pravin Singru for their continued efforts.

ON THE PROPERTIES OF NANODIMENSIONAL OBJECTS

A.V. Ankudinov[1], V.A. Eremeyev[1], E.A. Ivanova[1], and N.F. Morozov[1*]

ABSTRACT

In recent years, rapid development of nanotechnologies led to the necessity of constructing adequate physical models that make it possible to describe physico-mechanical properties of objects with a nanometersize (nanosize) scale. The interest to these problems is connected with the necessity of investigation of the mechanical deformation of nanotube devices, which are used intensively in the recent years in nanotechnology developments. The presented paper is devoted to problem of experimental determination of the stiffness characteristics of nano-objects. A theoretical foundation for the experimental determination of the bending stiffness of nano-shells is carried out. Proposed experimental method is based on measuring of the eigenfrequencies of nano-objects. If we try to measure eigenfrequencies of nano-objects by using the optical methods, we will have very many difficulties. The fact, that the laser ray is a spot of certain diameter (not a point), is the main problem. If the size of investigated object is smaller than the laser ray diameter, then the results of measuring does not have a sense. So, using of optical methods do not able to measure eigenfrequencies of nano-objects, but we can measure eigenfrequencies of the system, consisting of micro-substrate and regular structure of identical nano-objects. Two problems, topics of which are in the intersection of mechanics and experimental physics are discussed in this paper. The first problem is determination of elastic modulus of nano-objects in the case when the eigenfrequencies of the system consisting of micro-substrate and nano-objects and the elastic characteristics of the micro-substrate are known. The second problem is analysis of possibility to extract eigenfrequencies of nano-objects from the spectrum of the system consisting of micro-substrate and nano-objects. Success of solving of both these problems depends on conditions of experiment. That is why mechanics problem is not only interpretation of experimental data but and development of theoretical foundation of the experimental methods in the form of recommendations concerning mechanical aspects of the methods. At the present time investigation of properties of nano-objects is carried out by

[1]Institute for Problems in Mechanical Engineering of the Russian Academy of Sciences, St. Petersburg, Russia

[*]E-mail: *morozov@NM1016.spb.edu*

using of atomic force microscope (or others sound microscopes). Cantilever of AFM acts on the investigated object and it can change properties of the investigated object. Using AFM we measure not eigenfrequencies of investigated object but eigenfrequencies of the system consisting of AFM cantilever and investigated object. That is why studying of the problem of interaction of nano-object and AFM cantilever is very important. In fact, it is necessary to solve two problems. The first problem is determination of mechanical characteristics of nano-objects in the case when the eigenfrequencies of the system consisting of nano-objects and cantilever are known. The second problem is elaboration of such experiment conditions, which give possibility to extract eigenfrequencies of nano-objects from the spectrum of the system consisting of the nano-objects and AFM cantilever.

Key words: nanomechanics, natural vibrations, nanotubes, bending stiffness, shells, electric elasticity, finite elements method

1. EXPERIMENTAL METHOD OF DETERMINATION OF BENDING STIFFNESS OF NANOSHELLS

The problem of the experimental determination of elastic moduli of nanoscale objects is of present interest. The determination of the elastic moduli of thin macroscopic shells is usually based on experiments with plates. It is known that, when grown using certain techniques, nanoobjects are obtained only in the form of shells. Therefore, it is necessary to develop a method for determining the elastic moduli of nanoobjects on the basis of experiments with shells. Experimental determination of the bending stiffness of nanosize shells presents a serious problem, because for such widespread nanoobjects as nanotubes and fullerenes under arbitrary deformation, the material is subjected to both bending and tension. Therefore, all parameters (e.g., natural frequencies) that can be measured directly are complicated functions of both bending and tension stiffness. In recent years, together with nanotubes and fullerenes, nanoobjects of a more intricate configuration have been obtained [1–3]. Nanosize cylindrical helices [1] (see Fig. 1) are of particular interest in connection with the possible experimental determination of bending stiffness.

This is due to the fact that in helical shells under arbitrary deformation, the material is mainly bent, so that the material tension effect can be neglected when interpreting experimental data; and the natural oscillation shapes of helical shells are much more easily observed than those of cylindrical shells associated with pure bending of the material. The latter statement is illustrated in Fig. 2, which presents the first four helical shell oscillation shapes. The analysis of helical shell dynamics may be a theoretical foundation for experimental testing of the applicability of the continuum theory to (a) the calculation of mechanical characteristics of nanoobjects and (b) the experimental determination of the bending stiffness of nanoshells [4].

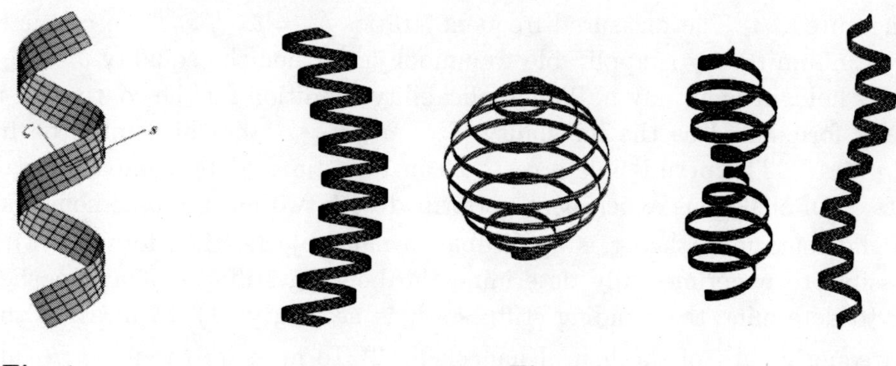

Fig. 1 **Fig. 2**

Let us take into consideration dimensionless frequency

$$\Omega = \sqrt{\frac{\rho R^4}{D}}\omega, \tag{1}$$

where, ρ is the surface mass density, R is radius of cylindrical helical shell, D is ending stiffness, and ω is eigenfrequency. It is shown, that dimensionless frequencies Ω_n depend on the number of frequency n and three dimensionless parameters

$$\Omega_n = \Omega_n\left(\alpha, \frac{l}{R}, \frac{a}{R}\right), \tag{2}$$

where $\alpha$is helix angle, $l$is helix-forming band length, a is helix-forming band width.

We will consider two thin helical shells with different physical and geometric characteristics but the same dimensionless parameters α, l/R, and a/R. We will assume that both shells are fixed at corners. In this case, the spectra of the dimensionless frequencies of shells under consideration coincide with (2). Then, in accordance with (1), the frequency ratio $\omega_n^{(1)}/\omega_n^{(2)}$ is independent of their ordinal number n

$$\frac{\omega_n^{(1)}}{\omega_n^{(2)}} = \sqrt{\frac{D_1 \rho_2 R_2^4}{D_2 \rho_1 R_1^4}}. \tag{3}$$

Relation (3) may serve as a theoretical basis for the experimental investigation of the applicability of the continuum theory to nanoobjects and, if the answer is affirmative, for experimental determination of the bending stiffness of nanoshells.

To test the applicability of the continuum theory to nanoobjects, the following measurements can be performed: 1) Several first eigenfrequencies of a helical nanoshell are measured. 2) The eigenfrequencies of a macroscopic helical shell with the same dimensionless parameters α, l/R, a/R, and the same fixation conditions are measured. 3) The measured frequency ratios $\delta_n = \omega_n^{(1)}/\omega_n^{(2)}$ are calculated. If the continuum theory is applicable to nanoobjects, then the equality $\delta_n = \delta_1$ theoretically holds true for any n. The applicability condition for the continuum theory is really formulated as the inequality $|\delta_n - \delta_1|/\delta_1 \leq \varepsilon_N$, which must be fulfilled for $\forall n \leq N$. The permissible error ε_N can be estimated by comparing with the results of an analogous experiment performed with two macroscopic helical shells.

If the continuum theory is applicable to nanoobjects, then formula (3) makes it possible to experimentally determine the bending stiffness of a nanoshell. In order to determine the bending stiffness, it is necessary: 1) To measure the first eigenfrequency $\omega_1^{(1)}$ of the helical nanoshell. 2) To measure the mass m_1 and the geometric dimensions l_1, a_1, and R_1 of the nanoshell and to calculate its surface density $\rho_1 = m_1/(l_1 a_1)$. 3) To determine the characteristics $\omega_1^{(2)}$, D_2, ρ_2, and R_2 of a compared macroscopic helical shell with the same dimensionless parameters α, l/R, a/R, and the same fixation conditions as those of the nanoshell under study. 4) To calculate the bending stiffness of the nanoshell D_1, using formula (3).

2. ON THE DETERMINATION OF THE RIGIDITY PARAMETERS OF NANOOBJECTS BY USING ATOMIC FORCE MICROSCOPE

Interpretation of experimental data is a challenging scientific problem. This problem becomes still more acute in studying nanometer-scale objects. This is because experimental conditions for nanoobjects differ radically from those for macroobjects. In the latter case, the dimensions of measuring devices (e.g., strain gages) are, as a rule, much less than those of an object being studied. Therefore, in experiments with macroobjects, measuring equipment has a negligible effect, if any, on the object and measurements give its true characteristics. In the case of nanoobjects, the situation is basically different, since microinstruments used in these experiments may strongly influence the object, changing its properties up to restructuring or even causing destruction. Eventually, measurements may represent the characteristics of the modified nanoobject or of the nanoobject–instrument system. Thus, to study the interplay between nanoobjects and measuring equipment is of great importance. Below, this problem is exemplified in determining the elastic properties of nanoobjects with an atomic force microscope (AFM).

The problem of finding the elastic moduli of nanometer-scale objects has become topical. Many researchers note that the values of elastic moduli derived from micro- and macroexperiments differ. In [5, 6], the Young's modulus and flexural rigidity of a 2D singlecrystal sheet was studied as a function of the number of atomic layers in the sheet. The results obtained in [5, 6] indicate that three expressions for the flexural rigidity of the sheet (the expression known from the continuum theory, the expression obtained by substituting Young's modulus into the expression from the continuum theory, and that based on a discrete model [6]) give results differing considerably for a small number of atomic layers. This means that the formulas from the continuum theory, which ignore the discreteness of the material properties across the thickness of the sheet, may yield erroneous results. Clearly, the discreteness along the length of the sheet, where the number of layers is large, is insignificant and the continuum expressions seem to apply well in this case. The same is true for analysis of nanorods and nanosheaths. Thus, it is important to devise techniques allowing for directly finding the elastic properties of thin-walled nanoobjects, i.e., without using formulas relating the elastic moduli of a nanoobject to its thickness and the Young's modulus of the material. Specifically, a burning issue is experimental determination of the mechanical properties of nanoobjects [7]. An efficient way of determining elastic moduli, which are used in macromechanics, consists in measuring the eigenfrequencies of an object under test (the resonance method). Below, details of applying the resonance method to nanoobjects are discussed and another (antiresonance) method is suggested.

Today, the properties of nanoobjects, including their eigenfrequencies, are studied with probe microscopy, specifically, with an AFM, which interacts with a test object through an electric field [8, 9]. The major component of the AFM is a scanning probe (cantilever) [10, 11]. Mechanically, the cantilever represents an elastic beam one end of which is tightly fixed and the other is free and bears a nanometer tip with a radius of curvature of 10–50 nm. Standard commercial cantilevers have resonance frequencies ranging from 10 to 400 kHz. The AFM operates in three (contact, contactless, and tapping) modes. Note that all the modes are contact from the mechanical point of view. As the tip approaches a test object, they come into interaction, no matter whether this is interaction is contact or is accomplished through force fields. The dependence of the force of interaction on the tip–object distance is akin to that observed in the case of interaction with the Lenard–Jones potential. Depending on the tip–object distance, the force of interaction is either repulsive (contact mode), attractive (contactless modes), or alternating (tapping mode). All the three modes can be used to examine the surface relief, and the tapping mode is also applied to determine the eigenfrequencies. In this mode, the clamped end of the cantilever executes vertical vibrations with a given frequency. Measuring the vibration amplitudes of cantilever points at different frequencies, one fixes resonances. If the tip is away from the surface, the resonance frequencies are close to the eigenfrequencies of the cantilever. As the tip approaches the surface and comes into interaction with it, the resonance frequencies vary. From the dependence of the resonance frequency on the tip–surface distance, one can judge the properties of the test object. Application of an AFM to measure eigenfrequencies is not free of disadvantages and limitations, which do not allow making good use of the

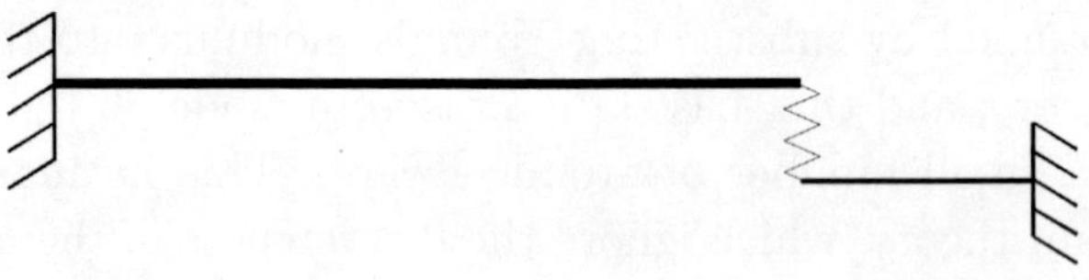

Fig. 3

potential of this instrument. The problems associated with this method are the following. 1) The frequency range accessible to measurements is limited. To extend it to higher frequencies requires a decrease in the dimensions (weight) of the cantilever and/or an increase in its rigidity. 2) Imperatively, the top of the tip must be much smaller than the test object; therefore, the radius of curvature of the tip should be decreased as the test object gets smaller. 3) The nanoobject should be mounted so that the support has no influence on its eigenfrequencies. Otherwise, extra difficulties in interpreting measurement results and gaining information on the nanoobject may arise. 4) The tip is in contact with the object and inevitably influences it somehow or other. As a result, the eigenfrequencies of the nanoobject–cantilever system, rather than of the nanoobject itself, are measured. The last-mentioned problem is related to the well known mechanical effect: the redistribution of the vibration eigenfrequencies of the cantilever–object system among the cantilever and object [12]. In this case, the change in the frequency spectrum considerably depends on the tip–surface distance, since such a redistribution is equivalent to a change in the "rigidity" of field interaction coupling. Physically, two problems lying at the interfaces between mechanics and experimental physics arise here: (i) determination of the nanoobject's elastic moduli from the eigenfrequencies of the system and (ii) design of experiments making it possible to separate out the eigenfrequencies of a nanoobject from the frequency spectrum of the system.

The problem stated here and the cantilever–nanoobjects test system elaborate upon work [10]. The basic difference is that here test nanoobjects possess intrinsic dynamics. Examples are the elements of thin-walled (nanodimensional) structures, such as rods, sheaths, and helices. Below, we set forth theoretical grounds for determining the vibration eigenfrequencies of rodlike structures with an AFM. Consider the mechanical model of a test object shown in Fig. 3. The rod at the left of the figure symbolizes a cantilever. Its left end is firmly fixed, while the right end interacts with the object.

Experimentally, one can detect not only a sharp buildup of the vibration amplitude but also vanishing of the amplitude. In many-body systems with distributed parameters, the latter fact may be observed in two cases: (i) when the point where the amplitude is measured is a node of a given waveform of vibration or (ii) when the vibrations of one body are dynamically damped at the partial frequency of another (so-called antiresonance). Let us pose the question: whether there exist forced vibration frequencies at which the cantilever's right-hand end, which is contact with the nanoobject, remains stationary at any time instant? It can be shown that these frequencies Ω should be solution of following equation

$$\begin{aligned} &1+\cos(\mu L)\cosh(\mu L)+\frac{C}{D\mu^3}[\sin(\mu L)\cosh(\mu L)-\cos(\mu L)\sinh(\mu L)]=0, \\ &\mu^2=\sqrt{\rho/D}\,\Omega. \end{aligned} \tag{4}$$

The problem of finding the rigidity of nanodimensional objects was considered in [4]. In our case, the flexural rigidity of the nanorod can be found from both the resonance frequencies, and antiresonance frequencies. Having measured two (resonance or antiresonance) frequencies, one substitutes them into the respective equation, and thus reduces the problem of finding the flexural rigidity of the nanorod to the solution of transcendental equations in two unknowns. But determination of the flexural rigidity of the nanorod from antiresonance frequencies is computationally easier. However, it makes sense to apply both approaches and compare the obtained values of flexural rigidity to improve the reliability of the results.

3. METHOD OF DETERMINING THE EIGENFREQUENCIES OF AN ORDERED SYSTEM OF NANOOBJECTS

In this study, we suggest a method of determining the eigenfrequencies of nanostructures (nanotubes and nanocrystals) from the measured eigenfrequencies of a large system comprising a highly ordered array of identical nanotubes or nanocrystals grown on a substrate (Fig. 4) as a result, e.g., of self-organized growth [13, 14]. Nanoobjects constituting such an array are usually of the same size; therefore, one can use the macroscopic sizes of the array to study the properties of nanoobjects by finding the first eigenfrequencies of the nanoobjects–substrate system. It is shown that, from the spectrum of the large (nanoobjects–substrate) system, the eigenfrequencies of an individual nanoobject can be derived; that is, the eigenfrequencies of a single nanoobject can be determined from experimental data obtained for the large system. The main problem arising in measuring the frequencies of nanoobjects mounted on an elastic substrate is that the eigenfrequencies of the system as a whole are distributed among its components — a phenomenon well known in mechanics [12]. The shift of the spectrum strongly depends on the parameters of the substrate and object. It is known that, in many-body systems with distributed parameters, the vibration of one body may be dynamically damped at the partial frequency of another body (antiresonance).

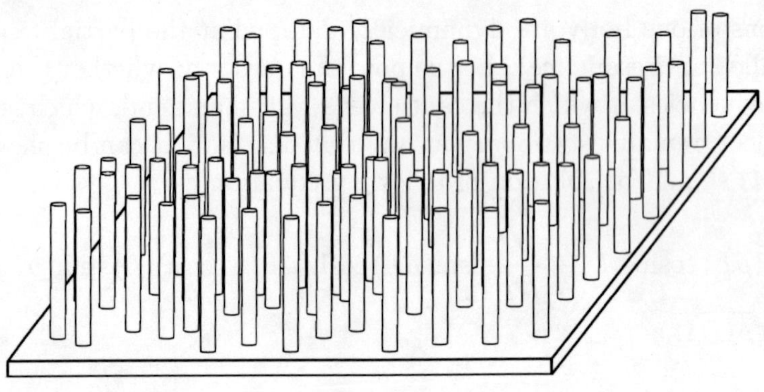

Fig. 4

It will be shown below that systems of highly ordered arrays of identical nanotubes or nanocrystals grown on a substrate (Fig. 4) may also exhibit antiresonance and that this effect can be used to extract the eigenfrequencies of the nanoobjects from the spectrum of the large system. The normal modes of the system shown in Fig. 4 can hardly be analyzed in terms of the 3D elasticity theory. Therefore, we will first consider a rod model of the large system where a horizontal rod represents a substrate and vertical rods simulate nanoobjects. In terms of this model, the normal modes of an array of nanocrystals are analyzed and the feasibility of extracting their eigenfrequency spectrum from the respective spectrum of the large system is demonstrated. The array of nanocrystals assumed in this work is to a great extent similar to micro- and nanocrystals of semiconducting zinc oxide. Owing to their high mechanical and physical performance, such crystals are of considerable interest for nanomechanics and nanophotonics and can be fabricated by different techniques, pulsed laser evaporation among them [15–17]. Numerical analysis of the dynamics of such an array is the second stage of this study. The eigenfrequencies of an array of zinc oxide micro- or nanocrystals on a sapphire substrate are calculated in terms of the 2D and 3D problems of the elasticity theory. The calculation results also indicate the feasibility of extracting the spectrum of nanoobjects from the spectrum of the large system.

4. ON EFFECTIVE PROPERTIES OF NANOSIZED PLATES

In this section we discuss the application of the so-called direct approach to the modeling the nanosized plates [18–22]. From the direct approach point of view a plate or a shell is modeled as a material surface each particle of which has five degrees of freedom (three displacements and two rotations, the rotation about the normal to plate is not considered as a kinematically independent variable). Note that the application of the direct approach to the thin-walled nanostructures has advantages because it is free from the consideration of the nanostructure on the base of the three-dimensional continuum mechanics. The equations of motion are formulated as the Euler's laws of dynamics

$$\nabla \cdot \mathbf{T} + \mathbf{q} = \rho \ddot{\mathbf{u}} + \rho \boldsymbol{\Theta}_1 \cdot \ddot{\boldsymbol{\varphi}}, \quad \nabla \cdot \mathbf{M} + \mathbf{T}_\times + \mathbf{m} = \rho \boldsymbol{\Theta}_1^T \cdot \ddot{\mathbf{u}} + \rho \boldsymbol{\Theta}_2 \cdot \ddot{\boldsymbol{\varphi}}. \tag{5}$$

Here $\mathbf{T}$, $\mathbf{M}$ are the tensors of forces and moments, $\mathbf{q}$, $\mathbf{m}$ are the vectors of surface loads (forces and moments), $\mathbf{T}_\times$ is the vector invariant of the force tensor, ∇ is the nabla (Hamilton) operator, $\mathbf{u}$, $\boldsymbol{\varphi}$ are the vectors of displacements and rotations, $\boldsymbol{\Theta}_1$, $\boldsymbol{\Theta}_2$ are the first and the second tensor of inertia, ρ is the density (effective property of the deformable surface), $(\ldots)^T$ denotes transposed and $(\ldots)^\bullet$ the time derivative. The geometrical equations are given as

$$\boldsymbol{\mu} = (\nabla \mathbf{u} \cdot \mathbf{a})^{\mathrm{sym}}, \quad \boldsymbol{\gamma} = \nabla \mathbf{u} \cdot \mathbf{n} + \mathbf{c} \cdot \boldsymbol{\varphi}, \quad \boldsymbol{\kappa} = \nabla \boldsymbol{\varphi}, \tag{6}$$

where $\mathbf{a}$ is the first metric tensor (plane tensor), $\mathbf{n}$ is the unit outer normal vector at the surface, $\mathbf{c}$ is the discriminant tensor ($\mathbf{c} = -\mathbf{a} \times \mathbf{n}$), $\boldsymbol{\mu}$, $\boldsymbol{\gamma}$, and $\boldsymbol{\kappa}$ are the strain tensors (the tensor of in-plane strains, the vector of transverse shear strains and the tensor of the out-of-plane strains), while $\mathbf{t}^{\mathrm{sym}}$ denotes the symmetric part. The boundary conditions are given by the relations

$$\boldsymbol{\nu} \cdot \mathbf{T} = \mathbf{f}, \quad \boldsymbol{\nu} \cdot \mathbf{M} = \mathbf{l} \quad (\mathbf{l} \cdot \mathbf{n} = 0) \quad \text{or} \quad \mathbf{u} = \mathbf{u}^0, \quad \boldsymbol{\varphi} = \boldsymbol{\varphi}^0 \quad \text{along } S. \tag{7}$$

Here $\mathbf{f}$ and $\mathbf{l}$ are external force and moment vectors acting along the boundary of the plate S, while $\mathbf{u}^0$ and $\boldsymbol{\varphi}^0$ are given functions describing the displacements and rotations of the plate boundary, respectively. $\boldsymbol{\nu}$ is the unit outer normal vector to the boundary S ($\boldsymbol{\nu} \cdot \mathbf{n} = 0$). The relations (3) are the static and the kinematic boundary conditions. Other types of boundary conditions are also possible. The constitutive equations are given by the following relation for the strain energy of the deformable surface W

$$W(\boldsymbol{\mu}, \boldsymbol{\gamma}, \boldsymbol{\kappa}) = \frac{1}{2} \boldsymbol{\mu} \cdot\cdot \mathbf{A} \cdot\cdot \boldsymbol{\mu} + \boldsymbol{\mu} \cdot\cdot \mathbf{B} \cdot\cdot \boldsymbol{\kappa} + \frac{1}{2} \boldsymbol{\kappa} \cdot\cdot \mathbf{C} \cdot\cdot \boldsymbol{\kappa} + \frac{1}{2} \boldsymbol{\gamma} \cdot \boldsymbol{\Gamma} \cdot \boldsymbol{\gamma} + \boldsymbol{\gamma} \cdot (\boldsymbol{\Gamma}_1 \cdot\cdot \boldsymbol{\mu} + \boldsymbol{\Gamma}_2 \cdot\cdot \boldsymbol{\kappa}), \tag{8}$$

where $\mathbf{A}$, $\mathbf{B}$, $\mathbf{C}$ are 4th rank tensors, $\boldsymbol{\Gamma}_1$, $\boldsymbol{\Gamma}_2$ are 3rd rank tensors, $\boldsymbol{\Gamma}$ is a 2nd rank tensor of the effective stiffness properties. They depend on the material properties and the cross-section geometry. In the general case the tensors contain 36 different values — a reduction is possible assuming some symmetries. In [18–22] a new approach to the determination of the effective stiffness tensors was developed. The idea of this approach is based on the comparison of the exact solutions of the three-dimensional elasticity and the corresponding solutions of Eqs (5)–(8) for several test problems of

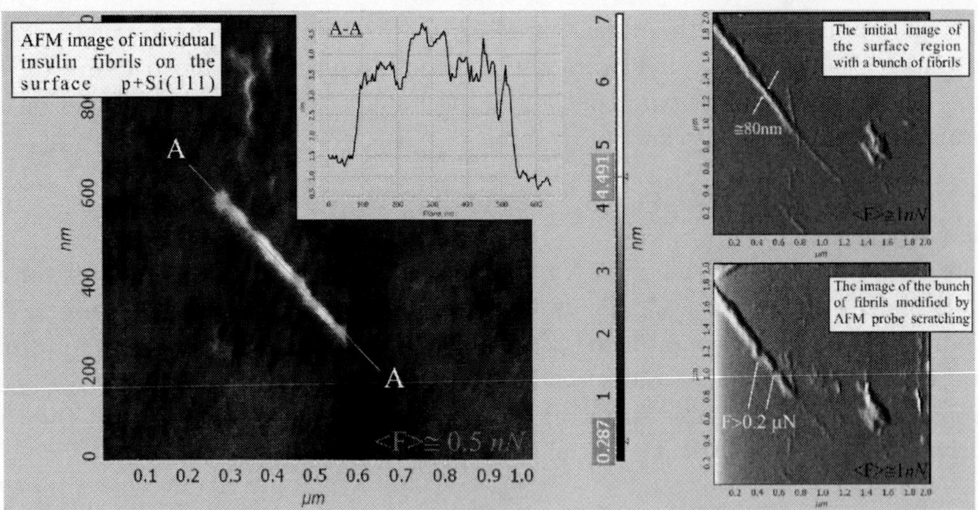

Fig. 5

the elastostatics such as tension and bending, plane shear, and torsion. The method was applied to isotropic plates as well as to orthotropic and transversally isotropic material behavior. Now it is well-known that the nanocomposites demonstrate mechanical properties which may be quite different from its macro-analogues, see e.g. [5, 13, 14]. One of the reasons of such difference is the surface tension, because these materials may have a fractal-like surface. The influence of the surface tension on the elastic moduli was investigated by several authors, see for example [23]. From this point of view they may consider the nanocomposite based on an oriented array of nanocrystals or nanotubes as a transversally isotropic material. We propose to extend the procedure [18, 19, 22] to the case of such materials. Finally, one obtains the components of the effective stiffness tensors which depend not only on the bulk elastic properties and the thickness of the plate but also on the surface tension.

5. EXPERIMENTAL RESULTS

Study of surface relief, tensile and adhesion strength of insulin fibrils deposited on p+Si(111) surface, has been carried out. Results of the investigation are presented in Fig. 5. Results of investigation of surface relief and elasticity of insulin fibrils deposited on p+Si(111) surface are presented in Fig. 6. Study of internal structure and elastic properties of individual insulin fibrils is presented in Fig. 7. Study of ordering processes of insulin fibrils is presented in Fig. 8.

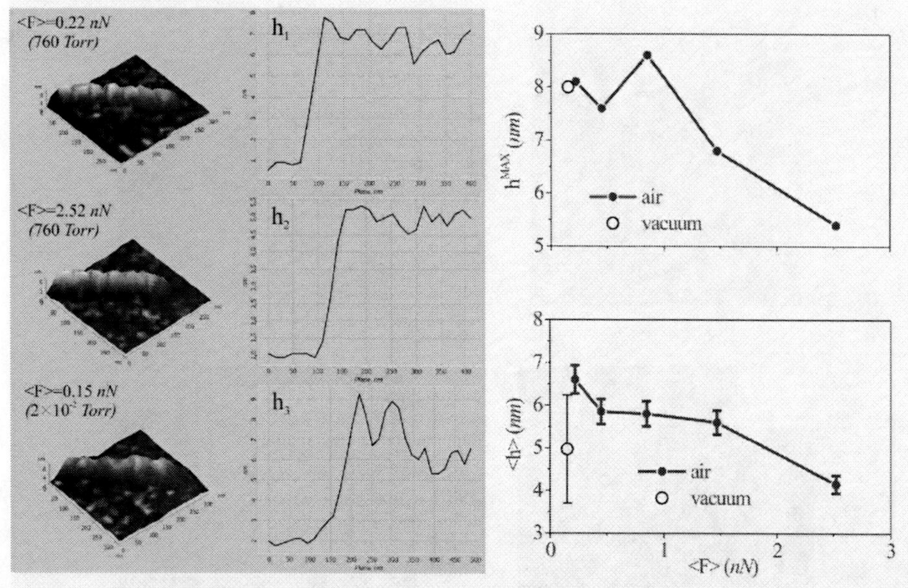

Fig. 6

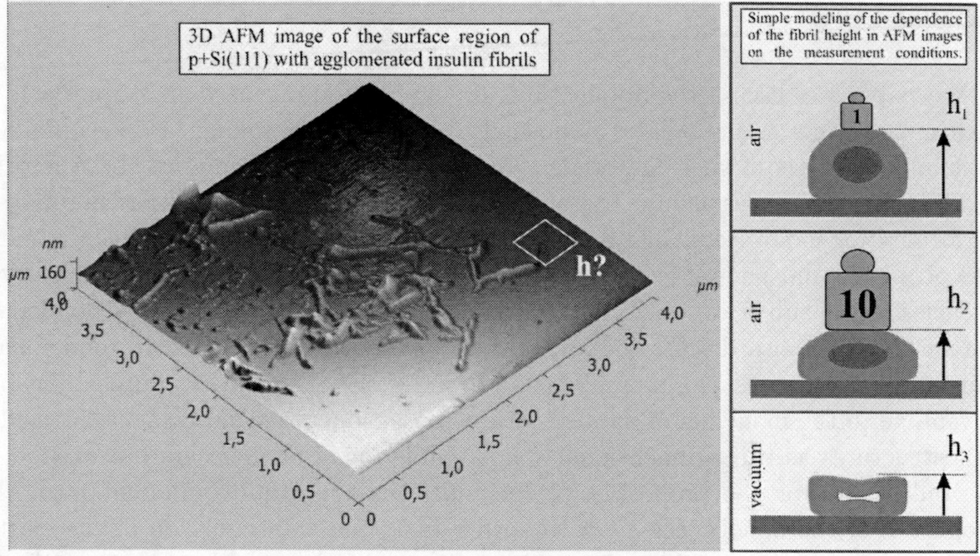

Fig. 7

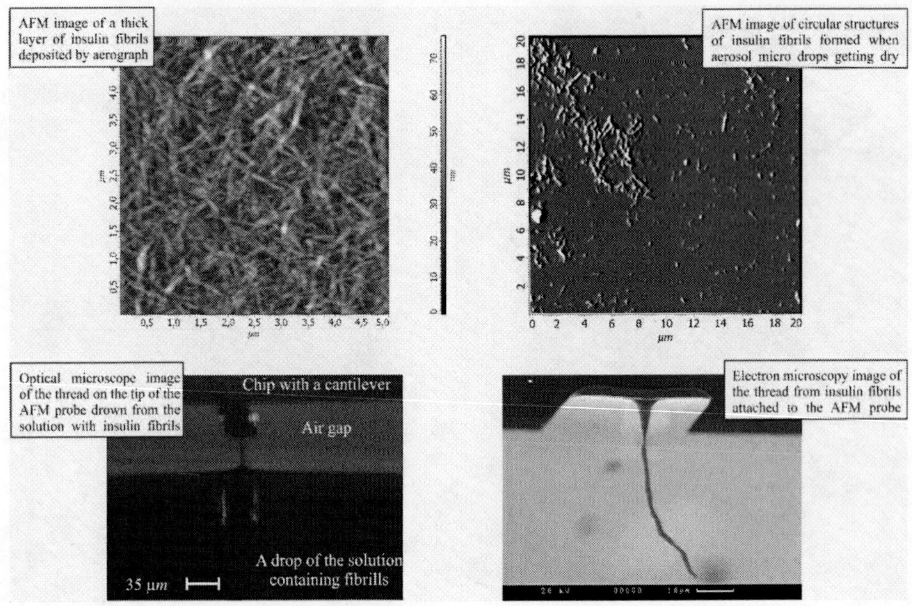

Fig. 8

CONCLUSION

In the paper we discuss the applications of the continuum mechanics approach to such nanoobjects as nanoplates nanoshells and nanofilms taking into account additional properties of such materials. We consider the mathematical base for the possible physical experiments. In particular, the theoretical and numerical analysis performed for nanobjects, shows that it is possible to determine the eigenfrequencies of micro- and nanoobjects in experiments using the measurement of eigenfrequencies of the substrate with an array of such nanodimensional objects bonded to the surface. We also discuss the application of the direct approach to model such thin-walled structures as a nanoplates and nanoshells. The direct approach has an advantage to the modeling of nanostructures because it is not necessary to consider nanostructures as three-dimensional solids, while many of them are not existed in the bulk phase or the properties of the bulk phase are quite different from the properties of the nanoobjects. On the other hand the determination of the elastic stiffness tensors is essential for the applicability of the theory. The effective stiffness may be obtained by consideration of experimental data on eigenfrequencies. Here we discussed a possible method of determination of the eigenfrequencies of single nanoobjects by investigation the spectrum of the large system consisting of an array of identically nanoobjects grown on a substrate.

ACKNOWLEDGMENTS

The research was financially supported by the Council of the President of the Russian Federation for Support of Young Scientists and Leading Scientific Schools (projects Nos. MD–4829.2007.1 and NSh–4518.2006.1).

REFERENCES

1. S.V. Golod, V.Ya. Prinz, V.I. Mashanov, and A.K. Gutakovsky, "Fabrication of Conducting GeSi/Si Micro- and Nanotubes and Helical Microcoils," Semicond. Sci. Technol. **16** (3), 181–185 (2001).
2. A.B. Vorob'ev and V.Ya. Prinz, "Directional Rolling of Strained Heterofilms," Semicond. Sci. Technol. **17** (6), 614–616 (2002).
3. V.Ya. Prinz, "A New Concept in Fabricating Building Blocks for Nanoelectronic and Nanomechanic Devices," Microelectron. Eng. **69** (2/4), 466–475 (2003).
4. E.A. Ivanova and N.F. Morozov, "An Approach to the Experimental Determination of the Bending Stiffness of Nanosize Shells," Dokl. Phys. **50** (2), 83–87 (2005).
5. A.M. Krivtsov and N.F. Morozov, "Anomalies in Mechanical Characteristics of Nanometer-Size Objects," Dokl. Phys. **46** (11), 825–827 (2001).
6. E.A. Ivanova, A.M. Krivtsov, and N.F. Morozov, "Peculiarities of the Bending-Stiffness Calculation for Nanocrystals," Dokl. Phys. **47** (8), 620–622 (2002).
7. M.S. Dunaevskii, J.J. Grob, A.G. Zabrodskii, et al., "Atomic-Force-Microscopy Visualization of Si Nanocrystals in SiO_2 Thermal Oxide Using Selective Etching," Semiconductors **38** (11), 1254–1259 (2004).
8. G. Binnig, C.F. Quate, and C. Gerber, "Atomic Force Microscope," Phys. Rev. Lett. **31** (9), 930–933 (1986).
9. B. Gotsmann and H. Fuchs, "Dynamic Force Spectroscopy of Conservative and Dissipative Forces in an Al-Au(111) Tip-Sample System," Phys. Rev. Lett. **86** (12), 2597–2600 (2001).
10. U. Rabe, K. Janser, and W. Arnold, "Vibrations of Free and Surface-Coupled Atomic Force Microscope Cantilevers: Theory and Experiment," Rev. Sci. Instrum. **67** (9), 3281–3293 (1996).
11. C.T. Gibson, D.A. Smith, and C.J. Roberts, "Calibration of Silicon Atomic Force Microscope Cantilevers," Nanotechnology **16** (2), 234–238 (2005).
12. S.H. Gould, *Variational Methods for Eigenvalue Problems. An Introduction to the Weinstein Method of Intermediate Problems*, 2 ed. (Oxford Univ. Press, Oxford, 1966; Mir, Moscow, 1970).
13. B. Bhushan (Editor), *Springer Handbook of Nanotechnology* (Springer, Berlin, 2004).
14. W.A. Goddard, D.W. Brenner, S.E. Lyshevsky, and G.J. Iafrate (Editors), *Handbook of Nanoscience, Engineering, and Technology* (CRC, Boac Ratoon, 2003).

15. M. Lorenz, J. Lenzner, E.M. Kaidashev, et al., "Cathodoluminescence of Selected Single ZnO Nanowires on Sapphire," Ann. Phys. (Leipzig) **13** (1/2), 39–42 (2004).
16. E.M. Kaidashev, M. Lorenz, H. Wenckstern, et al., "High Electron Mobility of Epitaxial ZnO Thin Films on *c*-plane Sapphire Grown by Multistep Pulsed-Laser Deposition," Appl. Phys. Lett. **82** (22), 3901–3903 (2003).
17. M. Lorenz, H. Hochmuth, R. Schmidt-Grund, et al., "Advances of Pulsed Laser Deposition of ZnO Thin Films," Ann. Phys. (Leipzig) **13** (1/2), 59–60 (2004).
18. H. Altenbach, "An Alternative Determination of Transverse Shear Stiffnesses for Sandwich and Laminated Plates," Int. J. Solids Struct.**37** (25), 3503–3520 (2000).
19. H. Altenbach, "On the Determination of Transverse Shear Stiffnesses of Orthotropic Plates," ZAMP **51** (4), 629–649 (2000).
20. H. Altenbach and V.A. Eremeyev, "Direct Approach-Based Analysis of Plates Composed of Functionally Graded Materials," Arch. Appl. Mechanics **78** (10), 775–794 (2008).
21. H. Altenbach and P. Zhilin, "General Theory of Elastic Simple Shells," Uspekhi Mekhaniki **11** (4), 107–148 (1988).
22. P.A. Zhilin, *Applied Mechanics. Foundations of the Theory of Shells* (St. Petersburg State Polytechnical University, St. Petersburg, 2006) [in Russian].
23. H.L. Duan, J. Wang, B.L. Karihaloo, and Z.P. Huang, " Nanoporous Materials Can Be Made Stiffer Than Non-Porous Counterparts by Surface Modification," Acta Materialia **54** (11), 2983–2990 (2006).

MECHANICAL BEHAVIOR OF ULTRAFINE GRAINED/NANOSTRUCTURED MATERIALS

R. Jayaganthan[1*] and S.K. Panigrahi[2]**

ABSTRACT

The mechanical properties and microstructural characteristics of ultrafine grained Al 6063 produced by cryorolling (CR) with different stain levels is reported in the present work. The effect of rolling temperature and post CR ageing treatment on mechanical properties of the Al 6063 alloy was investigated. The microstructural characteristics of ultrafine grained Al 6063 alloy were characterized by using SEM/EBSD and transmission electron microscopy (TEM). The hardness and tensile strength of the ultrafine grained Al alloys were measured. The severe strain induced during cryorolling of Al 6063 alloys in the solid solutionised state produces ultrafine microstructures with improved mechanical properties such as strength and hardness. It is observed that the strength and hardness of the cryorolled materials (CR) at different percentage of thickness reductions are higher as compared to the room temperature rolled (RTR) materials at the same strain due to suppression of dynamic recovery and accumulation of higher dislocations density in the CR materials. The ageing treatment of cryorolled Al 6063alloys has improved its strength and ductility significantly due to the precipitation hardening and grain coarsening mechanisms, respectively.

Key words: ultrafine grain, Al 6063 alloy, cryorolling, microstructural characterization

INTRODUCTION

Polycrystalline materials with ultrafine-grained microstructure exhibit enhanced mechanical properties as compared to the conventional bulk materials [1]. With the ever-growing demand for superior mechanical properties of Al alloys, a new processing route, namely severe plastic deformation technique (SPD) has been used extensively to produce this alloy with the ultrafine-grained microstructures owing to the

[1]Department of Metallurgical and Materials Engineering, Indian Institute of Technology Roorkee, Roorkee — 247667, Uttarakhand, India
[*]E-mail: *rjayafmt@iitr.ernet.in*
[**]E-mail: *susant02@yahoo.co.in*

limitation of conventional processing. The processing of bulk aluminium alloys to grain sizes of $\approx 1\ \mu m$ through the conventional route is very difficult due to its high stack fault energy. The SPD techniques have been widely used to produce ultrafine-grained Al alloys [2–3]. However, majority of these methods produces a relatively small quantity of materials [4]. To overcome these difficulties, cryorolling has been identified as one of the potential routes to produce nanostructured/ultrafine grained pure metals [5–7] and Al alloys [8–9] from its bulk counterparts. The rolling of pure metals and alloys in cryogenic temperature suppresses dynamic recovery which in turn results in accumulating the higher density of dislocations, reaching a higher steady state level as compared to room temperature rolling. These higher densities of dislocations act as driving force for the initiation of large number of nucleation sites during annealing, forming the sub-microcrystalline or ultrafine grain structures (ufg) [5–7]. The literature on ultrafine grained Al alloys produced by cryorolling is very limited.

In the present work, ultrafine grained (ufg) Al 6063 alloy was produced by cryorolling technique. In order to enhance its mechanical properties further, the cryorolled alloys were aged at different temperature. The objectives of the present study are as follows: (a) To study the effect of cryorolling strain on mechanical properties and microstructure of ufg Al 6063 alloy; (b) To investigate the influence of ageing treatment on their mechanical properties.

1. EXPERIMENTAL

The 6063 Al alloy plates (9.6 mm×30 mm×30 mm) of T4 temper treated, with the composition of 0.45% Si, 0.3% Mg, 0.015% Cu, 0.013% Mn, 0.058% Fe, 0.022% Zn, 0.02% Cr (wt%) were procured from the Hindalco Industries Ltd, Aditya Birla Group, Renukoot, India for the present work. These plates were solutionised at 520°C for 60 minutes and then quenched in water. These plates were solution treated, quenched, and rolled at both room and liquid nitrogen temperature up to a true strain (e) of 3.6. The details of cryorolling are discussed elsewhere [10–14]. In order to improve the mechanical properties, the cryorolled (CR) and the uncryorolled samples (Un CR) upon solutionising treatment were subjected to ageing treatment at 100°C, 150°C and 175°C respectively.

The microstructural features of starting material solution treated, quenched (ST) and cryorolled (CR) sheets were characterized using a scanning electron microscope (FEI, Quanta 200F)/EBSD analysis. The dislocation network and microstructure of the cryorolled sheets after 97% reduction ($e = 3.6$) were characterized by using TEM. Microhardness and tensile tests were conducted to evaluate the strength and ductility of the CR and room temperature rolled (RTR) Al 6063 alloys subjected to different thickness reductions and ageing conditions. The tensile specimens were machined as per ASTM E-8 sub-size specifications parallel to the rolling direction with gauge length of 25 mm. The tensile tests were performed after polishing the samples in air at room temperature using a S-Series, H25K-S materials testing machine operated at a constant crosshead speed with an initial strain rate of 5×10^{-4}S-1. The tensile samples after rolling had the same length and width but a different thickness in accordance with rolling reductions.

2. RESULTS AND DISCUSSION

2.1. *Effect of cryorolling strain on microstructure and mechanical properties*

The microstructural evolution [12] in the Al 6063 alloy cryorolled over a strain range of 0, 0.4, 0.8, 2.6, and 3.6 was investigated by EBSD and shown in Fig. 1. The ST material exhibits equiaxed grain morphology with an average grain size of 70 μm (Fig 1 *a*). At low strains, bands of elongated cells, aligned at $5 - 15°$ to the rolling direction were observed in Figs. 1 *b* and 1 *c*. After the CR strain of 2.3, the EBSD map reveals the long fibrous grains parallel to the rolling direction or the lamella structure with high angle grain boundaries (HAGB) with the average spacing of about 3 μm (Fig. 1 *d*). When the strain is > 3.6, the lamella HAGB spacing has decreased in the transverse direction and the remaining long fibrous grains were fragmented and reduced in thickness to submicron scale (Fig. 1 *e*). Fig. 1 *f* shows the elongated sub grains, which are in the range of $100 - 500$ nm.

The histograms of the boundary misorientation distribution of the CR materials as a function of strain are shown in Fig. 2. The ST material shows the HAGB as evident from this figure. At low strains (0.4 and 0.8), the misorientation distribution of CR materials is heavily skewed to low angle boundaries (LAB). When the CR strain is between $2.6 - 3.6$, the proportion of LAB reduces drastically and the formation of higher fraction of HAGB is noticed, which is due to dynamic recrystallization.

The TEM micrograph [12] of the CR Al 6063 alloy after the strain of 3.6 (97% reduction) is shown in Fig. 3 *a*. The size of individual grains produced during cryorolling lies in the range of $100 - 400$ nm. The heavily strained grains with dislocations are evident from this figure. The electron diffraction pattern of the TEM micrograph exhibits diffraction rings (Fig. 3 *b*), which indicates a polycrystalline structure.

The hardness values [10] of the solution treated alloys rolled at cryogenic temperature and room temperature (RT), respectively as a function of true strain are shown in Fig. 4 *a*. The true strain corresponds to different percentage of thickness reduction in the samples. The hardness of the cryorolled samples has increased from 58 to 85.5 Hv (nearly 47% increase) after 55 percent thickness reduction ($e = 0.79$). Subsequent thickness reductions of the samples increased its hardness further and

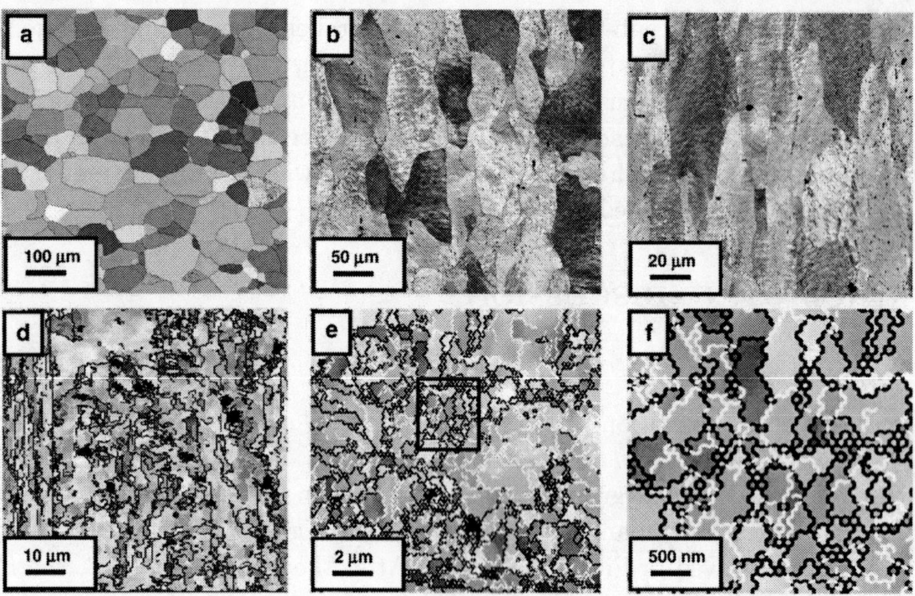

Fig. 1. [12] EBSD micrographs of original and cryorolled 6063 Al alloy at different strains (a) Starting solution treated material (ST) (b) True strain $e = 0.4$, (c) $e = 0.8$, (d) $e = 2.6$, (e) $e = 3.6$, (f) $e = 3.6$

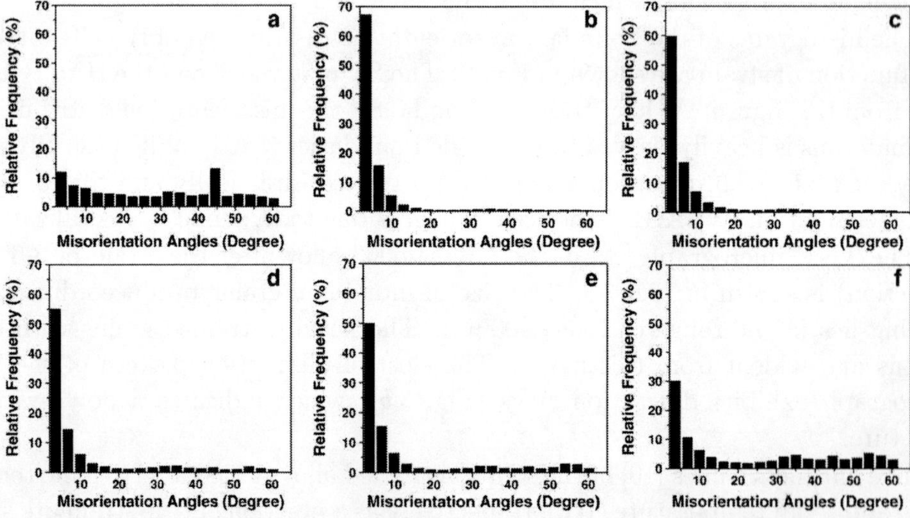

Fig. 2. The frequency histograms of misorientation angles of original and cryorolled 6063 Al alloy at different strains. (a) Starting solution treated material (ST), (b) True strain $e = 0.4$, (c) $e = 0.8$, (d) $e = 1.2$, (e) $e = 2.6$, (f) $e = 3.6$

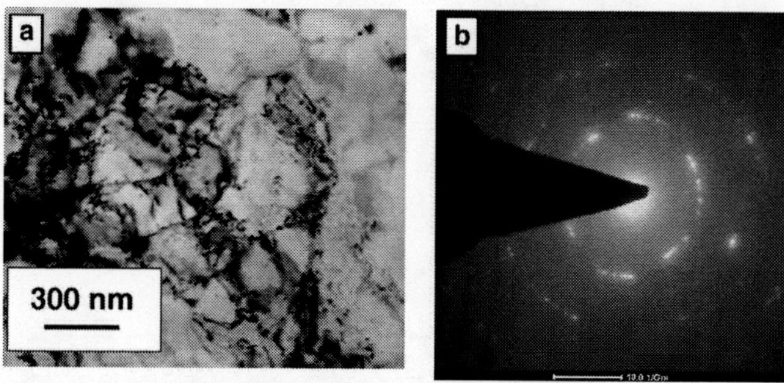

Fig. 3. [12] TEM micrograph and SAD pattern of the cryorolled Al 6063 alloy after 97% reduction ($e = 3.6$) (*a*) TEM microstructure, (*b*) SAD pattern

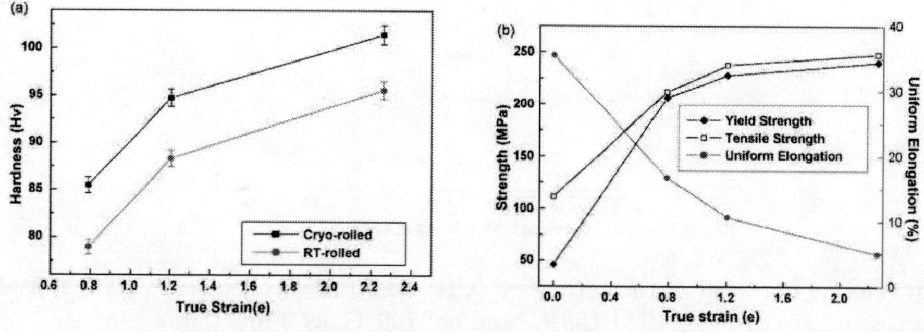

Fig. 4. Hardness and tensile properties 6063 Al alloy as a function of total amount of deformation during cryorolling: (*a*) Hardness property, (*b*) Tensile property

after 90 percent thickness reduction ($e = 2.262$), it has increased about 75% . An enhancement of hardness for the cryorolled Al 6063 alloys could be directly attributed to the considerable substructure refinement, which occurs during severe plastic deformation. It is observed that the hardness of cryorolled alloys is higher than that of RT rolled materials at different strains. It can be explained based on the mechanism that dynamic recovery was effectively suppressed during cryorolling leading to a higher dislocation density as known in the earlier literature [10–14]. The ultrafine-grained microstructures formed during of Al 6063 alloy obeys the Hall-Petch effect to substantiate the improved hardness observed cryorolling of Al 6063 alloy obeys the Hall-Petch effect to substantiate the improved hardness observed in the present work. With the formation of ultrafine grains in the cryorolled alloys, hardness would increase due to the restricted dislocation mobility imposed by grain boundaries as well as misorientation of the grains.

Fig. 4 *b* shows [10] the tensile properties of Al 6063 alloy cryorolled at different

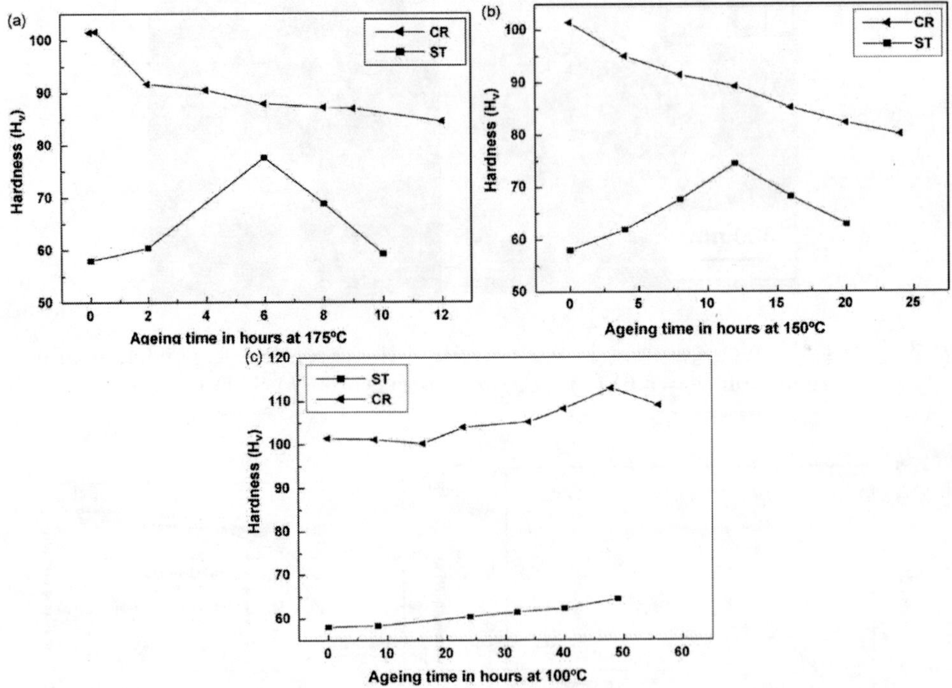

Fig. 5. Vickers hardness of the cryorolled (CR) and uncryorolled (Un-CR) materials after aging at (*a*) 175°C, (*b*) 150°C, and (*c*) 100°C, as a function ageing time

percentage of thickness reduction. It is shown that the tensile strength (UTS) has increased from 111 to 248 MPa (nearly 123% increase) for the cryorolled samples with 90 percent reduction ($e = 2.262$) but for the room temperature rolled samples with the same thickness reduction, it showed a relatively lower tensile strength of 223 MPa. An enhancement of tensile strength of the Al 6063 alloy upon cryorolling treatment could occur due to the fact that the cryogenic temperature can effectively suppress the dynamic recovery and builds up a higher dislocation density in the samples. The influence of cryorolling treatment is well pronounced in yield strength (YS) as compared to the tensile strength of the samples due to effective grain refinement. However, an uniform elongation of the samples decreases with cryorolling deformation as shown in Fig. 4 *b*.

2.2. *Effect of ageing on mechanical properties of cryorolled Al 6063 alloy*

A post CR-ageing treatment was given to the samples in order to increase the strength and ductility of the cryorolled Al alloys, which has been discussed in detail in our earlier work [10]. Three ageing treatments were selected for CR and Un-

CR materials, 100°C, 150°C, and 175°C, respectively. The hardness data of the Al 6063 alloys (CR and Un-CR) measured after ageing at 175°C as a function of ageing time up to 12 hours is plotted in Fig. 5 *a*. In the uncryorolled materials after solution treatment, there is a significant increase in hardness up to 6 hours (35% increase) and than it decreases. On the other hand, the cryorolled materials exhibit an opposite trend; hardness decreases with time. This is due to the recovery or grain coarsening, which might have occurred during ageing of the severely strained microstructure with time. The effect of recovery or grain coarsening may be well pronounced as compared to precipitate hardening for the severely deformed samples. Fig. 5 *b* shows the effect of ageing treatment at 150°C up to 24 h on the ST and CR materials. The hardness of the CR materials decreases with ageing time but for the ST materials, it has increased up to 12 h (28%) and then decreases. It is evident that the observed trend of hardness behavior of both ST and CR materials are similar to the materials aged at 175°C.

In order to minimize the softening effect during post-CR treatment of Al-Mg-Si alloys, an ageing was carried out at a lower temperature of 100°C. The Fig. 5 *c* shows the effect of ageing on hardness value of both CR and Un-CR materials as a function of time at a temperature of 100°C. It is evident that the hardness of cryorolled materials decreases initially up to few hours and then it increases, approximately 10 – 15% , after 48 hours. The precipitation hardening effect contributes to the improved hardness observed for the post-ageing of the cryorolled samples. The prolonged ageing of cryorolled samples after 48 hours decreases its hardness because of the substructure coarsening. An increase in hardness is also observed in the un-CR solution treated materials but the effect is much smaller than that of the un-CR material aged at 175°C. An 11% increase in hardness was observed for the un-CR solution treated materials subjected to ageing at 100°C for 48 hours as compared to 35% increase when they were aged at 175°C for 6 hours.

Upon optimizing the peak aging condition of the cryorolled material [10] as shown in Fig. 5, the tensile test was carried out for the Un-CR and the CR materials with the peak aging condition at 100°C for 48 hours. The tensile properties of the Un-CR and CR materials with peak ageing condition are shown in Fig. 6. The combination of cryorolling with low temperature aging treatment (100°C — 48 hours) results in a further increase in UTS, YS and ductility of the CR materials as observed in Fig. 6 which is tandem with the trend of hardness data plotted in Fig. 5 *c*. The YS and UTS of the cryorolled materials after peak aging have increased from 240 to 259 MPa and 248 to 277 MPa, respectively, as compared to the CR materials. The uncryolled materials with the post aging treatment (100°C for 48 hours) show the YS and UTS of 87 MPa and 159 MPa, respectively, which are very less when compared to the peak aged cryorolled materials.

Along with the YS and UTS, there is a large enhancement of ductility of the CR

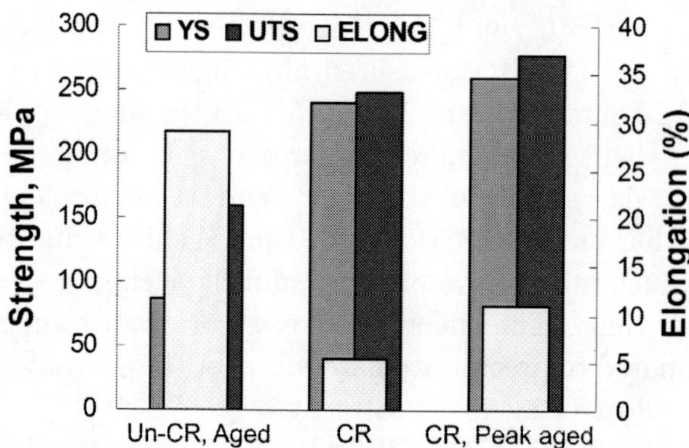

Fig. 6. Tensile properties of the Al 6063 alloy of Un-CR material with ageing, Cryorolled (CR) materials and Cryorolled with peak aged materials after 90% reduction

materials after post aging treatment. As compared to the peak aged (T6) treated commercial 6063 Al alloys (YS of 214 MPa, UTS of 241 MPa, and ductility of 12%), the cryorolled samples after peak ageing shows a significant increase in YS and UTS with the same ductility. These results clearly indicate that pre-CR solid solution treatment combined with post-CR low temperature ageing of Al 6063 alloy show a much higher strength and ductility due to the combined effect of recovery, grain refinement, and precipitation hardening. Kim et al. [15] and Zheng et al. [16] investigated the Al alloys processed through ECAP and observed that pre-ECAP solid solution treatment combined with post-ECAP ageing of the alloys exhibited a significant improvement in strength but with less ductility. However, in the present work, the cryorolled Al 6063 alloy after peak ageing shows a significant increase in YS and UTS with the improved ductility. The improved YS and UTS of CR materials, subjected to peak ageing, are better than that of the commercial T6 treated 6063 Al alloys.

The XRD peaks of Un-CR, CR and CR with peak-aged materials [10] are in Fig. 7. The starting material (Un-CR) subjected to solutionizing and then quenching prior to cryorolling shows the disappearance of the Mg_2Si precipitates as seen from Fig. 7 A. This may be due to dissolution of the coarse precipitates into the material and diffused throughout the whole structure. In the case of cryorolled materials after 90% reduction shown in Fig. 7 B, the precipitation peak is not observed because of the dissolution of Mg_2Si precipitate. Fig. 7 C shows a peak of CR material after peak aging, indicating the presence of Mg_2Si precipitates. Hence, a precipitation hardening is the main strengthening mechanism for the post aged cryorolled materials.

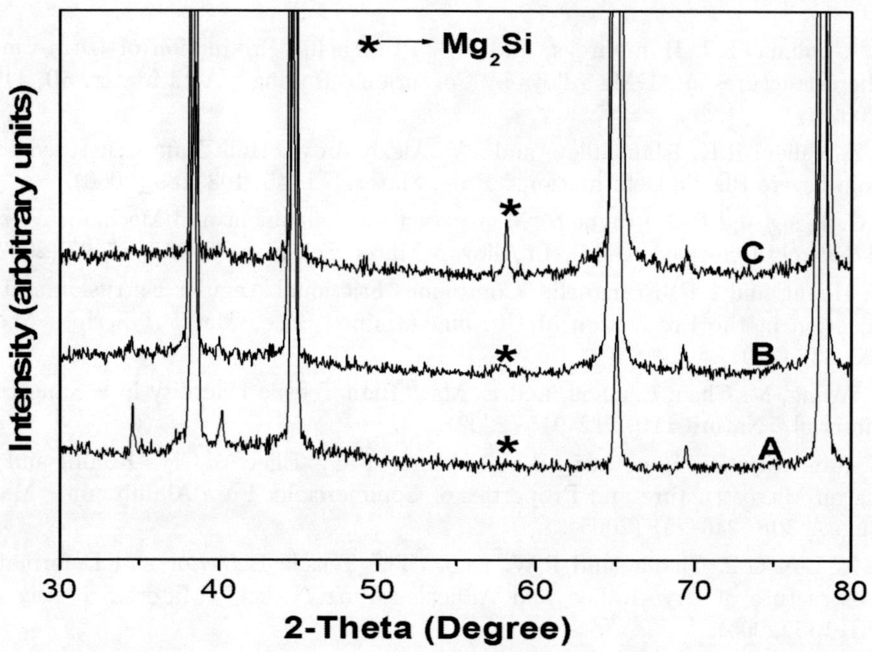

Fig. 7. XRD patterns of Un-CR, CR and CR with peak aged Al 6063 alloys A) the starting material (Un-Cr), annealed at 520°C for 45 min; B) Cryorolled (CR) with 90% reduction; C) CR with 90% reduction and peak aged material

CONCLUSIONS

The mechanical properties and microstructural characteristics of the ultrafine grained Al alloys produced by cryorolling have been investigated in the present work. The strength and hardness of the cryorolled materials (CR) at different percentage of thickness reductions are higher as compared to the room temperature rolled (RTR) materials at the same strain as observed in the present work. The suppression of dynamic recovery and accumulation of higher dislocations density in the CR materials contribute to its improved mechanical properties. The strength and ductility of ultrafine grained Al 6063 alloy, upon ageing treatment, has significantly improved due to the precipitation hardening and grain coarsening mechanisms, respectively.

REFERENCES

1. A. Gholinia, F.J. Humphreys, and P.B. Prangnell, “Production of Ultra-Fine Grain Microstructures in Al–Mg Alloys by Coventional Rolling,” Acta Mater. **50**, 4461–4476 (2002).
2. R.Z. Valiev, R.K. Islamgaliev, and I.V. Alexandrov, “Bulk Nanostructured Materials from Severe Plastic Deformation,” Prog. Mater. Sci. **45**, 103–189 (2000).
3. Z.C. Wang and P.B. Pragnell, “Microstructure Refinement and Mechanical Properties of Severely Deformed Al–Mg–Li Alloys,” Mater. Sci. Eng. A. **328**, 87–97 (2002).
4. Y. Huang and P.B. Prangnell, “Continuous Frictional Angular Extrusion and Its Application in the Production of Ultrafine-Grained Sheet Metals,” Scripta Mater. **56**, 333–336 (2007).
5. Y. Wang, M. Chen, F. Zhou, and E. Ma, “High Tensile Ductility in a Nanostructured Material,” Nature **419**, 912–915 (2002).
6. N. Rangaraju , T. Raghuram, B.V. Krishna, et al., “Effect of Cryo-Rolling and Annealing on Microstructure and Properties of Commercially Pure Aluminium,” Mater. Sci. Eng. A **398**, 246–251 (2005).
7. T.R. Lee, C.P. Chang, and P.W. Kao, “The Tensile Behavior and Deformation Microstructure of Cryo-Rolled and Annealed Pure Nickel,” Mater. Sci. Eng. A **408**, 131–135 (2005).
8. Y.H. Zhao, X.Z. Liao, S. Cheng, et al., “Simultaneously Increasing the Ductility and Strength of Nanostructured Alloy,” Adv. Mater. **18**, 2280–2283 (2006).
9. S. Cheng, Y.H. Zhao, Y.T. Zhu, and E. Ma, “Optimizing the Strength and Ductility of Fine Structured 2024 Al Alloy by Nano-Precipitation,” Acta Mater. **55**, 5822–5832 (2007).
10. S.K. Panigrahi and R. Jayaganthan, “A Study on Mechanical Properties of Cryorolled Al-Mg-Si Alloy,” Mater. Sci. Eng. A **480**, 299–305 (2008).
11. S.K. Panigrahi and R. Jayaganthan, “Effect of Rolling Temperature on Microstructure and Mechanical Properties of 6063 Al Alloy,” Mater. Sci. Eng. A **492**, 300–305 (2008).
12. S.K. Panigrahi, R. Jayaganthan, and V. Chawla, “Effect of Cryorolling on Microstructure of Al-Mg-Si Alloy,” Mater. Lett. **62**, 2626–2629 (2008).
13. R. Jayaganthan and S.K. Panigrahi, “Effect of Cryorolling Strain on Precipitation Kinetics of Al 7075 Alloy,” Mater. Sci. Forum. **584**, 911–916 (2008).
14. S.K. Panigrahi, R. Jayaganthan, and V. Pancholi, “Effect of Plastic Deformation Conditions on Microstructural Characteristics and Mechanical Properties of Al 6063 Alloy,” Mater. Design (2008) (in press).
15. J.K. Kim, H.G. Jeong, S.I. Hong, et al., “Effect of Aging Treatment on Heavily Deformed Microstructure of a 6061 Aluminum Alloy after Equal Channel Angular Pressing,” Scripta Mater. **45**, 901–907 (2001).
16. L.J. Zheng, H.X. Li, M.F. Hashmi, et al., J. Mater. Proc. Technology **171**, 100–107 (2006).

NON-RIGIDITY OF MATTER IN NANO-DOMAINS: POSSIBILITY OF ABRIDGING CLASSICAL TO QUANTUM

N.N. Sharma[1]

ABSTRACT

An analytical model for Brownian motion of nano-size particles has been obtained and explored by author using Langevin equation and considering non-rigidity (elasticity) and dissipative capabilities of matter. A systems-level abstraction of Brownian motion of nanoparticle attributable to thermal agitation has been done using a non-rigid nanoparticle model. The extension of classical lumped-parameter systems model of Brownian motion of nanoparticle to abstract power dissipated per unit area for nanoparticle radiating at temperature greater than absolute zero has been obtained for vacuum and non-vacuum condition and attempts to explain quantum phenomenon with classical principles. The analytical model developed in the work has been based on synergism of local deformation and global motion of nanoparticle. The work suggests the hypothesis of explaining a quantum phenomenon from a classical approach. The model is validated with the well established with Planck's radiation law.

Key words: nanoparticle, non-rigidity, Brownian motion, radiation law

INTRODUCTION

The concept of absolute rigidity of matter is debatable in nano-domains. The attempt to capture non-rigidity of matter has been done with classical approach using a systems-level abstraction of Brownian motion of nanoparticle attributable to thermal agitation [1–4]. Study of motion of small-size particles has interested many researchers and has been analyzed with novel thoughts and abstract mathematics. The English botanist Robert Brown discovered the small sized particle motion in 1827 as a physical phenomenon and it was named after him as Brownian motion.

[1]Mechanical Engineering Group, Birla Institute of Technology and Science, Pilani — 333031, Vidya Vihar, India. E-mail: *nns@bits-pilani.ac.in*

Albert Einstein derived a mathematical description of the phenomenon from the laws of physics in 1905.

The Brownian motion of particles has been extensively studied considering surrounding medium as Newtonian fluid with Stokes approximation of the drag and with Gaussian-Markovian driving forces as input [5–8]. The drag term from surrounding medium has been modified considering Boussineq-Basset correction [9–11]. The consideration of Non-Newtonian fluids leading to Non-Markovian process [12–16] and analysis considering the non-homogenous surrounding medium [17] has been done. The effects of small-time [18–19] and Non-Gaussian driving forces [20] have been explored. The two common approaches to obtain the Brownian motion models present in the literature are (i) diffusion equation method and (ii) spectral analysis using Langevin equation. In the Langevin equation approach [3], correlation technique is used to obtain variance. The Langevin equation is formulated considering the mass of the Brownian-particle and the damping from surrounding medium. The analysis of motion assumes particle to be rigid in the entire literature available on Brownian motion. The rigid body assumption implies that the motion of the body is not influenced by the deformations of the particles caused by the applied forces.

It has been observed in nano-domains that the nanoparticles have low values of spring constant k [21]. This suggests that in nano-domains, local deformations are of appreciable magnitude and must affect overall motion. Therefore, the additional property like elasticity represented by spring constant should be included in nanoparticle motion modeling. The influence of nanoparticle properties has been explored [1] by modeling the impact transfer due to Gaussian white noise input using variance model approach. In another work [2], a model of Brownian motion has been developed considering elastic and dissipative properties of nanoparticle using a system-modeling approach.

The physics of radiation has been also investigated over last one and half century and has been instrumental in development of Modern Physics [22, 23]. The mathematical abstractions of radiation phenomenon of black and grey bodies are present in literature at quantum levels [22, 24] and classical/semi-classical levels. In the classical attempts [25–36], radiation-matter interactions are based on implicit assumption that zero point thermal radiation also known as background radiation is present at $T = 0\,\mathrm{K}$. The subject of thermal radiation from small particles as well as clusters and molecules is recently investigated both by experimentalists and theoreticians [37–40] and suggests a departure from Planck's radiation law. Classically, radiation is an equilibrium interaction between matter and electromagnetic waves in absence of medium for which a classical solution has been done recently using finite difference time domain (FDTD) stochastic differential equation [41]. In a classical attempt [42], modeling of radiation from nanoparticle in vacuum has been done as a simple extension of associated Brownian motion problem at system level.

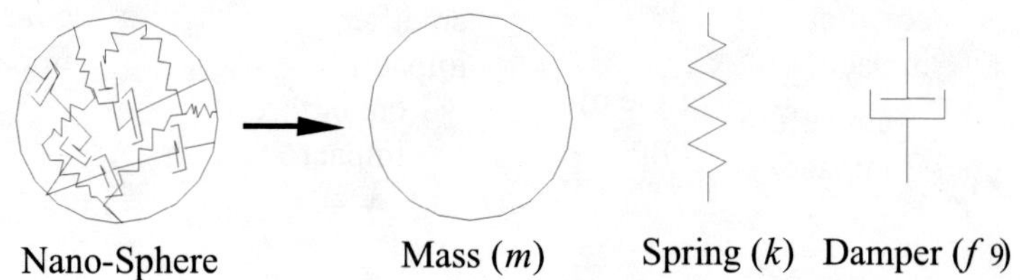

Fig. 1. Model of Nanoparticle and its Lumped parameters

The approach is direct, simple and presents a possible explanation for simultaneous coexistence of blackbodies and gray bodies in nature. The universality of the thermal fluctuation making it consistent with Planck's formula and Grey body radiation has been achieved using instantaneous white noise and non-rigid nanoparticle model considering dissipative and absorptive parameters.

In the present work, the theory in [42] has been extended to model and analyze the synergism between radiation and Brownian motion of nanoparticle under non-vacuum conditions. The work also discusses the possibility of abridging classical theory to explain the quantum phenomenon of radiation and elucidates the relevance of non-rigidity of matter in nano-domains in nanorobotic propulsions. The work is organized in three more sections: In section 1, a model for radiation and Brownian motion of nanoparticle under non-vacuum condition has been developed; Section 2 presents the simulation results using the mathematical model developed in section 1, and in section 3 the simulation results are discussed and scope for further work is presented.

1. POWER DISSIPATION MODEL OF NANOPARTICLE IN NON-VACUUM CONDITION

The development of radiation models of nanoparticle is done using lumped parameter model of nanoparticle [1] and the systems model of free nanoparticle exchanging thermal energy with surrounding [2].

The nanoparticle is considered as sphere of radius r and lumped parameters mass m, damping coefficient f', spring constant k representing inertia, dissipation and absorption properties of matter respectively as shown in Fig. 1.

At equilibrium, a free nanoparticle exchanges thermal energy with surrounding medium and executes Brownian motion in accordance with fluctuation-dissipation theorem. The motion is synergism of local deformation and global motion shown in Fig. 2. The surrounding medium is modeled by a dissipative parameter f representing damping coefficient in resistance to motion and as provider of motive stochastic force [5].

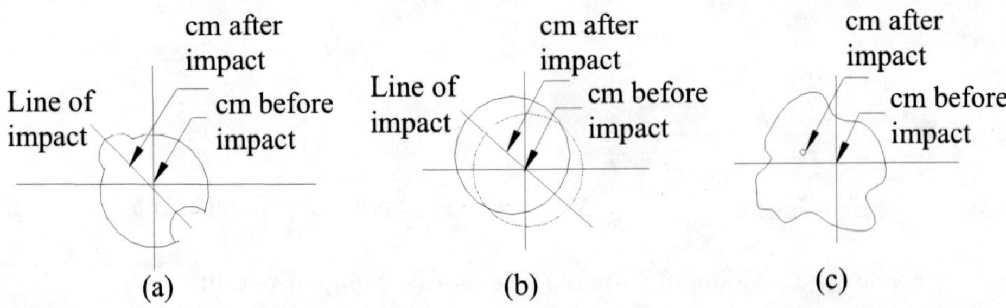

Fig. 2. Deformation and motion of nanoparticle (*a*) Local motion due to single photon/particle impact (*b*) Global motion (rigid body motion) due to single photon/particle impact (*c*) Plausible snapshot of local motion due to random number of random impacts from random directions

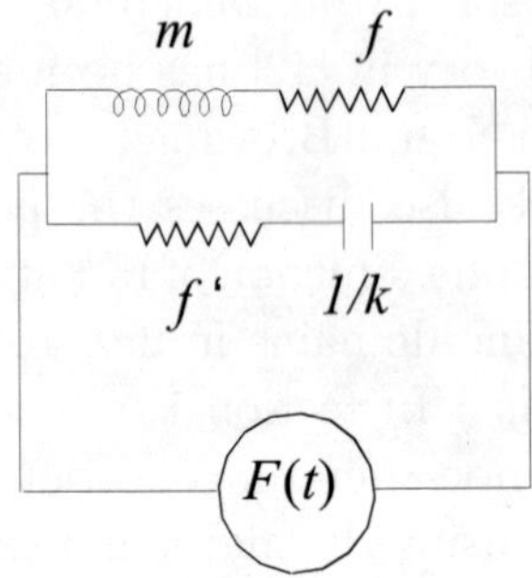

Fig. 3. Equivalent electric analog circuits for three models of nanoparticle and surrounding medium

A well-argued and verified systems lumped parameter model for nanoparticle, as represented in Fig. 1, surrounded by medium for Brownian motion of nanoparticle has been developed [2] in the form of equivalent analog electric circuits, presented in Fig. 3. The model was explored to include non-Brownian motion of nanoparticle [45], analyze impact spectrum [46], include Non-Newtonian fluids surrounding nanoparticle [47–49] and to predict dynamics of 1-dof nanorobot [50]. The model in Fig. 3 for non-rigid nanoparticle was also explored for development of possible models for radiation from nanoparticle in vacuum and verifying it with Planck's radiation law [42].

Possible radiation model of nanoparticle is obtained considering the total dissipation from the Brownian motion model for non-rigid nanoparticle subjected to stochastic force $F(t) = 2\kappa T f_{\text{eq}}$ with Boltzman's constant 1.3807×10^{-23} J/K, absolute temperature T. The total damping from medium and nanoparticle f_{eq}, in classical domain is given as:

$$f_{\mathrm{eq}} = \frac{(ff' + mk)(f + f') + (\omega m f' - fk/\omega)(\omega m - k/\omega)}{(f + f')^2 + (\omega m - k/\omega)^2}. \tag{1}$$

In case of radiation in vacuum, equating $f = 0$ captures the absence of medium. The total power dissipated by the nanoparticle P modeled by electrical analogue circuit in Fig. 3 considering $f = 0$ is:

$$P = \dot{x}^2 f'. \tag{2}$$

For stochastic input $F(t)$, the expected value of power dissipated for the proposed model was obtained as [42]:

$$u_{\mathrm{proposed}}(\upsilon) = \frac{2\kappa T f'^3}{4\pi r^2 m^2 (f'^2 + k^2/4\pi^2\upsilon^2)^2} \frac{mkf' + 2\pi\upsilon m f'[2\pi\upsilon m - k/(2\pi\upsilon)]}{f'^2 + [2\pi\upsilon m - k/(2\pi\upsilon)]^2}, \tag{3}$$

where $\omega = 2\pi\upsilon$. The nanoparticle under consideration is a classical particle with radiation losses $\dot{x}^2 f'$ (mechanical equivalent of i^2R losses in electrical circuits). At thermal equilibrium, in order that the temperature of nanoparticle remains constant at T, the replenishment of energy to compensate the $\dot{x}^2 f'$ losses from the nanoparticle is provided by photon (particle dual of electromagnetic waves) bombardments from all directions randomly. Photons, constituting the surrounding medium of nanoparticle, are identical particles with zero spin and cannot be distinguished from one another and come under the category of bosons as compared to distinguishable surrounding medium molecules in Brownian motion of nanoparticle. The photons obey exclusion principle and Bose–Einstein energy distribution against Maxwell–Boltzman's distribution of distinguishable molecules [43] bombarding nanoparticle in Brownian motion case. Therefore, the energy distribution for photons following Bose–Einstein distribution given by $h\upsilon/(e^{h\upsilon/\kappa T} - 1)$ replaces classical energy distribution κT for distinguishable molecules in (3), where $h = 6.624 \times 10^{-34}$ J·s is Planck's constant. Substituting $h\upsilon/(e^{h\upsilon/\kappa T} - 1)$ for κT in (3), the modified model is given as

$$u(\upsilon) = \frac{h\upsilon f'^3 (mkf' + (2\pi\upsilon m f')[2\pi\upsilon m - k/(2\pi\upsilon)])}{2\pi r^2 m^2 (e^{h\upsilon/\kappa T} - 1)[f'^2 + k^2/(4\pi^2\upsilon^2)]^2 (f'^2 + [2\pi\upsilon m - k/(2\pi\upsilon)]^2)}. \tag{4}$$

For well-known Planck's radiation law, spectral distribution of emissive power per unit area of blackbody $u_{\mathrm{planck}}(\upsilon)$ given as [43]

$$u_{\mathrm{planck}}(\upsilon) = \frac{8\pi h\upsilon^3}{c^2(e^{h\upsilon/\kappa T} - 1)}, \tag{5}$$

where $c = 3 \times 10^8$ m/s is the speed of light. The radiation energy distributions given by semi-classical model developed in (4) as an extension of Brownian motion model were simulated [42] and found to validate with the established Planck's radiation law (5). In case of presence of medium (non-vacuum condition), the medium particle as well as photon particles bombard the nanoparticle. The exchanged energy at equilibrium is obtained as equivalent of the total power dissipated by the nanoparticle P modeled by electrical analogue circuit in Fig. 3. The total power dissipated P is given as

$$P = \dot{x}_1^2 f + \dot{x}_2^2 f', \tag{6}$$

where $\dot{x}_1$ and $\dot{x}_2$ are mechanical equivalents of current flowing through f and f' respectively and are obtained using Kirchoff's law. For stochastic input $F(t)$, the expected value of power dissipated is obtained from equation (6) as

$$E\{P\} = E\{F^2(t)\}\left\{\frac{f^3}{(f^2+\omega^2 m^2)^2} + \frac{f'^3}{(f'^2+k^2/\omega^2)^2}\right\}. \tag{7}$$

The value $E\{F^2(t)\}$ is the autocorrelation of the input noise and is $2\kappa T f_{\text{eq}}$ [2]. From (1) and (7), the power dissipated per unit area at equilibrium $u'(\omega)$ for the nanoparticle in presence of surrounding medium is obtained as

$$u'(\omega) = \left\{\frac{(2\kappa T)[(ff'+mk)(f+f') + (\omega m f' - fk/\omega)(\omega m - k/\omega)]}{4\pi m^2 r^2\{(f+f')^2 + (\omega m - k/\omega)^2\}}\right\} \times \left\{\frac{f^3}{(f^2+\omega^2/k^2)^2} + \frac{f'^3}{(f'^2+k^2/\omega^2)^2}\right\}. \tag{8}$$

Keeping κT for surrounding medium particles and replacing κT by $h\upsilon/(e^{h\upsilon/\kappa T}-1)$ for photons (bosons); the equation (8) is modified as given in (9). For $f = 0$, (9) reduces to (4). The power dissipated per unit area of nanoparticle as given in (9) is simulated next.

$$\begin{aligned} u'(\upsilon) = &\left\{\frac{(2\kappa T)\{(ff'+mk)(f+f') + [2\pi\upsilon m f' - fk/(2\pi\upsilon)][2\pi\upsilon m - k/(2\pi\upsilon)]\}}{\{4\pi^2 m^2 r^2\}\{(f+f')^2 + [2\pi\upsilon m - k/(2\pi\upsilon)]^2\}}\right\} \\ &\times \left\{\frac{f^3}{(f^2+4\pi^2\upsilon^2/k^2)^2}\right\} \\ &+ \left\{\frac{(2\pi h\upsilon)\{(ff'+mk)(f+f') + [2\pi\upsilon m f' - fk/(2\pi\upsilon)][2\pi\upsilon m - k/(2\pi\upsilon)]\}}{\{4\pi^2 m^2 r^2 (e^{h\upsilon/\kappa T}-1)\}\{(f+f')^2 + [2\pi\upsilon m - k/(2\pi\upsilon)]^2\}}\right\} \\ &\times \left\{\frac{f'^3}{[f'^2 + k^2/(4\pi^2\upsilon^2)]^2}\right\}. \end{aligned} \tag{9}$$

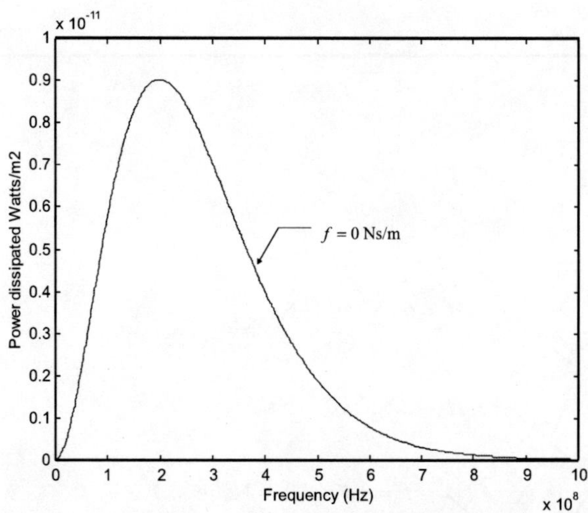

Fig. 4. Power dissipated per unit area from Proposed Model in absence of Medium (vacuum) as a function of frequency

2. SIMULATION

The proposed power dissipation model given by (9) is simulated for obtaining the power dissipated per unit area as a function of frequency (υ) for different parametric values of f. The variation in f represents the variation in presence of medium surrounding nanoparticle and is captured in terms of bulk properties viscosity and size of nanoparticle. The simulation is done using Matlab®. Polystyrene nanoparticle for which Brownian motion simulation was done in [2] is chosen for simulation with parameters as: radius r = 500 nm, density ρ = 1060 kg/m^3, at T = 296.01 K. For $f = 0$ implying absence of medium, simulation was done in [42] and results were obtained. It was observed from simulation in [42] that the variation of k has no effect on emissive power per unit area in the chosen range. The simulation results from radiation model (4) for frequency 1 MHz$\leq \upsilon \leq$ 100 MHz varied in steps of 1 MHz are plotted for $f' = 3.02 \times 10^7$ N·s/m and $k = 2.3$ N/m in Fig. 4. For these parametric values, results of proposed model were found to be validating with the results obtained from Planck's model given in (5). The plots in Fig. 4 have emissive power dissipated per unit area on Y-axis and frequency on X-axis.

Further, simulation results for variation of f from $9.42 \times 10^{-11} \leq f \leq 5.93 \times 10^{-9}$ N·s/m for $f' = 3.02 \times 10^7$ N·s/m, and $k = 2.3$ N/m are plotted in Fig. 5 to explore variation of power dissipation using model in (9). The viscosity is varied in geometric progression of two. The frequency is varied from 1 MHz to 1 GHz in steps

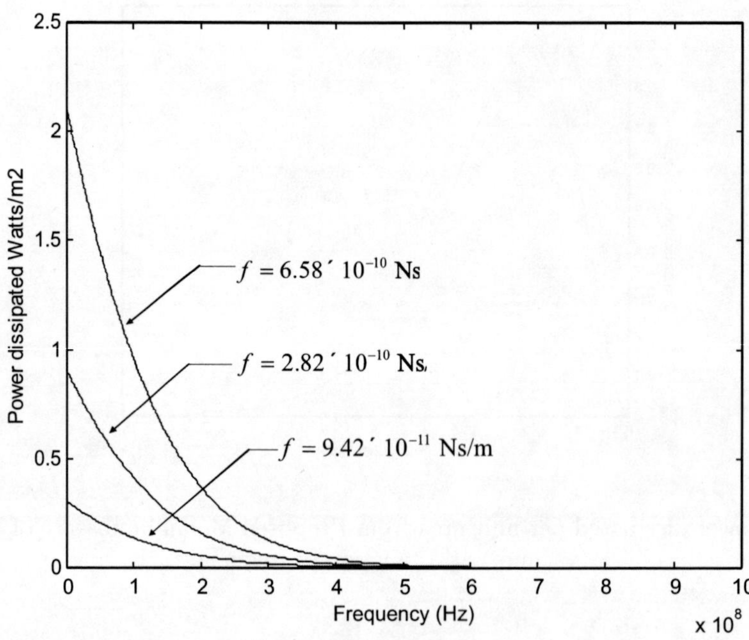

Fig. 5. Power dissipated per unit area from Proposed Model in presence of medium as a function of frequency

of 1 MHz and $f' = 3.02 \times 10^7$ N·s/m. It is observed from Fig. 5 that power dissipated in presence of medium is very high in comparison to power dissipated in vacuum. Further, power dissipated is high at low frequency and reduces asymptotically to zero at very high frequency. The trend is same for all values of f in the simulated regime. The power dissipation is more for higher values of f, which is obvious as presence of more viscous medium (high f) cause higher dissipation of energy. The relevance of proposed modeling is discussed in next section.

3. DISCUSSION

Radiation model given by (9) has been obtained from Brownian motion model [1, 2] considering a non-rigid nanoparticle. The Brownian motion model using system-modeling approach physically considers non-rigid nanoparticle parameterized by lumped parameters m, f', k against m only in case of rigid body model of nanoparticle. It is noted that neglecting f' and k, as is done in present literature of Brownian motion based on rigid-body model, using Langevin equation; and in the absence of medium ($f = 0$), radiation phenomenon cannot be accounted in the classical framework of Brownian motion model.

The physical nature of the model is hypothesized considering the non-rigid nanoparticle receiving only one impact from one surrounding medium molecule from some random direction. The nanoparticle being non-rigid will get deformed, i.e. the nanoparticle will be compressed at the impact site and bulge at the other end, resulting in a shift of the center of mass. This is characterized as local motion of nanoparticle, as shown in Fig. 2 *a*. The nanoparticle will also move (get displaced) from its location due to the impact and finite values. This is characterized as global motion of nanoparticle, shown in Fig. 2 *b*. If the nanoparticle is considered as rigid, there is no local motion and the global motion thus is unable to account for radiation in classical framework. The inclusion of f' and k as free parameters characterizing dissipation and absorption of nanoparticle makes it possible to account for radiation phenomenon in (4).

In real time, a number of single impacts (surrounding medium molecule impacts in case of Brownian motion and photon impacts bombarding nanoparticle in case of absence of medium) will take place on the nanoparticle from all directions randomly. At any instant, a plausible snapshot of the deformed and moved nanoparticle is shown in Fig. 2 *c*. The center of mass will get displaced due to local and global motions and the process will continue in time continuum. In the absence of medium, there will be local motion of the nanoparticle due to photon impacts. The shift of center of mass due to local motion of nanoparticle is stochastic with zero mean and normal distribution.

Moreover, the non-rigid nanoparticle will experience resistance to both global and local motion. The resistance to global motion is offered by the damping characteristics (parameterized by damping coefficient $f = 6\pi r\eta$ in Stokes regime) of the surrounding medium, while the resistance to local motion is due to dissipative properties parameterized by f' of nanoparticle. It is logical to assume that there is a coupling of resistance offered by the medium and nanoparticle to global and local motions of nanoparticle and the synergistic model of the two resistances along with parameters has been obtained using systems-modeling approach.

The present work is important from the point of view that an attempt to model power dissipation and radiation phenomenon in classical framework of Brownian motion has been done. The developed nanoparticle radiation model can be further refined using distributed parameter modeling and using non-instantaneous fluctuation-dissipation relation. Further, the non-rigidity is an issue in propulsion in nano-domains. With the advancement in pursuit of autonomous NEMS like nanorobots [51], non-rigidity and Brownian motion becomes relevant in context to equilibrium energy exchange model modeling done in the present work. The motion by diffusion may be more effective choice in comparison to convective motion in nanorobots. In order to achieve higher Peclect number at low Reynolds number, the diffusion of the nanorobot will play a vital role. Moreover in nano-domains, the non-rigid nature of the moving links of nanorobots is also to be taken into account. The future may see realization of nanorobots where design is done to enhance movements based on diffusion in addition to convection thereby increasing the efficiency.

REFERENCES

1. N.N. Sharma, M. Ganesh, and R.K. Mittal, "Non-Brownian Motion of Nanoparticle — An Impact Process Model," IEEE Tr. Nanotech. **3** (1), 180–186 (2004).
2. N.N. Sharma and R.K. Mittal, "Brownian Motion of Nanoparticle Considering Non Rigidity of Matter–Systems Modeling Approach," IEEE Tr. Nanotech. **4** (2), 180–186 (2005).
3. R. Kubo, "Brownian Motion and Nonequilibrium Statistical Mechanics," Science, **233**, 330–334 (1986).
4. M.D. Haw, "Colloidal Suspensions, Brownian Motion, Molecular reality: A Short Story", J. Phys: Condens. Matter **14**, 7769–7779 (2002).
5. S. Chandrasekhar, "Stochastic Problems in Physics and Astronomy," Rev. Mod. Phys. **15** (1), 1–89 (1943).
6. G.E. Uhlenbeck and L.S. Ornstein, "On the Theory of Brownian Motion," Phys. Rev. **36**, 823–841 (1930).
7. M.C. Wang and G.E. Uhlenbeck, "On the Theory of Brownian Motion II," Rev. Mod. Phys. **17**, 323–342 (1945).
8. N. Wax (Editor), *Selected papers on noise and stochastic process* (Dover, New York, 1954).
9. L.D. Landau and E.M. Lifschitz, *Fluid Mechanics* (Addison–Wesley, Reading MA, 1960).
10. K.M. Case, "Velocity Fluctuations of a Body in Fluid," Phy. Fluids, **14** (10), 2091–2095 (1971).
11. J.W. Dufty, "Gaussian Model for Fluctuation of a Brownian Particle," Phys. Fluids **17** (2), 328–333 (1974).
12. B.J. Berne, J.P. Boon, and S.A. Rice, "On the Calculation of Autocorrelation Functions of dynamical Variables," J. Chem. Phys. **45** (4), 1086–1096 (1966).
13. R. Zwanzig and M. Bixon, 'Compressibility Effects in the Hydrodynamic Theory of Brownian Motion," J. Fluid Mech. **69** (1), 21–25 (1975).
14. H. Kato, M. Tachibana, and K. Oikawa, "On the Drag of a Sphere in Polymer Solutions," Bull. JSME **15** (90), 1556–1567 (1972).
15. T.S. Chow and J.J. Hermans, "Effect of Inertia on the Brownian Motion of Rigid Particles in a Viscous Fluid," J. Chem. Phys. **56** (6), 3150–3154 (1972).
16. R.F. Rodriguez, "Brownian Motion and Correlation Function in a Viscoelastic Fluid," J. Phys. A: Math. Gen. **21**, 2121–2130 (1988).

17. J. Piasceck, "A Model of Brownian Motion in an Inhomogeneous Environment," J. Phys.: Condens. Matter **14**, 9265–9273 (2002).
18. R.A. Sack, "A Modification of Smoluchowski's Diffusion Equation," Physica **22**, 917–918 (1956).
19. W.T. Coffey, "On the Application of a Modification of the Smoluchowski Equation to Rotational Brownian Motion," J. Phys. D: Appl. Phys. **10**, L83–L86 (1977).
20. Carolyn M. Van Vliet, "Macroscopic and Microscopic methods for noise in Device," IEEE Tr. Electron Devices **41** (11), 1902–1915 (1994).
21. M.L. Roukes, "Nano Electromechanical Systems," in *Tech. Digest of 2000 Solid-State Sensor and Actuator Workshop* (Hilton Isl., SC, 6/4-8/2000, 2000), pp. 1–10.
22. M. Planck, *Warmestrahlung (1914)*, translated in *The Theory of Heat Radiation* (Dover, New York, 1991).
23. T.H. Boyer, *Foundation of Radiation Theory and Quantum Electrodynamics*, Ed. by A.O. Banet (Plenum, New York, 1980), pp. 49–63.
24. T.S. Kuhn, *Blackbody Theory and the Quantum Discontinuity 1894–1912* (Oxford University Press, New York, 1978).
25. T.H. Boyer, "Derivation of the Blackbody Radiation Spectrum without Quantum Assumption," Phys. Rev. **182** (5), 1374–1383 (1969).
26. O. Theimer, "Derivation of the Blackbody Radiation Spectrum by Classical Statistical Mechanics," Phys. Rev. D **4** (6), 1597–1601 (1971).
27. O. Theimer and P.R. Peterson, "Statistics of Classical Blackbody Radiation with Ground State," Phys. Rev. D **10** (12), 3962–3971 (1974).
28. J.P. Gordon, "Neoclassical Physics and Blackbody Radiation," Phys. Rev. A **12** (6), 2487–2497 (1975).
29. T.H. Boyer, "Equilibrium of Classical Electromagnetic Radiation in the Presence of Non-relativistic Nonlinear Electric Dipole Oscillator," Phys. Rev. D **13** (10), 2832–2845 (1976).
30. T.H. Boyer, "Derivation of Planck Radiation Spectrum as an Interpolation Formula in Classical Electrodynamics with Classical Electromagnetic Zero Point Radiation," Phys. Rev. D **27** (2), 2906–2911 (1983).
31. O. Theimer and P.R. Peterson, "Semiclassical Stochastic Radiation Theory", Phy. Rev. A **16** (5), 2055–2067 (1977).
32. T.H. Boyer, "Equilibrium Distribution for Relativistic Free Particles in Thermal Radiation with Classical Electrodynamics," Phys. Rev. A **20** (3), 1246–1259 (1979).
33. T.H. Boyer, "Thermal Effects of Acceleration through Random Classical Radiation," Phys. Rev. D **21** (8), 2137–2148 (1980).
34. T.H. Boyer, "Derivation of the Blackbody Radiation Spectrum from the Equivalence Principle in Classical Physics with Classical Electromagnetic Zero- Point Radiation," Phys. Rev. D **29** (6), 1906–1908 (1984).
35. D.C. Cole, "Reinvestigation of the Thermodynamics of Blackbody Radiation via Classical Physics," Phys. Rev. A **45** (12), 8471–8489 (1992).

36. C. Tsallis and F.C. SaBarreto, “Generalization of the Planck Radiation Law and Application to the Cosmic Microwave Background Radiation,” Phys. Rev. E **52** (2), 1447–1451 (1995).

37. E.A. Rohlfing, "Optical Emission Studies of Atomic, Molecular and Particulate Carbon produced from a Laser Vaporization Cluster Source", J. Chem. Phys., **89** (10), 6103–6112 (1988).

38. R. Scholl and B. Weber, *Physics and Chemistry of Finite Systems; From Clusters to Crystals*, 374 (NATO Advanced Study Institute Series C, Physics, Kluwer Academics, New York, 1992), p. 1275.

39. R. Mitzner and E.E.B. Campbell,"Optical Emission Studies of Laser Desorbed C_{60}", J. Chem. Phys. **103**, 2445–2453 (1995).

40. K.Hansen and E.E.B. Campbell, “Thermal Radiation from Small Particles,” Phys. Rev. E **58** (5), 5477–5482 (1998).

41. C. Luo, A. Narayanswamy, G. Chen, and J.D. Joannopoulos, “Thermal Radiation from Photonic Crystals: A Direct Calculation,” Phys. Rev. Lett. **93** (21), 213905 (2004).

42. Niti Nipun Sharma, “Radiation model for Nanoparticle:extension of classical Brownian motion,” J. Nanoparticle Res. **10** (2), 330–340 (2008).

43. A. Beiser, *Concepts of Modern Physics* (McGraw-Hill, New York, 1987).

44. E.M. Purcell, *Electricity & Magnetism.* Part 9 (McGraw-Hill, New York, 1984).

45. N.N. Sharma and R.K. Mittal, “On the Theory of Non-Brownian Motion in Nano-Regimes,” in *Proc. of Int. INAE Conference on Nanotechnology ICON-2003* (CSIO, Chandigarh, India, 2003), pp 586-596.

46. N.N. Sharma, M. Ganesh, and R.K. Mittal, “Nano-Electromechanical System Impact Spectrum Modeling and Clubbing of Structural Properties,” IE (I) Journal-MC, **85**, 88–193 (2005).

47. N.N. Sharma, *Need for Experimental Investigation for Validation of Relationship between Rigid and Non-Rigid Parameters in Nano-Domains*, Tata Institute of Fundamental Research (TIFR), Department of Nuclear and Atomic Physics (Mumbai, India, 2005).

48. N.N. Sharma and R.K. Mittal, “Non-Rigidity: Vital Link between Dynamics of Nanoparticle and Bio-species in Nano-domains,” in *Proc. of III Int. Conference on From Solid State to Biophysics* (Dubrovnik, Croatia, 2006).

49. N.N. Sharma and R.K. Mittal, “Brownian Motion System Models of Non-rigid Nanoparticle in Viscoelastic Medium,” in *National Symposium on Recent Trends in Physics. Nov. 4–5, 2005* (BITS, Pilani, India, 2005).

50. N.N. Sharma and R.K. Mittal, “Brownian Motion for 1-DOF Nanorobot”, in *Proc. Int. Conference on Emerging Mechanical Technology-Macro to Nano, EMTM2N-2007*, Ed. by R.K. Mittal and N.N. Sharma (BITS, Pilani, India, 2007), pp 35-38.

51. N.N. Sharma and R.K. Mittal, “Nanorobot Movement: Challenges and Biologically inspired solutions,” Int. J. Smart Sensing and Intelligent Systems **1** (1), 87–109 (2008).

A MODEL FOR THE PHONON TRANSPORT DUE TO LASER EXCITATION

D.A. Indeitsev[1], V.N. Naumov[1], B.N. Semenov[1], and A.K. Belyaev[1]

ABSTRACT

A mechanical two-component model of the solid of complex structure is presented. The suggested structural-rheological model is governed by a system of equations describing deformation of thermoelastic body of complex structure and accounting for the force and energy exchange between the material components of the body. Propagation of mechanical and temperature disturbances due to a pulse laser excitation is considered. The influence of the material structure on the character of transfer of the temperature disturbances is studied. The possibility of transfer of temperature disturbances with the velocity close to that of the mechanical pulse propagation is shown.

Key words: two-component model, solid, force and energy exchange, complex structure

INTRODUCTION

The rational mechanics of the continuous media ignores such an important physical property of a material as a discrete structure of real bodies. Such approach demonstrates an obvious discrepancy with known experiments, cf. [1–3]. It is clear that the model of a solid within the framework of rational mechanics should have complex structure in order to reflect the properties of discrete structure of the matter. Such a complex structure of a material is determined by presence of the internal degrees of freedom and influences its dynamics. The presence of these degrees of freedom can result in change of the basic macroparameters which are usually used for description of the material by means of the classical equations of continuum mechanics.

As shown in [4, 5], one of the approaches to description of behavior of the continuous media is introducing the two-component models. The latter allows one

[1]Institute for Problems in Mechanical Engineering of the Russian Academy of Sciences, St. Petersburg, Russia

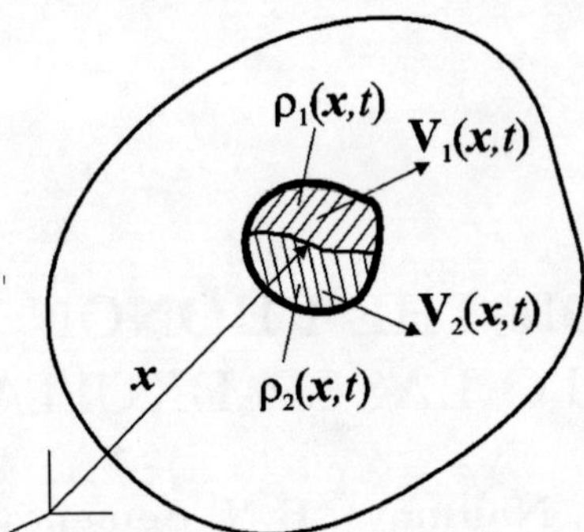

Fig. 1. A schematics of the two-component model

to explain some physical phenomena which have not been properly understood. In particular, this is the question of distribution of mechanical and temperature pulses in solids under short-term laser excitation [1].

The classical approaches usually assume introducing additional parameters in the constitutive equations. In this case the required thermodynamic relations serve allows one to determine these new variables. The basic equations for the two-component model introduced in the present paper point out an essential role of the internal structure of the material and allow us to describe the above-mentioned physical phenomena.

1. THE BASIC ASSUMPTIONS AND EQUATIONS

We postulate a model of the material with a carrying medium whose components are particles described by the displacement vector $\mathbf{u}_1(\mathbf{x}, t)$. An additional set of particles interacting with each other and with the carrying medium is attached to the carrying medium. The absolute displacement of particles of this additional media is given by vector $\mathbf{u}_2(\mathbf{x}, t)$. Both sets are supposed to be mutually penetrating continuous media. In other words, we introduce the concept of the material point that has a complex structure and consists of two components. In the expressions for displacements the argument $\mathbf{x}$ is the position vector of the material point in actual configuration, i.e. Euler's description is taken, see Fig. 1.

Physically, the different components of the material occupy different spatial volumes. In this regard there arises a question of the conditions of interaction of parts of a material on their internal boundaries. An axiomatic construction of the model reduces to assignment of interaction force $\mathbf{R}$ and heat exchange $\kappa(T_1 - T_2)$ between the components. (If needed, account for the mass exchange, chemical interactions between the components etc. can be modeled, too.) Realization of these representations results in structural-rheological two-component model. In essentially non-equilibrium processes the velocities, temperatures of the components can considerably differ from each other, so the effects of elastic and viscous interactions as well as the heat exchange between components are observable.

The law of mass conservation in the local form for each component and for the entire material is supposed to hold true

$$\frac{d_i\rho_i}{dt} + \nabla \cdot (\rho_i \mathbf{V}_i) = 0, \quad i = 1,\, 2 \tag{1}$$

$$\frac{d\rho}{dt} + \nabla \cdot (\rho \mathbf{V}) = 0. \tag{2}$$

Here

$$\frac{d_i}{dt} = \frac{\partial}{\partial t} + \mathbf{V}_i(\mathbf{x},t) \cdot \nabla, \quad \frac{d}{dt} = \frac{\partial}{\partial t} + \mathbf{V}(\mathbf{x},t) \cdot \nabla \tag{3}$$

denote the material derivatives, while ρ_1, ρ_2, ρ are densities of components and the entire material, respectively.

By virtue of the law of conservation of momentum we have

$$\rho \mathbf{V}(\mathbf{x},t) = \rho_1 \mathbf{V}_1(\mathbf{x},t) + \rho_2 \mathbf{V}_2(\mathbf{x},t). \tag{4}$$

Here and in what follows we assume the following expression for density of the material $\rho = \rho_1 + \rho_2$.

Velocities of the components and the center of mass of the material point are expressed as follows

$$\mathbf{V}_i(\mathbf{x},t) = \frac{d_i \mathbf{u}_i(\mathbf{x},t)}{dt}, \quad \mathbf{V}(\mathbf{x},t) = \frac{d\mathbf{u}(\mathbf{x},t)}{dt}. \tag{5}$$

The motion of the entire material point is governed by the law of dynamics in the local form

$$\nabla \cdot \tau + \rho \mathbf{F} = \rho \frac{d\mathbf{V}}{dt}. \tag{6}$$

The mass external force $\mathbf{F}$ can be given by

$$\rho \mathbf{F} = \rho_1 \mathbf{F}_1 + \rho_2 \mathbf{F}_2. \tag{7}$$

The equation of dynamics is convenient to rewrite in the form of two equations

$$\nabla \cdot \tau_1 + \rho_1 \mathbf{F}_1 + \mathbf{R} = \rho_1 \frac{d\mathbf{V}_1}{dt}, \quad \nabla \cdot \tau_2 + \rho_2 \mathbf{F}_2 - \mathbf{R} = \rho_2 \frac{d\mathbf{V}_2}{dt}, \tag{8}$$

where $\mathbf{R}$ is the force of interaction of two components of the material of complex structure. This interaction force $\mathbf{R}$ has an expression which is explicitly determined by the specific structure of the medium under consideration. In addition to this, the overall stress tensor of the material point is supposed to be the sum of the stress tensors of separate components

$$\tau = \tau_1 + \tau_2. \tag{9}$$

Let us note that the above equations for the two-component body (the two-component medium) are in agreement with the equations of mechanics of continuous heterogeneous media developed for modeling diverse mixtures [6–7].

The content of the present section follows the materials of Ref. [8]. The equation of the energy balance which usually refers to as the first law of thermodynamics, is widely used in continuum mechanics. The first law of thermodynamics deals with the various forms of energy and ways of exchange of energy between these forms. The basic (ideological) difficulty of the further analysis is the correct introduction of the concept of internal energy and total energy for a representative volume of the multi-component media. The total energy of particles of the media in the representative volume is understood as the sum of kinetic and internal energies. It is necessary to take into account that the kinetic energy is an additive function of mass and consequently can be represented by an integral over the mass. Generally speaking, the internal energy is additive with respect to the medium particles, some of the particles being inertialess. However in what follows we take advantage of the traditional view of additivity of the internal energy with respect to mass density $\rho_1 U_1$, $\rho_2 U_2$. In this case the equation of balance of energy for the representative (material) volume of the two-component media in the local form is given by

$$\begin{aligned}\rho_1\left(\frac{d_1 U_1}{dt} + \mathbf{V}_1 \cdot \nabla U_1\right) + \rho_2\left(\frac{d_2 U_2}{dt} + \mathbf{V}_2 \cdot \nabla U_2\right)\\ = \tau_1 \cdot\cdot \nabla \mathbf{V}_1 + \tau_2 \cdot\cdot \nabla \mathbf{V}_2 + \mathbf{R} \cdot (\mathbf{V}_2 - \mathbf{V}_1) - \nabla \cdot \mathbf{h} + \rho q.\end{aligned} \tag{10}$$

According to the idea of two-component medium the heat fluxes (energy fluxes of the non-mechanical character) $\mathbf{h}_1$, $\mathbf{h}_2$ in each material component are introduced into consideration. We also enter in consideration the temperatures θ_1, θ_2 of the first and the second component.

We split the equation of energy balance (10) into following two equations

$$\rho_1\left(\frac{d_1 U_1}{dt} + \mathbf{V}_1 \cdot \nabla U_1\right) = \tau_1 \cdot\cdot \nabla \mathbf{V}_1 + \frac{1}{2}\mathbf{R} \cdot (\mathbf{V}_2 - \mathbf{V}_1) - \nabla \cdot \mathbf{h}_1 + \rho_1 q_1 - Q, \tag{11}$$

$$\rho_2\left(\frac{d_2 U_2}{dt} + \mathbf{V}_2 \cdot \nabla U_2\right) = \tau_2 \cdot\cdot \nabla \mathbf{V}_2 + \frac{1}{2}\mathbf{R} \cdot (\mathbf{V}_2 - \mathbf{V}_1) - \nabla \cdot \mathbf{h}_2 + \rho_2 q_2 + Q, \tag{12}$$

where Q denotes the heat exchange between the components and $\mathbf{h} = \mathbf{h}_1 + \mathbf{h}_2$ is the vector of the overall heat flux.

Traditional form of the second law of thermodynamics is usually given as the Clausius-Duhem inequality: The rate of change of the internal entropy of the medium is not less than the rate of influx of the entropy in this media from outside, [9]. This general formulation concerns with a classical solid. In our case of two-component media, it is possible to introduce two absolute temperatures $\theta_1 > 0$, $\theta_2 > 0$, attributed to each component, two internal entropies S_1, S_2 and formulate the second law of thermodynamics for each component (that is, in the form of two inequalities)

$$\frac{d}{dt}\int_{(V)} \rho_1 S_1 \, dV \geq \int_{(V)} \left(\frac{\rho_1 q_1}{\theta_1} + \frac{Q}{\theta_2}\right) dV - \int_{(S)} \mathbf{n} \cdot \left(\frac{\mathbf{h}_1}{\theta_1} - \rho_1 \mathbf{V}_1 S_1\right) dS, \quad (13)$$

$$\frac{d}{dt}\int_{(V)} \rho_2 S_2 \, dV \geq \int_{(V)} \left(\frac{\rho_2 q_2}{\theta_2} + \frac{Q}{\theta_1}\right) dV - \int_{(S)} \mathbf{n} \cdot \left(\frac{\mathbf{h}_2}{\theta_2} - \rho_2 \mathbf{V}_2 S_2\right) dS. \quad (14)$$

Here V denotes the representative volume containing particles of both components and S is the boundary surface.

2. TWO-COMPONENT ONE-DIMENSIONAL MODEL OF A THERMO-ELASTIC MATERIAL OF COMPLEX STRUCTURE

2.1. *Basic equations of one-dimensional model*

Using the one-dimensional equations for analysis of distribution of wave pulses in a specimen is substantiated for extended specimens. Assuming smallness $\partial\rho_1/\partial x$, $\partial\rho_2/\partial x$, u_1, u_2, T_1, T_2 and their derivatives with respect to time and spatial coordinate, we carry out linearization of the above three-dimensional equations of thermoelasticity

$$\begin{aligned}
&\frac{\partial \sigma_1}{\partial x} - \rho_1 \frac{\partial^2 u_1}{\partial t^2} + \rho_1 F_1^e - R = 0, \\
&\frac{\partial \sigma_2}{\partial x} - \rho_2 \frac{\partial^2 u_2}{\partial t^2} + \rho_2 F_2^e + R = 0, \\
&\lambda_1 \frac{\partial^2 T_1}{\partial x^2} - \rho_1 c_1 \frac{\partial T_1}{\partial t} = E_1 \alpha_1 \theta_0 \frac{\partial^2 u_1}{\partial x \partial t} - \rho_1 b_1 - \kappa(T_1 - T_2), \\
&\lambda_2 \frac{\partial^2 T_2}{\partial x^2} - \rho_2 c_2 \frac{\partial T_2}{\partial t} = E_2 \alpha_2 \theta_0 \frac{\partial^2 u_2}{\partial x \partial t} - \rho_2 b_2 + \kappa(T_1 - T_2).
\end{aligned} \quad (15)$$

As $\sigma_k = E_k[\varepsilon_k - \alpha_k(\theta_k - \theta_0)]$ $(k = 1, 2)$, Eq. (14) takes the form, cf. [10]

$$\begin{aligned}
&E_1 \frac{\partial^2 u_1}{\partial x^2} - \rho_1 \frac{\partial^2 u_1}{\partial t^2} - E_1 \alpha_1 \frac{\partial T_1}{\partial x} + \rho_1 F_1^e - R = 0, \\
&E_2 \frac{\partial^2 u_2}{\partial x^2} - \rho_2 \frac{\partial^2 u_2}{\partial t^2} - E_2 \alpha_2 \frac{\partial T_2}{\partial x} + \rho_2 F_2^e + R = 0,
\end{aligned} \quad (16)$$

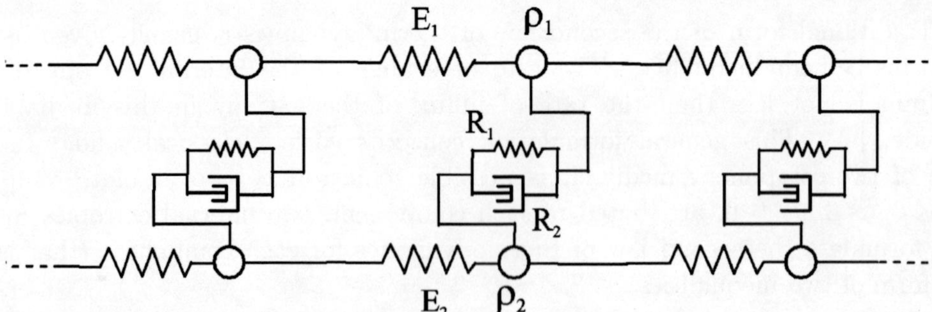

Fig. 2. The general form of the structural-rheological model

where $R = R_1(u_1 - u_2) + R_2(\partial u_1/\partial t - \partial u_2/\partial t$ is the force of a elastic-viscous interaction of the components and $T_k = \theta_k - \theta_0$, θ_0 being the reference temperature of model (material).

The equations in Eq. (16) represent the coupled equations of dynamics of one-dimensional two-component thermoelastic model, whereas equations in (15) describe propagation of thermal perturbations in the first and second components. So, according to Eqs. (15) and (16) the interaction between the components of the one-dimensional two-component model is carried out in terms of the force of mechanical interaction R and the thermal interaction $Q = \kappa(T_1 - T_2)$.

2.2. *Features of behavior of two-component model under non-stationary loading*

Equations (15) and (16) should be subjected to initial and boundary conditions. A structural-rheological model is suggested for visualization of the approach and displayed in the Fig. 3.

A linear elastic-viscous force interaction of the material components is supposed, that is, $R = R_1(u_1 - u_2) + R_2(\partial u_1/\partial t - \partial u_2/\partial t$.

This two-component model can reflect the properties of various real thermoelastic bodies. In the present paper the model of a two-component body is used for analysis of the phenomena related to intensive ultrashort laser excitation of the material surface. The above construction allows us to consider material as a system of two continua, each possessing own properties and the cumulative properties due to coexistence and mutual interaction of the material components. The equations for each material component have equal rights. Physical distinctions in components are introduced by compulsory assignment of the parameters which are essentially distinguished from each other.

In all cases of the present analysis the first material component plays the role of a rigid skeleton of a body. This component models a crystal lattice of a solid consisting of the massive (nearly motionless) particles (atoms, ions). In the first (rough) approximation the first material component is supposed to be a physically and geometrically linear thermoelastic continuum.

The second material component models a set of small particles (set of electrons) interacting with the first component. The character of interaction of particles of the second component with the first one and each other depends on the physical properties of the considered specimen. The peculiarity of interaction of each particles with the particles of both components influences the overall rheological properties of specimen.

Actually the specimen has a discrete structure which cannot be ignored under the conditions of a laser pulse excitation. The two-component model suggested here is an essentially continuum model. In this regard, there exist some questions about statement of the boundary conditions for the formulated problems, in particular, as to how to describe the mechanism of transfer of energy of a laser pulse to a tested specimen. It is obvious, that from the point of view of a continuum model the laser excitation influence results either in force or temperature loading. One has to neglect micromechanical interaction of radiation and boundary particles of a specimen, that is, it is necessary to ignore the fact of influence of the oscillating electromagnetic field on the charged particles of a specimen. Clearly, the micromechanics of the energy transfer should be considered separately, and the result of the analysis should be formulated for the continuum model as a set of certain postulates.

2.3. *Pulse temperature loading of a semi-conductor crystal (a Si specimen)*

It is known, that under certain conditions the thermal perturbation in crystals can propagate on macroscopical distances with a velocity close to the sound velocity. So at very low temperatures (≈ 4.2 K) generation of the wave fluctuations of temperature is possible, cf. [11].

In the considered linear two-component model, such phenomenon is not feasible. However, it could be simulated under the condition of practical absence of heat conductivity in both components, $\lambda_1 \approx 0$, $\lambda_2 \approx 0$ and vanishingly small thermal capacities of both components.

Let us proceed to Eqs. (15)–(16). A plenty of the physical parameters in these equations allows us to model the various phenomena due to the pulse laser excitation.

As an example we consider the propagation of a wave thermo-mechanical pulse in a crystal semi-conductor silicon specimen.

Let us estimate parameters of the two-component model:

The specific thermal capacity (at $20 \div 100^\circ$C) ≈ 800 J/(kg·K). On example of other substances it is possible to assume, that the thermal capacity of a crystal lattice of silicon at cryogenic temperatures is hundreds times lower, i.e. we take

$c_1 \approx 0.01$ J/(kg·K), $c_2 \approx 1$ J/(kg·K). This value of c_1 assumes practically full absence of fluctuations of atoms (ions) of the crystal lattice of the specimen, value c_2 testifies that for the modeled material the electrons are not free and do not determine the thermal capacity of the second components (that is, the second component does not affect the thermal capacity).

Heat conductivity (at 250°C) $\approx 84 \div 126$ W/(m·K). Obviously, at low (cryogenic) temperatures the heat conductivity of silicon should be much lower, i.e. for modeling we take $\lambda_1 \approx 0$ W/(m·K), $\lambda_2 \approx 0$ W/(m·K), that promotes the wave distribution of thermal perturbations (together with mechanical pulses).

The temperature coefficient of linear expansion for the silicon at normal temperature $\alpha \approx 2.33 \cdot 10^{-6}$ 1/K. At temperatures lower than 120 K this coefficient becomes negative. For modeling we take $\alpha_1 \approx 2.33 \cdot 10^{-6}$ 1/K, $\alpha_2 \approx 2.33 \cdot 10^{-9}$ 1/K.

The elasticity module of silicon at normal temperature is $E = 109 \cdot 10^9$ N/m^2. For two-component model we accept $E_1 = 109 \cdot 10^9$ N/m^2, $E_2 = 3.4 \cdot 10^7$ N/m^2. The very fact $E_2 \neq 0$ means presence of own elasticity arising from polarization of the electric fields under displacement of electronic clouds of the silicon atoms.

Density of silicon is $\rho \approx 2.33 \cdot 10^3$ kg/m^3. The atom of silicon has 14 electrons. They settle down on 3 levels: 2–8–4, the last level having 4 electrons. Most likely, these last electrons "form" the second deformable component of the considered two-component model. Probably, this statement will demand some correction in the future. Generally speaking, it is possible to think of formation of the second component by amount of electrons from 1 up to 14. The mass of a nucleus of atom of silicon is $M \approx 4.676 \cdot 10^{-26}$ kg. The mass of the electrons involved in formation of the second components is $9.11 \div 1.275 \cdot 10^{-29}$ kg. Hence the densities of components of the model could be taken as $\rho_1 \approx 2.33 \cdot 10^3$ kg/m^3, $\rho_2 \approx 0.0454 \div 0.635$ kg/m^3. At the rate of 4 electrons, forming the second component, we have the density of the component $\rho_2 \approx 0.182$ kg/m^3. The moduli of elasticity and the mass density define the following velocities of propagation of wave pulses in both components $v_2 \approx 13700 > v_1 \approx 6850$, $v_2 = 2v_1$.

As we have no information about the temperature exchange between the components we accept the condition of the absence of free electrons, i.e. factor κ is close to zero $\kappa \approx 0$ W/(m^3·K).

While determining the factor of force interaction between the components R_1, it is necessary to be guided by the value of frequency of partial fluctuations of the second component relative to the first one. If we accept $\rho_2 \approx 0.182$ kg/m^3 and $f \approx 1.18 \cdot 10^8$ Hz then we obtain the following factor of the elastic interaction $R_1 \approx 1 \cdot 10^{17}$ N/m^4. We also put $R_2 \approx 10^3$ N·s/m^4.

The accepted parameters for two-component models are close enough to the real physical values. It is supposed the one-dimensional specimen $0 < x < l \approx 5.5 \cdot 10^{-3}$ m

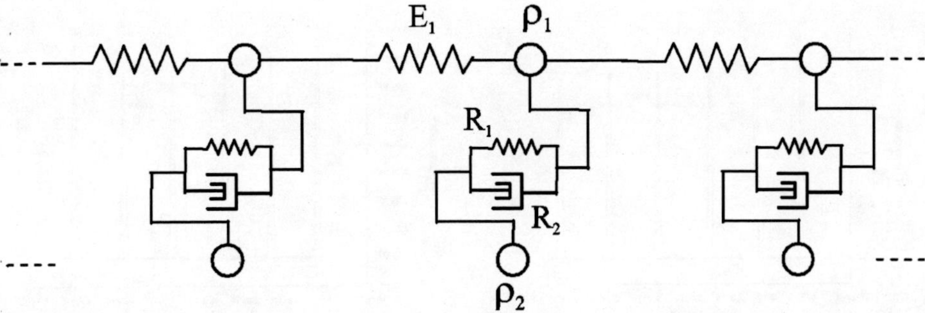

Fig. 3. A structural-rheological model for one-dimensional two-component nonmetal body

is loaded on the end face $x = 0$ in terms of a short-term temperature pulse

$$T_1\Big|_{x=0} = T_{10}[H(t) - H(t-\tau)], \quad \frac{\partial T_2}{\partial x}\Big|_{x=0} = 0$$

or

$$T_2\Big|_{x=0} = T_{20}[H(t) - H(t-\tau)], \quad \frac{\partial T_1}{\partial x}\Big|_{x=0} = 0.$$

That is, both ends $x = 0, l$ of the specimen are assumed to be free of loading.

For modeling of behavior of nonmetals (at cryogenic temperatures the semi-conductor properties are close to properties of nonmetals) the following structural-rheological model is applied (Fig. 3).

The semi-conductor material at cryogenic temperatures behaves as an isolator.

The thermal phenomena due to oscillatory micromovements are "frozen" at low temperatures. It can be treated as reduction of the thermal capacities c_1, c_2 and the factors of heat conductivity λ_1, λ_2 in comparison with their values at normal temperatures. For numerical modeling we take $c_1 = 0.001$, $c_2 = 1$, $\lambda_1 = \lambda_2 = 0$. The specific thermal capacities are accepted to be non-zero in order to avoid absurdity of physical sense. Besides, it is accepted $c_2 > c_1$ as the second component (electrons near atoms and ions of a crystal lattice of the first component) possesses a greater micromobility than that of the first component.

Acceptance of values $\lambda_1 = \lambda_2 = 0$ assumes that at cryogenic temperatures the heat on a specimen (on components) is not transferred "independently" as a thermal flux. Since transfer of heat (temperature perturbation) is carried out by means of the mechanical wave pulses.

We also make the following assumption in regard of the factors of temperature expansion. Clearly, reducing the specimen temperature affects the values of this factor. It was not possible to find out these changes in reference books. It is obvious that the factor of temperature expansion of the first component (representing a crystal lattice) can be higher than that for the second component (representing a

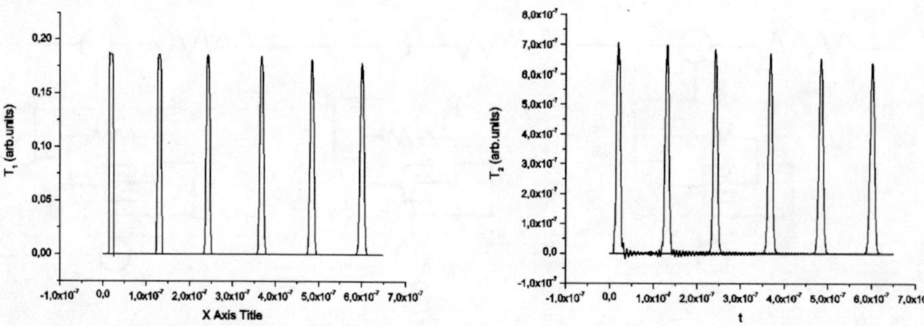

Fig. 4. Temperature of the first and second components versus time for the cross-sections model $x \approx 0$, $x \approx 1$, $x \approx 2$, $x \approx 3$, $x \approx 4$, $x \approx 5$ mm, respectively

"friable" set of electrons). For modeling we take $\alpha_1 \approx 2.33 \cdot 10^{-6}$, $\alpha_2 \approx 2.33 \cdot 10^{-9}$. As the system of equations is linear only the relations between the parameters are important rather than the absolute vales of the temperature factors. The numerical calculations show that the final results weakly depend on sizes and the ratio of temperature expansion factors.

Under the given conditions the second component in the micromovements is constrained (in the semiconductor the free electrons are practically absent) it is accepted $\kappa \approx 0$. This implies absence of the heat exchange between components. The other parameters remain unchanged.

As the described two-component model is based upon the principles (laws) of rational mechanics its components cannot transfer anything other but mechanical wave pulses and temperature indignations. For the considered pulse loading of the process of excitation of the model is as follows. The temperature pulse $T_1|_{x=0} = T_{10}(H(t) - H(t-\tau))$, $\partial T_2/\partial x|_{x=0} = 0$ applied at end $x = 0$, heats up the first component and practically does not affect propagation of the temperature perturbation. The heating generates temperature deformations in the first component which propagate as wave pulses along the specimen. Thermomechanical indignation of first component through force interaction generates as well in the second component a thermomechanical wave pulse. Since we use the equations of the coupled thermoelasticity Eqs. (17)–(19) the temperature perturbation propagates along the specimen also in the form of the mechanical wave pulse.

Let us notice once again that we use a model of rational mechanics. The question as to how to treat the described phenomena from a perspective of the physical processes real semi-conductor crystal, e.g. a silicon specimen, remains to be tackled.

The results below are represented in Figs. 4 and 5 in terms of the non-dimensional values.

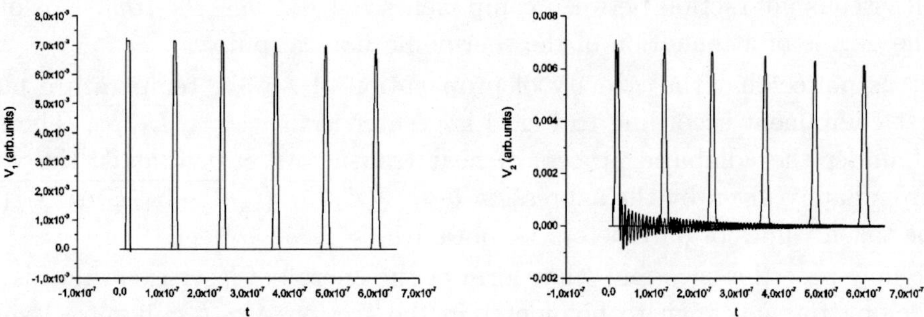

Fig. 5. Velocities of the first and second components versus time for the cross-sections $x \approx 0$, $x \approx 1$, $x \approx 2$, $x \approx 3$, $x \approx 4$, $x \approx 5$ mm, respectively

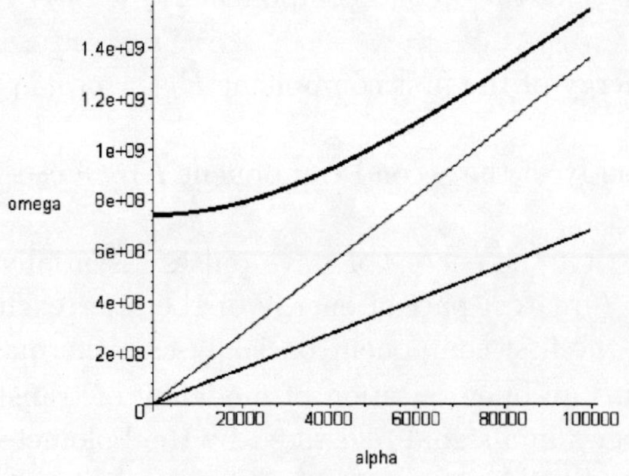

Fig. 6. Dispersion curves

The figures for T_2, V_2 clearly demonstrate oscillations with frequency which is equal to frequency of partial fluctuations of the second component on elastic constraint R_1. It is approximately estimated as $f = 1/(2\pi)\sqrt{R_1}/\rho_2 \approx 1/(2\pi) \cdot 0.741 \cdot 10^9 \approx 1.18 \cdot 10^8$ Hz. They can be understood as confirmation of existence of the second (electronic) component. The feasibility of existence of high-frequency wave pulses is confirmed by the dispersion curves shown in Fig. 6.

Let us note that the temperature pulse of the first component weakly attenuates as the pulse propagates along the specimen. This attenuation is partly caused by that the specimen has a small (however non-zero) thermal capacity. Besides, there is a small viscous interaction between components $R_2(\partial u_1/\partial t - \partial u_2/\partial t)$. It is obvious that the degree of attenuation of the thermomechanical pulse.

Let us notice that the velocity of propagation of a wave temperature pulse in the first component according to Fig. 4 is greater than $v_1 = \sqrt{E_1/\rho_1}$. The reason is that under the adiabatic process of heat transfer $\lambda_1 \approx 0$, that is, the velocity is approximately given by the expression $\tilde{v}_1 \approx \sqrt{E_1/\rho_1} \times \sqrt{1 + \alpha_1(E_1\alpha_1\theta_0)/(\rho_1 c_1)}$. For the taken values of parameters we obtain $\tilde{v}_1 \approx 1.23v_1$.

We now pose the question, what form of the energy of laser excitation is transferred along the specimen to bolometer in the section $x = l \approx 5$ mm. Using the above result for the considered pulse temperature loading, we obtain the following maximal values

- the kinetic energy of the first component $K_1 = \frac{1}{2}\rho_1 V_1^2 \approx 0.57 \cdot 10^{-1}$,
- the kinetic energy of the second component $K_2 = \frac{1}{2}\rho_2 V_2^2 \approx 0.33 \cdot 10^{-5}$,
- the thermal energy of the first component $E_{T1} = c_1\rho_1 T_1 \approx 0.42$,
- the thermal energy of the second component $E_{T2} = c_2\rho_2 T_2 \approx 0.13 \cdot 10^{-6}$.

One can observe that the energy of wave pulses essentially differs for two components. The major (greater) part of energy of the laser excitation is delivered to the bolometer along the first component basically as a thermal pulse.

There still remains an open question of modeling of transformation of thermal and mechanical energies in a signal registered by the bolometer.

Certainly, the obtained data are result of the simulation, however, they are not too far from real ones. We shall also notice that in the present research the second (electronic) component does not practically affect transport of the wave pulse excited by the laser along the specimen. The influence of the second component is exposed in terms of high-frequency oscillations of velocity and temperature.

The result of the above simulation leads us to the conclusion about an opportunity of updating of considered model by introducing a single temperature. However, this conclusion seems to be valid only for the considered material and the above loading conditions.

The further result is shown for a lower value of the force interaction between the components, namely $R_1 \approx 10^{16}$ instead of $R_1 \approx 1 \cdot 10^{17}$.

It is observable that all characteristic features of distribution of wave pulses are preserved. The partial frequency of fluctuations of the second component decreases by factor $\sqrt{10}$.

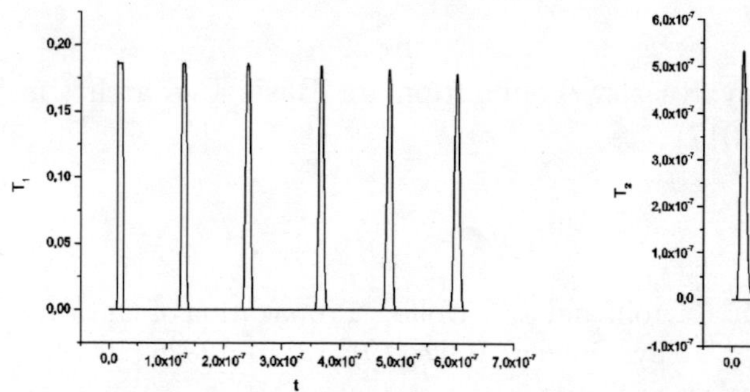
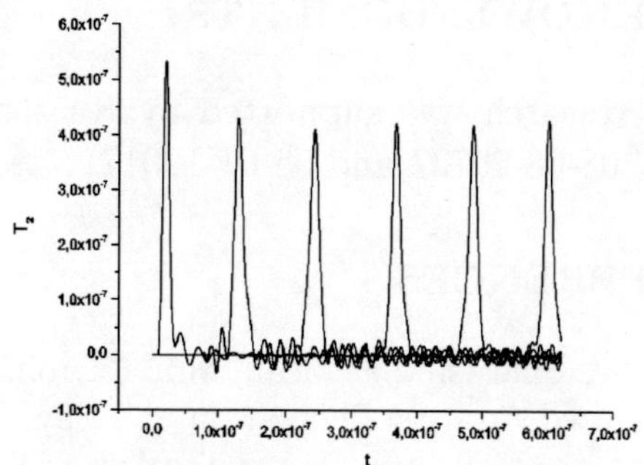

Fig. 7. Temperature of the first and second components versus time for the cross-sections $x \approx 0$, $x \approx 1$, $x \approx 2$, $x \approx 3$, $x \approx 4$, $x \approx 5$ mm, respectively

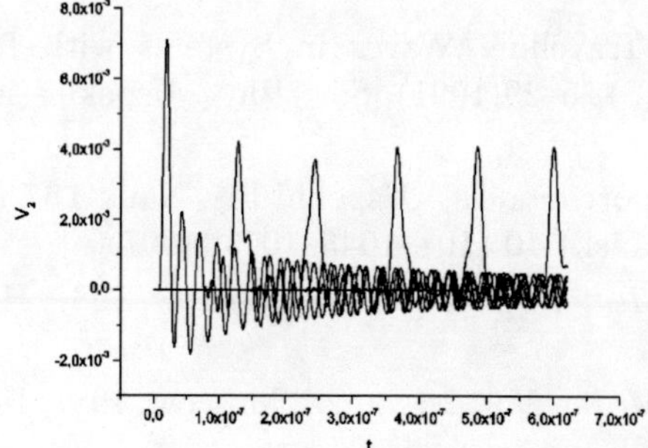
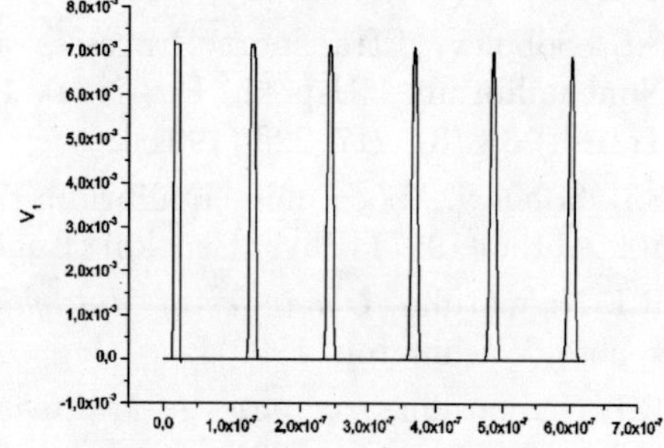

Fig. 8. Velocities of the first and second components versus time for the cross-sections $x \approx 0$, $x \approx 1$, $x \approx 2$, $x \approx 3$, $x \approx 4$, $x \approx 5$ mm, respectively

CONCLUSION

The present paper is concerned with an attempt of using a two-component model for treatment of features of propagation of thermal and mechanical pulses caused by non-stationary loading from a perspective of the rational mechanics The essential role coupled thermoelasticity becomes evident. However we also obtain the results which confirm the well-known fact of weak influence of effect of coupling thermal and mechanical processes on the temperature distribution. In this case, the second (electronic) component practically does expose itself in the process of transfer of the thermomechanical pulse through the semi-conductor specimen. Practically the whole energy of laser excitation is transferred along the first component.

ACKNOWLEDGMENTS

The research was supported by Russian Foundation for Basic Research (projects Nos. 08-08-00737 and 08-01-12017).

REFERENCES

1. J.A. Shields, M.E. Msall, M.S. Carroll, and J.P. Wolfe, "Propagation of Optically Generated Acoustic Phonons in Si," Phys. Review B **47** (19), 12510–12526 (1993).
2. Yu.V. Sud'enkov and A.I. Pavlishin, "Nanosecond Pressure Pulses Propagating at Anomalously High Velocities in Metal Foils," Pis'ma Zh. Tekh. Fiz. **29** (12), 14–20 (2003) [Tech. Phys. Lett. (Engl. Transl.) **29** (6), 491–493 (2003)].
3. Yu.G. Gurevich, G. González de la Cruz, G.N. Logvinov, and M.N. Kasyanchuk, "Effect of Electron-Phonon Energy Exchange on Thermal Wave Propagation in Semiconductors," Fiz. Tekh. Poluprovodn. **32** (11), 1325–1330 (1998) [Semiconductors (Engl. Transl.) **32** (11), 1063–7826 (1998)].
4. S.L. Sobolev, "Transport Processes and Travelling Waves in Systems with Local Nonequilibrium," Uspekhi Fiz. Nauk **161** (3), 5–29(1991) [Sov. Phys. Uspekhi (Engl. Transl.) **34** (3), 217–229 (1991)].
5. S.L. Sobolev, "Local non-equilibrium transport models," Uspekhi Fiz. Nauk **167** (10), 1095–1106 (1997) [Phys. Uspekhi (Engl. Transl.) **40** (10), 1043–1053 (1997)].
6. R.I. Nigmatulin, *Dynamics of Multiphase Media.* Vol. 1 (Nauka, Moscow, 1987; Hemisphere, Washington, 1990).
7. R.I. Nigmatulin, *Dynamics of Multiphase Media.* Vol. 2 (Nauka, Moscow, 1987; Hemisphere, Washington, 1991).
8. P.A. Zhilin, *Advanced Problems in Mechanics. Selection of articles.* Vol. 2 (Edition of IPME RAS, St. Petersburg, 2006).
9. V.A. Palmov, *Vibrations of Elasto-Plastic bodies* (Springer, Heidelberg, 1998).
10. W. Nowacki, *Theory of Elasticity* (PWN, Warsaw, 1970; Mir, Moscow, 1975).
11. K.P. Zol'nikov, R.I. Kadyrov, I.I. Naumov, et al., "Possible Nonlinear Heat-Pulse Propagation in Solids at Debye Temperatures," Pis'ma Zh. Tekh. Fiz. **25** (6), 55–59 (1999) [Tech. Phys. Lett. (Engl. Transl.) **25** (3), 230-232 (1999)].

LASER INDUCED EXCITONIC AND BI-EXCITONIC PROPERTIES OF CuCl THIN FILMS ON Si

A. Mitra[1*], L. O'Reilly[2], O.F. Lucas[2], G. Natarajan[3], A.L. Bradley[4], P.J. McNally[2], and S. Daniels[3]

ABSTRACT

Recently, study of alternative wide band gap semiconductors which can replace the GaN and its family has drawn lot of attentions among the researchers. CuCl is another kind of wide band gap semiconductor with band gap 3.4 eV which has a potential of replacing GaN. Major advantage of CuCl is that it has a large excitonic binding energy of 190 meV compare to other wide band gap semiconductors like ZnO (60 meV) and GaN (25 meV). In this paper we present a report on study of excitonic and bi-excitonic mediated optical properties of vacuum deposited CuCl thin films on Si substrate with excellent lattice matching. Properties of luminescence from CuCl thin films on Si have been studied using temperature dependent photoluminescence spectroscopy. We have been identified four peaks attributed to free exciton (Z_3) (3.203 eV), bound exciton (I_1) (3.181 eV), bi-exciton (M) (3.159 eV) and bound bi-exciton (N_1) (3.134 eV) from PL spectrum at 10 K. A strong free excitonic peak at room temperature has been observed at 3.230 eV. Binding energies for bound exciton, bi-exciton and bound bi-exciton have been calculated. Parameters like PL peak intensity, energy and line–width have been extracted from the temperature dependence of the free excitonic Z_3 peak and compared to CuCl films on other substrates and in single crystal form. It has also been observed that temperature dependent variation of band gap of CuCl with temperature is just opposite to that of GaN and ZnO. The luminescence properties of CuCl films on Si system are found to be compared well with reports for single crystal CuCl.

[1]Department of Physics, Indian Institute of Technology Roorkee, Roorkee — 247667, Uttarakhand, India

[2]Nanomaterials Processing Laboratory, Research Institute for Networks and Communications Engineering (RINCE), School of Electronic Engineering, Dublin City University, Dublin 9, Ireland

[3]Nanomaterials Processing Laboratory, National Centre for Plasma Science & Technology (NCPST), School of Electronic Engineering, Dublin City University, Dublin 9, Ireland

[4]Semiconductor Photonics, Physics Department, Trinity College, Dublin 2, Ireland

[*]E-mail: *anmitra@yahoo.com*. Phone: +911332285652. Fax: +911332286662

Key words: CuCl, exciton, bi-exciton, wide band gap semiconductor, photoluminescence, ultra-violet emission

INTRODUCTION

Wide band gap materials have drawn lot of attraction among the researchers for a range of applications such as UV light emitting diodes, diode lasers and detectors [1]. Efforts have mainly focused on II-VI and III-Nitride materials systems. The latter have been the most successful to date, though a fundamental problem with this material system is the large lattice mismatch ($\sim$ 13% [2]) between the GaN epitaxial layers and suitable compatible substrates (e.g. SiC, α-Al_2O_3). ZnO is also the subject of extensive research, particularly for applications wishing to exploit the high exciton binding energy, such as room temperature cavity polariton physics [3]. We propose an alternative; a direct wide band gap semiconductor material, emitting in the UV, which is closely lattice matched to the substrate, γ-CuCl on Si. γ-CuCl has band gap energy of 3.40 eV and a lattice mismatch with cubic Si of $< 0.4\%$ at room temperature [4]. It has been extensively studied in the form of micro-sized crystals embedded in various host matrices [5–8] and exhibits interesting properties such as large exciton binding energy of 190 meV compared to 60 meV in ZnO and 25 meV in GaN. Due to the large exciton binding energy strong exciton lasing action [9–10] has been observed at low temperatures. It also exhibits a large bi-exciton binding energy of 34 meV [11–12] and bi-exciton lasing action from CuCl quantum dots embedded in NaCl matrix [8, 13–14] has also been reported. CuCl thin films deposited on various substrates such as Al_2O_3, CaF_2, quartz, TiO_2 and GaAs have been previously studied [4, 11, 15–16]. However these material systems are not easily compatible with current electronic or optoelectronic technologies. γ-CuCl is closely lattice matched to both silicon and GaAs and is an ideal candidate for the development of hybrid electronic-optoelectronic platforms. In order to make an assessment of the luminescence properties the measurements of γ-CuCl thin films on Si are placed in the context of previously reported measurements on single crystal CuCl and thin film of CuCl on other substrates.

1. EXPERIMENTAL TECHNIQUE

CuCl thin film samples with typical layer thicknesses of $\sim$ 500 nm were grown on Si(100) at room temperature using an Edwards Auto 306A vacuum deposition system at a base pressure of $\sim 1 \times 10^{-6}$ mbar. Details of the sample preparation and growth can be found elsewhere [17–18]. The crystallinity of the films was characterized using the x-ray diffraction (XRD) technique. The optical properties of the films were studied using temperature dependent photoluminescence (PL)

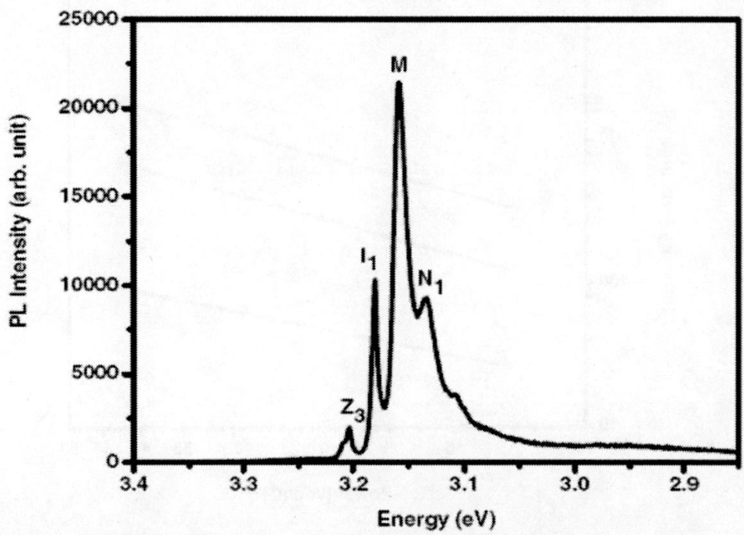

Fig. 1. PL spectrum of CuCl thin film on Si(100) at 10 K

in the range 10 K to room temperature. Due to the high hygroscopicity of the CuCl material the samples were held under vacuum at all times. Photoexcitation, at 244 nm, was provided via frequency doubling of the 488 nm line from a CW Innova Ar ion laser using a BBO crystal. A Jobin Yvon-Horiba, Triax 190 spectrometer with a spectral resolution of 0.3 nm, coupled to a liquid nitrogen cooled CCD, was used to record the photoluminescence spectra.

2. RESULTS AND DISCUSSION

We have been optically characterized the vacuum deposited CuCl films on Si using power and temperature dependent photoluminescence. A typical low temperature PL spectrum, recorded at 10 K, for CuCl on Si(100) is shown in Fig. 1. Four main peaks are observed and identified. The free exciton peak, Z_3, occurs at 3.203 eV. The peak at 3.181 eV is attributed to the bound exciton peak, I_1, in agreement with the literature [19]. The bound exciton emission may be associated with an impurity such as a Cu^+ vacancy [11]. The energies for the free and bound exciton peaks agree exactly with previous reported measurements for CuCl bulk crystal at 8 K [19] and thin films on Al_2O_3 [11]. The peak occurring at 3.159 eV, is close to what is expected for the well-known free bi-exciton PL band (M). The fourth strong peak at 3.134 eV is attributed to the bound bi-exciton (N_1) [11].

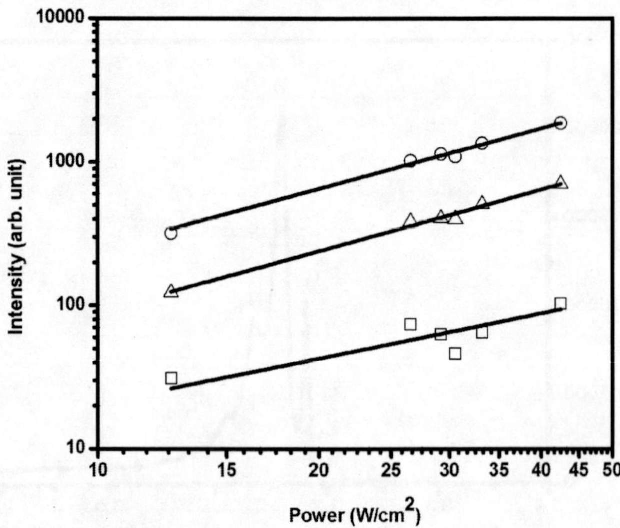

Fig. 2. Variation of free exciton (Z_3), bi-exciton (M) and bound bi-exciton (N_1) peak intensities with laser power. The theoretical simulation (solid curve) to the experimental data points for free exciton (□), bi-exciton (○) and bound bi-exciton (△) are obtained using the equations $y = 45.22x^{0.86}$, $y = 631.8x^{1.45}$, $y = 237x^{1.42}$

The bi-exciton peak assignments were further tested by performing power dependent PL measurements, undertaken at 10 K. Theoretically a bi-exciton peak intensity increases as the square of the incident laser power while exciton peak intensity increases linearly with the incident optical power, though this is rarely seen experimentally [7–8]. An approximate factor of two between both power dependencies is a typical signature of bi-excitonic behavior, even if it is only evident over a small range of excitation power [8]. The peak intensities are plotted as a function of the power density in Fig. 2. This power range is almost 100 times smaller than that used in other studies reporting bi-excitonic features. However, we are pumping at 244 nm where the absorption coefficient is higher than the 355 nm and 337 nm sources used in the other studies [20]. We measure power dependencies of 1.45 ± 0.08 and 1.4 ± 0.1 for the bi-exciton and bound bi-exciton respectively, in contrast with 0.86 ± 0.38 for the free exciton. This behavior is consistent with a bi-excitonic nature of the two lower energy peaks.

Based on these assignments we can estimate the binding energies of the bound exciton (E^b_{bX}), the free bi-exciton (E^b_{XX}) and the bound bi-exciton (E^b_{bXX}) from the following equations based on the corrected energy calculation scheme in accordance with reference [11].

$$E^b_{bX} = E_X - E_{bX}, \tag{1}$$
$$E^b_{XX} = E_X - E_{XX}, \tag{2}$$
$$E^b_{bXX} = 2E_X - E_{bXX} - E_{bX} - E^b_{XX} \tag{3}$$

where E_X (3.203 eV), E_{XX} (3.159 eV), E_{bX} (3.181 eV), and E_{bXX} (3.134 eV) are free exciton, bi-exciton, bound exciton and bound bi-exciton energies respectively taken from the PL spectrum at 10 K, as shown in Fig. 2. The estimated binding energy for the bound exciton of 22 ± 2 meV is in exact agreement previously reported values for vacuum deposited CuCl thin film on Al_2O_3 and bulk CuCl. However using this analysis the free bi-exciton binding energy would be 44 ± 2 meV, approximately 11 meV higher than for bulk CuCl [8].

The increase of band gap energy as a function of temperature is in contrast to other semiconductors, which generally follow the Varshini or Einstein model [21–22]. Similar results have been previously reported for vacuum evaporated thin films on fused quartz substrates [23]. To explain the behavior Göbel et al. proposed a two harmonic oscillator model to describe the renormalization of the CuCl band gap by electron-phonon interaction [24]. Due to the relatively large mass difference between Cu and Cl, one oscillator describes purely chlorine-like vibrations at high (optic) frequencies and the other purely copper-like vibrations at low (acoustic) frequencies. The following expression describes the mass and temperature dependence of the fundamental gap:

$$E_0(T, M) = E_0 + \frac{A_{\mathrm{Cu}}}{\omega_{\mathrm{Cu}} M_{\mathrm{Cu}}}\left[n(\omega_{\mathrm{Cu}}, T) + \frac{1}{2}\right] + \frac{A_{\mathrm{Cl}}}{\omega_{\mathrm{Cl}} M_{\mathrm{Cl}}}\left[n(\omega_{\mathrm{Cl}}, T) + \frac{1}{2}\right], \tag{4}$$

where, $n(\omega, T) = [\exp(\hbar\omega/kT) - 1]^{-1}$ is Bose-Einstein occupation number of the phonon. $M_{\mathrm{Cu/Cl}}$ is the atomic mass of Cu/Cl, E_0 is the unrenormalized band gap, and $A_{\mathrm{Cu/Cl}}$ is an effective electron-phonon interaction parameter. As in references of 24 we take an average optical, purely Cl-like, phonon frequency of $\omega_{\mathrm{Cl}} = 6$ THz and an average acoustic, purely Cu-like, phonon frequency of $\omega_{\mathrm{Cu}} = 1$ THz. This equation results in an excellent fit with our experimental data as shown in Fig. 3. Values for E_0 and the $A_{\mathrm{Cu/Cl}}$ parameters were determined: $E_0 = 3.233 \pm 0.002$ eV, $A_{\mathrm{Cu}} = 0.0032 \pm 0.0001$ eV^2 amu, $A_{\mathrm{Cl}} = -0.057 \pm 0.004$ eV^2 amu. By looking separately at the contributions of each of the two oscillators to the renormalization of the band gap energy we determine the optical, chlorine-like vibration reduces the unrenormalised band gap at 0 K by 33 meV while the acoustic, copper-like vibration increases the gap by 6 meV resulting in an overall 0 K band gap renormalization of 27 meV.

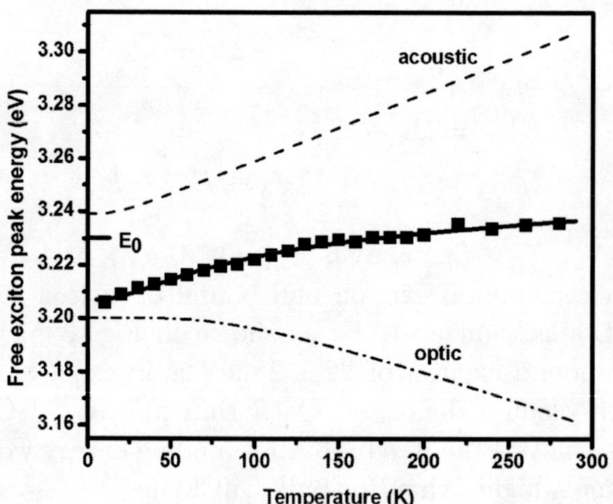

Fig. 3. Variation of free exciton (Z_3) energy for the CuCl thin film on Si(100) with temperature. Experimental data points (!) are fitted (solid line) using equation (5). The separate contributions arising from the acoustic and optical phonon interactions are also shown.

CONCLUSIONS

CuCl thin films grown on Si(100) substrates have been optically characterized. Strong exciton and bi-exciton features have been identified at low temperature, and free exciton emission is observed at room temperature. Power dependent PL confirmed the peak assignments. The binding energies of free and bound excitons and bi-excitons in the CuCl on Si material system have been determined.

ACKNOWLEDGMENTS

The research was financially supported by the Science Foundation Ireland Grant No. RFP/ENE027.

REFERENCES

1. S. Nakamura and G. Fasol, *The Blue Laser Diode: GaN based Light Emitters and Lasers* (Springer, Berlin, 1997).
2. O. Ambacher, J. Phys. D: Appl. Phys. **31**, 2653 (1998).

3. M. Zamfirescu, A. Kakovin, B. Gil, and G. Malpuech, Phys. Stat. Sol. (A) **195**, 563 (2003).
4. N. Nishida, K. Saiki, and A. Koma, Surf. Sci. **324**, 149 (1995).
5. T. Itoh, Y. Iwabuchi, and T. Kirihara, Phys. Stat. Sol. (B) **146**, 531 (1988).
6. H. Kurisu, K. Nagoya, N. Nakayama, et al., J. Lumin. **87–89**, 390 (2000).
7. S. Yano, T. Goto, T. Itoh, and A. Kasuya, Phys. Rev. B **55**, 1667 (1997).
8. Y. Masumoto, T. Kawamura, and K. Era, Appl. Phys. Lett. **62**, 225 (1993).
9. K. Reimann and St. Rubenacke, J. Appl. Phys. **76**, 4897 (1994).
10. M. Nagai, F. Hoshino, S. Yamamoto, et al., Optics Lett. **22**, 1630 (1997).
11. M. Nakayama, H. Ichida and H. Nishimura, J. Phys.: Condens. Matter. **11**, 7653 (1999).
12. Y. Masumoto, S. Okamoto, and S. Katayangi, Phys. Rev. B **50**, 18658 (1994).
13. Y. Kagotani, K. Miyajima, G. Oohata, et al., J. Lumin. **112**, 113 (2005).
14. G. Oohata, Y. Kagotani, K. Miyajima, et al., Physica. E **26**, 347 (2005).
15. D. K. Shuh, R. S. Williams, Y. Segawa, et al., Phys. Rev. B **44**, 5827 (1991).
16. A. Kawamori, K. Edamatsu, and T. Itoh, J. Cryst Growth. **237**, 1615 (2002).
17. L. O'Reilly, G. Natarajan, O.F. Lucas, et al., J. Appl. Phys. **98**, 113512 (2005).
18. L. O'Reilly, G. Natarajan, P.J. McNally, et al., J. Mater. Sci.: Mater. Electron. **16**, 415 (2005).
19. T. Goto, T. Takahashi, and M. Ueta, J. Phys. Soc. Japan **24**, 314 (1968).
20. Y. Kondo, Y. Kuroiwa, N. Sugimoto, et al., J. Opt. Soc. Am. B. **17**, 548 (2000).
21. Y. P. Varshini, Physica. E **34**, 149 (1967).
22. X.T. Zhang, Y.C. Liu, Z.Z. Zhi, et al., J. Lumin. **99**, 149 (2002).
23. Y. Kaifu and T. Komatsu, Phys. Stat. Solidi. (B) **48**, k125 (1971).
24. A. Göbel, T. Ruf, M. Cardona, et al, Phys. Rev. B **57**, 15183 (1998).

THE DIFFERENTIAL FACTORIZATION METHOD IN PROBLEMS FOR CONTINUOUS MEDIA

V.A. Babeshko[1*], O.V. Evdokimova[1], and O.M. Babeshko[1]**

ABSTRACT

The application of the differential factorization method in boundary-value problems for continuous media is described, including its first application, i.e. to dynamic problems of the elasticity theory. These problems are convenient when comparing different solution methods and they demonstrate high potential of the differential factorization method. They are also convenient because they include main cases of the types of zero determinants of characteristic equations of differential equations of boundary-value problems, namely single and repeated zeros. Hence, it is not difficult to use the method when studying anisotropic materials.

Key words: factorization, automorphism, boundary value problem, differential Lamé equation, pseudodifferential equations, Fourier transformation functional equation, transformation groups

1. The differential factorization method recently developed at Kuban State University [1, 2] is intended for obtaining an integral representation of solutions to systems of partial differential equations with constant coefficients in domains of complex geometry. The method is designed primarily for seismology problems.

The following principles of topological algebra underlie the method. The domain of the boundary value problem in question is viewed as a topological manifold with boundary. An automorphism, i.e., a topological mapping of this manifold into itself generates transformation groups that are isomorphic to some groups of nonsingular matrices. The latter generate representations of these groups described in the general case by composite special functions. The partial differential expression in the statement of the boundary value problem is treated as a differentiable mapping to

[1]Kuban State University, Krasnodar, Russia

[*]E-mail: *babeshko@kubsu.ru*

[**]E-mail: *evdokimova.lga@mail.ru*

the manifold of a vector field defined on the same manifold. This mapping leads to a functional equation.

To ensure an automorphism, the functional equation has to be examined by the factorization method. When the special functions generated by the automorphism are invariant under the differentiable mapping, the functional equation is especially easy to study, since the boundary conditions are globally stated on coordinate surfaces. In the general case, to ensure an automorphism, we have to use local coordinates and apply a topological partition of unity.

2. Various versions of applying the factorization method to boundary value problems in various statements can be found in [1, 2] (see also the references therein). As a result of these studies, algorithms were developed for applying the differential factorization method to the study and solution of boundary value problems involving systems of partial differential equations with constant coefficients. Below, a new algorithm is demonstrated as applied to a rather general boundary problem.

Consider a fairly general boundary value problem system of P partial differential equations of an arbitrary order with constant coefficients written in operator form in a convex three-dimensional domain Ω:

$$\mathbf{K}(\partial x_1, \partial x_2, \partial x_3)\boldsymbol{\varphi} = \sum_{m=1}^{M}\sum_{n=1}^{N}\sum_{k=1}^{K}\sum_{p=1}^{P} A_{spmnk}\varphi_{p,x_1x_2x_3}^{(m)(n)(k)} = 0, \tag{1}$$

$$s = 1, 2, \ldots, P, \quad A_{sqmnk} = \text{const}, \quad \boldsymbol{\varphi} = \{\varphi_1, \varphi_2, \ldots, \varphi_P\},$$

$$\boldsymbol{\varphi} = \{\varphi_s\}, \quad \boldsymbol{\varphi}(\mathbf{x}) = \boldsymbol{\varphi}(x_1, x_2, x_3), \quad \mathbf{x} = \{x_1, x_2, x_3\}, \quad \mathbf{x} \in \Omega.$$

On the boundary $\partial\Omega$ we set the boundary conditions

$$\mathbf{R}(\partial x_1, \partial x_2, \partial x_3)\boldsymbol{\varphi} = \sum_{m=1}^{M_1}\sum_{n=1}^{N_1}\sum_{k=1}^{K_1}\sum_{p=1}^{P} B_{spmnk}\varphi_{p,x_1x_2x_3}^{(m)(n)(k)} = f_s, \tag{2}$$

$$s = 1, 2, \ldots, s_0 < P, \quad \mathbf{x} \in \partial\Omega, \quad M_1 < M, \quad N_1 < N, \quad K_1 < K.$$

Note that, like in the integral factorization method described above, in the differential factorization method, the boundary value problem is solved exactly if Ω is a half lem is reduced to a system of normally solvable pseudodifferential equations. space. If Ω is a convex domain, the problem is reduced to a system of normally solvable pseudodifferential equations.

To give a systematic description of the differential factorization method, we divide it into several steps.

2.1. *Reduction of the differential equations to a functional equation by applying the Fourier transform*

The tree-dimensional Fourier transform

$$\Phi_n(\boldsymbol{\alpha}) = \iiint_{\Omega} \varphi_n(x) e^{i\langle \boldsymbol{\alpha}\mathbf{x}\rangle}\, d\mathbf{x} \equiv F\varphi_n, \quad \Phi_m = F\varphi_m$$

is applied to the system to reduce it to a functional equation of the form

$$\mathbf{K}(\boldsymbol{\alpha})\boldsymbol{\Phi} = \iint_{\partial\Omega} \boldsymbol{\omega}, \quad \mathbf{K}(\boldsymbol{\alpha}) \equiv -\mathbf{K}(-i\alpha_1, -i\alpha_2, -i\alpha_3) = \|k_{nm}(\alpha)\|. \tag{3}$$

Here $\mathbf{K}(\boldsymbol{\alpha})$ is a polynomial matrix function of order P.

The components of the vector of exterior forms $\boldsymbol{\omega}$ are two-dimensional functions of the form

$$\begin{aligned} &\boldsymbol{\omega} = \{\omega_s\}, \quad s = 1, 2, \ldots, P, \\ &\omega_s = P_{12s}\, dx_1 \wedge dx_2 + P_{13s}\, dx_1 \wedge dx_3 + P_{23s}\, dx_2 \wedge dx_3. \end{aligned} \tag{4}$$

The exterior-form operations are defined as

$$\begin{aligned} dx_1 \wedge dx_2 &= dx_1^1\, dx_2^2 - dx_1^2\, dx_2^1, \\ dx_1 \wedge dx_3 &= dx_1^1\, dx_2^3 - dx_1^3\, dx_2^1, \\ dx_2 \wedge dx_3 &= dx_1^2\, dx_2^3 - dx_1^3\, dx_2^2. \end{aligned}$$

Here, we introduced vectors of an arbitrary coordinate system lying in the coverings of the tangent bundle of the body surface. In a Cartesian coordinate system, we used the following notation for the tangent vectors of an arbitrary element of a covering:

$$\begin{aligned} x_1 &= \{x_1^1, x_1^2, x_1^3\}, \\ x_2 &= \{x_2^1, x_2^2, x_2^3\}. \end{aligned}$$

The coefficients of the exterior forms are give by

$$
\begin{aligned}
P_{12s} &= \sum_{m=1}^{M}\sum_{n=1}^{N}\sum_{k=1}^{K}\sum_{p=1}^{P} A_{spmnk}(-i\alpha_1)^m(-i\alpha_2)^n \sum_{p_3=1}^{k}(-i\alpha_3)^{(p_3-1)}\varphi_{px_3}^{(k-p_3)}e^{i\langle\boldsymbol{\alpha}\mathbf{x}\rangle},\\
P_{13s} &= -\sum_{m=1}^{M}\sum_{n=1}^{N}\sum_{k=1}^{K}\sum_{p=1}^{P} A_{spmnk}(-i\alpha_1)^m(-i\alpha_3)^k \sum_{p_2=1}^{n}(-i\alpha_2)^{(p_2-1)}\varphi_{px_2}^{(n-p_2)}e^{i\langle\boldsymbol{\alpha}\mathbf{x}\rangle}\\
&\quad -(-i\alpha_1)^m\sum_{p_2=1}^{n}\sum_{p_3=1}^{k}(-i\alpha_2)^{(p_2-1)}(-i\alpha_3)^{(p_3-1)}\frac{\partial}{\partial x_3}(\varphi_{x_2x_3}^{(n-p_2),(k-p_3)}e^{i\langle\boldsymbol{\alpha}\mathbf{x}\rangle}),\\
P_{23s} &= \sum_{m=1}^{M}\sum_{n=1}^{N}\sum_{k=1}^{K}\sum_{p=1}^{P} A_{spmnk}(-i\alpha_2)^n(-i\alpha_3)^k \sum_{s_1=1}^{m}(-i\alpha_1)^{(s_1-1)}\varphi_{px_1}^{(m-s_1)}e^{i\langle\boldsymbol{\alpha}\mathbf{x}\rangle}\\
&\quad +(-i\alpha_2)^n\sum_{s_1=1}^{m}\sum_{p_3=1}^{k}(-i\alpha_1)^{(s_1-1)}(-i\alpha_3)^{(p_3-1)}\frac{\partial}{\partial x_3}(\varphi_{px_1x_3}^{(m-s_1),(k-p_3)}e^{i\langle\boldsymbol{\alpha}\mathbf{x}\rangle})\\
&\quad +(-i\alpha_3)^k\sum_{s_1=1}^{m}\sum_{s_2=1}^{n}(-i\alpha_1)^{(s_1-1)}(-i\alpha_2)^{(s_2-1)}\frac{\partial}{\partial x_2}(\varphi_{px_1x_2}^{(m-s_1),(n-s_2)}e^{i\langle\boldsymbol{\alpha}\mathbf{x}\rangle})\\
&\quad +\sum_{s_1=1}^{m}\sum_{s_2=1}^{n}\sum_{s_3=1}^{k}(-i\alpha_1)^{(s_1-1)}(-i\alpha_2)^{(s_2-1)}(-i\alpha_3)^{(s_3-1)}\\
&\quad \times\frac{\partial}{\partial x_2}\frac{\partial}{\partial x_3}(\varphi_{px_1x_2x_3}^{(m-s_1),(n-s_2),(k-s_3)}e^{i\langle\boldsymbol{\alpha}\mathbf{x}\rangle}).\\
&\langle\boldsymbol{\alpha}\mathbf{x}\rangle = \alpha_1x_1+\alpha_2x_2+\alpha_3x_3,\quad \boldsymbol{\varphi}=\{\varphi_n\},\quad \boldsymbol{\alpha}=\{\alpha_1,\alpha_2,\alpha_3\},\quad \boldsymbol{\Phi}=\{\Phi_m\}.
\end{aligned}
\tag{5}
$$

2.2. *Fulfillment of given boundary conditions (2)*

To achieve this, the solution $\varphi(\partial\Omega)$ and its normal derivatives on $\partial\Omega$ taken from the boundary conditions are introduced into the representations of the exterior forms. The tangent derivatives are not taken into account. The exterior forms contain the solution φ_n and its derivatives on $\partial\Omega$. The functions or normal derivatives on the boundary are found by fitting and inverting the nonsingular matrix from boundary conditions (4) and are introduced into the corresponding representations of $\boldsymbol{\omega}$. The remaining functions or normal derivatives have to be found from the pseudodifferential equations obtained by transformations of the functional equations.

The following steps are to be performed to determine the remaining unknowns in the representation of the solution.

2.3. *Factorization of the matrix function* $\mathbf{K}(\alpha)$ *in the functional equation*

Let λ_+ denote a domain containing all the zeros z_{s+}^v with Im $z_{s+}^v > 0$ and z_{s-}^v with Im $z_{s-}^v < 0$ ($s\pm = 1, 2, \ldots, G_\pm$) of the determinant $K(\alpha_3^\nu) = \det \mathbf{K}(\alpha_3^\nu)$, and let λ_- denote its complement to the entire plane with the boundary Γ separating the domains. The location of the contour will be specified later. By using the results of [3], the matrix function $\mathbf{K}(\alpha_3)$ can be factorized as

$$\mathbf{K}(\alpha_3^\nu) = \mathbf{K}(\alpha_3^\nu, -)\mathbf{K}_r(\alpha_3^\nu). \tag{6}$$

Here, $\mathbf{K}(\alpha_3^\nu, -)$ is a regular matrix function in X_- and its determinant has no zeros in this domain. The elements of the matrix function $\mathbf{K}_r(\alpha_3^\nu)$ are polynomials in α_3, and its determinant is independent of this parameter. Thus, all the zeros of the determinant of $\mathbf{K}(\alpha_3^\nu)$ with respect to α_3 coincide with the zeros of the determinant of $\mathbf{K}(\alpha_3^\nu, -)$ that lie in λ_+.

The elements of the matrix function $\mathbf{K}^{-1}(\alpha_3^\nu, -)$ can be represented in integral form.

To derive them, we introduce the adjoint $\mathbf{K}^*(\alpha_3^\nu)$ of $\mathbf{K}(\alpha_3^\nu)$ by setting

$$\mathbf{K}^*(\alpha_3^\nu) = \|M_{pn}(\alpha_3^\nu)\|.$$

Consider a matrix function $\mathbf{K}^*(\alpha_3^\nu, m)$ of order $P-1$ obtained from $\mathbf{K}^*(\alpha_3^\nu)$ by deleting the mth row and column and such that the zeros ξ_n^ν of its determinant $Q(\alpha_3^\nu) = \det \mathbf{K}(\alpha_3^\nu, m)$ do not coincide with z_{s+}^v, z_{s-}^v.

The elements of the inverse matrix function are denoted by

$$[\mathbf{K}^*(\alpha_3^\nu, m)]^{-1} = \|Q^{-1}Q_{ps}\|.$$

Then the elements of the matrix function $\mathbf{K}^{-1}(\alpha_3^\nu, -)$ given by

$$\mathbf{K}^{-1}(\alpha_3^\nu, -) = \left\| \begin{matrix} 1 & & & & & & 0 \\ & 1 & & & & & \\ & & \ddots & & & & \\ S_{m1} & S_{m2} & \cdots & S_{mm} & \cdots & & S_{mN} \\ & & & & & \ddots & \\ 0 & & & & & & 1 \end{matrix} \right\|, \tag{7}$$

have an integral representation of the form

$$S_{mp}(\alpha_3^\nu) = \frac{1}{2\pi i}\oint_{\Gamma_\mp}\sum_{s=1}^{N}{}'\frac{Q_{ps}(u_3)M_{sm}(u_3)\,du_3}{Q(u_3)K(u_3)(u_3-\alpha_3^\nu)} - \left(\frac{1}{2}\mp\frac{1}{2}\right)\frac{R_{mp}(\alpha_3^\nu)}{K(\alpha_3^\nu)},$$

$$m\neq p,\quad \frac{R_{mp}(\alpha_3^\nu)}{K(\alpha_3^\nu)} = \frac{Z_{mp}(\alpha_3^\nu)}{Q(\alpha_3^\nu)K(\alpha_3^\nu)} + \sum_n \frac{Z_{mp}(\xi_n^\nu)}{Q'(\xi_n^\nu)K(\xi_n^\nu)(\xi_n^\nu-\alpha_3^\nu)}, \tag{8}$$

$$S_{mm}(\alpha_3^\nu) = K^{-1}(\alpha_3^\nu),\quad \alpha_3^\nu\in\lambda_\mp,$$

$$Z_{mp}(\alpha_3^\nu) = \sum_{s=1}^{N}{}'Q_{ps}(\alpha_3^\nu)M_{sm}(\alpha_3^\nu).$$

Here, Γ_+ is a closed contour such that the domain λ_+ contains only the zeros z_{s+}^ν, z_{s-}^ν and the domain λ_- contains only the zeros ξ_n^ν. The closed contour Γ_- encloses a domain containing all the zeros z_{s+}^ν, z_{s-}^ν and ξ_n^ν. This representation implies that the elements of $\mathbf{K}^{-1}(\alpha_3^\nu,-)$ are rational functions with their only singularities occurring at the zeros z_{s+}^ν, z_{s-}^ν, and the term containing them, $\mathbf{K}^{-1}(\alpha_3^\nu)$ is explicitly expressed.

2.4. *Reduction of the functional equation to a system of pseudodifferential equations*

The contour Γ is deformed so that it encloses an infinite strip with the real line and still surrounds the zeros z_{s+}^ν, z_{s-}^ν and ξ_n^ν. Consider the functional equation on the real line assuming that it contains no zeros z_{s+}^ν, z_{s-}^ν. Otherwise, we have to use the techniques described in [4] in order to proceed to a curved real line. Obviously, the zeros z_{s+}^ν, z_{s-}^ν lie in the upper and lower half-planes, respectively.

In what follows, we use a local system of Cartesian coordinates $\mathbf{x}^\nu = \{x_1^\nu, x_2^\nu, x_3^\nu\}$ where the first two components lie in the tangent plane to the boundary $\partial\Omega$ and divided according to the following rule: the third component lies on the outward normal. In each local coordinate system, we perform an operation that ensures an automorphism of Ω. To this end, we perform factorization (6) and represent functional equation (6) in the form

$$\mathbf{\Phi} = \mathbf{K}_r^{-1}(\alpha_3^\nu)\mathbf{K}^{-1}(\alpha_3^\nu,-)\iint_{\partial\Omega}\omega. \tag{9}$$

Applying the inverse three-dimensional Fourier transform to this functional matrix equation, we require that the original vector function ô vanish for $x_3 > 0$, i.e., outside Ω. Dropping the intermediate rearrangements, we obtain the relations

$$\sum_{p=1}^{P} \iint_{\partial\Omega} \omega_p Z_{mp}(z^{\nu}_{s-}) = 0, \quad s- = 1, 2, \ldots, G_-, \qquad (10)$$
$$Z_{mm}(\alpha^{\nu}_3) = -Q(\alpha^{\nu}_3).$$

This system consists of pseudodifferential equations.

2.5. *Derivation of a representation of the solution to the boundary value problem*

In view of Section 2.2, assume that system (10) has been solved. Introducing the determined components into the vector of exterior forms (9) and applying the three-dimensional Fourier transform to $\mathbf{\Phi}(\alpha)$, we obtain

$$\boldsymbol{\varphi}(\mathbf{x}^{\nu}) = \frac{1}{8\pi^3} \iiint_{-\infty}^{\infty} \mathbf{K}_r^{-1}(\alpha^{\nu}_3)\mathbf{K}^{-1}(\alpha^{\nu}_3, -) \iint_{\partial\Omega} \boldsymbol{\omega} e^{-i\langle\alpha^{\nu}_3 x^{\nu}_3\rangle}\, d\alpha^{\nu}_1\, d\alpha^{\nu}_2\, d\alpha^{\nu}_3,$$
$$\mathbf{x}^{\nu} \in \Omega. \quad (11)$$

Due to formulas (8), the solution can be made more visual if we evaluate the integral with respect to α^{ν}_3 by using residue theory. As a result, we have

$$\boldsymbol{\varphi}(\mathbf{x}^{\nu}) = \frac{1}{4\pi^2} \iint_{-\infty}^{\infty} \sum_{s} e^{-i(\alpha^{\nu}_1 x^{\nu}_1 + \alpha^{\nu}_2 x^{\nu}_2)} [\mathbf{K}_r^{-1}(i\frac{\partial}{\partial x^{\nu}_3})\mathbf{T}_+(\alpha^{\nu}_1, \alpha^{\nu}_2, z^{\nu}_{s+})e^{-iz^{\nu}_{s+}x^{\nu}_3}$$
$$- \mathbf{K}_r^{-1}(i\frac{\partial}{\partial x^{\nu}_3})\mathbf{T}_-(\alpha^{\nu}_1, \alpha^{\nu}_2, z^{\nu}_{s-})e^{-iz^{\nu}_{s-}x^{\nu}_3}]\, d\alpha^{\nu}_1\, d\alpha^{\nu}_2, \qquad (12)$$
$$t_{m\pm}(\alpha^{\nu}_1, \alpha^{\nu}_2, z^{\nu}_{s\pm}) = -\sum_{p=1}^{P} \iint_{\partial\Omega_{\pm}} \frac{\omega_p Z_{mp}(z^{\nu}_{s\pm})}{Q(z^{\nu}_{s\pm})K'(z^{\nu}_{s\pm})}$$
$$\mathbf{T}_{\pm} = \{0, 0, \ldots, 0, t_{m\pm}, 0, \ldots, 0\}.$$

Here, the boundary $\partial\Omega$ for chosen $x^{\nu}_3 < 0$, $\mathbf{x}^{\nu} \in \Omega$ is divided according to the following rule:

$$\iint_{\partial\Omega} \omega = \iint_{\partial\Omega_+} \omega + \iint_{\partial\Omega_-} \omega,$$
$$\iint_{\partial\Omega_+} \boldsymbol{\omega} \exp(-i\alpha^{\nu}_3 x^{\nu}_3) \to 0, \quad \operatorname{Im} \alpha^{\nu}_3 \to \infty,$$
$$\iint_{\partial\Omega_-} \boldsymbol{\omega} \exp(-i\alpha^{\nu}_3 x^{\nu}_3) \to 0, \quad \operatorname{Im} \alpha^{\nu}_3 \to -\infty.$$

In the case of a half-space or a layered medium, the pseudodifferential equations in (10) degenerate into algebraic ones. By inverting them, the solution is constructed in a finite form.

The problems considered in [5, 6] show that both factorization methods do not iterate and supplement each other, providing the possibility of analyzing a wider circle of problems.

3. Consider the homogeneous differential Lamé equation in conventional form :

$$\begin{aligned}&(\lambda+\mu)\operatorname{grad}\operatorname{div}\mathbf{u}+\mu\Delta\mathbf{u}-\delta\mathbf{u}=0,\\&\mathbf{u}=\{u_1,u_2,u_3\}.\end{aligned}\tag{13}$$

Here, $\delta=-\rho\omega^2$ in vibration problems and $\delta=\rho p^2$ in nonstationary problems, where ω is the vibration frequency, p is the Laplace transform parameter, and ρ is the density of the material. In the boundary-value problem, certain boundary conditions have to be set, which will be discussed below.

After applying a three-dimensional Fourier transform over all the coordinates x_1, x_2, x_3 substituting the $-ia_k$ parameters of the Fourier transform for the corresponding derivatives; and multiplying by -1, the above system of equations takes the following form:

$$\mathbf{KU}=\iint_{\partial\Omega}\boldsymbol{\omega},\quad \mathbf{U}=\{U_1,U_2,U_3\},\quad \mathbf{U}=\mathbf{F}_3(\alpha_1,\alpha_2,\alpha_3)\mathbf{u},\tag{14}$$

$$\mathbf{K}=\begin{Vmatrix}(\lambda+2\mu)\alpha_1^2+\mu\alpha_2^2+\mu\alpha_3^2+\delta & (\lambda+\mu)\alpha_1\alpha_2 & (\lambda+\mu)\alpha_1\alpha_3\\ (\lambda+\mu)\alpha_1\alpha_2 & \mu\alpha_1^2+(\lambda+2\mu)\alpha_2^2+\mu\alpha_3^2+\delta & (\lambda+\mu)\alpha_2\alpha_3\\ (\lambda+\mu)\alpha_1\alpha_3 & (\lambda+\mu)\alpha_2\alpha_3 & \mu\alpha_1^2+\mu\alpha_2^2+(\lambda+2\mu)\alpha_3^2+\delta\end{Vmatrix}.$$

Let us consider a tangent bundle of the boundary $\partial\Omega$ system $\mathbf{x}^\nu$ such that the x_1^ν, x_2^ν axes lie in the tangent plane and the x_3^ν axis is aligned with the outward normal to the boundary. The Fourier transform parameters corresponding to them are denoted as $\boldsymbol{\alpha}^\nu$. Formulas for the passage from one local system to another are given by the well-known transformation relationships:

$$\mathbf{x}^\nu=\mathbf{c}_\nu^\tau\mathbf{x}^\tau+\mathbf{x}_0^\tau,\quad \boldsymbol{\alpha}^\nu=\mathbf{c}_\nu^\tau\boldsymbol{\alpha}^\tau,\tag{15}$$

where $\mathbf{x}_0^\tau$ are the coordinates of the origin of the new coordinate system in the initial one.

Using similar expressions, let us pass to new unknown quantities defined by the following formulas:

$$\mathbf{u}^\nu=\mathbf{c}_\nu^\tau\mathbf{u}^\tau.\tag{16}$$

Lemma. On passage to the new local coordinate system, the images and preimages of the Fourier transform in Eqs. (13) are transformed according to formulas (15) and (16).

The lemma is proved by direct substitution of the transform into Eqs. (13), after which differential equations (13) should be written in each local coordinate system $\mathbf{x}^\nu$ with $\mathbf{u}^\nu = \{u_1^\nu, u_2^\nu, u_3^\nu\}$.

4. In functional equations (14), the vector of exterior forms $\mathbf{w}$ has the following components [7–10]:

$$\begin{aligned} \omega_{sk} &= R_{sk}\, dx_1 \wedge dx_2 + Q_{sk}\, dx_1 \wedge dx_3 + P_{sk}\, dx_2 \wedge dx_3, \\ \boldsymbol{\omega} &= \{\omega_{s1}, \omega_{s2}, \omega_{s3}\}, \end{aligned} \tag{17}$$

where subscript s indicates the group of exterior forms and k is the number of the row of the Lame equations. Transformation of the components $\mathbf{R}_3 = \{R_{31}, R_{32}, R_{33}\}$ of the obtained vector of the exterior form yields the following representation:

$$\begin{aligned} R_{31} &= [\sigma_{13} - i\mu\alpha_3 u_1 - i\lambda\alpha_1 u_3] e^{i\langle\boldsymbol{\alpha}\mathbf{x}\rangle}, \\ R_{32} &= [\sigma_{23} - i\mu\alpha_3 u_2 - i\lambda\alpha_2 u_3] e^{i\langle\boldsymbol{\alpha}\mathbf{x}\rangle}, \\ R_{33} &= [\sigma_{33} - i(\lambda + 2\mu)\alpha_3 u_3 - i\mu(\alpha_1 u_1 + \alpha_2 u_2)] e^{i\langle\boldsymbol{\alpha}\mathbf{x}\rangle}. \end{aligned} \tag{18}$$

Taking into account that an element of the tangent bundle is described by the oriented area $dx_1 \wedge dx_2$, we conclude that unknown quantities at the boundary can be set in terms of various combinations, using either stresses, or displacements, or mixed conditions.

Thus, for an isotropic body, the functional equations of the boundary-value problem under consideration in one of the local coordinate systems can be presented in the following form:

$$\mathbf{K}(\boldsymbol{\alpha}^\nu)\mathbf{U}^\nu = \iint_{\partial\Omega} \boldsymbol{\omega}^\nu = \sum_\tau \iint_{\partial\Omega} \varepsilon_\tau \boldsymbol{\omega}^\nu(\boldsymbol{\xi}^\tau, \boldsymbol{\alpha}^\nu). \tag{19}$$

where ε_τ are the elements of unity partition [5]. and introduce a local rectangular Cartesian coordinate.

5. For application of the differential factorization method to construction of the pseudodifferential equations for a matrix function, we use the approach developed in [7]. As a result, the factorizing matrix functions take the following form:

$$\mathbf{Q}_1 = \left\| \begin{matrix} \frac{1}{\alpha_3 - \alpha_{31-}} & \frac{-\alpha_1}{\alpha_2(\alpha_3 - \alpha_{31-})} & 0 \\ 0 & 1 & 0 \\ 0 & 0 & 1 \end{matrix} \right\|, \quad \mathbf{Q}_2 = \left\| \begin{matrix} \frac{1}{\alpha_3 - \alpha_{31-}} & 0 & \frac{-\alpha_2}{\alpha_{31-}(\alpha_3 - \alpha_{31-})} \\ 0 & 1 & 0 \\ 0 & 0 & 1 \end{matrix} \right\|,$$

$$\mathbf{Q}_3 = \left\| \begin{matrix} 1 & 0 & 0 \\ 0 & 1 & 0 \\ \frac{\alpha_1}{\alpha_{32-}(\alpha_3-\alpha_{32-})} & \frac{\alpha_2}{\alpha_{32-}(\alpha_3-\alpha_{32-})} & \frac{1}{\alpha_3-\alpha_{32-}} \end{matrix} \right\|.$$

Here, the double roots of the determinant $\det \mathbf{K}$ are expressed as

$$\alpha_{31+} = i\sqrt{\alpha_2^2 + \alpha_1^2 + \frac{\delta}{\mu}}, \quad \alpha_{31-} = -\alpha_{31+}, \tag{20}$$

where the signs at the subscripts indicate that the roots belong to the upper (plus) or lower (minus) half-planes of the complex plane.

The simple roots are expressed as

$$\alpha_{32+} = i\sqrt{\alpha_2^2 + \alpha_1^2 + \frac{\delta}{\lambda + 2\mu}}, \quad \alpha_{32-} = -\alpha_{32+}.$$

In order to obtain the required pseudodifferential equations, it is necessary to equate the corresponding Leray residue forms to zero. Calculating these residue forms in the neighborhood of the local coordinate system, we obtain the following relationships:

$$\begin{aligned} &\lim_{\alpha_3 \to \alpha_{31-}} (\alpha_3 - \alpha_{31-})\mathbf{Q}_m \mathbf{F}_2 \mathbf{R}_3 = 0, \quad m = 1,\, 2, \\ &\lim_{\alpha_3 \to \alpha_{32-}} (\alpha_3 - \alpha_{32-})\mathbf{Q}_m \mathbf{F}_2 \mathbf{R}_3 = 0, \quad m = 3, \end{aligned} \tag{21}$$

where $\mathbf{F}_2 = \mathbf{F}_2(a_1, a_2)$ is the two-dimensional Fourier transform (with respect to the parameters x_1 and x_2) of functions defined in the same neighborhood of the local coordinate systems.

In the matrix form, system (21) can be presented as

$$\mathbf{L}\mathbf{F}_2\mathbf{u} = \mathbf{D}\mathbf{F}_2\mathbf{t}, \quad \mathbf{t} = \{t_1, t_2, t_3\}, \quad t_1 = \sigma_{13}, \quad t_2 = \sigma_{23}, \quad t_3 = \sigma_{33}, \tag{22}$$

$$\mathbf{L} = \left\| \begin{matrix} \alpha_1\sqrt{\tau_1^2 - v^2} & \alpha_2\sqrt{\tau_1^2 - v^2} & s \\ \alpha_2\sqrt{\tau_2^2 - v^2} & -\alpha_1\sqrt{\tau_2^2 - v^2} & 0 \\ 2s + \alpha_2^2 & -\alpha_1\alpha_2 & -2\alpha_1\sqrt{\tau_2^2 - v^2} \end{matrix} \right\|, \quad \mathbf{D} = \frac{i}{\mu} \left\| \begin{matrix} \frac{\alpha_1}{2} & \frac{\alpha_2}{2} & \frac{\sqrt{\tau_1^2 - v^2}}{2} \\ \alpha_2 & -\alpha_1 & 0 \\ \sqrt{\tau_2^2 - v^2} & 0 & -\alpha_1 \end{matrix} \right\|,$$

$$\tau_1 = -\frac{\delta}{\lambda + 2\mu}, \quad \tau_2 = -\frac{\delta}{\mu}, \quad v = \sqrt{\alpha_1^2 + \alpha_2^2}, \quad s = 0.5\tau_2^2 - v^2.$$

Calculation of the determinant of matrix $\mathbf{L}$ yields

$$\det \mathbf{L} = 2\alpha_1\sqrt{\tau_2^2 - v^2}\left[v^2\sqrt{\tau_1^2 - v^2}\sqrt{\tau_2^2 - v^2} + s^2\right] = \Delta_2,$$

$$\alpha_{310} = -i\sigma_1 = \sqrt{\tau_1^2 - v^2}, \quad \alpha_{320} = -i\sigma_2 = \sqrt{\tau_2^2 - v^2}, \quad \operatorname{Im} \alpha_{3n0} \le 0, \quad n = 1,\, 2.$$

Multiplying system (22) by the matrix-function $\mathbf{L}^{-1}$ on the left and applying the two-dimensional inverse Fourier transform, we get the following representation:

$$\mathbf{F}_2^{-1}\mathbf{K}_0\mathbf{F}_2\mathbf{t} = \mathbf{u}, \tag{23}$$

where the matrix function can be written as

$$\mathbf{K}_0 = -\frac{1}{2\mu}\left\|\begin{array}{ccc} \alpha_1^2 M + \alpha_2^2 N & \alpha_1\alpha_2(M-N) & i\alpha_1 P \\ \alpha_1\alpha_2(M-N) & \alpha_1^2 N + \alpha_2^2 M & i\alpha_2 P \\ -i\alpha_1 P & -i\alpha_2 P & R \end{array}\right\|,$$

$$M(v) = \frac{-0.5\tau_2^2\sigma_2}{v^2\Delta_0}, \quad N(v) = \frac{2}{v^2\sigma_2}, \quad P(v) = \frac{v^2 - 0.5\tau_2^2 - \sigma_1\sigma_2}{\Delta_0},$$

$$R(v) = \frac{-0.5\tau_2^2\sigma_1}{\Delta_0}, \quad \Delta_0 = (v^2 - 0.5\tau_2^2)^2 - v^2\sigma_1\sigma_2.$$

Using similar formulas, one can calculate the Leray residue forms in the right-hand side of functional equations (19) for the remaining τ after the change of variables $\boldsymbol{\alpha}^\tau = \mathbf{c}_\tau^\nu\boldsymbol{\alpha}^\nu$. An analysis of expression (23) shows that the formulas coincide with those for the case where the body is a half-space. However, it should be borne in mind that there is a significant distinction consisting in the fact that functions $\mathbf{u}$ and $\mathbf{t}$ are defined in , the neighborhood of the local coordinate systems generated by the tangent bundle of the boundary. Taking into account that the unity partition leads to coverage of the boundary by disjoint neighborhoods, we conclude that the set of given and unknown functions for the system of pseudodifferential equations under consideration will contain functions defined in the neighborhoods of the local coordinate systems.

6. For further investigation, let us write the system of pseudodifferential equations (23) constructed after calculating the Leray residue forms as follows:

$$\iint_{\partial\Omega_\nu} \boldsymbol{\omega}_0^\nu(\xi^\nu, \alpha_1^\nu, \alpha_2^\nu, \alpha_{3r-}^\nu(\alpha_1^\nu, \alpha_2^\nu)) + \sum_\tau \iint_{\partial\Omega_\tau} \boldsymbol{\omega}_0^\tau(\xi^\tau, \alpha_1^\nu, \alpha_2^\nu, \alpha_{3r-}^\nu(\alpha_1^\nu, \alpha_2^\nu)) = 0,$$

$\nu = 1, 2, \ldots, T.$

Here, $\boldsymbol{\omega}_0^\nu$, $\boldsymbol{\omega}_0^\tau$ are no longer the exterior forms; these quantities are given by expressions obtained multiplying equations by the factorizing matrix functions and calculating the Leray residue forms. This system of equations can be rewritten in the following form:

$$\begin{aligned}&\mathbf{L}^\nu(\alpha_1^\nu, \alpha_2^\nu, \alpha_{3r-}^\nu(\alpha_1^\nu, \alpha_2^\nu))\mathbf{U}_0^\nu(\alpha_1^\nu, \alpha_2^\nu, \alpha_{3r-}^\nu(\alpha_1^\nu, \alpha_2^\nu)) \\ &\quad - \mathbf{D}^\nu(\alpha_1^\nu, \alpha_2^\nu, \alpha_{3r-}^\nu(\alpha_1^\nu, \alpha_2^\nu))\mathbf{T}^\nu(\alpha_1^\nu, \alpha_2^\nu, \alpha_{3r-}^\nu(\alpha_1^\nu, \alpha_2^\nu))\end{aligned}$$

$$+\sum_{\tau=1}^{T}\left[\mathbf{L}^{\tau}(\alpha_1^{\nu},\alpha_2^{\nu},\alpha_{3r-}^{\nu}(\alpha_1^{\nu},\alpha_2^{\nu}))\mathbf{U}_0^{\tau}(\alpha_1^{\nu},\alpha_2^{\nu},\alpha_{3r-}^{\nu}(\alpha_1^{\nu},\alpha_2^{\nu}))\right.$$
$$\left.-\mathbf{D}^{\tau}(\alpha_1^{\nu},\alpha_2^{\nu},\alpha_{3r-}^{\nu}(\alpha_1^{\nu},\alpha_2^{\nu}))\mathbf{T}^{\tau}(\alpha_1^{\nu},\alpha_2^{\nu},\alpha_{3r-}^{\nu}(\alpha_1^{\nu},\alpha_2^{\nu}))\right]=0. \qquad (24)$$

where the primed sum symbol implies that the term with $\tau = \nu$ in this sum is missing. The obtained pseudodifferential equations make it possible to formulate various boundary-value problems for elastic bodies. For example, let us assume that the displacement vector $\mathbf{u}^{\nu}$ is set at the boundary. Then, the system of equations can be rewritten as follows:

$$(\mathbf{L}^{\nu})^{-1}\mathbf{D}^{\nu}\mathbf{T}^{\nu}+\sum_{\tau}(\mathbf{L}^{\nu})^{-1}\mathbf{D}^{\tau}\mathbf{T}^{\tau}=\mathbf{U}_0^{\nu}+\sum_{\tau}(\mathbf{L}^{\nu})^{-1}\mathbf{L}^{\tau}\mathbf{U}_0^{\tau}. \qquad (25)$$

We obtained the system of integral equations with respect to stresses, which can be written in a more explicit form by introducing, for example, the following notation

$$\mathbf{K}^{\nu}(\alpha_1^{\nu},\alpha_2^{\nu})=(\mathbf{L}^{\nu})^{-1}\mathbf{D}^{\nu},\quad \mathbf{K}^{\nu\tau}(\alpha_1^{\nu},\alpha_2^{\nu})=(\mathbf{L}^{\nu})^{-1}\mathbf{D}^{\tau},\quad \mathbf{B}^{\nu\tau}(\alpha_1^{\nu},\alpha_2^{\nu})=(\mathbf{L}^{\nu})^{-1}\mathbf{L}^{\tau}.$$

As a result, applying the inverse Fourier transform $\mathbf{F}_2^{-1}(x_1^{\nu},x_2^{\nu})$ with respect to parameters $\alpha_1^{\nu},\alpha_2^{\nu}$, we arrive at the following system of integral equations:

$$\iint_{\partial\Omega_{\nu}}\mathbf{k}^{\nu}(x_1^{\nu}-\xi_1^{\nu},x_2^{\nu}-\xi_2^{\nu})\mathbf{t}^{\nu}(\xi_1^{\nu},\xi_2^{\nu})\,d\xi_1^{\nu}\,d\xi_2^{\nu}$$
$$+\sum_{\tau=1}^{T}{}'\iint_{\partial\Omega_{\tau}}\mathbf{k}^{\nu\tau}(x_1^{\nu},\xi_1^{\tau},x_2^{\nu},\xi_2^{\tau})\mathbf{t}^{\tau}(\xi_1^{\tau},\xi_2^{\tau})\,d\xi_1^{\tau}\,d\xi_2^{\tau}$$
$$=\mathbf{u}^{\nu}(x_1^{\nu},x_2^{\nu})+\sum_{\tau=1}^{T}{}'\iint_{\partial\Omega_{\tau}}\mathbf{b}^{\nu\tau}(x_1^{\nu},\xi_1^{\tau},x_2^{\nu},\xi_2^{\tau})\mathbf{u}^{\tau}(\xi_1^{\tau},\xi_2^{\tau})\,d\xi_1^{\tau}\,d\xi_2^{\tau}, \qquad (26)$$

$$x_1^{\nu},\ x_2^{\nu}\in\partial\Omega_{\nu},\quad 1\le\nu\le T,$$
$$\mathbf{k}^{\nu}(x_1^{\nu},x_2^{\nu})=\mathbf{F}_2^{-1}\mathbf{K}^{\nu}(\alpha_1^{\nu},\alpha_2^{\nu}),$$
$$\mathbf{k}^{\nu\tau}(x_1^{\nu},\xi_1^{\tau},x_2^{\nu},\xi_2^{\tau})=\mathbf{F}_2^{-1}\mathbf{K}^{\nu\tau}(\alpha_1^{\nu},\alpha_2^{\nu})\exp i<\mathbf{c}_{\tau}^{\nu}\boldsymbol{\alpha}^{\nu},\quad \boldsymbol{\xi}^{\tau}>,$$
$$\mathbf{b}^{\nu\tau}(x_1^{\nu},\xi_1^{\tau},x_2^{\nu},\xi_2^{\tau})=\mathbf{F}_2^{-1}\mathbf{B}^{\nu\tau}(\alpha_1^{\nu},\alpha_2^{\nu})\exp i<\mathbf{c}_{\tau}^{\nu}\boldsymbol{\alpha}^{\nu},\quad \boldsymbol{\xi}^{\tau}>,$$

where T is the number of local coordinate systems for the tangent bundle of the boundary. Similarly, one can derive the system of integral equations for a boundary-value problem with preset stresses. The following theorem is valid.

Theorem. The operator $\mathbf{K}^{\nu}(\alpha_1^{\nu},\alpha_2^{\nu})$ in system (14) is principal, corresponding to the boundary-value problem for a half-space; the remaining operators are subordinate, being completely continuous in spaces where the principal operator is invertible.

This theorem determines plenty of methods for the analytical and numerical investigation into systems of integral equations of the type under consideration.

ACKNOWLEDGMENTS

This work was partially supported by the Russian Foundation for Basic Research (grants Nos. 06-01-00295, 08-08-00468, 06-08-00671, and 08-08-00669), the program "The South of Russia" (projects Nos. 08-01-99012, 08-01-99013, 08-01-99016, 08-08-99090, 08-08-99091, 07-01-12028, and 06-01-96634–6-01-96638), the Presidential Program of Support for Leading Scientific Schools in Russia (project No. NSh-4839.2006.1), and the programs of the OMMPU Department and the Presidium of the Russian Academy of Sciences performed at the Southern Scientific Center of the Russian Academy of Sciences.

REFERENCES

1. V.A. Babeshko and O.M. Babeshko, "Method of Factorizing a Solution to Some Boundary-Value Problems," Dokl. Akad. Nauk **389** (2), 184–188 (2003) [Dokl. Phys. (Engl. Transl.) **48** (3), 134–137 (2003)].
2. V.A. Babeshko and O.M. Babeshko, "Integral Transformations and the Factorization Method in Boundary Value Problems," Dokl. Akad. Nauk **403** (6), 28–32 (2005) [Dokl. Math. (Engl. Transl.) **72** (1), 630–633 (2005)].
3. V.A. Babeshko and O.M. Babeshko, "Factorization Formulas for Some Meromorphic Matrix Functions," Dokl. Akad. Nauk **399** (1), 26–28 (2004) [Dokl. Math. (Engl. Transl.) **70** (3), 963–965 (2004)].
4. I.I. Vorovich and V.A. Babeshko, *Dynamic Mixed Elasticity Problems for Nonclassical Domains* (Nauka, Moscow, 1979) [in Russian].
5. V.A. Babeshko, O.M. Babeshko, and O.V. Evdokimova, "Addressing the Problem of Investigating Coated Materials," Dokl. Akad. Nauk **410** (1), 49–52 (2006) [Dokl. Phys. (Engl. Transl.) **51** (9), 509–512 (2006)]
6. V.A. Babeshko, O.M. Babeshko, and O.V. Evdokimova, "On the Problem of Estimating the State of Coated Materials," Dokl. Akad. Nauk **409** (4), 481–485 (2006) [Dokl. Phys. (Engl. Transl.) **51** (8), 423–428 (2006)].
7. O.V. Evdokimova, Ekolog. Vestnik Nauchn. Tsentr. ChES, No. 2, 51 (2007).
8. V.A. Babeshko, O.M. Babeshko, and O.V. Evdokimova, "On Integral and Differential Factorization Methods," Dokl. Akad. Nauk **410** (2), 168–172 (2006) [Dokl. Math. (Engl. Transl.) **74** (2), 762–766 (2006)].
9. O.V. Evdokimova, Ekolog. Vestnik Nauchn. Tsentr. ChES, No. 4, 32 (2006).

10. V.A. Babeshko, O.V. Evdokimova, and O.M. Babeshko, “Differential Factorization Method in Block Structures and Nanostructures,” Dokl. Akad. Nauk **415** (5), 596–599 (2007) [Dokl. Phys. (Engl. Transl.) **76** (1), 614–617 (2007)].
11. B.V. Shabat, *An Introduction to the Complex Analysis*, Parts 1, 2 (Nauka, Moscow, 1985) [in Russian].
12. O.V. Evdokimova, Ekolog. Vestnik Nauchn. Tsentr. ChES, No. 2, 8 (2007).

MODELS OF ADHESIVE INTERACTION OF ELASTIC SOLIDS

I.G. Goryacheva[1*] and Yu.Yu. Makhovskaya[1]**

ABSTRACT

An approach to solving problems of the interaction of axisymmetric elastic bodies in the presence of adhesion is developed. The different natures of adhesion, i.e. capillary adhesion, or molecular adhesion described by the Lennard-Jones potential are examined. The effect of additional loading of the interacting bodies outside the contact zone is also investigated. The approach is based on the representation of the pressure outside the contact zone arising from adhesion by a step function. The analytical solution is obtained and is used to analyze the influence of the form of the adhesion interaction potential, of the surface energy of interacting bodies or the films covering the bodies, their shapes (parabolic, higher power exponential function), volume of liquid in the meniscus, density of contact spots, of elastic modulus and the Poisson ratio on the characteristics of the interaction of the bodies in the presence of adhesion.

Key words: contact interaction, adhesion, capillary forces, roughness

INTRODUCTION

Approaches to solving the problems on adhesion of elastic bodies can be divided into two classes. The first comprises numerical methods in which integral equations of the contact problem are solved numerically for a given form of the adhesion interaction potential (such as the Lennard-Jones potential [1]). The second presents approximate models (among them the classical Johnson–Kendall–Roberts theory and the classical Derjaguin–Muller–Toporov theory [2]) that yield asymptotic solutions to the adhesion problem in the case of two contacting elastic spheres. There

[1]Ishlinsky Institute for Problems in Mechanics of the Russian Academy of Sciences, Moscow, Russia

[*]E-mail: *goryache@ipmnet.ru*

[**]E-mail: *makhovskaya@mail.ru*

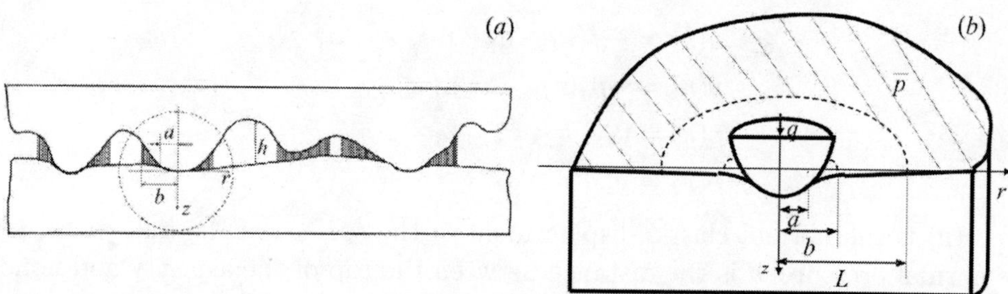

Fig. 1. Adhesive interaction between a rough surface and an elastic half-space (a). Adhesive interaction between an asperity and an elastic half-space, the effect of other asperities is replaced by the uniform pressure $\bar{p}$ applied in the region $r \geq L$ (b)

are a number of approximate methods [3–6] developed for solving this problem in a wide range of variation of problem parameters. These methods are based on the approximation by given functions of adhesion pressure arising on the surfaces of interacting bodies.

In this study, we propose a more general approach using the representation of the adhesion pressure in the form of a piecewise-constant multistep function. This provides the possibility of considering arbitrary forms of the adhesion interaction potential (including the case of capillary adhesion), as well as of taking into account the presence of another additional load, in particular, the effect of neighboring asperities on adhesion of both rough bodies (Fig. 1) and bodies with a regular surface relief.

1. PROBLEM FORMULATION

Consider the interaction of an elastic half-space with a rough rigid surface (Fig. 1a). Rough surface is pressed to the elastic half-space by the external normal pressure $\bar{p}$. Let the origin of the local cylindrical system of coordinates (r, z, φ) coincide with the point at which the elastic half-space contacts with an asperity before deformation. The axis z is directed inside the half-space. The shape of top of each asperity is the same and is described by the function $f(r) = Ar^{2n}$, where n is an integer.

To determine the stress-strain state near this asperity, we replace the influence of other asperities by the additional uniform pressure $\bar{p}$, applied in the region $r \geq L$ (Fig. 1b).

The boundary conditions for the elastic half-space for $z = 0$ have the form

$$\begin{aligned} u(r) &= -f(r) - d, \quad 0 < r < a, \\ p(r) &= -p_a(r), \quad a \le r \le b, \\ p(r) &= 0, \quad b < r \le L, \\ p(r) &= \bar{p}, \quad r > L, \end{aligned} \tag{1}$$

where $u(r)$ is the normal elastic displacement of the surface of the half-space, $p(r)$ is the normal pressure, d is the distance between the top of the asperity and nondeformed half-space. The first condition of (6) specifies contact between the asperity and the half-space over the circular region $0 < r < a$. In the case of no contact ($a = 0$) this condition is not used. The second condition of (6) specifies the adhesive attraction between the surfaces. Adhesive pressure $p_a(r)$ in the region $a \le r \le b$ is represented by the stepwise function:

$$p_a(r) = \begin{cases} p_1, & b_0 \le r \le b_1, \\ p_2, & b_1 \le r \le b_2, \\ \dots \\ p_N, & b_{N-1} \le r \le b_N, \end{cases} \tag{2}$$

where $b_0 = a$ and $b_N = b$.

The relationship between the normal displacement $u(r)$ and pressure $p(r)$ is defined by the familiar equation for the axisymmetric loading of the elastic half-space [7]

$$\begin{aligned} &u(r) = A[p(r), \infty], \\ &A[p(r), c] = \frac{4}{\pi E^*} \int_0^c p(r')\mathbf{K}\left(\frac{2\sqrt{rr'}}{r + r'}\right)\frac{r'dr'}{r + r'}, \quad E^* = \frac{E}{1 - \nu^2}, \end{aligned} \tag{3}$$

where E and ν are the Young modulus and Poisson's ratio for the elastic half-space, $\mathbf{K}(x)$ is the complete elliptic integral of the first kind.

The equilibrium equation for the asperity is satisfied

$$q = 2\pi \int_0^b rp(r)\, dr, \tag{4}$$

where $q$is the external normal force acting on the asperity.

2. METHOD OF SOLUTION

Introduce the new function $p_*(r)$ such that the function of contact pressure $p(r)$ in the region $0 \le r \le a$ is represented as

$$p(r) = p_*(r) - p_1. \tag{5}$$

To obtain the relationship between the new function $p_*(r)$ and elastic displacement $u(r)$, we substitute Eq. (5) into Eq. (3). Taking into account Eqs. (6) and (2) we have

$$u(r) - \sum_{k=1}^{N}(p_{k+1} - p_k)\chi(r, b_k) + \bar{p}\chi(r, R) - \bar{p}A[1, \infty] = A[p_*(r), a], \qquad (6)$$

where the function $\chi(r, c)$ is specified by the relation [7]

$$\chi(r,c) = A[1,c] = \frac{4}{\pi E^*}\begin{cases} c\mathbf{E}\left(\frac{r}{c}\right), & r \le c, \\ r\left[\mathbf{E}\left(\frac{c}{r}\right) - \left(1 - \frac{c^2}{r^2}\right)\mathbf{K}\left(\frac{c}{r}\right)\right], & r > c, \end{cases}$$

where $\mathbf{E}(x)$ is the complete elliptic integral of the second kind. In Eq. (6) and in what follows, we assume that $p_{N+1} = 0$.

In the case where there is no contact between the asperity and elastic half-space, and only the adhesive attraction exists ($a = 0$), the solution is defined by Eq. (6), the right-hand side of which is zero. This equation defines the elastic displacement $u(r)$ caused by the application of the adhesion pressure $-p_a(r)$ specified by (2) inside the circular region $0 \le r < b$ and additional load $\bar{p}$ in the region $r \ge L$.

In the case of contact of the surfaces, from Eq. (6) taking into account the contact condition (the first condition of (6)), we obtain the integral equation for the determination of the function $p_*(r)$

$$\begin{aligned} &A[p_*(r), a] = -f_*(r) - d_*, \quad r \le a, \\ &f_*(r) = f(r) + \sum_{k=1}^{N}(p_{k+1} - p_k)\chi(r, b_k) - \bar{p}\chi(r, L), \\ &d_* = d + \bar{p}A[1, \infty]. \end{aligned} \qquad (7)$$

Note that the distance d between the top pf the asperity and the surface of the nondeformed half-space is infinite, since the half-space is loaded over the infinite region. But the value of d_* is finite. We call this quantity d_* as the additional distance between the asperity and the half-space.

The equilibrium equation (4) taking into account (2) and (5) takes the form

$$q + \pi a^2 p_1 + \pi \sum_{k=1}^{N} p_k (b_k^2 - b_{k-1}^2) = 2\pi \int_0^a r p_*(r)\, dr. \qquad (8)$$

Since $p_*(a) = 0$ in virtue of (5), the integral equation (7) is similar to the equation for the problem about an axisymmetric punch of the shape $f_*(r)$ pressed to the elastic half-space by the force specified by the left-hand side of (8). In this case, relation (6) for $a < r \leq b$ defines the elastic displacement $u(r)$ in the region of adhesive interaction. The solution of this problem obtained in [8] gives the following relation for the contact pressure

$$p(r) = \frac{AE^* a^{2n+1}}{\pi} \left[\frac{(2n)!!}{(2n-1)!!} \right]^2 \sqrt{1 - \frac{r^2}{a^2}} \sum_{j=1}^{n} \frac{(2j-3)!!}{(2j-2)!!} \left(\frac{r}{a} \right)^{2(n-j)} - p_1$$
$$- \frac{2}{\pi} \sum_{k=1}^{N} (p_{k+1} - p_k) \arctan \sqrt{\frac{a^2 - r^2}{b_k^2 - a^2}} + \frac{2}{\pi} \bar{p} \arctan \sqrt{\frac{a^2 - r^2}{L^2 - a^2}}, \quad r \leq a. \quad (9)$$

A similar relation is obtained for the elastic displacement $u(r)$ for $a < r \leq b$. The additional distance d_* between the bodies is specified by the equation

$$d_* = - \frac{(2n)!!}{(2n-1)!!} A a^{2n} - \frac{2}{E^*} \sum_{k=1}^{N} (p_{k+1} - p_k) b_k \sqrt{1 - \frac{a^2}{b_k^2}} + \frac{2}{E^*} \bar{p} L \sqrt{1 - \frac{a^2}{L^2}}. \quad (10)$$

And the load q has the form

$$q = - \frac{(2n)!!}{(2n+1)!!} 4E^* A n a^{2n+1} - 2 \sum_{k=1}^{N} (p_{k+1} - p_k) b_k^2 \left(\arcsin \frac{a}{b_k} - \frac{a}{b_k} \sqrt{1 - \frac{a^2}{b_k^2}} \right)$$
$$+ 2 \bar{p} L^2 \left(\arcsin \frac{a}{L} - \frac{a}{L} \sqrt{1 - \frac{a^2}{L^2}} \right) - \pi a^2 p_1 - \pi \sum_{k=1}^{N} p_k (b_k^2 - b_{k-1}^2). \quad (11)$$

The solution obtained allows one to model various types of adhesive interaction. Examples of such models are given in the subsequent sections.

3. MOLECULAR ADHESION

First we consider models which do not take into account the interaction between asperities ($\bar{p} = 0$). Consider the interaction of two elastic asperities in the presence of adhesive attraction specified by the function $p_{\mathrm{ad}}(h)$, where h is the value of the gap between the surfaces. In particular, it may be the Lennard-Jones function describing the molecular attraction of surfaces.

In order to solve this problem, we divide the region of adhesive interaction $a \leq r \leq b$ into rings by b_j, $j = 1, 2, \ldots, N$. The values of the pressure p_j in each ring are assumed to be known. The value of the gap between the surfaces is given by the relation

$$h(r) = f(r) - f(a) + u(r) - u(a). \quad (12)$$

The elastic displacement $u(r)$ in the region $a \leq r \leq b$ is determined in accordance with the solution obtained in the previous section. After this, the values p_j can be found from the system of equations

$$p_j = p_{\text{ad}}(h(b_{j-1})), \quad j = 1, 2, \dots, N.$$

To determine the unknown values a and b, we need to additional equations — the condition of continuity of the pressure at the boundary of the contact region $p(a) = p_1$ and Eq. (11). After the determination of the adhesive pressure p_j and the values a and b, we can calculate the contact pressure distribution $p(r)$, and distance d between the bodies in accordance with Eqs. (9) and (10).

The graphs of the external force q versus the distance d between the bodies are presented in Fig. 2 for the cases of one-step and linear functions $p_{\text{ad}}(h)$. The shape of the interacting bodies is described by the parabolic function $f(r) = r^2/(2R)$, i.e. $A = 1/(2R)$, $n = 1$. The parameters of the functions $p_{\text{ad}}(h)$ are chosen such that the surface energies for these functions:

$$\gamma = \int_0^\infty p_{\text{ad}}(h)\, dh,$$

and the adhesive pressure at zero space between the surfaces $p_{\text{ad}}(0) = p_0$ are the same for both functions — linear and step-wise. We use the dimensionless notation suggested in [3], for which the solution depends on the only parameter

$$\lambda = p_0 \left(\frac{9R}{2\pi\gamma E^{*2}} \right)^{1/3}.$$

The results indicate that the form of the function $p_{\text{ad}}(h)$ have the most influence on the solution obtained for the negative force q in the case where the contact region is absent or small in comparison with the region of adhesive interaction. The dependence of the force q on distance d is nonmonotonic and ambiguous. This means that if the surfaces are separated under the controlled force q, the contact breaks in the instant corresponding to minimum of the force q in Fig. 2. If the surfaces are brought together and then separated under the controlled distance d, a loss of energy takes place, with corresponds to the area of the hatched region in Fig. 2.

Note that the results for the one-step function of adhesion pressure coincide with the solution obtained in [3]. The detailed analysis of the energy dissipation in an approach-separation cycle for this case is presented in [9].

4. CAPILLARY ADHESION

One more model based on the approach presented is the model of interaction of two elastic asperities in the presence of meniscus of liquid. Let two elastic asperities be

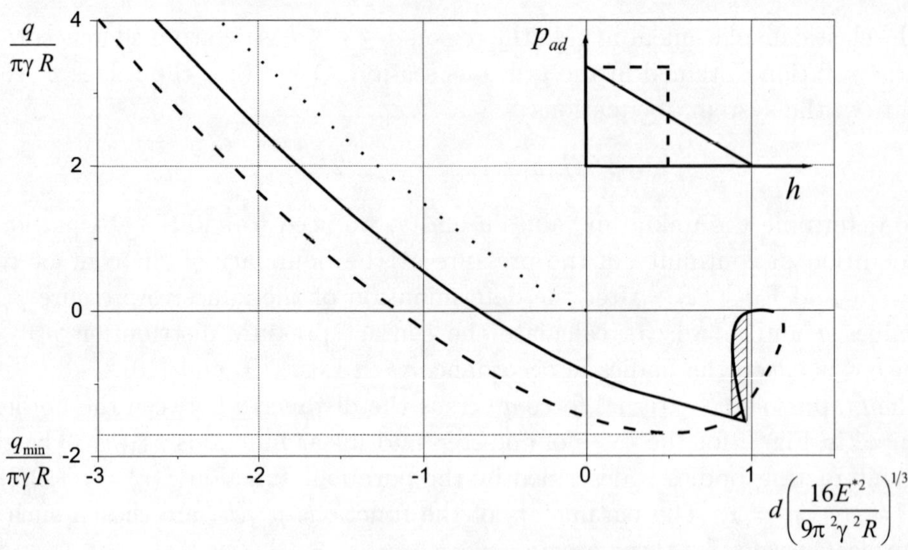

Fig. 2. The dimensionless load versus the dimensionless distance between the bodies for the linear (solid line) and one-step (dashed line) functions $p_{\rm ad}(h)$ for $\lambda = 0.6$. Dotted line corresponds to the case of no adhesion (Hertzian contact)

in contact over the region $r < a$ and the meniscus occupies the region $a \leq r \leq b$. Liquid in the meniscus produces a negative pressure $-p_0$ on the surfaces, which is specified by the relation $p_0 \approx 2\sigma/h(b)$ following from the Laplace formula under the assumption that the value of the gap $h(b)$ is small in comparison with the radius of the meniscus b. Here, σ is the surface tension in the liquid. Imposing also the condition of the constant volume v of the liquid in the meniscus,

$$v = 2\pi \int_a^b r h(r)\, dr$$

we arrive at the additional relation for determining the value of p_0.

The analysis of the solution to this problem presented in [8,9] showed, in particular, that an approach and separation of elastic bodies in the presence of a meniscus is accompanied by an energy loss, as is the case for adhesion defined by the function of adhesion pressure $p_{\rm ad}(h)$. For parabolic bodies ($n = 1$), the dependencies of this energy loss $w$and the pull-off force $q_{\min}$, on the surface tension in the liquid are shown in Fig. 3 (for different volumes of the liquid in the meniscus). The results indicate that the greater the surface tension σ of the liquid and the smaller the amount of the liquid v, the greater the energy loss. The pull-off force depends virtually linearly on the surface tension and weakly depends on the volume of liquid in the meniscus.

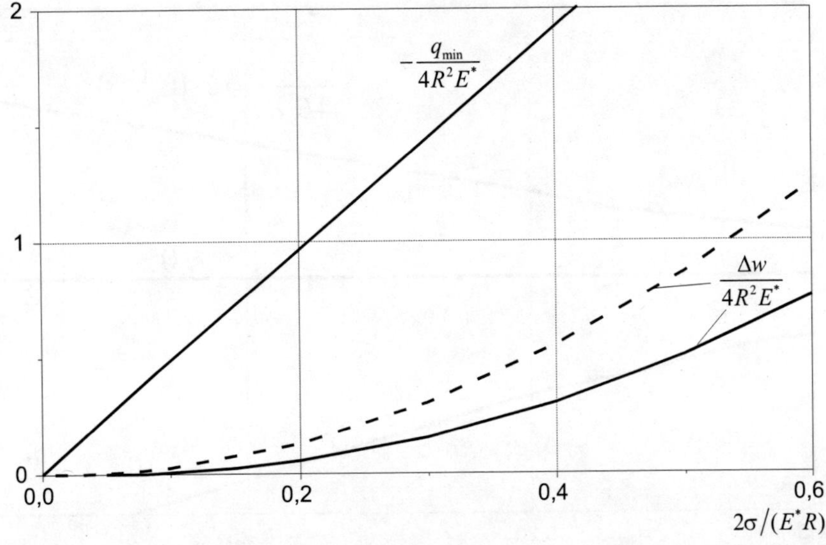

Fig. 3. Dimensionless pull-off force and dimensionless energy loss as functions of the dimensionless surface tension in the liquid in the case of capillary adhesion for $v/R^3 = 0.40$ (solid line) and 0.08 (dashed line)

5. ADHESION OF BODIES WITH REGULAR MICROGEOMETRY

The above method is also applicable for analyzing the adhesion interaction between elastic bodies with a regular surface microgeometry. Suppose that an elastic half-space interacts with a periodic system of axisymmetric identical asperities whose shape is described by the function $f(r)$. The asperities are assumed to be situated at the sites of a hexagonal lattice with a step l. The boundary conditions in the neighborhood of each of the asperities correspond to the adhesion attraction of the surfaces with a given function $p_{\text{ad}}(h)$ or to the case that each asperity is surrounded by the meniscus of a liquid.

To solve this problem, we used the localization method of [10]. In the simplest variant of this method, only the interaction between the half-space and a single asperity in the presence of an additional load in the form of uniform pressure $\bar{p}$ acting in the region $r \geq L$ (Fig. 1*b*) is taken into account. The mean pressure is calculated by the formula $\bar{p} = 2q/(\sqrt{3}l^2)$, and the value of L is found from the equality condition of the mean pressure inside and outside of the region $r \leq L$, i.e., $\bar{p} = q/(\pi L^2)$.

As a result, we have to solve the axisymmetric problem for the half-space with boundary conditions similar to those described in Section 2, in which the region

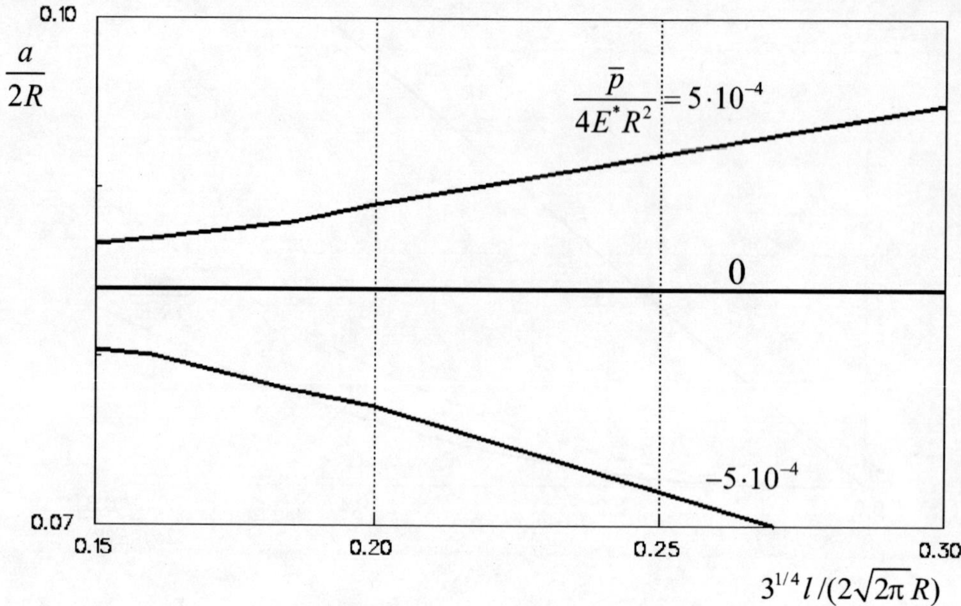

Fig. 4. Dependence of the dimensionless radius of the contact area on the dimensionless spacing between neighboring asperities in the case of the adhesion of surfaces with a regular microgeometry $\gamma/(RE^*) = 2 \times 10^{-4}$, $p_0/E^* = 0.02$

of loading by an additional step pressure is infinitely large. The results of solving this problem for the step function of the adhesion attraction $p_{\text{ad}}(h)$ and $n = 1$ (parabolic asperities) are presented in [11]. Fig. 4 shows the dependencies of the contact radius a on the spacing between the neighboring asperities at a fixed mean pressure $\bar{p}$ onto the half-space. The results indicate that, for the positive mean pressure $\bar{p}$, a decrease in the spacing l reduces the size of the contact area. On the contrary, for negative pressures, provided that the surfaces are still in contact, a decrease in l leads to increasing radius a of the contact area. This implies that the character of the dependence of the actual contact area on the density of the asperities on a rough surface is determined by the sign of the external nominal pressure applied to the interacting bodies.

ACKNOWLEDGMENTS

The authors are grateful to the Leverhulme Trust for funding the International Network Adhesint and to the Russian Foundation for Basic Research (grants Nos. 05-08-18204-a and 07-01-00282-a).

REFERENCES

1. J.A. Greenwood, "Adhesion of Elastic Spheres," Proc. Roy. Soc. London Ser. A **453**, 1277–1297 (1961) (1997).
2. D. Tabor, "Surface Forces and Surface Interactions," J. Colloid Interface Sci. **58** (1), 2–13 (1977).
3. D. Maugis, "Adhesion of Spheres: The JKR-DMT Transition Using a Dugdale Model," J. Colloid Interface Sci. **150** (1), 243–269 (1992).
4. D. Maugis and M. Barquins, "Fracture Mechanics and Adherence of Viscoelastic Bodies," J. Phys. D: Appl. Phys. **11** (14), 1989–2033 (1978).
5. J.A. Greenwood and K.L. Johnson, "An alternative to the Maugis Model of Adhesion between Elastic Spheres," J. Phys. D: Appl. Phys. **31** (22), 3279–3290 (1998).
6. E. Barthel, "On the Description of the Adhesive Contact of Spheres with Arbitrary Interaction Potentials," J. Colloid Interface Sci. **200** (1), 7–18 (1998).
7. K.L. Johnson, *Contact Mechanics* (Cambridge University Press, Cambridge, 1985).
8. I.G. Goryacheva and Yu.Yu. Makhovskaya, "Adhesion Interaction of Elastic Bodies," J. Appl. Math. Mech. **65** (2), 273–282 (2001).
9. Yu.Yu. Makhovskaya and I.G. Goryacheva, "The Combined Effect of Capillarity and Elasticity in Contact Interaction," Tribology Internat. **32** (9), 507–515 (1999).
10. I.G. Goryacheva, "The Periodic Contact Problem for an Elastic Half-Space," J. Appl. Math. Mech. **62** (6), 959–966 (1998).
11. Yu.Yu. Makhovskaya, "Discrete Contact of Elastic Bodies in the Presence of Adhesion," Mech. Solids **38** (2), 39–48 (2003).

EFFICIENT CALCULATION OF TRACTION COEFFICIENT IN LUBRICATED CONCENTRATED ELLIPTICAL CONTACTS

R.K. Pandey[1*] and N.K. Gupta[1]**

ABSTRACT

Accurate and efficient prediction of the traction coefficient at lubricated concentrated point/elliptical contact is of great importance in design of rolling element bearings, gears, cam and followers, and continuous variable transmissions (CVT) etc. In order to minimize energy loss/maximize power output and avoiding surface failures at the lubricated contacts, designers try to optimize traction coefficient. Hence, rapid calculation of traction at any general lubricated concentrated contact is an important step in design. Thus, an efficient thermal analysis of traction coefficient using much simpler mathematical modeling has been dealt in this paper. The results are presented for wide range of loads [P_H(Hertzian pressures)= 1.0 GPa to 4.0 GPa] and rolling speeds (5 m/s to 30 m/s) considering reasonable slips. Significant reduction in traction coefficient has been observed at higher rolling/sliding speeds with increase in loads. An empirical equation is developed for the prediction of traction coefficient in the contact zone in terms of operating performance parameters.

Key words: concentrated contact, rheological model, empirical formula, thermal effects, contact pressure, accuracy, EHL

INTRODUCTION

Many machine elements such as gears, rolling element bearings and traction drives etc. transmit large loads through the thin film thickness of lubricant present at the rolling/sliding concentrated contacts. The maximum Hertzian contact pressures and rolling velocities in hard working elliptical lubricated contacts may possibly reach some where in between 1.0 to 4.0 GPa and 5 to 30 m/s, respectively. Thus,

[1]Indian Institute of Technology Delhi, New Delhi — 110016, Hauz Khas, India
[*]E-mail: *rkpandey@itmmec.iitd.ernet.in*
[**]E-mail: *nkgupta@am.iitd.ernet.in*

an accurate and efficient prediction of the traction coefficient at the lubricated concentrated contacts is of great importance as it determines the power loss/efficiency and life of the contacts. Authors [1–2] have nicely reported works pertaining to EHL of concentrated contacts in their review papers. Thickness of the lubricating film separating two matting surfaces of a concentrated contact plays vital role in its operation. A lot of works exist in the open literature pertaining to studies and predictions of minimum film thickness in EHL area. Compared to the amount of research works on film thickness in elliptical EHL contacts, very little publications have been done by researchers [3–19] pertaining to the development of formulas for the prediction of traction coefficients. The reasons for this may be attributed to; (i) lack of understanding of the rheological behavior of lubricant in film at the central zone of the EHL contact at elevated operating conditions, and (ii) involved task in incorporation of non-Newtonian rheology in the modeling of elliptical EHL contact.

Based on the literature review, it is observed that few references have reported thermal EHL analysis of point/elliptic contacts, yet a thermal EHL analysis has not been presented with high rolling speeds (between 5 and 30 m/s) considering sets of heavy loads in the range of 1.0 GPa to 4.0 GPa. Moreover, it is also noticed that the biggest difficulty in the full-solution model (coupled solution of Reynolds equation, energy equation, elastic deformation incorporated film thickness relation, and rheological relations of lubricant) of the elliptical EHL contact is the lack of robustness in numerical method. For heavy loads and high rolling speeds of operating conditions, no work has been seen dealing with traction modeling. Thus, the objective of this research paper is to investigate traction coefficient in EHL elliptic contact at elevated operating conditions by considering heating effect in lubricating film. In the present analysis, much simpler and faster solution approach has been adopted. A reasonably modified Hertzian pressure profile has been assumed at the elliptical contact in order to avoid numerical instability due to high-pressure gradient (refer the base of pressure profile). Temperature variation across the film in energy equation has been approximated by parabolic polynomial. Non-Newtonian rheology of lubricant has been considered in modeling. Surface roughness and transient effects have been neglected in this work. It is, therefore, essentially an analysis for traction coefficient in the contact zone of a full film lubrication problem of smooth elliptically contacting surfaces at high loads and high rolling/sliding speeds. Results indicate that due to sliding, traction coefficient reduces significantly at elevated operating conditions.

1. GOVERNING EQUATIONS

The method of traction calculation presented in this paper incorporates some key features of the full numerical solution methods and approximation method. This has

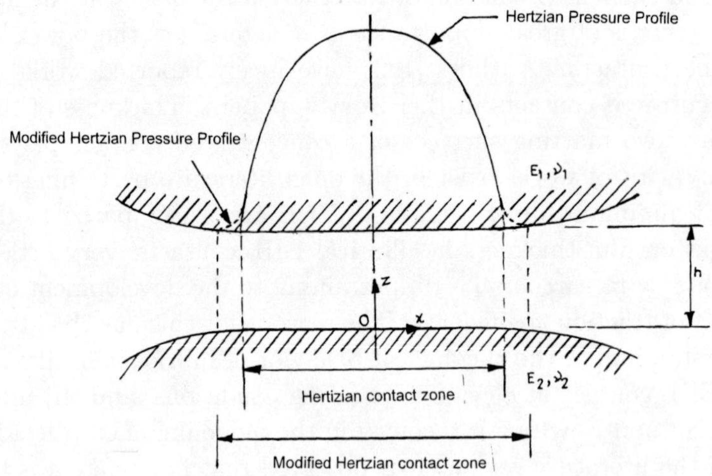

Fig. 1. View of Hertzian and modified pressure profiles

been done to avoid numerical instability. Numerical instability is primarily trigger in achieving the converged solution of Reynolds equation for pressure. Since pressure distribution in an EHL contact is similar to the corresponding Hertzian pressure for large practical problems, thus Hertzian pressure may be directly assumed at the EHL contact without loosing much accuracy in traction calculation.

Expression for pressure distribution

Hertzian pressure distribution at the EHL elliptic contact is modified as illustrated in Fig. 1 in order to avoid higher-pressure gradient near the base. In Fig. 1, the pressure profile is modified by blending the Hertzian surface and the surface formed by the revolution of the curve with respect to the z-axis towards the outside of base. Stitching of the pressure surfaces are carried out in such a way that discontinuity is not arising. The order of the polynomial for pressure curve is chosen to be three. The higher order (more than 3) polynomial is not chosen because it resulted in bending of the pressure curve.

Rheological relations for lubricant

Following relations for viscosity variation has been chosen in modelling:

$$\dot{\gamma}_{xz} = \frac{\tau_{xz}\tau_0}{\tau_e\eta}\sinh\left(\frac{\tau_e}{\tau_0}\right), \tag{1}$$

$$\dot{\gamma}_{yz} = \frac{\tau_{yz}\tau_0}{\tau_e\eta}\sinh\left(\frac{\tau_e}{\tau_0}\right), \tag{2}$$

$$\tau_e = \sqrt{\tau_{xz}^2 + \tau_{yz}^2}, \tag{3}$$

$$\tau_0(x,y) = \frac{1}{3}[0.095 - 0.00035T_m(x,y)]p(x,y), \tag{4}$$

$$\eta = \eta_0 \exp\{[\ln(\eta_0) + 9.67][-1 + (1 + 5.1e - 9p)^z] - \gamma(T - T_0)\}. \tag{5}$$

Density-pressure relation of lubricant

Following density relation provided by Dowson and Higginson is taken in modeling:

$$\overline{\rho} = \rho_0\Big(1 + \frac{\beta_1 p}{1 + \beta_2 p}\Big)[1 - \beta(T - T_0)]. \tag{6}$$

Energy equation

Under thin-film flow conditions, the general expression of the energy equation for a non-Newtonian fluid is expressed as:

$$\rho C_p\Big(u\frac{\partial T}{\partial x} + v\frac{\partial T}{\partial y}\Big) = \frac{\partial}{\partial z}\Big(k_f\frac{\partial T}{\partial z}\Big) + \beta T\Big(u\frac{\partial P}{\partial x} + v\frac{\partial P}{\partial y}\Big) + \tau_e\dot{\gamma}_e,$$

where

$$\dot{\gamma}_e = \sqrt{\gamma_{xz}^2 + \gamma_{yz}^2}.$$

Parabolic profile of temperature across the film is considered to transform the three-dimensional energy equation into two dimensions. The simplified form of the energy equation for a non-Newtonian fluid can thus be written as:

$$\begin{aligned}
&6T_1 + 6T_2 - 12T_m - \frac{\rho C_p h^4}{120k_f\eta_x}\frac{\partial p}{\partial x}\left(\frac{\partial T_1}{\partial x} + \frac{\partial T_2}{\partial x} - 12\frac{\partial T_m}{\partial x}\right)\\
&- \frac{\rho C_p h^4}{120k_f\eta_y}\frac{\partial p}{\partial y}\left(\frac{\partial T_1}{\partial y} + \frac{\partial T_2}{\partial y} - 12\frac{\partial T_m}{\partial y}\right) - \frac{\rho C_p u h^2}{k_f}\frac{\partial T_m}{\partial x} - \frac{\rho C_p u S h^4}{12k_f}\left(\frac{\partial T_2}{\partial x} - \frac{\partial T_1}{\partial x}\right)\\
&+ \frac{\beta h^4}{120k_f\eta_x}(T_1 + T_2 - 12T_m)\Big(\frac{dp}{dx}\Big)^2 + \frac{\beta h^4}{120k_f\eta_y}(T_1 + T_2 - 12T_m)\Big(\frac{dp}{dy}\Big)^2\\
&+ \frac{h^2\tau_{x1}\tau_0}{k_f\eta}\sinh\left(\frac{\tau_{x1}}{\tau_0}\right) - \frac{\beta u_e h^2}{k_f}\frac{dp}{dx}\Big[\frac{1}{12}(T_1 - T_2)S - T_m\Big] = 0,
\end{aligned} \tag{7}$$

where

$$\eta_x = \frac{\eta}{\cosh(\tau_{x1}/\tau_0)}, \quad \eta_y = \frac{\eta(\tau_{x1}/\tau_0)}{\sinh(\tau_{x1}/\tau_0)}, \quad \text{and} \quad \tau_{x1} = \tau_{xz}\big|_{z=0}. \tag{8}$$

Traction model

$$F = \pm \iint_{\Omega} \tau_{0,h} \, dx \, dy, \tag{9}$$

where

$$\begin{aligned} F &= \text{Traction Force}, \\ \Omega &= \text{Domain of Lubrication}, \\ \tau_{0,h} &= \frac{\eta}{h}\Delta U + \frac{h}{2}\frac{\partial p}{\partial x}, \\ \Delta U &= U_1 - U_2. \end{aligned}$$

2. NUMERICAL METHOD

Computation comprises of following steps:

- Pressure is determined using the following manner:

  if $(\sqrt{x^2+y^2} < 0.9)$ then $P(x,y) = \sqrt{x^2+y^2}$,
  elseif $((\sqrt{x^2+y^2} > 0.9)$ and $(\sqrt{x^2+y^2} < 1.4))$ then
  $r = \sqrt{x^2+y^2},\ P(x,y) = a_0 + a_1 r + a_2 r^2 + a_3 r^3$,
  else $P(x,y) = 0.0$ endif.

- Film thickness for relatively higher load ($P_{\max} > 1$ GPa) would be flat in the contact zone. Employing film thickness formula given by Hamrock and Dowson for evaluation of film thickness in domain.

- Calculate the shear stress and directional viscosities (η_x, η_y by solving the constitutive relationships.

- Calculate the mean temperature field by solving the energy equation and estimate the surface temperature field using the BCs.

- Estimate the new film thickness for modified temperature field employing the thermal reduction approach.

- Iterate the procedure from steps 3 to 5 till both the thermal film thickness and temperature field converge to the relative error of 1×10^{-5}.

- Calculate the traction coefficient of the lubricated contact.

Table 1. Properties of bounding solids and lubricants

Property	Value
Equivalent radius in x-direction (R_x), m	0.014
Modulus of elasticity of solids (E_1,E_2), N/m^2	2×10^{11}
Poisson's ratio of solids (ν_1, ν_2)	0.3
Specific heat capacity of solids (C_1, C_2), J/(kg·K)	460
Density of bounding solids (ρ_1, ρ_2), kg/m^3	7865
Inlet/ambient temperature for oil (T_0), °C	50
Inlet viscosity of oil (η_0), Pa·s	0.058
Inlet density of oil (ρ_0), kg/m^3	875
Coefficient for pressure-viscosity Index (Z_1)	0.72
Temperature-viscosity index (S_0)	1.14
Coefficient of compressibility (C_A), Pa^{-1}	0.326×10^{-9}
Coefficient of compressibility (C_B), Pa^{-1}	0.238×10^{-9}
Thermal expansivity (D_T), °C^{-1}	0.35×10^{-3}
Thermal conductivity (K_L), Wm^{-1}K^{-1}	0.13
Specific heat capacity (C_L), J·kg^{-1}K^{-1}	2010

3. RESULTS AND DISCUSSIONS

Based on the model discussed in this paper, traction coefficient in EHL elliptical contact is evaluated for numerous load and speed conditions. Input data of lubricant and bounding solids are provided in Table 1.

Calculated traction results are compared to those of Dama and Chang [11]. For a sensible comparison, identical parameters are used in calculation. Good correlation is visible at high slips.

In Figs. 3 to 6, traction coefficients values are plotted against slip for different values of G, k_e, u, and P_H. It is visible in these figures that traction coefficient is reducing with increase in slip for any values of G, k_e, u, and P_H. Based on these traction results, an empirical relation for the prediction of traction coefficient is developed as follows:

$$\mu_{\text{trac}} = 3.442281 \times 10^{-9} (P_H)^{(4.92)} (k_e)^{(-0.165)} (G)^{(2.355)} (U)^{(-0.736)} (S)^{(-0.543)}.$$

CONCLUSIONS

This paper presents a robust and accurate methodology to calculate the traction coefficient in EHL elliptical contact. The key points of this work are the use of the modified Hertzian pressure distribution and a formula based calculated film thickness. Closed form approximation of the pressure and film thickness largely

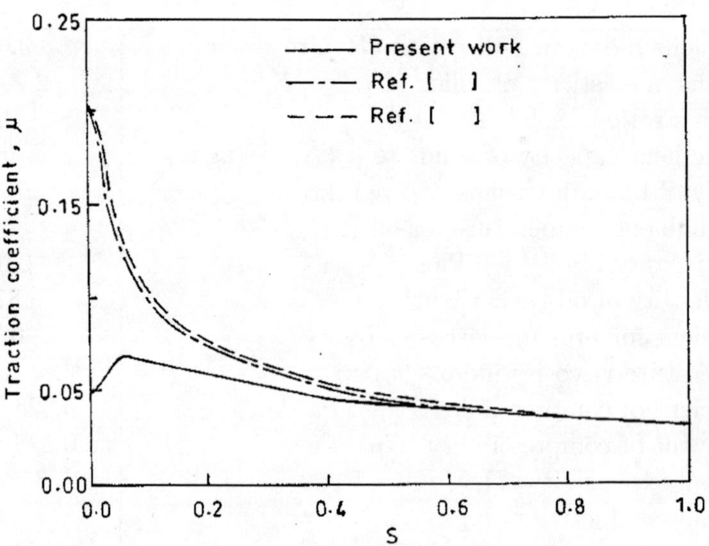

Fig. 2. Comparison of present results with other published works [11]

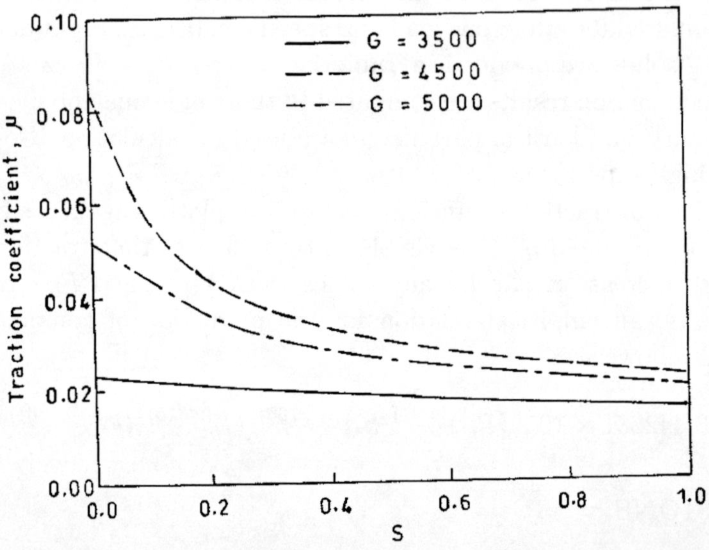

Fig. 3. Variation of traction coefficients with slips for different G

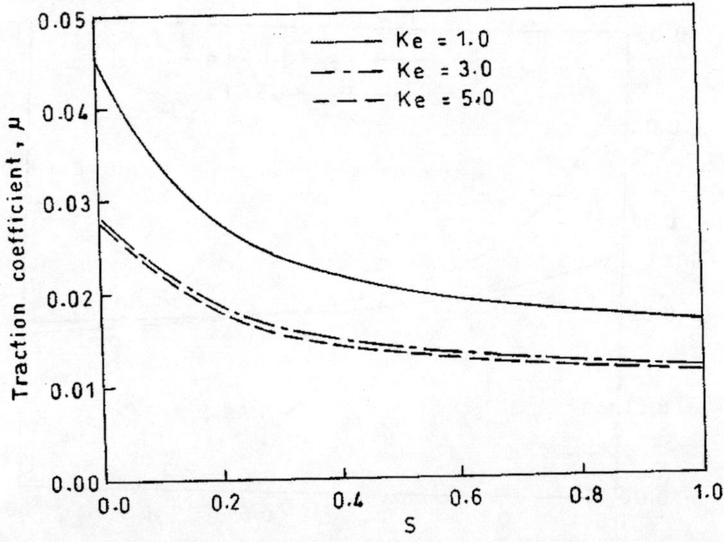

Fig. 4. Variation of traction coefficients with slips at different k_e

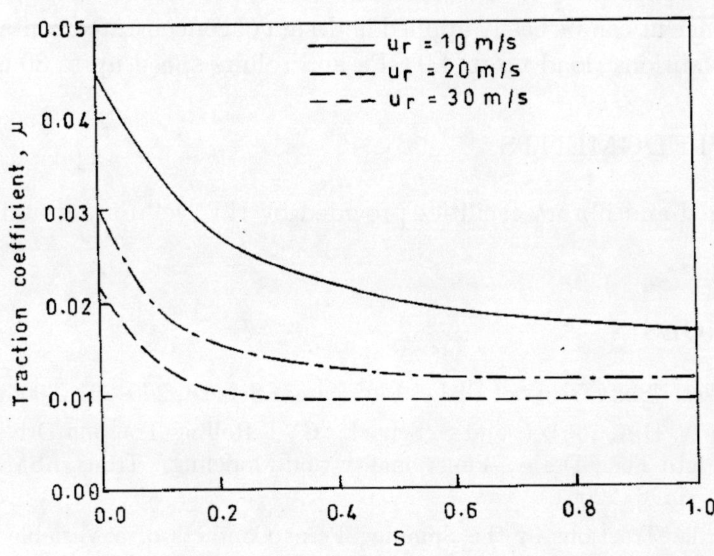

Fig. 5. Variation of traction coefficients with slips at different rolling speeds

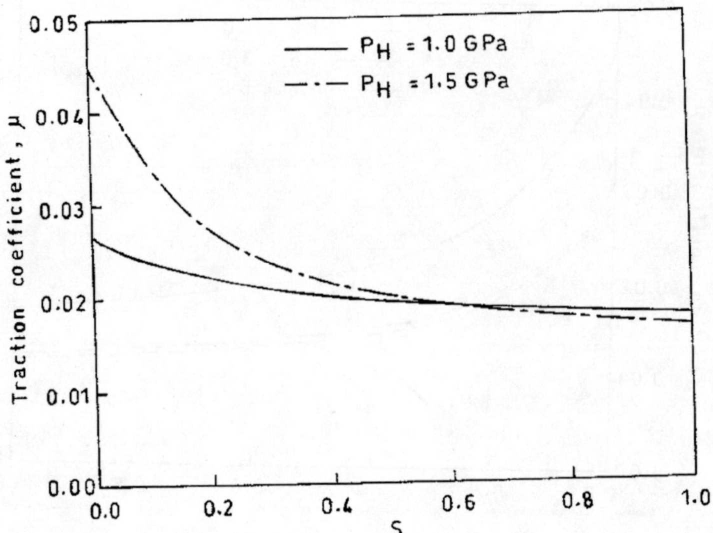

Fig. 6. Variation of traction coefficient with slip at two loads

enhances the robustness of the solution. Newly developed empirical relation of traction coefficient can be easily applied in design of concentrated contact at elevated operating conditions (load up to 4.0 GPa and rolling speed up to 30 m/s).

ACKNOWLEDGMENTS

Computational and library facilities provided by IIT Delhi are gratefully acknowledged.

REFERENCES

1. H.A. Spikes, "Sixty Years of EHL," Lubrication Sci. **18**, 265–291 (2006).
2. S. Akehurst, D.A. Parker, and S. Schaaf, "CVT Rolling Traction Drives-A Review of Research into Their Design, Functionality, and Modeling," Trans. ASME. J. Tribology **128**, 1165–1176 (2006).
3. S. Lingard, "Tractions at the Spinning Point Contacts of a Variable Ratio Friction Drive," Tribology Int. **7**, 228–234 (1974).
4. R. Kunz and W.O. Winer, "Prediction of Traction in Sliding EHD Contacts," Trans. ASME. J. Lubrication Technol. **98**, 362–366 (1976).
5. J.L. Tevaarwerk and K.L. Johnson, "The Influence of Fluid Rheology on the Performance of Traction Drives," Trans. ASME. J. Lubrication Technol. **101**, 266–274 (1979).

6. T.F. Conry, "Thermal Effects on Traction in EHD Lubrication," Trans. ASME. J. Lubrication Technol. **103**, 533–538 (1981).
7. I. Andersson, "Traction measurements in the Thermal Region of the Traction Curve," Tribology Int. **15**, 97–101 (1982).
8. W. Hirst and J.W. Richmond, "Traction in Elastohydrodynamic Contacts," Proc. Instn. Mech. Engrs. **202**, 129–144 (1988).
9. S. Wang, C. Cusano, and T.F. Conry, "Thermal Analysis of Elastohydrodynamic Lubrication of Line Contacts Using the Ree-Eyring Fluid Model," Trans. ASME. J. Tribology **113**, 232–244 (1991).
10. D.W. Dareing, "Traction Coefficients for Coated Bearing Races Lubricated with Teflon Transfer Films," Trans. ASME. J. Tribology **113**, 343–348 (1991).
11. R. Dama and L. Chang, "An Efficient and Accurate Calculation of Traction in Elastohydrodynamic Contacts," Wear **206**, 113–121 (1997).
12. A.V. Olver and H.A. Spikes, "Prediction of Traction in Elastohydrodynamic Lubrication," Proc. Instn. Mech. Engrs. Part J **212**, 321–332 (1998).
13. T. Nonishi, S. Oda, K. Miyachika, and T. Koide, "Limit Transmissible Torque in the Traction Derive of a Concave and Convex Roller Pair," Tribology Int. **33**, 233–240 (2000).
14. B. Jacod, C.H. Venner, and P.M. Lugt, "A Generalised Traction Curve for EHL Contacts," Trans. ASME. J. Tribology **123**, 248–253 (2001).
15. S. Blair and W.O. Winer, "Prediction of Traction in Elastohydrodynamic Lubrication," Proc. Instn. Mech. Engrs. Part J **215**, 309–310 (2001).
16. P. Ge and Z. Liu, "Experimental and computational Investigation of the Traction Coefficient of a Ball Traction Drive Device," Tribology Int. **35**, 219–224 (2002).
17. O.S. Cretu and R.P. Glovnea, "Traction Drive with Reduced Spin Losses," Trans. ASME. J. Tribology **125**, 507–512 (2003).
18. Y.S. Wang, B.Y. Yang, and L.Q. Wang, "Investigation into the Traction Coefficient in Elastohydrodynamic Lubrication," Tribotest J. **11**, 113–124 (2004).
19. X. Liu, M. Jiang, P. Yang, and M. Kaneta, "Non-Newtonian Thermal Analysis of Point EHL Contacts Using the Eyring Model," Trans. ASME. J. Tribology **127**, 70–81 (2005).

CONFORMAL CONTACT BETWEEN FOUNDATIONS AND PUNCHES

A.V. Manzhirov[1*] and K.E. Kazakov[1]**

ABSTRACT

We study the contact interaction between rigid punches and viscoelastic foundations with thin coatings for the cases in which the punch and coating surfaces are conformal (mutually repeating). Such problems can arise, for example, when the punch immerses into a solidificating coating before its complete solidification; as a result, the surface takes the shape of the punch base. Examples of such coatings can be a layer of glue, concrete at its young age, many polymeric materials. We consider plane contact problems for aging viscoelastic basements in the case of their conformal contact with rigid punches. We present the statements of the problems and derive their basic mixed integral equation. The solution of this equation is constructed by using the generalized projection method. We present numerical computations of model problems, including the problem in which the shape of the punch base is described by a rapidly oscillating function.

Key words: conformal contact, foundation, punch, settlement, tilt angle, contact stresses, mixed integral equation, rapidly oscillating function, projective method, analytic solution

1. STATEMENT OF THE PROBLEM

We assume that a viscoelastic layer with a coating lies on a rigid basis. At time τ_0, the force $P(t)$ with eccentricity $e(t)$ starts to indent a smooth rigid punch of width $2a$ (Fig. 1) into the surface of such a foundation. A specific characteristic of this contact interaction is the fact that the coating shape (the shape of the surface of the layer packet) coincides with the punch base shape. Such a contact interaction will be called conformal. The coating is assumed to be thin compared with the contact

[1]Ishlinsky Institute for Problems in Mechanics of the Russian Academy of Sciences, Moscow, Russia

[*]E-mail: *manzh@ipmnet.ru*

[**]E-mail: *kazakov@ipmnet.ru*

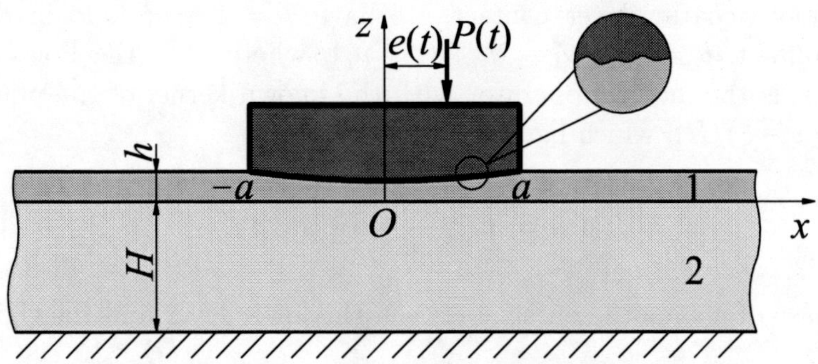

Fig. 1. Plane contact problem

area, i.e., its thickness satisfies the condition $h(x) \ll 2a$. Both the thin coating and the lower layer of an arbitrary thickness H are made of viscoelastic materials. We denote the moments of their production by τ_1 and τ_2, respectively. We assume that the coating rigidity is less than the rigidity of the lower layer or they are of the same order of magnitude [1–3]. We consider the case of plane strain.

We note that the simple case of conformal contact is the contact of a punch with a plane base and a plane part of a solid (including basements with coating of constant thickness).

To derive the integral equation of the problem, we replace the punch by some normally distributed load $p(x,t) = -q(x,t)$ acting on the same region $(-a \le x \le a)$ and equal to zero outside this region. Then the vertical displacement of the upper face of the foundation described above under the action of the normal force $q(x,t)$ can be written as [1–3]

$$u_z(x,t) = (\mathbf{I} - \mathbf{V}_1)\frac{\theta q(x,t)h(x)}{E_1(t-\tau_1)} + (\mathbf{I} - \mathbf{V}_2)\mathbf{F}\frac{2(1-\nu_2^2)q(x,t)}{\pi E_2(t-\tau_2)}, \tag{1.1}$$

$$\mathbf{F}f(x,t) = \int_{-a}^{a} k_{\mathrm{pl}}\Big(\frac{x-\xi}{H}\Big) f(\xi,t)\,d\xi,$$

$$\mathbf{V}_k f(x,t) = \int_{\tau_0}^{t} K^{(k)}(t-\tau_k, \tau-\tau_k) f(x,\tau)\,d\tau,$$

$$K^{(k)}(t,\tau) = E_k(\tau)\frac{\partial}{\partial\tau}\Big[\frac{1}{E_k(\tau)} + C^{(k)}(t,\tau)\Big], \quad k = 1,\, 2,$$

where $E_k(t)$ are Young's moduli of the coating $(k = 1)$ and the lower layer $(k = 2)$ and ν_2 is the Poisson ratio of the lower layer; $\mathbf{I}$ is the identity operator; $\mathbf{V}_k$ are the Volterra integral operators with tensile creep kernels $K^{(k)}(t,\tau)$ $(k = 1,\, 2)$; $C^{(k)}(t,\tau)$ $(k = 1,\, 2)$ are the tensile creep functions; θ is a dimensionless coefficient depending on the contact conditions between coating and lower layer; in the case of a smooth coating-layer contact, we have $\theta = 1 - \nu_1^2$, and in the case of an perfect contact, $\theta = (1 - \nu_1 - 2\nu_1^2)/(1 - \nu_1)$, where ν_1 is the Poisson ratio of the coating; $\mathbf{F}$ is the integral operator with the known kernel of the plane contact problem $k_{\text{pl}}[(x - \xi)/H]$, which has the form [3, 4]

$$k_{\text{pl}}(s) = \int_0^\infty \frac{L(u)}{u} \cos(su)\, du,$$

and, in the case of a smooth contact between the lower layer and the rigid base,

$$L(u) = \frac{\cosh 2u - 1}{\sinh 2u + 2u},$$

and in the case of a perfect contact,

$$L(u) = \frac{2\varkappa \sinh 2u - 4u}{2\varkappa \cosh 2u + 4u^2 + 1 + \varkappa^2}, \qquad \varkappa = 3 - 4\nu_2.$$

By equating the vertical displacements of the upper face of the coating with the displacement of the rigid punch and taking into account (1.1) and the fact that the contact interaction is conformal, we obtain the integral equation of our problem in the form

$$(\mathbf{I} - \mathbf{V}_1)\frac{\theta q(x,t)h(x)}{E_1(t - \tau_1)} + (\mathbf{I} - \mathbf{V}_2)\mathbf{F}\frac{2(1 - \nu_2^2)q(x,t)}{\pi E_2(t - \tau_2)} = \delta(t) + \alpha(t)x$$
$$(-a \leqslant x \leqslant a), \quad (1.2)$$

where $\delta(t)$ is the punch settlement and $\alpha(t)$ is its tilt angle.

We supplement Eq. (1.2) with the condition of the punch equilibrium on the foundation:

$$\int_{-a}^{a} q(\xi, t)\, d\xi = P(t), \quad \int_{-a}^{a} \xi q(\xi, t)\, d\xi = M(t). \qquad (1.3)$$

Here $M(t) = e(t)P(t)$ denotes the moment of application of the force $P(t)$.

In (1.2) and (1.3), we make the change of variables by the formulas

$$x^* = x/a, \quad \xi^* = \xi/a, \quad t^* = t/\tau_0, \quad \tau^* = \tau/\tau_0,$$
$$\tau_1^* = \tau_1/\tau_0, \quad \tau_2^* = \tau_2/\tau_0, \quad \lambda = H/a,$$
$$\delta^*(t^*) = \frac{\delta(t)}{a}, \quad \alpha^*(t^*) = \alpha(t), \quad c^*(t^*) = \frac{E_2(t - \tau_2)}{E_1(t - \tau_1)},$$
$$m^*(x^*) = \frac{\theta}{1 - \nu_2^2}\frac{h(x)}{2a}, \quad q^*(x^*, t^*) = \frac{2(1 - \nu_2^2)q(x,t)}{E_2(t - \tau_2)},$$

$$
\begin{aligned}
&P^*(t^*) = \frac{2P(t)(1-\nu_2^2)}{E_2(t-\tau_2)a}, \quad M^*(t^*) = \frac{2M(t)(1-\nu_2^2)}{E_2(t-\tau_2)a^2}, \\
&\mathbf{V}_k^* f(x^*,t^*) = \int_1^{t^*} K_k(t^*,\tau^*) f(x^*,\tau^*)\,d\tau^*, \quad k=1,2, \\
&K_1(t^*,\tau^*) = \frac{E_1(t-\tau_1)}{E_1(\tau-\tau_1)}\frac{E_2(\tau-\tau_2)}{E_2(t-\tau_2)} K^{(1)}(t-\tau_1,\tau-\tau_1)\tau_0, \\
&K_2(t^*,\tau^*) = K^{(2)}(t-\tau_2,\tau-\tau_2)\tau_0, \\
&\mathbf{F}^* f(x^*,t^*) = \int_{-1}^{1} k_{\rm pl}^*(x^*,\xi^*) f(\xi^*,t^*)\,d\xi^*, \\
&k_{\rm pl}^*(x^*,\xi^*) = \frac{1}{\pi} k_{\rm pl}\Big(\frac{x-\xi}{H}\Big) = \frac{1}{\pi} k_{\rm pl}\Big(\frac{x^*-\xi^*}{\lambda}\Big).
\end{aligned}
\tag{1.4}
$$

Then, omitting the asterisks, we obtain a mixed integral equation in the form

$$
c(t)m(x)(\mathbf{I}-\mathbf{V}_1)q(x,t) + (\mathbf{I}-\mathbf{V}_2)\mathbf{F}q(x,t) = \delta(t) + \alpha(t)x \quad (-1 \leqslant x \leqslant 1) \tag{1.5}
$$

with the additional conditions

$$
\int_{-1}^{1} q(\xi,t)\,d\xi = P(t), \quad \int_{-1}^{1} \xi q(\xi,t)\,d\xi = M(t). \tag{1.6}
$$

Now we divide Eq. (1.5) by $\sqrt{m(x)}$ and introduce the notation

$$
\begin{aligned}
&Q(x,t) = \sqrt{m(x)}q(x,t), \quad k(x,\xi) = \frac{k_{\rm pl}(x,\xi)}{\sqrt{m(x)}\sqrt{m(\xi)}}, \\
&\mathbf{A}Q(x,t) = \int_{-1}^{1} k(x,\xi)Q(\xi,t)\,d\xi.
\end{aligned}
$$

Then the integral equation (1.5) can be reduced to the following integral equation with the Hilbert–Schmidt kernel $k(x,\xi)$ (see, e.g., [5]):

$$
c(t)(\mathbf{I}-\mathbf{V}_1)Q(x,t) + (\mathbf{I}-\mathbf{V}_2)\mathbf{A}Q(x,t) = \frac{\delta(t)}{\sqrt{m(x)}} + \frac{\alpha(t)x}{\sqrt{m(x)}} \quad (-1 \leqslant x \leqslant 1). \tag{1.7}
$$

The additional conditions (1.6) take the form

$$
\int_{-1}^{1} \frac{Q(\xi,t)}{\sqrt{m(\xi)}}\,d\xi = P(t), \quad \int_{-1}^{1} \frac{Q(\xi,t)}{\sqrt{m(\xi)}}\xi\,d\xi = M(t). \tag{1.8}
$$

In what follows, we construct the solution of the two-dimensional equation (1.7), which contains integral operators with constant as well as variable limits of integration, with the additional conditions (1.8) taken into account.

There exist four different versions of the substitution: 1) the settlement and tilt angle of the punch are given (i.e., the right-hand side of the equation is given); 2) the punch settlement and the moment of the load application are given; 3) the tilt angle of the punch and the force of the load application are given; 4) the force and the moment of the load application are given. Each of these statements is a separate problem with its specific integral operator, and hence it is required to construct its own system of functions for each of the four problems.

We note that earlier no solutions of the problem with incomplete information about the right-hand side and additional conditions (cases 2)–3)) have been obtained even in the case of an elastic material. This could be done only by using the generalized projection method. Moreover, this methods allowed us to construct a new effective solution of the classical problem for the Fredholm integral equation of second kind with the Schmidt kernel for a given right-hand side, which differs from the classical solution presented in Goursat's book [5].

The solution for case 4) is given in [6]. Here we will consider solutions for problems 1)–3).

2. SOLUTION FOR A GIVEN FORCE AND TILT ANGLE

Consider the following statement of a problem: it is required to define a change rule of eccentricity $e(t)$ of the force $P(t)$, to provide the given law of change of an tilt angle $\alpha(t)$. It is considered that the force $P(t)$ is known.

We assume that the function $\alpha(t)$ in (1.7) is given, first additional condition (1.8) remains valid, and second additional condition (1.8) gives the formula for function

$$e(t) = \frac{1}{P(t)} \int_{-1}^{1} \frac{Q(\xi,t)}{\sqrt{m(\xi)}} \xi \, d\xi. \tag{2.1}$$

We seek a solution of Eq. (1.7) under first condition (1.8) using (2.1) in the class of functions continuous in time t in the Hilbert space $L_2[-1,1]$ (e.g., see [1]). To this end, we first construct an $L_2[-1,1]$-orthonormal system of functions such that it contains $1/\sqrt{m(x)}$ and the remaining basis functions can be represented as products of functions depending on x by the weight function $1/\sqrt{m(x)}$. A system of functions satisfying the above conditions can be constructed by the formulas [7]:

$$\int_{-1}^{1} p_i(\xi) p_j(\xi) \, d\xi = \delta_{ij}, \quad p_n(x) = \frac{P_n(x)}{\sqrt{m(x)}},$$

$$P_0(x) = \frac{1}{\sqrt{J_0}}, \quad J_n = \int_{-1}^{1} \frac{\xi^n}{m(\xi)} \, d\xi,$$

$$P_n(x) = \frac{1}{\sqrt{\Delta_{n-1}\Delta_n}} \begin{vmatrix} J_0 & J_1 & \cdots & J_n \\ J_1 & J_2 & \cdots & J_{n+1} \\ \vdots & \vdots & \ddots & \vdots \\ 1 & x & \cdots & x^n \end{vmatrix}, \tag{2.2}$$

$$\Delta_{-1} = 1, \quad \Delta_n = \begin{vmatrix} J_0 & J_1 & \cdots & J_n \\ J_1 & J_2 & \cdots & J_{n+1} \\ \vdots & \vdots & \ddots & \vdots \\ J_n & J_{n+1} & \cdots & J_{2n} \end{vmatrix}.$$

Note that if $m(x) = \text{const}$, then the polynomials $p_n(x)$ are the orthonormal Legendre polynomials.

The Hilbert space $L_2[-1,1]$ can be represented as the direct sum of orthogonal subspaces $L_2[-1,1] = L_2^{(1)}[-1,1] \oplus L_2^{(2)}[-1,1]$, where $L_2^{(1)}[-1,1]$ is the Euclidean space with basis $\{p_0(x)\}$ and $L_2^{(2)}[-1,1]$ is the Hilbert space with basis $\{p_1(x), p_2(x), p_3(x), \ldots\}$. The integrand and the right-hand side of (1.7) can also be represented as the algebraic sum of functions continuous in time t and ranging in $L_2^{(1)}[-1,1]$ and $L_2^{(2)}[-1,1]$, respectively, i.e.,

$$Q(x,t) = Q_1(x,t) + Q_2(x,t), \quad f(x,t) = f_1(x,t) + f_2(x,t),$$

$$Q_1(x,t) = z_0(t)p_0(x), \quad f_1(x,t) = \frac{\delta(t)}{\sqrt{m(x)}} = \sqrt{J_0}\delta(t)p_0(x), \quad f_2(x,t) \equiv \frac{\alpha(t)x}{\sqrt{m(x)}}.$$

Note that the representation for $Q(x,t)$ contains the known term $Q_1(x,t)$, which is determined by the first additional condition (1.8):

$$z_0(t) = \frac{P(t)}{\sqrt{J_0}},$$

and the term $Q_2(x,t)$ is to be found. Conversely, for the right-hand side, one should find $f_1(x,t)$, while $f_2(x,t) \equiv 0$. These peculiarities permit one to class the resulting problem as a specific case of the generalized projection problem stated and solved in [8,9].

Following [8,9], we can introduce the orthogonal projection operator, mapping the space $L_2[-1,1]$ onto subspace $L_2^{(1)}[-1,1]$:

$$\mathbf{P}_1\phi(x,t) = \int_{-1}^{1} \phi(\xi,t)[p_0(x)p_0(\xi)]\,d\xi.$$

Obviously, the orthoprojector $\mathbf{P}_2 = \mathbf{I} - \mathbf{P}_1$ maps the space $L_2[-1,1]$ onto $L_2^{(2)}[-1,1]$. In addition, the following relations hold:

$$\mathbf{P}_i f(x,t) = f_i(x,t), \quad \mathbf{P}_i Q(x,t) = Q_i(x,t), \quad i = 1, 2.$$

Using [8], we apply the orthogonal projection operator $\mathbf{P}_2$ to Eq. (1.7). As a result, we obtain the equation for determining $Q_2(x,t)$ with a known right-hand side:

$$c(t)(\mathbf{I}-\mathbf{V}_1)Q_2(x,t)+(\mathbf{I}-\mathbf{V}_2)\mathbf{P}_2\mathbf{A}Q_2(x,t)=-(\mathbf{I}-\mathbf{V}_2)\mathbf{P}_2\mathbf{A}Q_1(x,t). \qquad (2.3)$$

It is necessary to construct its solution in the form of a series in the eigenfunctions of the operator $\mathbf{P}_2\mathbf{A}$, which, as one can show using the results obtained in [8, 9], is a compact strongly positive self-adjoint operator from $L_2^{(2)}[-1,1]$ into $L_2^{(2)}[-1,1]$. The system of eigenfunctions of such an operator is a basis in the space $L_2^{(2)}[-1,1]$, see [10]. The spectral problem for the operator $\mathbf{P}_2\mathbf{A}$ can be written in the form

$$\mathbf{P}_2\mathbf{A}\varphi_k(x)=\gamma_k\varphi_k(x),$$

$$\varphi_k(x)=\sum_{i=1}^{\infty}\varphi_i^{(k)}p_i(x), \quad k=1,\,2,\,\ldots, \quad k(x,\xi)=\sum_{m=0}^{\infty}\sum_{n=0}^{\infty}R_{mn}p_m(x)p_n(\xi),$$

$$R_{mn}=\int_{-1}^{1}\int_{-1}^{1}k(x,\xi)p_m(x)p_n(\xi)\,dx\,d\xi, \quad R_{nm}=R_{mn}, \quad m,\,n=0,\,1,\,\ldots,$$

$$\sum_{n=1}^{\infty}R_{mn}\varphi_n^{(k)}=\gamma_k\varphi_m^{(k)}, \quad k,\,m=1,\,2,\,\ldots$$

We expand the function $Q_2(x,t)$ with respect to the new basis functions $\varphi_k(x)$ $(k=1,\,2,\,\ldots)$ in $L_2^{(2)}[-1,1]$, i.e., $Q_2(x,t)=\sum\limits_{k=1}^{\infty}z_k(t)\varphi_k(x)$, substitute this representation into (2.3), and see that the unknown expansion functions $z_k(t)$ $(k=1,\,2,\,\ldots)$ can be determined by the formula

$$z_k(t)=-(\mathbf{I}+\mathbf{W}_k)\frac{-\alpha(t)g_k^{\alpha}+(\mathbf{I}-\mathbf{V}_2)z_0(t)K_k^{\alpha}}{c(t)+\gamma_k},$$

$$K_k^{\alpha}=\sum_{n=1}^{\infty}R_{0n}\varphi_n^{(k)}, \quad g_k^{\alpha}=\sqrt{\frac{J_0J_2-J_1^2}{J_0}}\int_{-1}^{1}p_1(\xi)\varphi_k(\xi)\,d\xi=\varphi_1^{(k)}\sqrt{\frac{J_0J_2-J_1^2}{J_0}},$$

$$\mathbf{W}_kf(x,t)=\int_1^t R_k^*(t,\tau)f(x,\tau)\,d\tau,$$

where $R_k^*(t,\tau)$ $(k=1,\,2,\,\ldots)$ is the resolvent of the kernel

$$K_k^*(t,\tau)=\frac{c(t)K_1(t,\tau)+\gamma_kK_2(t,\tau)}{c(t)+\gamma_k}.$$

Note that the resulting solution has the following structure

$$q(x,t)=\frac{1}{m(x)}\big[z_0(t)P_0(x)+\ldots\big],$$

i.e., one can explicitly single out the weight function $m(x)$, and hence the coating thickness function $h(ax)$ related to it by the change of variables (1.4). The formulas thus obtained permit deriving efficient analytic solutions for the layers with coatings of thickness described by complicated, in particular, rapidly oscillating functions, which can hardly be done by other well-known methods.

So, having defined the contact pressure $q(x,t)$ under the punch, using (2.1) we can find the change rule of eccentricity of the load:

$$e(t) = \frac{1}{P(t)}\left[\frac{J_1}{\sqrt{J_0}} z_0(t) + \sum_{i=1}^{\infty} g_i^{\alpha} z_i(t)\right] = \frac{J_1}{J_0} + \frac{1}{P(t)} \sum_{i=1}^{\infty} g_i^{\alpha} z_i(t).$$

In particular, it is possible to provide absence of a warp of a punch at any moment ($\alpha(t) \equiv 0$).

Determining the contact pressure under the punch, we can find the unknown punch settlement. To this end, we must apply the operator $\mathbf{P}_1$ to Eq. (1.7):

$$\delta(t) = \frac{1}{\sqrt{J_0}}\left\{-\alpha(t)\frac{J_1}{\sqrt{J_0}} + c(t)(\mathbf{I} - \mathbf{V}_1)z_0(t) + (\mathbf{I} - \mathbf{V}_2)\left[R_{00} z_0(t) + \sum_{k=1}^{\infty} K_k^{\alpha} z_k(t)\right]\right\}.$$

3. SOLUTION FOR A GIVEN SETTLEMENT AND MOMENT

One more possible statement is definition of a force $P(t)$, tilt angle $\alpha(t)$ and contact pressure $q(x,t)$ on known punch settlement $\delta(t)$ and the moment $M(t)$ (eccentricity $e(t)$).

We assume that the function $\delta(t)$ in (1.7) is given, second additional condition (1.8) gives the formula for the force

$$P(t) = \int_{-1}^{1} \frac{Q(\xi,t)}{\sqrt{m(\xi)}}\, d\xi, \tag{3.1}$$

and first additional condition (1.8) remains valid.

The Hilbert space $L_2[-1,1]$ can be represented as the direct sum of orthogonal subspaces $L_2[-1,1] = \tilde{L}_2^{(1)}[-1,1] \oplus \tilde{L}_2^{(2)}[-1,1]$, where $\tilde{L}_2^{(1)}[-1,1]$ is the Euclidean space with basis $\{\tilde{p}_0(x)\}$ and $\tilde{L}_2^{(2)}[-1,1]$ is the Hilbert space with basis $\{\tilde{p}_1(x), \tilde{p}_2(x), \tilde{p}_3(x), \ldots\}$. Here the basis functions can be fond by the formulas

$$\tilde{p}_0(x) = \frac{x}{\sqrt{J_2 m(x)}}, \quad \tilde{p}_1(x) = \frac{J_2 - J_1 x}{\sqrt{J_2(J_0 J_2 - J_1^2) m(x)}}, \quad \tilde{p}_k(x) \equiv p_k(x), \quad k = 2, 3, \ldots$$

For the integrand and the right-hand side of (1.7):

$$Q(x,t) = \tilde{Q}_1(x,t) + \tilde{Q}_2(x,t), \quad f(x,t) = \tilde{f}_1(x,t) + \tilde{f}_2(x,t),$$

where $\tilde{Q}_i(x,t)$, $\tilde{f}_i(x,t)$ are functions continuous in time t and ranging in $\tilde{L}_2^{(1)}[-1,1]$ and $\tilde{L}_2^{(2)}[-1,1]$, respectively.

The representation for $Q(x,t)$ contains the known first term, and the second term is to be found. Conversely, for the right-hand side, one should find $\tilde{f}_1(x,t) = \alpha(t)x/\sqrt{m(x)}$, while $\tilde{f}_2(x,t) = \delta(t)/\sqrt{m(x)}$ in unknown.

The orthogonal projection operator, mapping the space $L_2[-1,1]$ onto $\tilde{L}_2^{(1)}[-1,1]$ can be introduced by formulas:

$$\tilde{\mathbf{P}}_1\phi(x,t) = \int_{-1}^{1} \phi(\xi,t)\tilde{p}_0(x)\tilde{p}_0(\xi)\,d\xi.$$

The orthoprojector $\tilde{\mathbf{P}}_2 = \mathbf{I} - \tilde{\mathbf{P}}_1$ maps the space $L_2[-1,1]$ onto $\tilde{L}_2^{(2)}[-1,1]$.

We apply the orthogonal projection operator $\tilde{\mathbf{P}}_2$ to Eq. (1.7). As a result, we obtain the equation for determining $\tilde{Q}_2(x,t)$ with a known right-hand side. It is necessary to construct its solution in the form of a series in the eigenfunctions of the operator $\tilde{\mathbf{P}}_2\mathbf{A}$. The spectral problem for this operator can be written in the form

$$\tilde{\mathbf{P}}_2\mathbf{A}\tilde{\varphi}_k(x) = \tilde{\gamma}_k\tilde{\varphi}_k(x),$$

$$\tilde{\varphi}_k(x) = \sum_{i=1}^{\infty} \tilde{\varphi}_i^{(k)}\tilde{p}_i(x), \quad k = 1,\,2,\,\ldots, \quad k(x,\xi) = \sum_{m=0}^{\infty}\sum_{n=0}^{\infty} \tilde{R}_{mn}\tilde{p}_m(x)\tilde{p}_n(\xi),$$

$$\tilde{R}_{mn} = \int_{-1}^{1}\int_{-1}^{1} k(x,\xi)\tilde{p}_m(x)\tilde{p}_n(\xi)\,dx\,d\xi, \quad \tilde{R}_{nm} = \tilde{R}_{mn}, \quad m,\,n = 0,\,1,\,\ldots,$$

$$\sum_{n=1}^{\infty} \tilde{R}_{mn}\tilde{\varphi}_n^{(k)} = \tilde{\gamma}_k\tilde{\varphi}_m^{(k)}, \quad k,\,m = 1,\,2,\,\ldots$$

Final formulas for contact pressure under the punch take the form

$$q(x,t) = \frac{Q(x,t)}{\sqrt{m(x)}}, \quad Q(x,t) = \tilde{z}_0(t)\tilde{p}_0(x) + \sum_{k=1}^{\infty} \tilde{z}_k(t)\tilde{\varphi}_k(x),$$

$$\tilde{z}_0(t) = \frac{M(t)}{\sqrt{J_2}}, \quad \tilde{z}_k(t) = -(\mathbf{I} + \widetilde{\mathbf{W}}_k)\frac{-\delta(t)g_k^\delta + (\mathbf{I} - \mathbf{V}_2)\tilde{z}_0(t)K_k^\delta}{c(t) + \tilde{\gamma}_k},$$

$$K_k^\delta = \sum_{n=1}^{\infty} \tilde{R}_{0n}\tilde{\varphi}_n^{(k)}, \quad g_k^\delta = \sqrt{\frac{J_0J_2 - J_1^2}{J_2}}\int_{-1}^{1} \tilde{p}_1(x)\tilde{\varphi}_k(\xi)\,d\xi = \tilde{\varphi}_1^{(k)}\sqrt{\frac{J_0J_2 - J_1^2}{J_2}},$$

$$\widetilde{\mathbf{W}}_k f(x,t) = \int_1^t \tilde{R}_k^*(t,\tau)f(x,\tau)\,d\tau.$$

Kernels $\tilde{R}_k^*(t,\tau)$ ($k = 1,\,2,\,\ldots$) are the resolvents of the kernels

$$\tilde{K}_k^*(t,\tau) = \frac{c(t)K_1(t,\tau) + \tilde{\gamma}_kK_2(t,\tau)}{c(t) + \tilde{\gamma}_k}.$$

So, having defined the contact pressure $q(x,t)$ under the punch, using (3.1) we can find the force

$$P(t) = \frac{J_1}{\sqrt{J_2}}\tilde{z}_0(t) + \sum_{i=1}^{\infty} g_i^\delta \tilde{z}_i(t) = \frac{J_1}{J_2}M(t) + \sum_{i=1}^{\infty} g_i^\delta \tilde{z}_i(t).$$

Note that the decomposition factors $\tilde{R}_{mn}$ of the kernel $k(x,\xi)$ are expressed through factors R_{mn}. As it is possible to express basis functions $\tilde{p}_k(x)$ $(k = 0,\ 1)$ through $p_k(x)$ $(k = 1,\ 2)$,

$$\tilde{p}_0(x) = k_0^{(0)} p_0(x) + k_1^{(0)} p_1(x), \quad \tilde{p}_1(x) = k_0^{(1)} p_0(x) + k_1^{(1)} p_1(x),$$

$$k_0^{(0)} = -k_1^{(1)} = \frac{J_1}{\sqrt{J_0 J_2}}, \quad k_1^{(0)} = k_0^{(1)} = \sqrt{1 - \frac{J_1^2}{J_0 J_2}},$$

decomposition factors $\tilde{R}_{mn}$ can be presented in a form:

$$\tilde{R}_{mn} = \sum_{i,j=1}^{2} k_i^{(m)} k_j^{(n)} R_{ij} = k_0^{(m)} k_0^{(n)} R_{00} + [k_0^{(m)} k_1^{(n)} + k_1^{(m)} k_0^{(n)}] R_{01} + k_1^{(m)} k_1^{(n)} R_{11},$$
$$m,\ n = 0,\ 1,$$

$$\tilde{R}_{nk} = \tilde{R}_{kn} = \sum_{i=1}^{2} k_i^{(n)} R_{ik} = k_0^{(n)} R_{0k} + k_1^{(n)} R_{1k}, \quad n = 0,\ 1, \quad k = 2,\ 3,\ \ldots,$$

$$\tilde{R}_{kl} = R_{kl}, \quad k,\ l = 2,\ 3,\ \ldots$$

Determining the contact pressure under the punch, we can find the unknown tilt angle:

$$\alpha(t) = \frac{1}{\sqrt{J_2}}\left\{-\delta(t)\frac{J_1}{\sqrt{J_2}} + c(t)(\mathbf{I} - \mathbf{V}_1)\tilde{z}_0(t) + (\mathbf{I} - \mathbf{V}_2)\left[\tilde{R}_{00}\tilde{z}_0(t) + \sum_{k=1}^{\infty} K_k^\delta z_k(t)\right]\right\}.$$

4. SOLUTION FOR A GIVEN SETTLEMENT AND TILT ANGLE

This method allowed us to construct a solution of a problem with given right-hand side, i.e. when the settlement $\delta(t)$ and tilt angle $\alpha(t)$ of the punch are given. It is required to define change rules of force $P(t)$ and eccentricity $e(t)$ (or moment $M(t)$).

Additional conditions (1.8) give at once formulas for definition of functions $P(t)$ and $e(t)$ on function $Q(x,t)$:

$$P(t) = \int_{-1}^{1} \frac{Q(\xi,t)}{\sqrt{m(\xi)}}\,d\xi, \quad e(t) = \frac{1}{P(t)}\int_{-1}^{1} \frac{Q(\xi,t)}{\sqrt{m(\xi)}}\xi\,d\xi. \tag{4.1}$$

As the right side of the Eq. (1.7) is given, the orthogonal projection operator $\hat{\mathbf{P}}_1 = \mathbf{0}$. Obviously $\hat{\mathbf{P}}_2 = \mathbf{I}$. The spectral problem for the operator $\hat{\mathbf{P}}_2\mathbf{A}$ can be written in the form

$$\hat{\mathbf{P}}_2 \mathbf{A} \hat{\varphi}_k(x) = \hat{\gamma}_k \hat{\varphi}_k(x),$$

$$\hat{\varphi}_k(x) = \sum_{i=1}^{\infty} \hat{\varphi}_i^{(k)} p_i(x), \quad k = 0,\, 1,\, \ldots, \quad k(x,\xi) = \sum_{m=0}^{\infty} \sum_{n=0}^{\infty} R_{mn} p_m(x) p_n(\xi),$$

$$R_{mn} = \int_{-1}^{1} \int_{-1}^{1} k(x,\xi) p_m(x) p_n(\xi)\, dx\, d\xi, \quad R_{nm} = R_{mn}, \quad m,\, n = 0,\, 1,\, \ldots,$$

$$\sum_{n=1}^{\infty} R_{mn} \hat{\varphi}_n^{(k)} = \hat{\gamma}_k \hat{\varphi}_m^{(k)}, \quad k,\, m = 0,\, 1,\, \ldots$$

To setting aside the technical details of solution we state the final formulas for contact pressure under the punch:

$$q(x,t) = \frac{Q(x,t)}{\sqrt{m(x)}}, \quad Q(x,t) = \sum_{k=0}^{\infty} \hat{z}_k(t) \hat{\varphi}_k(x),$$

$$\hat{z}_k(t) = (\mathbf{I} + \widehat{\mathbf{W}}_k) \frac{\delta(t) \hat{g}_k^{\delta} + \alpha(t) \hat{g}_k^{\alpha}}{c(t) + \hat{\gamma}_k}, \quad \widehat{\mathbf{W}}_k f(x,t) = \int_1^t \hat{R}_k^*(t,\tau) f(x,\tau)\, d\tau,$$

$$\hat{g}_k^{\alpha} = \frac{J_1}{\sqrt{J_0}} \int_{-1}^{1} p_0(\xi) \hat{\varphi}_k(\xi)\, d\xi + \sqrt{\frac{J_0 J_2 - J_1^2}{J_0}} \int_{-1}^{1} p_1(\xi) \hat{\varphi}_k(\xi)\, d\xi$$

$$= \hat{\varphi}_0^{(k)} \frac{J_1}{\sqrt{J_0}} + \hat{\varphi}_1^{(k)} \sqrt{\frac{J_0 J_2 - J_1^2}{J_0}},$$

$$\hat{g}_k^{\delta} = \sqrt{J_0} \int_{-1}^{1} p_0(\xi) \hat{\varphi}_k(\xi)\, d\xi = \hat{\varphi}_0^{(k)} \sqrt{J_0}.$$

The functions $p_k(x)$ can be calculated under Eq. (2.2); $\hat{R}_k^*(t,\tau)$ ($k = 0,\, 1,\, \ldots$) is resolvent of the kernel

$$\hat{K}_k^*(t,\tau) = \frac{c(t) K_1(t,\tau) + \hat{\gamma}_k K_2(t,\tau)}{c(t) + \hat{\gamma}_k}.$$

So, having defined the contact pressure $q(x,t)$ under the punch, using (4.1) we can find the force $P(t)$ and the eccentricity $e(t)$:

$$P(t) = \sum_{i=0}^{\infty} \hat{g}_i^{\delta} \hat{z}_i(t), \quad e(t) = \frac{1}{P(t)} \sum_{i=0}^{\infty} \hat{g}_i^{\alpha} \hat{z}_i(t).$$

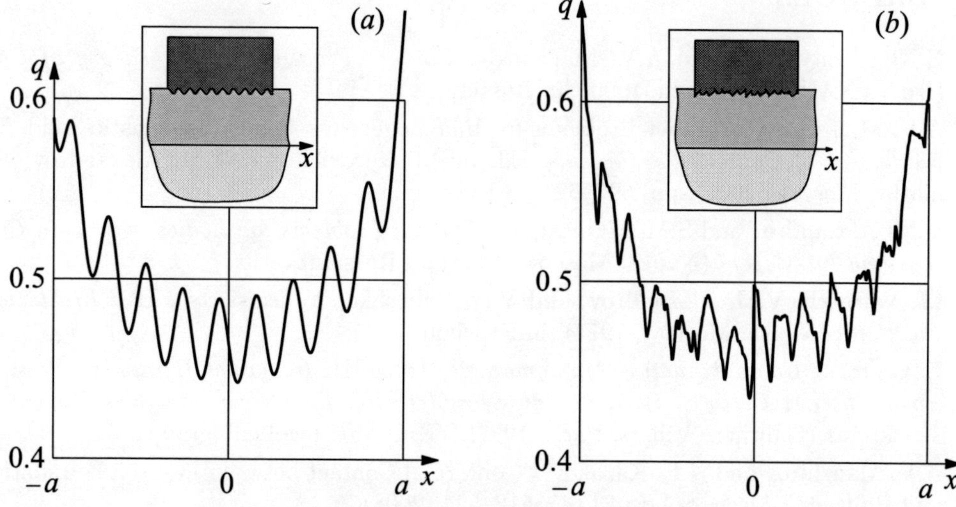

Fig. 2. Examples of solutions

5. MAIN RESULTS AND CONCLUSIONS

In the present paper, we introduce the notion of conformal contact interaction or a conformal contact of bodies, which is a generalization of the interaction between bodies and plane surface. We pose and solve plane problems of conformal contact between viscoelastic aging basements with coatings and rigid punches. We show that it is important to take the conformal contact into account. We also demonstrate the efficiency of the projection method for solving mixed integral equations in the case of plane contact problems and coated bodies with surfaces of complicated shape. The solution of the problem is obtained analytically, and, in the expressions for the contact stresses, the shape function of the foundation is distinguished explicitly, which allows one to perform computations for actual shapes of the coating surface, which are described by oscillating functions (Fig. 2). For the first time, the explicit formulas were obtained for the punch settlement and the tilt angle.

ACKNOWLEDGMENTS

The research was financially supported by the Russian Foundation for Basic Research (under grants Nos. 08-01-91302-IND_a, 08-01-00553-a, and 06-01-00521-a) and by the Department of Energetics, Mechanical Engineering, Mechanics and Control Processes of the Russian Academy of Sciences (Program No. 13 OE).

REFERENCES

1. N.Kh. Arutyunyan and A.V. Manzhirov, *Contact Problems in the Theory of Creep* (Izd-vo NAN RA, Erevan, 1999) [in Russian].
2. A.V. Manzhirov, "Contact Problems for Inhomogeneous Aging Viscoelastic Solids," in *Mechanics of Contact Interactions*, Ed. by I.I. Vorovich and V.M. Alexandrov (Fizmatlit, Moscow, 2001), pp. 607–621 [in Russian].
3. V.M. Alexandrov and S.M. Mkhitaryan, *Contact Problems for Bodies with Thin Coatings and Interlayers* (Nauka, Moscow, 1983) [in Russian].
4. I.I. Vorovich, V.M. Alexandrov, and V.A. Babeshko, *Nonclassical Mixed Problems of Elasticity* (Nauka, Moscow, 1974) [in Russian].
5. E. Goursat, *Cours d'Analyse Mathématique*, Tome III. *Intégrales Infiniment Voisines; Équations aux Dérivées Partielles du Second Ordre; Équations Intégrales; Calcul des Variations* (Gauthier–Villars, Paris, 1927; GTTI, Moscow–Leningrad, 1934).
6. A.V. Manzhirov and K.E. Kazakov, "Conformal Contact between Layered Foundations and Punches," Mech. Solids **43** (3), 512–524 (2008).
7. G. Szegö, *Orthogonal Polynomials* (Amer. Math. Soc, Providence, 1959; Fizmatgiz, Moscow, 1962).
8. A.V. Manzhirov, "Mixed Integral Equations of Contact Mechanics and Tribology," in *Mixed Problems of Mechanics of Solid. Proceedings of V Russian Conf. with Intern. Participation*, Ed. by Acad. N.F. Morozov (Izd-vo Saratov Univ., Saratov, 2005), pp. 221–224 [in Russian].
9. A.V. Manzhirov and A.D. Polyanin, *Handbook of Mathematics for Engineers and Scientists* (Chapman & Hall/CRC Press, Boac Ratoon–London, 2007).
10. A.N. Kolmogorov and S.V. Fomin, *Elements of the Theory of Functions and Functional Analysis* (Nauka, Moscow, 1976; Dover Publications, New York, 1999).

NATURAL VIBRATION PROBLEMS OF VISCOELASTIC AND ELECTROVISCOELASTIC BODIES AND THEIR APPLICATION TO OPTIMIZATION OF DYNAMIC CHARACTERISTICS

V.P. Matveenko[1*], E.P. Kligman[1], and N.A. Yurlova[1***]**

ABSTRACT

The paper presents solutions to the optimization problems considering the dynamic characteristics (resonance and dissipative properties) of viscoelastic and electroviscoelastic deformable systems, the behavior of which is described by the equations of linear elasticity and linear hereditary viscoelasticity. The examined electroviscoelastic systems are fitted with piezoelements whose electrodes are connected to the external RLC-circuits having resistance, capacitance and inductance. The parameters of the viscoelastic system liable to optimization are the object geometry, combination of elastic and viscoelastic materials and boundary conditions. For viscoelastic systems this group of parameters should be extended to include the properties of materials, location of the piezoelements, the conditions of their interaction, and the parameters of the external RLC-circuits. Optimization of the dynamic characteristics of the systems under study is realized in the context of the mechanical problem on natural vibrations.

Key words: natural vibrations, viscoelasticity, electroviscoelasticity, piezomaterials, dissipative properties, optimization, finite element method

---

[1]Institute of Continuous Media Mechanics of the Ural Branch of the Russian Academy of Sciences, Perm, Russia

[*]E-mail: *mvp@icmm.ru*

[**]E-mail: *kligman@icmm.ru*

[***]E-mail: *yurlova@icmm.ru*

INTRODUCTION

Damping ability of materials plays an important role in the dynamic behavior of different structures. There is a variety of damping mechanisms. Damping properties of viscoelastic materials can be defined in the framework of the hereditary continuum theories taking into account the time factor. Among the well-known hereditary theories of continuum are the Maxwell theory of viscous resistance, the Kelvin–Voigt theory of viscous friction, the Boltzman–Volterra theory of heredity, the Zener theory of thermal diffusion, etc. In this study we use the most general Boltzman–Volterra theory, which reflects practically all peculiarities of the quasi-static and dynamic behavior of viscoelastic materials [1]. Application of new technologies in production of composite materials led to creation of SMART-materials. The specific feature of these materials is the ability to change their properties in response to variations of external conditions. Generally, SMART-composites include as a component part the special sensor elements to monitor the environmental variations or changes in the thermomechanical state of the materials (sensors) and the active elements (actuators) to change accordingly the mechanical properties of SMART-materials. Any material capable of changing its thermomechanical state under non-mechanical action (electric, magnetic, temperature) can be used as an actuator. In production of SMART-composites the most common choices are piezo-materials. Preference for these materials can be explained by the fact they have a direct and inverse piezoeffect, which allows designers to use piezoelements both as the sensors and actuators. A side benefit of SMART-materials with piesoelements is the possibility of realizing the additional damping mechanisms. In particular, as an alternative to such modification we consider the structure, in which the electrodes of the piezoelements located at prescribed points are connected with each other or with the point of the zero potential by passive RLC-circuits. In this case, damping of vibrations occurs by transformation of the mechanical energy into the electric energy followed by dissipation of the latter in the external RLC-circuits in the form heat and electromagnetic radiation. Qualitative estimation of the dissipative properties of the structure can be derived form the solution of two problems. One of the problems is concerned with the analysis of free vibrations. Here, dissipation of the system manifests itself as damping of vibrations and the damping rate is the variable that estimates quantitatively the dissipative properties of the system. The second problem considers the forced steady-state vibrations of the system, in which the dissipative properties are manifested in restriction of the resonance amplitudes.

A search for structures with optimal damping properties by methods of numerical simulation is a complicated problem involving time consuming computations. On the one hand, in the prescribed parameter range one should investigate the parameters, which control the damping properties of the structure. On the other hand, for each combination of these parameters it is necessary to analyze the behavior of the examined structure within a certain spectrum of dynamic action. In the case of free vibrations the optimization process necessitates construction of solutions to the dynamic problems under different boundary conditions whereas in the problems on forced vibrations — construction of solutions at different frequencies of disturbing actions.

From the viewpoint of practical applications, including treatment of optimization problems, the statement of the problem on natural vibrations, in which damping properties of the system can be estimated without reference to external forces, kinematic and other factors, is of considerable interest. In these problems, frequencies are complex quantities. The real part of these quantities is the frequency and the imaginary part is the damping coefficient (damping rate) of natural vibrations. This problem is used as a basis for constructing the effective numerical procedures for optimization of the spectrum of natural frequency modes. These procedures use as control parameters the values of mechanical characteristics of the materials composing SMART-materials, geometries, locations of piezoelements, and quantities defining the boundary conditions. The presence of the external RLC-circuits connecting the electrodes of piezoelements increases the number of optimization parameters.

1. MATHEMATICAL FORMULATION OF THE NATURAL VIBRATION PROBLEM

We consider a piecewise-homogeneous composite body of a volume V involving N homogeneous elastic and viscoelastic parts having volumes V_k $(k = 1, 2, ...N)$. For part of the surface, which bounds the volume, the following kinematic and force boundary conditions are given:

$$u_i = \bar{U}_i(\mathbf{x}, t) \text{ on } \Omega_u, \quad \sigma_{ij} n_j = \bar{P}_i(\mathbf{x}, t) \text{ on } \Omega_\sigma, \tag{1}$$

where n_j are the components of the external normal vector to the surface, and $\bar{U}_i$, $\bar{P}_i$ are the kinematic and external forces. Strains are related to displacements by the linear Cauchy equations. The elastic parts of the relation are governed by Hooke's law for isotropic or anisotropic materials. In the viscoelastic parts of the composite body, stresses and strains are related by the linear Bolzmann-Volterra relations with integral difference kernels for isotropic material

$$\begin{aligned} \sigma_{ij} - \sigma\delta_{ij} &= 2\mu_0^{(k)} \left\{ \varepsilon_{ij} - \frac{1}{3}\vartheta\delta_{ij} - \int_0^t R^{(k)}(t-\tau) \left[\varepsilon_{ij}(\tau) - \frac{1}{3}\vartheta(\tau)\delta_{ij} \right] d\tau \right\}, \\ \sigma &= K_0^{(k)} \left[\vartheta - \int_0^t T^{(k)}(t-\tau)\vartheta(\tau)\, d\tau \right] \sigma_{ij} n_j = \bar{P}_i(\mathbf{x}, t) \text{ on } \Omega_\sigma, \end{aligned} \tag{2}$$

or anisotropic material

$$\sigma_{ij} = C^{(k)}_{ijkl} - \int_0^t R^{(k)}_{ijkl}(t-\tau)\varepsilon_{kl}(\tau)\, d\tau, \tag{3}$$

where $\mu_0^{(k)}$ is the instantaneous shear modulus, $K_0^{(k)}$ is the instantaneous volume modulus, $R^{(k)}, T^{(k)}$ are the shear and volume relaxation kernels, $\rho^{(k)}$ is the material density, $\sigma = \sigma_{jj}/3$ is the mean stress, $\vartheta = \varepsilon_{jj}$ is the volume strain, and are the instantaneous tensors of elastic constants and $C^{(k)}_{ijkl}$, $R^{(k)}_{ijkl}$ relaxation kernels.

The system state components are derived from the variational Lagrangian equation by supplementing it, according to the Dalembertian principle, with the work done by inertial forces:

$$\sum_{k=1}^{N}\left\{-\int_{V_k}\sigma_{ij}\,\delta\varepsilon_{ij}\,dV + \int_{\Omega_\sigma} P_i\,\delta u_i\,d\Omega - \int_{V_k}\rho^{(k)}\frac{\partial^2 u_i}{\partial t^2}\,\delta u_i\,dV\right\} = 0. \tag{4}$$

Let us consider the stationary vibrations of anisotropic system with vibrational frequency p in the absence of transient processes. In this case, the solution can be written in the complex form as

$$u_i(\mathbf{x}, t) = \bar{u}_i(\mathbf{x})e^{-ipt}, \tag{5}$$

where $\bar{u}_i$ are the complex amplitudes of the displacement vector components. In our solution, the lower limit of integration in equation (3) can be changed for $-\infty$. Then, after replacing the variables $t - \tau = \zeta$ in (3) and reducing the exponential form of the complex number to a trigonometric form, we have

$$\sigma_{ij} = \left[C_{ijkl} - \int_0^\infty R_{ijkl}(\zeta)\cos(p\zeta)\,d\zeta - i\int_0^\infty R_{ijkl}(\zeta)\sin(p\zeta)\,d\zeta\right]\bar{\varepsilon}_{kl}e^{-ipt}. \tag{6}$$

Integration of equation (6) yields the following physical relation:

$$\sigma_{ij} = \left[C_{ijkl} - Q^c_{ijkl}(p) - iQ^s_{ijkl}(p)\right]\varepsilon_{kl} = \left[C^{(r)}_{ijkl}(p) + iC^{(i)}_{ijkl}(p)\right]\varepsilon_{kl} = C^*_{ijkl}(p)\varepsilon_{kl}, \tag{7}$$

where C^*_{ijkl} is the tensor of complex dynamic moduli, and $Q^c_{ijkl}(p)$, $Q^s_{ijkl}(p)$ are the cosine and sine Fourier-images of relaxation kernels. The moduli $C^{(r)}_{ijkl}(p)$ and $C^{(i)}_{ijkl}(p)$ are sometimes called the storage and loss moduli, respectively.

Let us now formulate the problem on natural vibrations of viscoelastic bodies.

Its solution under homogeneous boundary conditions ($\bar{U}_i = 0$, $\bar{P}_i = 0$) is written as

$$u_i(\mathbf{x}, t) = \bar{u}_i(\mathbf{x})e^{-i\omega t}, \tag{8}$$

where $\omega = \omega_R + i\omega_I$ is the complex natural frequency of vibrations.

It should be noted that the integral terms in the hereditary relations (6) that define the dissipative properties of the material are generally smaller than the instantaneous elastic terms. Hence, solutions (8) can be considered as a vibration process with weakly changing amplitude. Then, by considering that $\tilde{u}_i(\mathbf{x},t) = \bar{u}_i(\mathbf{x})e^{\omega_I t}$ is a slowly changing time function, we can use the assumption of "freezing" of the function $e^{\omega_I t}$. This makes it possible to remove it from the sign of the integral in equations (2) and (3).

The second assumption suggests that the lower limit of integration is changed for $-\infty$. One of the factors supporting the validity of this replacement is natural vibration independence of the initial conditions. In view of these assumptions physical relations (3) can be rewritten as:

$$\sigma_{ij} \approx \left[C_{ijkl}e^{-i\omega_R t} - \int_{-\infty}^{t} R_{ijkl}(t-\tau)e^{-i\omega_R \tau}\, d\tau \right] \bar{\varepsilon}_{kl}e^{\omega_I t}. \tag{9}$$

After accomplishing the procedures used for transition from relations (3) to relations (7), we obtain for the problem on natural vibrations of viscoelastic bodies the approximate analogue of equations (3) written as physical relations with complex moduli (7).

In the absence of information regarding relaxation kernels, the loss tangent $\chi(p)$ can be used. Then, $C_{ijkl}^{(r)}(p)$ are the instantaneous elastic characteristics of the material and $C_{ijkl}^{(i)}(p) = C_{ijkl}^{(r)}(p)\chi(p)$. For isotropic bodies, the complex moduli are represented as

$$\mu^* = \mu_0[1 - R^c(p) - iR^s(p)], \tag{10}$$

$$K^* = K_0[1 - T^c(p) - iT^s(p)]. \tag{11}$$

To study the dynamic behavior of SMART-constructions, we consider the electrovicoelastic problem for a piecewise-homogeneous body composed of elastic and viscoelastic elements of volume V_1 and piezoelectric elements of volume V_2. The piezoelectric elements can be connected through the electroded surface to the external circuit consisting of resistances, inductances, and capacitances.

The variational equation of motion for the body consisting of elastic and piezoelectric elements can be obtained using the liner elastic relations and quasi-static Maxwell equations [2–4].

For isothermal process, the variational equation can be written as:

$$\int_{V_1} (\sigma_{ij}\delta\varepsilon_{ij} + \rho\ddot{u}_i\delta u_i)\, dV + \int_{V_2} (\sigma_{ij}\delta\varepsilon_{ij} - D_i\delta E_i + \rho\ddot{u}_i\delta u_i)\, dV$$

$$- \int_{\Omega_\sigma} \delta u_i P_i \, d\Omega - \int_{\Omega_p} q_e \delta\phi \, d\Omega = 0, \quad (12)$$

where D, E are the vectors of electric induction and intensity of the electric field, P is the vector of loads, Ω_p is the surface bounding the piezoelectric element, and q_e and ϕ are the surface density of charges and the electric potential.

We assume that the electric filed is potential, i.e., the condition $\phi_{,i} = -E_i$ is fulfilled. The isothermal processes in linear electrovisoelastic media are governed by the following physical relations:

$$\sigma_{ij} = C_{ijkl}\varepsilon_{kl} \qquad \text{for } V_1, \quad (13)$$

$$\left.\begin{aligned} \sigma_{ij} &= C_{ijkl}\varepsilon_{kl} - \beta_{ijk}E_k, \\ D_k &= \beta_{ijk}\varepsilon_{ij} + e_{ki}E_i \end{aligned}\right\} \quad \text{for } V_2, \quad (14)$$

where C_{ijkl} is the tensor of elastic constants, β_{ijk} and e_{ki} are the tensors of piezoelectric and dielectric coefficients.

If the element of the body V_1 has viscoelastic properties, then the tensor of elastic constants C_{ijkl} should be replaced by the corresponding complex analogue C^*_{ijkl} (7).

Let us consider the processes described by the time function (8). In the absence of force action, the variational equation (12) takes the form:

$$\left[\int_{V_1} (\sigma_{ij}\, \delta\varepsilon_{ij} - \rho\omega^2 \bar{u}_i\, \delta u_i)\, dV + \int_{V_2} (\sigma_{ij}\, \delta\varepsilon_{ij} - D_i\, \delta E_i - \rho\omega^2 \bar{u}_i\, \delta u_i)\, dV \right] e^{-i\omega t} = \int_{\Omega_p} q_e\, \delta\phi\, d\Omega \quad (15)$$

We examine the case when one of the piezoelements with the electroded surface (electrode) Ω_0 is connected with the zero potential point by a conductor of resistance R, capacitance C and inductance L. To describe mathematically the boundary conditions of this type, we use Ohm's law for alternating current. Thus we have

$$\phi = \frac{Q}{C} + RI + L\dot{I} = \frac{Q}{C} + R\dot{Q} + L\ddot{Q}, \quad Q = \int_{\Omega_0} q\, d\Omega, \quad (16)$$

where Q is the charge on the electrode, and $I = \dot{Q}$ is the current in the conductor.

Since we are dealing here with the quasi-harmonic process, i.e., $\phi(t) = \bar{\phi}e^{-i\omega t}$, then we can rewrite the differential equation (16) to solve for Q:

$$Q(t) = \frac{\bar{\phi}e^{-i\omega t}}{C^{-1} - \omega^2 L - i\omega R}. \quad (17)$$

Taking into account the surface equipotentiality Ω_0 and the fact that RLC is the external circuit, the following condition will be satisfied:

$$\delta\phi \int_{\Omega_0} q_e \, d\Omega = -\frac{\delta\phi\bar{\phi}_0 e^{-i\omega t}}{C^{-1} - \omega^2 L - i\omega R}, \tag{18}$$

where ϕ_0 is the potential on the electroded part of the piezoelement Ω_0.

In the absence of other surface forces, the equation of motion (15) will finally take the form:

$$\int_{V_1} (\sigma_{ij}\,\delta\varepsilon_{ij} - \rho\omega^2 \bar{u}_i\,\delta u_i)\, dV + \int_{V_2} (\sigma_{ij}\,\delta\varepsilon_{ij} - D_i\,\delta E_i - \rho\omega^2 \bar{u}_i \delta u_i)\, dV + \frac{\delta\phi\,\bar{\phi}_0}{C^{-1} - \omega^2 L - i\omega R} = 0. \tag{19}$$

Equation (19) is homogeneous and can be considered as a variational eigenvalue problem.

Thus, we have obtained the variational equation describing quasi-harmonic vibration for a piezoelectric body. This equation involves dissipative terms caused by the energy loss in the external electric circuits of resistance R. The inductance of the external circuit L and capacitance C are accepted as analogues of the mechanical mass and rigidity which can be used to control natural vibration frequencies.

2. NUMERICAL REALIZATION

For numerical realization of the problems under consideration, we use the finite element method (FEM).

With the FEM, the system with distributed parameters is transformed to the system with a finite number of degrees of freedom. In this case, the equation of motion (variational Lagrangian equation) for the dissipative system with a finite number of degrees of freedom has the form

$$\delta\{X\}^T([M]\{\ddot{X}\} + [K]\{X\}) = \{0\}, \tag{20}$$

where $\{X\}$ is the displacement vector with the dimension N, $[K] = [C] + i[C_I]$ is the complex stiffness matrix, and $[M]$ is the mass matrix.

For electroviscoelsticity problems, each point of the body is related to the state vector $\{u\} = \{u_1, u_2, u_3, \varphi\}^T$, where u_i are the components of displacement vector, and φ is the electric potential. In order to solve the vibration problems of systems with general-type RLC circuits, we suggest the following universal FEM scheme.

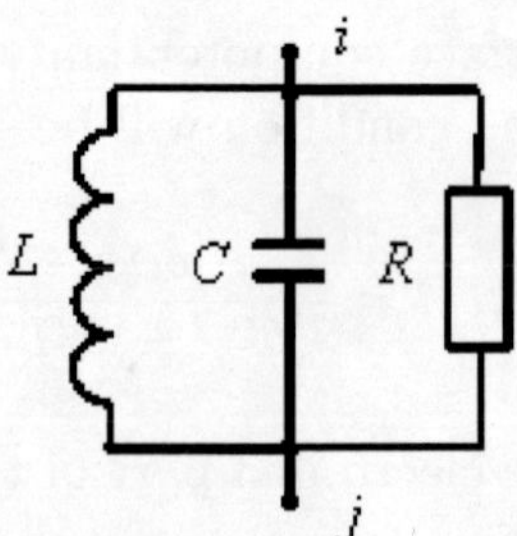

Fig. 1

In accord to the fundamentals of radio electronics, the behavior of passive R, L, and C elements in the alternating current circuits is described by the relations: $I = R^{-1}(\varphi_i - \varphi_j)$, $I = L^{-1} \int (\varphi_i - \varphi_j)\, dt$, and $I = C(\dot{\varphi}_i - \dot{\varphi}_j)$.

Here, I is the current of the conductor, R is the resistance, φ is the electric potential, L is the inductance, C is the capacity, and $q = \int I\, dt$ is the charge on the electrode.

Using the FEM, the behavior of one-dimensional R, L and C elements is defined as

$$\{I\} = R^{-1}[G]\{\varphi\}, \quad \{\dot{I}\} = L^{-1}[G]\{\varphi\}, \quad \{I\} = C[G]\{\dot{\varphi}\}. \tag{21}$$

Here I_i and I_j denote the current at the i-th and j-th nodes of the element

$$\{I\} = \left\{ \begin{array}{c} I_i \\ I_j \end{array} \right\}, \quad \{\varphi\} = \left\{ \begin{array}{c} \varphi_i \\ \varphi_j \end{array} \right\}, \quad [G] = \left[\begin{array}{cc} 1 & -1 \\ -1 & 1 \end{array} \right] \tag{22}$$

In electroelasticity, the work on varying the electric potential φ is done by the charge q and, therefore, relations (21) are transformed taking into consideration the identities $q = \int_{-\infty}^{t} I\, dt$:

$$\begin{aligned} \{q\} &= R^{-1}[G] \int \{\varphi\}\, dt = [K_R] \int \{\varphi\} dt, \\ \{q\} &= L^{-1}[G] \iint \{\varphi\}\, dt\, dt = [K_L] \iint \{\varphi\}\, dt\, dt, \\ \{q\} &= C[G]\{\varphi\} = [K_C]\{\varphi\}, \end{aligned} \tag{23}$$

where $\{q\} = \{q_i, q_j\}^T$.

For convenience in constructing arbitrary electric circuits (parallel, series, parallel-series etc.), the generalized element of the circuit is introduced according to the scheme shown in Fig. 1.

Combining the nodes of separate elements and using equation (23), we obtain

$$\{q\} = [K_C]\{\varphi\} + [K_R]\int\{\varphi\}\,dt + [K_L]\iint\{\varphi\}\,dt\,dt. \tag{24}$$

This expression describes the parallel RLC-circuit. By putting the corresponding elementary stiffness matrices equal to zero, it is possible to get any element of the scheme.

The work of the charge on variations in the electric potential can be written as

$$\delta\{\varphi\}^T\left([K_C]\{\varphi\} + [K_R]\int\{\varphi\}dt + [K_L]\iint\{\varphi\}dt\,dt\right) = \delta\{\varphi\}^T\{q\}. \tag{25}$$

To construct the matrices of network circuit coefficients, we use the congruent transformation approach.

After connecting the body to the circuit, we obtain the variational problem in the following form:

$$\delta\{\bar{X}\}^T\left([M]\{\ddot{\bar{X}}\}+([K]+[K_C])\{\bar{X}\}+[K_R]\int\{\bar{X}\}\,dt+[K_L]\iint\{\bar{X}\}dt\,dt\right)=\{0\}. \tag{26}$$

The solution to this problem is sought as $\{\bar{X}(t)\} = \{U\}e^{-i\omega t}$, which, in view of the arbitrariness of the variation, leads to the system of complex linear homogeneous equations

$$\left(-\omega^2[M] + [K] + [K_C] - \frac{1}{i\omega}[K_R] - \frac{1}{\omega^2}[K_L]\right)\{u\} = \{0\}. \tag{27}$$

Here, $[K]$ is the stiffness matrix, $[M]$ is the mass matrix, and $[K_C]$, $[K_R]$, $[K_L]$ are the matrices of coefficients of external RLC-circuits.

The condition for existence of the non-trivial solution to equation (27) is

$$D(\omega) = \det\left([K] + [K_C] - \omega^2[M] - \frac{1}{i\omega}[K_R] - \frac{1}{\omega^2}[K_L]\right) = 0. \tag{28}$$

Complex natural frequencies are obtained by the Muller method. The calculated eigen-numbers of equation (28) define the resonance frequencies $\mathrm{Re}(\omega)$ and the damping coefficients $\mathrm{Im}(\omega)$ of the system. Complex eigenfunctions can be used to determine the modes and phases of vibrations.

To solve numerically the stated problem, the displacement vector is represented in the form of a bounded series, whose basis involves the natural vibration modes of the electroelastic problem with the circuits including only the capacitance components.

$$\{X(t)\} \approx [\{u\}_1\ \{u\}_2\ \cdots\ \{u\}_j\ \cdots\ \{u\}_k]\cdot\{\vartheta(t)\} = [W]\cdot\{\vartheta(t)\}. \tag{29}$$

Here, $k \ll n$ is the number of terms in a series or the number of confined natural vibration modes in the corresponding electroelastic problem, $\vartheta_i(t)$ are the principal coordinates, and $[W]$ is the matrix of orthonormal eigenvectors of the corresponding electroelastic problem $(-\omega^2[M] + [C] + [K_C])\{U\} = \{0\}$.

Such representation permits using the potential of commercial software packages, in particular, ANSYS, for the solution of the problems under study. The ANSYS software package is applied to calculate the natural vibration modes of the corresponding conservative systems. Hence, substituting equation (29) in equation (26) yields

$$\delta\{\vartheta(t)\}^T\left[\{\ddot{\vartheta}(t)\}+(i\cdot[\alpha]+[\Omega])\{\vartheta(t)\}+[k_R]\int\{\vartheta(t)\}\,dt+[k_L]\iint\{\vartheta(t)\}dt\,dt\right]=\{0\}. \quad (30)$$

Here, $[\Omega] = \left\lfloor \omega_1^2 \ \omega_2^2 \ \cdots \ \omega_j^2 \ \cdots \ \omega_k^2 \right\rceil$ is the diagonal matrix which contains the squares of natural frequencies of the electroelastic problem, $[\alpha] = [W]^T[C_I][W]$, $[k_R] = [W]^T[K_R][W]$, $[k_L] = [W]^T[K_L][W]$.

Equation (30) is the system of linear homogeneous integro-differential equations, which can be formulated as the eigenvalue problem. The order of the system of equations (30) is significantly lower than the order of the initial problem (26), which provides the high computational efficiency of the numerical algorithms.

The partial solution of homogeneous equation (30) is sought as

$$\{\vartheta(t)\} = \{\theta\}e^{\lambda\cdot t}. \quad (31)$$

Substituting equation (31) in equation (30) in view of the arbitrariness of the variation, we have

$$\left([E]\lambda^2 + i\cdot[\alpha] + [\Omega] + \frac{1}{\lambda}[k_R] + \frac{1}{\lambda^2}[k_L]\right)\{\theta\} = \{0\}. \quad (32)$$

Here, $[E]$ is the unit matrix.

Equation (32) is a generalized algebraic eigenvalue problem for complex matrices. From the solution of this problem, the $2k$ complex eigen-numbers λ_j of the vector $\{\theta\}_j$ are obtained. The general solution of equation (30) is the sum of partial solutions

$$\{\vartheta(t)\} = \sum_j \{\theta\}_j e^{\lambda_j\cdot t}. \quad (33)$$

The final solution of the initial problem can be written as

$$\{X(t)\} \approx [\{u\}_1\ \{u\}_2\ \cdots\ \{u\}_j\ \cdots\ \{u\}_k]\cdot\sum_j\{\theta\}_j e^{\lambda_j\cdot t} = [W]\cdot\sum_j\{\theta\}_j e^{\lambda_j\cdot t}. \quad (34)$$

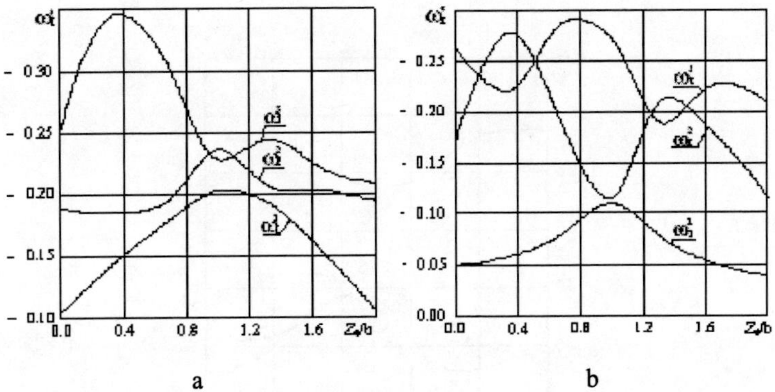

Fig. 2

3. RESULTS OF CALCULATIONS

Numerical realization of the problems under consideration is based on the finite element method. In order to illustrate the applicability of the proposed approach to determination and optimization of the dissipative properties of constructions, calculations of a few particular articles are made. As one of the applications of the problem on natural damping vibrations, we consider the problem of optimization of the damping properties of a two-layer cylinder during its transportation in vertical or horizontal positions in two rigid ring supports. The inner layer of the cylinder is made of the viscoelastic material, and its outer layer is made of the elastic material. The construction is defined by the following relations: $a/b = 0.3$, $L/b = 0.4$, $c/b = 1.01$, and $l_0/b = 0.08$. The parameters of the materials are $K/\mu_R = 24.7$, $\mu_I/\mu_R = 0.2$, $\mu_0/\mu_R = 100$, $\nu_0 = 0.3$, and $\rho_h = \rho_0$, where K, μ_R, μ_I, ρ_h are the volume modulus, imaginary and real parts of the complex shear modulus, specific density of the inner layer material and μ_0, ν_0, ρ_0 are the shear modulus, Poisson's ratio and specific density of the outer layer material. The components of the complex dynamic modulus are taken to be constant in the examined frequency range.

Suppose we have to achieve the maximum damping of the given number of the first vibration modes. In the case of symmetrically located supports, the part of the optimization parameter is played by z_0 (distance from the support to the edge of the cylinder).

Fig. 2 presents the dimensionless values ($\omega^* = \omega b\sqrt{\rho_h/\mu_R}$) of the imaginary parts of the first three natural frequencies at different positions of supports for two schemes of cylinder transportation, namely, in vertical (a) and horizontal (b) positions. The analysis of the obtained results led us to the conclusion that the optimal position of the supports corresponds to the relation $z_0/b = 1$.

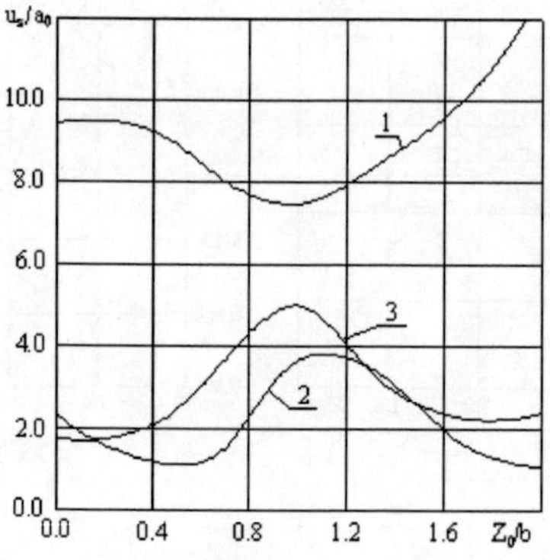

Fig. 3

As indicated earlier, the problem on natural damping vibrations was formulated under some assumptions leading to physical equations (10)–(11). Therefore, to verify the validity of the obtained solution, we consider the forced steady-state vibrations caused by the displacement of the supports according to the law $u_r = 0$, $u_z = a_0 \cos pt$.

This simulates the case when the cylinder is transported in a vertical position. The maximum displacement amplitudes at first three resonances are shown in Fig. 3 for different positions of the supports. As one can see, the optimal variant is at $z_0/b = 1$.

To support the possibility of suppressing mechanical vibrations with the aid of external circuits, a model experiment and the corresponding calculations in terms of this model were carried out. During the experiment, longitudinal vibrations (by second mode) are generated by piezoelement 2 in rectangular transversely polarized hinged-supported piezoceramic rod 1 (Fig. 4). A potential difference occurs on electrodes 3 located at the lateral surface. For additional suppression of mechanical vibrations, electrodes 3 are connected to the resistor R, where the electric energy dissipation takes place in the form of Joule heat.

By varying the resistance of the resistor R, the maximum suppression of vibrations can be obtained. The level of the amplitude of mechanical vibrations is defined indirectly in terms of the electrical potential U_1, appeared at the free end of the rod. The plot of the resonance amplitude U_1 [V] and the vibration frequency F [kHz] versus the resistance R [kOm] is given in Fig. 4.

It is seen that no additional suppression of vibrations arises at $R = 0$ and $R = \infty$. The maximum damping of vibrations takes place at $R \approx 2$ [kOm] and corresponds to a two-fold decrease in the resonance amplitude. Similar experiment is described

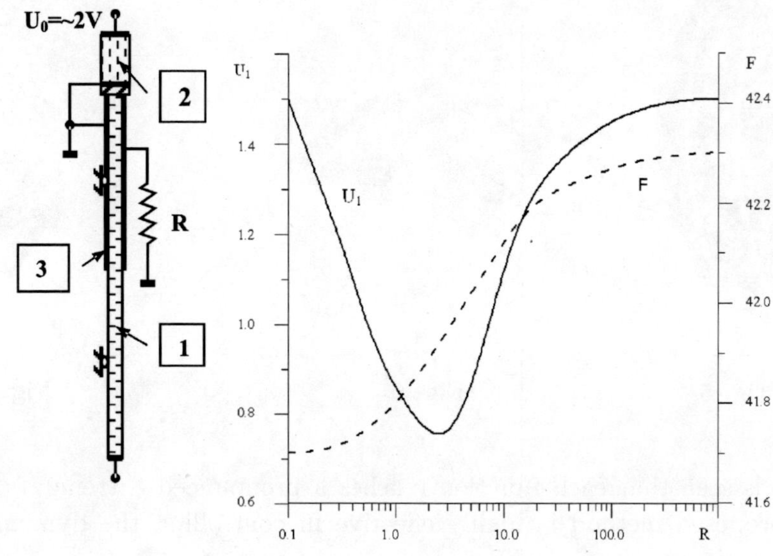

Fig. 4

in work [5], investigating the flexural vibrations of a gripped metal rod, which are damped with the aid of piezoelectric straps fixed at the rod base.

As another example, let us consider a parabolic shell (Fig. 5), having in plane the form of an ellipsoid with semi-axes $a = 1$ and $b = 0.75$ m, and 1 mm thick. The shell is made of aluminum alloy AD1M and reinforced by two pairs of stiffening ribs made of the unidirectional carboniferous fiber-based composite P-5-13N.

It is assumed that the material of the shell is elastic and is characterized by $E = 0.7 \cdot 10^{11}$ N/m^2, $\nu = 0.3$, and $\rho = 2600$ kG/m^3, and the stiffening ribs are viscoelastic with the loss tangent $\chi = 10^{-5}$ and $E = 1.5 \cdot 10^{11}$ N/m^2, $\nu = 0.2$, and $\rho = 1000$ kG/m^3.

The cross-sections of vertical and horizontal stiffening ribs are equal to 0.25×0.3, 0.25×0.2 mm, respectively. The white dots in Fig. 5 indicate the attaching points of the construction. We define the optimal coordinates for X and Y ribs to provide the maximum rate of vibration suppression in the system of the first four modes.

Figs. 6 and 7 illustrate the dependence of resonance frequencies (FREQ) and damping coefficients (DAMP) on rib positions. The intervals of $X-$ and $Y-$ coordinate variation were equal to $0.85a$ and $0.85b$, respectively, and their discreteness was 1/25.

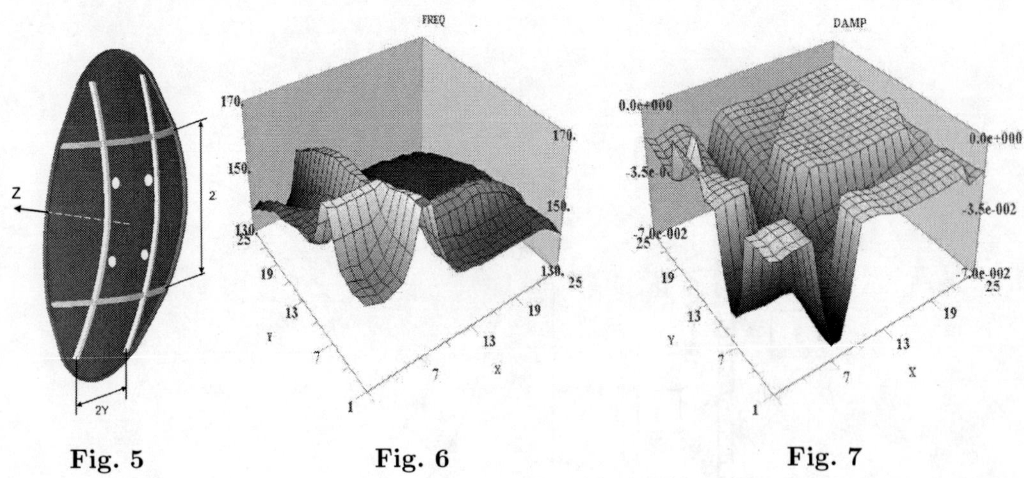

Fig. 5 **Fig. 6** **Fig. 7**

It is seen that each function reaches a pronounced extremum, suggesting that the proposed method is highly effective in controlling the dynamic properties of constructions.

The possibility of strengthening the damping effect with the use of SMART materials has been analyzed in a series of calculations made for the analogous parabolic shell with four piezoelectric zones in the central part (Fig. 8). The electrodes of these piezoelectric zones were connected to the external electric circuits (RLC) with a complex resistance $Z(\omega)$ (only one circuit is shown). The task is to define the circuit parameters, which can provide the maximum damping of the system.

Fig. 9 illustrates the dependence of the first complex mode of natural vibrations on the resistance value for circuits having a resistive component R, only. The obtained results have indicated that the external electrical circuits can provide an effective control of vibration damping.

The efficiency of the proposed approach for shell systems was verified in numerical calculations of the natural vibrations of hinged-supported two-layer cylindrical shells with the carbon plastic inner layer and the PZT-4 piezoceramics outer layer. The shells covered with piezoceramic material to full (a) or half (b) its length were examined (Fig. 10).

The cylinder had radius of 643 mm, length of 1286 mm, and thickness of 6.43 mm. The thickness of piezoelectric layer was 20% of the carrier layer. The characteristics of the materials entering the structure had the following values:

– carbon plastic: $E^1 = 4.3 \times 10^{10}$ N/m^2, $E^2 = 7.1 \times 10^{10}$ N/m^2, $E^3 = 6.0 \times 10^8$ N/m^2, $\nu_{21} = 0.4$, $\nu_{13} = \nu_{23} = 0.24$, $G_{12} = 2.0 \times 10^{10}$ N/m^2, $G_{13} = 1.4 \times 10^9$ N/m^2, $G_{23} = 2.1 \times 10^9$ N/m^2, $\rho = 1.47 \times 10^3$ kg/m^3;

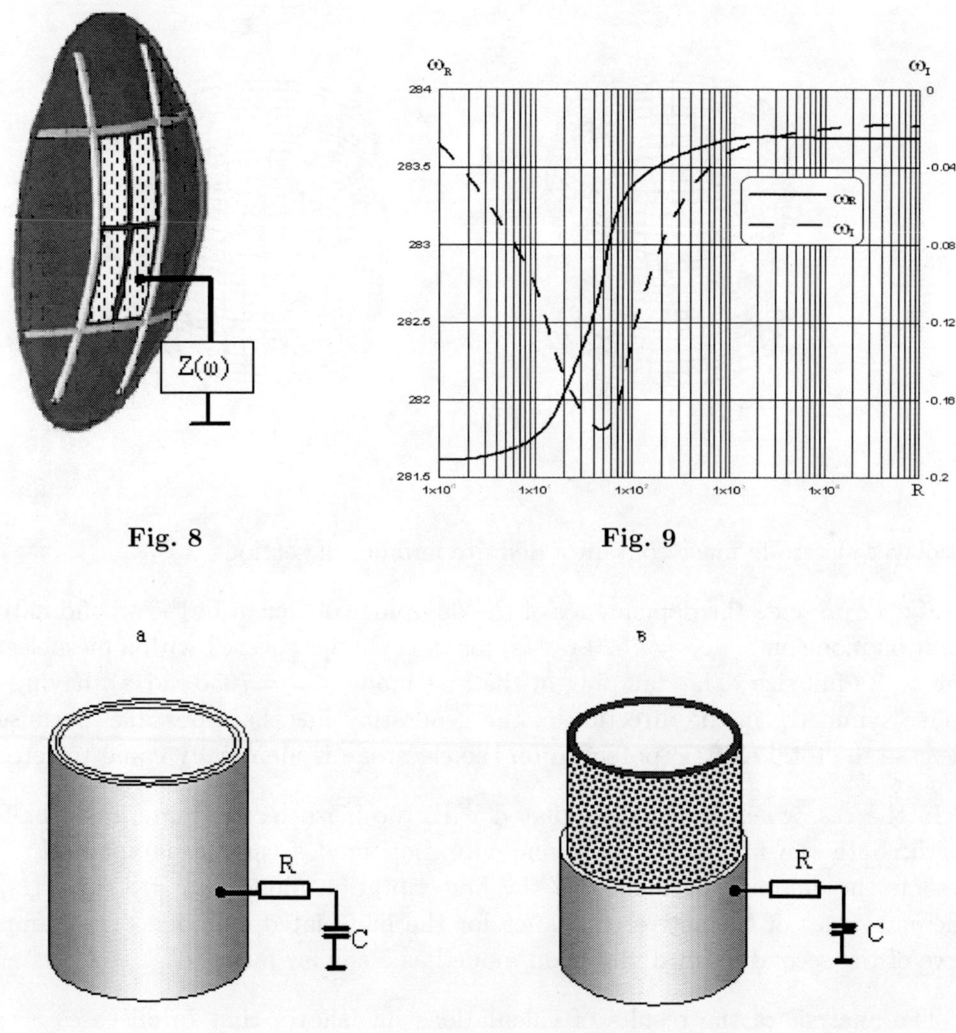

Fig. 8 **Fig. 9**

Fig. 10

– piezo-ceramics PZT-4: $\rho = 7500\,\mathrm{kg/m^3}$, $C_{11} = 13.9 \times 10^{10}\,\mathrm{N/m^2}$, $C_{12} = 7.78 \times 10^{10}\,\mathrm{N/m^2}$, $C_{13} = 7.43 \times 10^{10}\,\mathrm{N/m^2}$, $C_{33} = 11.5 \times 10^{10}\,\mathrm{N/m^2}$. As a damping element, we used the external RC-circuit of capacitance $C = 2\,\mathrm{nF}$.

The possibility of damping the first two modes of axisymmetric vibrations was examined (Fig. 11).

It should be noted that, at asymmetric modes of vibrations and in the case of full plating of the piezolayer, the electric potential on the electrode in a circumferential direction is identically equal to zero, which makes the damping of vibrations with the aid of external circuits impossible. For suppression of asymmetric modes, the

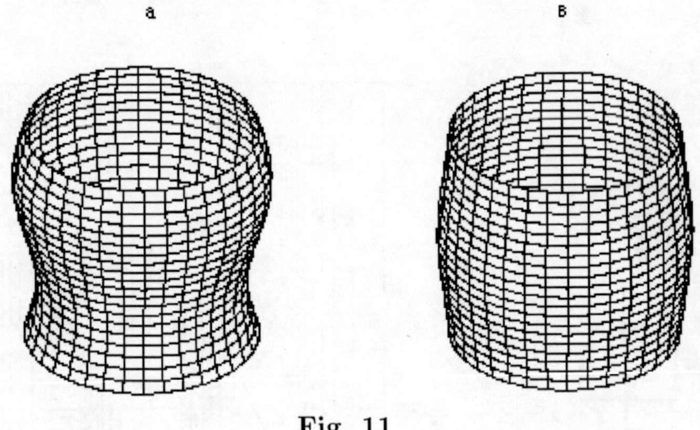

Fig. 11

piezolayer electrode must contain a definite number of sections.

Fig. 12 presents the dependence of the damping coefficient of the second natural vibration mode on R ($\omega_R = 7520$ rad/s) for the cylinder covered with a piezoelectric layer to its full size. The damping of the first mode ($\omega_R = 7050$ rad/s), having the inverse symmetry in the direction of the generating line, is impossible, because in this case the total electric potential on the electrode is identically equal to zero.

In the case when the shell is plated with the piezoelectric material to half its length, both the first and the second vibration modes can be damped. Fig. 13 presents the damping coefficient of the first vibration mode ($\omega_R = 8550$ rad/s) at different values of the active resistance for the half-plated cylinder. The damping curve of the second natural vibration mode has a similar form.

The analysis of the results of calculations has shown that in all cases a well-defined damping maximum is reached at certain values of the active resistance of the external circuit. The influence of the circuit capacitance C on the general damping factor of the system was considered.

Fig. 14 gives the dependence of the damping coefficient on the capacitance C of the external circuit for the second mode of natural vibrations of the cylinder fully covered with a piezolayer.

With an increase in the capacitance of the external RC-circuit up to 30 nF and more, the general damping coefficient of the system increases nearly 3 times, which indicates the dependence of the damping coefficient on the capacitance of the external circuit for the cylinder with the piezoelectric external layer.

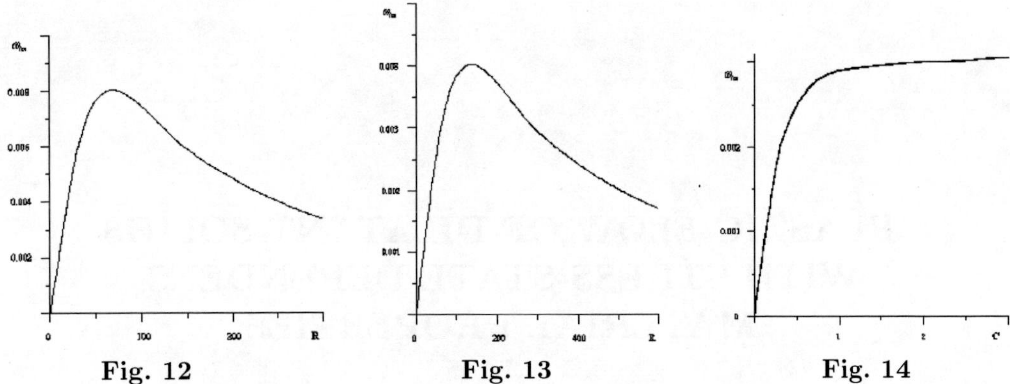

Fig. 12 Fig. 13 Fig. 14

CONCLUSION

In this paper, the mathematical formulation of the problem on natural vibrations of viscoelastic and electroviscoelastic bodies in the presence of external electric circuits including resistance, capacitance and inductance elements has been developed. The results of numerical experiments have indicated that the problem on natural vibrations of viscoelastic and electroviscoelastic bodies can be used for optimization of dissipative properties. Numerical experimentation has shown that the dynamic characteristics of the system can be effectively controlled by the elements of external electric circuits.

ACKNOWLEDGMENTS

This research was supported by the Russian Foundation for Basic Research (projects Nos. 07-08-12144 and 06-08-00405).

REFERENCES

1. A.D. Nashif, D.I.G. Jones, and J.P. Henderson, *Vibration Damping* (Wiley, New York, 1985; Mir, Moscow, 1988).
2. K. Washizu, *Variational methods in elasticity and plasticity*, 3rd ed. (Paragon, Oxford–New York, 1982; Mir, Moscow, 1987).
3. V.Z. Parton and B.A. Kudryavtsev, *Electromagnetoelasticity of piezoelectric and electrically conducting bodies*, (Nauka, Moscow, 1988) [in Russian].
4. V.G. Karnaukhov and I.F. Kirichok, *Electrothermoviscoelasticity*, (Naukova Dumka, Kiev, 1988) [in Russian].
5. D.J. Inman, "Smart structures solutions to vibration problems," in *Proceedings. International Conference on Noise and Vibration Engineering, Leuven, Belgium*, Vol. 1 (Leuven, 1998), pp. 1–12.

PLASTIC FLOW OF DILATANT SOLIDS WITH STRESS-STATE-DEPENDENT MATERIAL PROPERTIES

E.V. Lomakin[1]

ABSTRACT

In this study, a possible approach to the description of the dependence of the plastic properties of a dilatant medium on the type of loading-induced stress state is considered. The plasticity condition is accepted in corresponding generalized form where the parameter of stress state type is introduced. Adopting the flow rule associated with the proposed plasticity condition, the relations between the strain rates and stresses are obtained. It is shown that the processes of bulk and shear deformation are interrelated and the parameters of this relationship depend on the type of loading. The problem of the punch indentation is considered. The dependence of the limit state characteristics of the solids on the parameters characterizing the sensitivity of plastic properties of the materials to the alteration in the stress state type is studied for the plane strain conditions. The method for the determination of the displacement rate fields is proposed which permits one to investigate the process of the material dilatancy and to match the stress fields and the strain rate fields in the plastic regions.

Key words: dilatant materials, constitutive equations, plastic flow, stress state sensitivity, strain rate fields

INTRODUCTION

The process of plastic deformation of many materials involves not only the slip mechanism but also the development of existing and the formation of new discontinuity elements which are accompanied by the fracture and the displacement of the structural elements that causes the dilatation of the materials [1]. Theoretical

[1]Department of Mechanics and Mathematics, Lomonosov Moscow State University, Moscow, Russia. E-mail: *lomakin@mech.math.msu.su*

descriptions of the dilatancy process is commonly based on the use of the hydrostatic stress component in the plasticity condition [2–6]. Numerical simulation of the behavior of the porous materials with honeycomb structures shows that the characteristics of the plastic dilatation and the effective stress-strain curves depend not only on the magnitude of the hydrostatic component of the stress, but also on the stress state type parameter defined as the ratio of the hydrostatic component of the stress to the intensity of shear stresses named the triaxiality of the stress state [4–7].

Numerous experimental investigations of the plastic properties of the materials containing various kinds of structural inhomogeneity confirm these observations [11–13]. The plastic properties of rocks, graphites, refractory ceramics, cast iron, concrete, some particulate composites and many others are not invariant to the type of external forces. The condition of plastic incompressibility can not be accepted for these materials, too.

In this paper, the dependence of limit loads on the stress state sensitivity of the plastic properties of the materials is studied for plane strain conditions. The displacement fields are determined which describes the dilatation of a material in the plastic regions.

1. CONSTITUTIVE EQUATIONS

It was shown that the most proper stress state parameter is the ratio of the hydrostatic component of the stress, $\sigma = \frac{1}{3}\sigma_{kk}$ to the effective stress or the stress intensity $\sigma_0 = \sqrt{3/2 S_{ij} S_{ij}}$, where $S_{ij} = \sigma_{ij} - \sigma\delta_{ij}$ [12]. It must be noted that the parameter $\xi = \sigma/\sigma_0$ determines the average ratio of the normal stresses to the shear stresses, since the stress σ is the average normal stress at any point of a medium and the stress σ_0 is the average shear stress at the same point. This parameter can be used for the formulation of the constitutive equations.

The plasticity condition for a dilatant medium can be accepted in the following generalized form [12]:

$$F(\sigma_{ij}) = f(\xi)\sigma_0 = k. \tag{1}$$

One can assume without the loss of generality that $f(0) = 1$ for the pure shear ($\xi = 0$). In this case, it can be determined $k = \sqrt{3}\tau_s$, where τ_s is the yield stress for the pure shear. For an arbitrary stress state, the parameter ξ ranges from $-\infty$ (uniform triaxial compression) to ∞ (uniform triaxial tension).

By taking different analytical expressions for the function $f(\xi)$, we can obtain some of the well-known plasticity conditions for granular, porous and damaged media. The generalized Mohr-Coulomb condition $\sigma_0 + C\sigma = k$, which is widely used in soil mechanics, can be obtained if $f(\xi)$ is the linear function:

$$f(\xi) = 1 + C\xi. \tag{2}$$

The Green plasticity condition $\sigma_0^2 + \alpha\sigma^2 = k^2$ [4] suggested for porous media where the coefficients α and k are the functions of the porosity parameter of the medium can be obtained if the function $f(\xi)$ is accepted in the following form:

$$f(\xi) = \sqrt{1 + \alpha\xi^2} \quad (\alpha > 0). \tag{3}$$

For $f(\xi) = 1$, equation (1) coincides with Huber-Mises plasticity condition $\sigma_0 = k$. Other functions $f(\xi)$ have also been considered for various materials [12].

According to the flow rule $\dot{\varepsilon}_{ij} = h'\partial F/\partial\sigma_{ij}$ associated with plasticity condition (1), we obtain the following relation between the strain rates and the stresses for a rigid-plastic solid:

$$\dot{\varepsilon}_{ij} = h'\left[\frac{1}{3}\Lambda(\xi)\delta_{ij} + \frac{3}{2}\frac{\lambda(\xi)f(\xi)S_{ij}}{k}\right]. \tag{4}$$

The functions involved in this equation are $\Lambda(\xi) = f'(\xi)$, $\lambda(\xi) = f(\xi) - \xi f'(\xi)$, $h' = H/\chi(\xi)$, $H = \sqrt{\dot{\varepsilon}_{ij}\dot{\varepsilon}_{ij}}$, $\chi(\xi) = \sqrt{\frac{1}{3}\Lambda^2(\xi) + \frac{3}{2}\lambda^2(\xi)}$. The functions $\lambda(\xi)$, $\Lambda(\xi)$ and their derivatives are related by $\lambda(\xi) + \xi\Lambda(\xi) = f(\xi)$ and $\lambda'(\xi) + \xi\Lambda'(\xi) = 0$. Using (4), one can obtain the expressions for the intensity of plastic strain rates, $\Gamma = (\frac{2}{3}\dot{e}_{ij}\dot{e}_{ij})^{1/2} = h'\lambda(\xi)$, where $\dot{e}_{ij} = \dot{\varepsilon}_{ij} - \frac{1}{3}\dot{\varepsilon}\delta_{ij}$ and the bulk strain rate, $\dot{\varepsilon} = \dot{\varepsilon}_{ij}\delta_{ij} = h'\Lambda(\xi)$. These expressions imply that the function $\lambda(\xi)$ must be positive definite, while the rate of residual bulk strain must be proportional to the strain rate intensity:

$$\dot{\varepsilon} = \frac{\Lambda(\xi)}{\lambda(\xi)}\Gamma. \tag{5}$$

The proportionality factor $\Lambda(\xi)/\lambda(\xi)$ in general depends on the stress state parameter ξ and, hence, it has different values for different types of loading or different ratios of the normal to the shear stresses in a medium. If $f(\xi)$ is defined by equation (2), the coefficient in the relation (5) is constant.

In accordance with (4), the mechanical energy dissipation rate per unit volume is given by $D = \sigma_{ij}\dot{\varepsilon}_{ij} = kH/\chi(\xi)$. If $f'' \geq 0$, the yield surface in the stress space, defined by equation (1), is nonconcave.

2. GOVERNING EQUATIONS IN PLANE STRAIN

In the case of plane strain $\dot{\varepsilon}_{33} = \dot{\varepsilon}_{13} = \dot{\varepsilon}_{23} = 0$, it is possible to express the stress σ_{33} in terms of σ_{11}, σ_{22}, σ_{12} and eliminate it from the plasticity condition (1). As a result, we obtain:

$$\sigma_{33} = \sigma - \frac{2}{9}\sigma_0 \frac{\Lambda(\xi)}{\lambda(\xi)},$$
$$\sigma_0 = S_0 \left[1 - \frac{\Lambda^2(\xi)}{9\lambda^2(\xi)}\right]^{-1/2}, \tag{6}$$
$$\sigma = S - \sigma_0 \frac{\Lambda(\xi)}{9\lambda(\xi)}.$$

The new parameter $\zeta = S/S_0$, where $S_0 = \frac{\sqrt{3}}{2}\sqrt{(\sigma_{11} - \sigma_{22})^2 + 4\sigma_{12}^2}$, $S = \frac{1}{2}(\sigma_{11} + \sigma_{22})$, can be introduced and using (6), it is possible to express it in terms of ξ:

$$\zeta = \left[\xi + \frac{1}{9}\frac{\Lambda(\xi)}{\lambda(\xi)}\right]\left[1 - \frac{1}{9}\frac{\Lambda^2(\xi)}{\lambda^2(\xi)}\right]^{-1/2}. \tag{7}$$

This relation establishes a reciprocal correspondence between ξ and ζ provided that $3\lambda(\xi) > |\Lambda(\xi)|$, $\Lambda'(\xi) \geq 0$. The second inequality coincides with the condition for the yield surface (10) to be nonconcave. With reference to (6) and (7), one can represent the plasticity condition (1) in the following form:

$$f_1(\zeta)S_0 = k,$$
$$f_1(\zeta) = f(\xi(\zeta))\left[1 - \frac{1}{9}\frac{\Lambda_1^2(\zeta)}{\lambda_1^2(\zeta)}\right]^{-1/2}. \tag{8}$$

Here the functions are $\Lambda_1(\zeta) = \Lambda(\xi(\zeta))$, $\lambda_1(\zeta) = \lambda(\xi(\zeta))$. Since $\zeta = S/S_0 = Sf_1(\zeta)/k$, the parameter ζ and the stress S_0 can be expressed in terms of S. The stress S_0 is expressed as $S_0 = \sqrt{3}kF(S)$, where $F(S) = 1/[\sqrt{3}f_1(\zeta(S))]$. With reference to these relations, it is possible to express the stresses for plane strain conditions in the form:

$$\sigma_{11} = S - kF(S)\sin 2\theta,$$
$$\sigma_{22} = S + kF(S)\sin 2\theta, \tag{9}$$
$$\sigma_{12} = kF(S)\cos 2\theta.$$

Here θ is the angle between the x_1-axis and the normal to the plane on which the maximum tangential stress acts. Substituting the expression (9) into the equilibrium equations, we obtain the system of equations for the determination of the stress state in a medium:

$$\begin{aligned} &S_{,1} - kF'(S)(S_{,1}\sin 2\theta - S_{,2}\cos 2\theta) - 2kF(S)(\theta_{,1}\cos 2\theta + \theta_{,2}\sin 2\theta) = 0, \\ &S_{,2} + kF'(S)(S_{,1}\cos 2\theta + S_{,2}\sin 2\theta) - 2kF(S)(\theta_{,1}\sin 2\theta - \theta_{,2}\cos 2\theta) = 0. \end{aligned} \tag{10}$$

The form of these equations are similar to the equations of the general plane problem for an ideal incompressible plastic medium [14]. The prime denotes the differentiation with respect to S. The characteristics and the compatibility equations of the system (10), which we will denote by α and β, have the form:

$$\frac{dx_2}{dx_1} = \tan\varphi_{\alpha,\beta} = \frac{-\cos 2\theta \pm \sqrt{1-k^2F'^2}}{-kF' + \sin 2\theta}, \qquad dS \pm \frac{2kF\,d\theta}{\sqrt{1-k^2F'^2}} = 0. \tag{11}$$

Consider the case where the system (10) is the hyperbolic one, i. e. the inequality $|kF'| < 1$ holds. This implies certain constraints to be imposed on the degree of the sensitivity of the medium plastic properties to the change in the stress state type. These constraints are satisfied for a large number of materials [12]. The characteristics defined by (11) are orthogonal only if $f(\xi) \equiv$const, i. e. in the case where the condition (1) coincides with the Huber–Mises plasticity condition.

Using the relations (4), (8) and (9), we obtain the system of equations for the flow velocities:

$$(v_{1,2} + v_{2,1})\tan 2\theta + v_{1,1} - v_{2,2} = 0, \qquad (v_{1,1} + v_{2,2})\cos 2\theta + (v_{1,2} + v_{2,1})kF' = 0. \tag{12}$$

The characteristics of system (12) coincide with those of system (10), and along them Geiringer's compatibility relations [14] hold:

$$dv_\alpha - v_3\,d\varphi_\alpha = 0, \qquad dv_\beta - v_4\,d\varphi_\beta = 0. \tag{13}$$

Here v_α and v_β are the velocities components along the characteristics α and β, respectively, while v_3 and v_4 are those ones normal to these characteristics:

$$v_3 = \frac{v_\beta - v_\alpha\cos\varphi}{\sin\varphi}, \qquad v_4 = \frac{v_\beta\cos\varphi - v_\alpha}{\sin\varphi}, \qquad \varphi = \varphi_\beta - \varphi_\alpha. \tag{14}$$

The compatibility equations (11) can be integrated for some particular functions $f(\xi)$. For the function $f(\xi)$ defined by (2), the parameter ξ linearly depends on ζ:

$$\xi = \sqrt{1 - \frac{C^2}{9}}\zeta - \frac{C}{9}.$$

The function $F(S)$ is a linear one:

$$F(S) = m\Big(\frac{1}{C} - \frac{S}{k}\Big), \qquad m = \frac{\sqrt{3}C}{\sqrt{9 - C^2}}. \tag{15}$$

In this case, the characteristics and the compatibility relations have the form:

$$\tan\varphi_{\alpha,\beta} = \frac{-\cos 2\theta \pm \sqrt{1-m^2}}{m+\sin 2\theta}, \qquad \frac{\sqrt{1-m^2}}{2m}\ln\left(1 - C\frac{S}{k}\right) \pm \theta = \text{const.} \tag{16}$$

The system (10) is the hyperbolic one for $|m| < 1$, which corresponds to $|C| < \frac{3}{2}$. For positive C, the compressive yield stresses exceed the tensile yield stresses.

Let us consider the case of $f(\xi)$ defined by (3). The yield surface (1) is convex for $\alpha \geq 0$. The equation (7) has the form:

$$\xi = \zeta\left[\left(1+\frac{\alpha}{9}\right)^2 + \frac{\alpha^2\zeta^2}{9}\right]^{-1/2}.$$

The function $f_1(\zeta)$ in (8) can be represented in the form $f_1(\zeta) = \sqrt{1+\beta\zeta^2}$, where $\beta = 9\alpha/(9+\alpha)$, and the function $F(S) = \sqrt{1-\beta S^2/k^2}/\sqrt{3}$. The system (10) is a hyperbolic one under the condition $(S/k)^2 < 3/[\beta(3+\beta)]$. The characteristics and the compatibility relations have the form:

$$\tan\varphi_{\alpha,\beta} = \left[-\cos 2\theta \pm \sqrt{1 - \frac{(\beta S)^2}{3(k^2-\beta S^2)}}\right]\left[\frac{\beta S}{\sqrt{3(k^2-\beta S^2)}} + \sin 2\theta\right]^{-1},$$
$$\sqrt{\frac{3+\beta}{\beta}}\arcsin\left[\sqrt{\frac{\beta(3+\beta)}{3}}\frac{S}{k}\right] - \arcsin\left[\frac{\beta S}{\sqrt{3(k^2-\beta S^2)}}\right] \mp 2\theta = \text{const.} \tag{17}$$

Since the coefficients α and k in (3) are assumed to depend on the porosity of the medium, equation (17) can be used for the calculation of the limit states of the media with a certain value of the porosity.

It is not difficult to show that if the constitutive equations governing the behavior of the medium have the form of (4), a discontinuity of the velocity is impossible. Indeed, when passing through a discontinuity line, the strain rates are defined by the relations $\varepsilon_n = \partial v_n/\partial n$, $\varepsilon_t = 0$, $2\varepsilon_{tn} = \partial v_t/\partial n$, where the subscripts t and n denote the tangential and the normal directions to the discontinuity line. The physical considerations admit only a discontinuity in the tangential velocity component v_t. Equations (12) represented in the coordinates (t, n) imply that the discontinuity line of the tangential velocity component coincides with one of the characteristics of (11) and on this line holds the following relation:

$$\frac{\partial}{\partial n}(\pm\sqrt{1-k^2F'^2}v_n + 2kF'v_t) = 0. \tag{18}$$

Since the stresses are continuous on the characteristics and the values of the normal velocity components on both sides of the discontinuity line are equal ($v_n^+ = v_n^-$), from (18) it follows that the tangential velocity component is also continuous ($v_t^+ = v_t^-$).

3. ILLUSTRATIVE EXAMPLE OF DILATANT PLASTIC FLOW

The dependence of the medium plastic properties on the stress state type influences on the stress distribution in plastic regions. This influence, as well as the method for the determination of the velocity field characterizing the dilatancy of the medium, can be illustrated by the solution of simple problem of a punch indentation into a rigid-plastic half-space.

Using the relations (9), the boundary values of the normal and shear stresses can be obtained in the form:

$$\begin{aligned} \sigma_n &= S - kF(S)\sin[2(\theta - \varphi)], \\ \tau_n &= kF(S)\cos[2(\theta - \varphi)]. \end{aligned} \tag{19}$$

Here φ is the angle between the normal to the boundary and the x_1-axis. If the shear stress is equal zero, the relations (19) imply

$$\begin{aligned} \theta - \varphi &= \pm\frac{\pi}{4} + n\pi, \\ S &= \sigma_n \pm kF(S). \end{aligned} \tag{20}$$

The sign in the formulae (20) is determined by the sign of the tangential stress $\sigma_t = 2S - \sigma_n$ at the boundary.

To illustrate the dependence of the limit loads on the stress state sensitivity of plastic material properties and to study some features of the process of plastic dilatation of the material, the calculations can be performed for the most simple functions $f(\xi)$ and $F(S)$ defined by (2) and (15). Figure 1 illustrates a possible arrangement of the characteristics in the limit state for $C > 0$, which is similar to the perfect plasticity solution by Prandtl.

For the region BDE, we have $\theta = \pi/4$, while the stress S is determined from (20) with $\sigma_n = 0$:

$$S = -\frac{km}{C(1-m)}. \tag{21}$$

In this region, the uniform stress state of biaxial compression is realized. Using the relations (6), (9), and (21), we find $\sigma_{11} = -2km/[C(1-m)]$, $\sigma_{33} = -km(1+m)/[C(1-m)]$, $\sigma_{22} = \sigma_{12} = 0$.

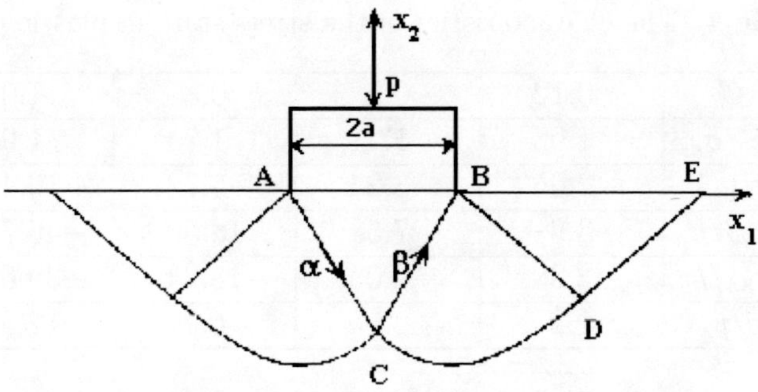

Fig. 1. Plastic flow region

In the region ABC, we have $\theta = -\pi/4$, and using (16) with reference to (22), we find $S=(k/C)[1-1/(1-m) \times \exp(\pi m/\sqrt{1-m^2})]$. The stresses in this region are expressed by $\sigma_{11}=(1-m)S+km/C$, $\sigma_{22}=(1+m)S-km/C$, $\sigma_{33}=(1+m^2)S-km^2/C$, $\sigma_{12}=0$. The limit load is defined by $p=2a\sigma_{22}$. For an incompressible material, the limit load is given by $p_*=2ak(2+\pi)/\sqrt{3}$. Table 1 presents the values of stresses in the region ABC and the ratio of the compressive yield stress, σ_s^-, to the tensile yield stress, σ_s^+, calculated in accordance with (1) and (2) for the different values of the coefficient C. These values indicate that in the region ABC, the stress state of nonuniform triaxial compression is realized. The last row of the table shows the ratios of the limit loads calculated on the basis of (1) and (2) to the load p_* for the perfect incompressible plastic medium. For the considered type of loading, even a very small difference between the compressive and tensile yield stresses noticeably influences on the value of the limit load. For example, if $C=0.12$, which corresponds to $\sigma_s^-/\sigma_s^+=1.08$, the limit load exceeds p_* by a factor 1.2.

Consider now the velocity fields in plastic regions. Since the plastic flow of the media under consideration is accompanied by dilatancy, the traditional scheme for the construction a discontinuous velocity field does not apply without additional conjectures about the behavior of the medium in the neighborhood of the hypothetical discontinuity surface. For this reason, we consider the construction of a continuous velocity field characterizing the dilatancy of the medium.

In the region ABC, the stresses are constant and, in accordance with (4), the strain rates are also constant. Hence, the velocities can be represented in the following form:

$$\begin{aligned} v_1 &= \alpha_0 + \alpha_1 x_1 + \alpha_2 x_2, \\ v_2 &= \beta_0 + \beta_1 x_1 + \beta_2 x_2. \end{aligned} \tag{22}$$

Since the distribution of the component v_1 is symmetric about the x_2-axis, we

Table 1. The characteristics of the stress state in plastic regions

C	0.12	0.5	0.8	1.0
σ_s^-/σ_s^+	1.08	1.41	1.74	1.98
σ_{11}/k	−2.05	−3.24	−5.72	−10.44
σ_{22}/k	−3.6	−7.58	−18.59	−46.71
σ_{33}/k	−2.88	−6.05	−15.24	−39.67
p/p_*	1.2	2.5	6.2	12.5

have $\alpha_0 = \alpha_2 = 0$. Using the conditions on the punch surface, $|x_1| \le a$, $x_2 = 0$, $v_2 = -V$, where V is the velocity of the punch indentation, we find $\beta_0 = -V$, $\beta_1 = 0$. Thus only two coefficients, α_1 and β_2 are to be determined and (22) is reduced to the following form:

$$\begin{aligned} v_1 &= \alpha_1 x_1, \\ v_2 &= -V + \beta_2 x_2. \end{aligned} \tag{23}$$

In accordance with (16), the slopes of the characteristics in the region ABC are defined by

$$\tan\varphi_{\alpha,\beta} = \mp\sqrt{\frac{1+m}{1-m}}. \tag{24}$$

Using (23) and (24), we find the expressions for the velocity components along the characteristics:

$$\begin{aligned} v_\alpha &= \alpha_1 x_1\sqrt{\frac{1-m}{2}} - (\beta_2 x_2 - V)\sqrt{\frac{1+m}{2}}, \\ v_\beta &= \alpha_1 x_1\sqrt{\frac{1-m}{2}} + (\beta_2 x_2 - V)\sqrt{\frac{1+m}{2}}. \end{aligned} \tag{25}$$

Since the velocities v_α and v_β are constant along the corresponding characteristics in the region ABC, by equating the values of v_β at the points B and C, we obtain the relation between α_1 and β_2: $\alpha_1 = -(1+m)/(1-m)\beta_2$. At the point C, the velocity components normal to the characteristics are continuous, i. e. $v_3 = v_4 = 0$. Using equations (14) and (25), we find $\beta_2 = \sqrt{(1-m)/(1+m)}V/a$. The components of the strain rate and the rate of the bulk strain can be written in the following form:

$$\begin{aligned} \dot{\varepsilon}_{11} &= \sqrt{\frac{1+m}{1-m}}\frac{V}{a}, \\ \dot{\varepsilon}_{22} &= -\sqrt{\frac{1-m}{1+m}}\frac{V}{a}, \\ \dot{\varepsilon} &= \frac{2m}{\sqrt{1-m^2}}\frac{V}{a}. \end{aligned} \tag{26}$$

On the base of equations (13), (14) and (25), we find the velocities in the region BCD:

$$v_\alpha = [\sqrt{2(1+m)}a - \rho\sqrt{1-m^2}]\exp\left[-\frac{m}{\sqrt{1-m^2}}\left(\varphi_\alpha + \arcsin\sqrt{\frac{1+m}{2}}\right)\right]\frac{V}{a},$$

$$v_\beta = 0, \tag{27}$$

$$\rho = \sqrt{(x_1-a)^2 + x_2^2}$$

In the region BDE, the slopes of characteristics are defined by

$$\tan\varphi_{\alpha,\beta} = \pm\sqrt{\frac{1-m}{1+m}}. \tag{28}$$

Using (27) and (28), we find v_α on the characteristic BD:

$$v_\alpha = [\sqrt{2(1+m)}a - \rho\sqrt{1-m^2}]\exp\left(-\frac{m\pi}{2\sqrt{1-m^2}}\right)\frac{V}{a}. \tag{29}$$

The values of v_α in the region BDE coincide with those on the characteristic BD. Thus, using the equations for α-characteristics passing through the points of BD, we find the velocities in the region BDE:

$$v_\alpha = \frac{1}{\sqrt{2}}[\sqrt{1-m}(a-x_1) + \sqrt{1+m}(x_2+2a)]\exp\left(-\frac{m\pi}{2\sqrt{1-m^2}}\right)\frac{V}{a}, \quad v_\beta = 0. \tag{30}$$

From (27) and (30), it follows that $v_\alpha = 0$ on the boundary characteristic CDE, and hence the velocities are distributed continuously. The instantaneous distribution of the velocity v_α on the part BE of the free surface is represented by a linear function.

The use of different functions $f(\xi)$ in the plasticity condition (1) leads to the similar results.

CONCLUSION

In this study a possible approach to the description of the dependence of the characteristics of the plastic deformation of a dilatant medium on the loading-induced stress state is considered. It is shown that the limit loads depend significantly on the value of the parameter characterizing the sensitivity of the plastic properties of the materials to the stress state type. The considered constitutive equations permit to determine on the base of the stress values the relative values of the strain rates. It is readily to show that this ratios coincide in various plastic regions with the values calculated on the base of the velocity field. This indicates the correspondence between the stress and velocity fields in the plastic regions. For each specific case of the characteristics distribution, the strain rates are determined by the shape of the plastic regions and the velocity on the boundary at which the loads are applied.

ACKNOWLEDGMENTS

This work was supported by the National Science Council, Taiwan, and the Russian Foundation for Basic Research (international project No. 08-01-92011).

REFERENCES

1. O. Reynolds, "On the Dilatancy of Media Composed of Rigid Particles in Contact," Philos. Mag. **20**, Ser. 5 (127), 469–481 (1885).
2. D.C. Drucker and W. Prager, "Soil Mechanics and Plastic Analysis or Limit Design," Quarterly Appl. Math. **10** (2), 157–165 (1952).
3. V.V. Novozhilov, "On Plastic Cavitation," J. Appl. Math. Mech. **29** (4), 811–819 (1965).
4. R.J. Green, "A Plasticity Theory for Porous Solids," Int. J. Mech. Sci. **14** (4), 215–226 (1972).
5. A.L. Gurson, "Continuum Theory of Ductile Rupture by Void Nucleation and Growth: Part 1 — Yield Criteria and Flow Rules for Porous Ductile Media," Trans. of ASME. J. Eng. Mater. Techn. **99**, 2–15 (1977).
6. V. Tvergaard and A. Needleman, "Analysis of the Cup-Cone Fracture in a Round Tensile Bar," Acta Metallurgica **32**, 157–169 (1984).
7. J. Koplic and A. Needleman, "Void Growth and Coalescence in Porous Plastic Solids," Int. J. Solids Structures **24** (8), 835–853 (1988).
8. R.M. McMeeking and C.L. Hom, "Finite Element Analysis of Void Growth in Elastic-Plastic Materials," Int. J. Fracture **42** (1), 1–19 (1990).
9. M.J. Worswick and R.J. Rick, "Void Growth and Constitutive Softening in a Periodically Voided Solids," J. Mech. Phys. Solids **38** (5), 601–625 (1990).
10. M. Kuna and D.Z. Sun, "Three-Dimensional Cell Model Analysis of Void Growth in Ductile Materials," Int. J. Fracture **81** (3), 235–258 (1996).
11. E.V. Lomakin, "Nonlinear Deformation of Materials whose Resistance Depends on the Form of the Stressed State," Mech. Solids **15** (4), 69–75 (1980).
12. E.V. Lomakin, "Dependence of the Limit State of Composite and Polymer Materials on the Type of the Stress State. 1. Experimental Dependencies and Determining Equations," Mech. Comp. Mater. **24** (1), 1–7 (1988).
13. E.V. Lomakin, "Constitutive Relations of Deformation Theory for Dilatant Media," Mech. Solids **26** (6), 64–72 (1991).
14. A.M. Freudental and H. Geiringer, "The Mathematical Theories of the Inelastic Continuum," in *Handbuch der Physik.* Bd. VI. *Elastizität und Plastizität*, Ed. by S. Flügge (Springer, Berlin, 1958), pp. 229–433.

SUPERPLASTIC FORMABILITY OF Al-Li 8090 ALLOY

V. Pancholi[1] and B.P. Kashyap[2]

ABSTRACT

Superplastic grade Al-Li 8090 alloy was subjected to constant pressure superplastic bulge forming at three different forming pressures of low, intermediate and high - corresponding to initial strain rates of 4×10^{-4}, 1×10^{-3} and 5×10^{-3} s^{-1}, respectively. Al-Li 8090 alloy contain distinct microstructures at different location along the thickness. To understand effect of such microstructure separated sheets were obtained and formed superplastically till failure. Formability was measured by parameters like thinning factor (measure of thickness variation) and bulge profile. The microstructure was characterized using optical microscopy and micro-texture (electron backscattered diffraction). It was found that the layer which contains superplastically favorable microstructure exhibited lower thickness variation and bulge shape then the layer which has unfavorable microstructure. However, the microstructure of layer with unfavorable microstructure evolved through dynamic recrystallization into factorable one but even then it could not improve the formability.

Key words: superplastic forming, Al-Li 8090 alloy, thickness variation, EBSD

1. STATEMENT OF THE PROBLEM

Superplastic forming (SPF) is usually carried out by applying inert gas pressure on one side of blank, while holding the sheet at its periphery into a die [1]. SPF through gas blowing has made it possible to form near net shapes and complex parts in one go. However, thickness variation in the formed component is an inherent problem; whereby the thickness is more near the region where the die holds the blank (edge) than the region away from it (center). High strain rate sensitivity index is shown

[1]Department of Metallurgical and Materials Engineering, Indian Institute of Technology Roorkee, Roorkee — 247667, Uttarakhand, India

[2]Department of Metallurgical Engineering and Materials Science, Indian Institute of Technology Bombay, Mumbai — 400076, Maharashtra, India

both experimentally and analytically [2–6], to reduce the thickness variation in the superplastically formed component.

The AA8090 Al–Li alloy is known to have microstructural gradient in the through thickness direction; with the outer surface layers having fully recrystallized and nearly equiaxed grains, and the middle layer having unrecrystallized elongated grains. The unrecrystallized elongated grains in the middle layer are reported to evolve into recrystallized microstructure through deformation induced continuous recrystallization (DICR) [7,8]. The microstructural evolution profoundly affect the strain rate sensitivity of the material in the middle layer [9]. In the present work effect of starting microstructure and microstructural evolution in Al-Li 8090 alloy and its formability was studied.

2. EXPERIMENTAL

The commercial superplastic grade AA 8090 Al–Li alloy was obtained in the form of 3 mm thick sheet. The composition in wt% is 2.7Li–1.4Cu–0.56Mg–0.12Zr-balance Al. The initial microstructure is shown in Fig. 1, which contains distinct microstructural gradient along thickness direction; fully recrystallized pancake shaped grains, elongated in longitudinal and transverse directions in the surface layer and elongated unrecrystallized microstructure in the middle layer. To understand the effect of starting microstructure on forming properties separated layers containing only one kind of microstructure were obtained by removing the material unwanted material. In this way separated surface and middle layers of approximately 1 mm thickness were obtained and formed superplastically into free flowing bulges at constant temperature of 530°C, using constant argon gas pressure. The three different pressures, corresponding to three different strain rates, used in the present study were—low ($\dot{\varepsilon} \sim 4 \times 10^{-4}$ s^{-1}), intermediate ($\dot{\varepsilon} \sim 1 \times 10^{-3}$ s^{-1}) and high ($\dot{\varepsilon} \sim 5 \times 10^{-3}$ s^{-1}) forming pressures. The forming was continued to the level of crack formation whereupon the tests were stopped. After forming, the strain was measured in the tangential, circumferential and thickness directions. The measurements were made from the apex of the bulge to its edge at five different locations. Microtexture measurements were carried out using a TSL OIM system on a FEI Quanta 200 HV SEM.

3. RESULTS AND DISCUSSION

3.1. *Effect of starting microstructure on bulge shape*

During superplastic forming, pressure was kept nearly constant through out the forming cycle. Figs. 2 *a*, *b*, and *c* show the hemispherical bulges formed using sur-

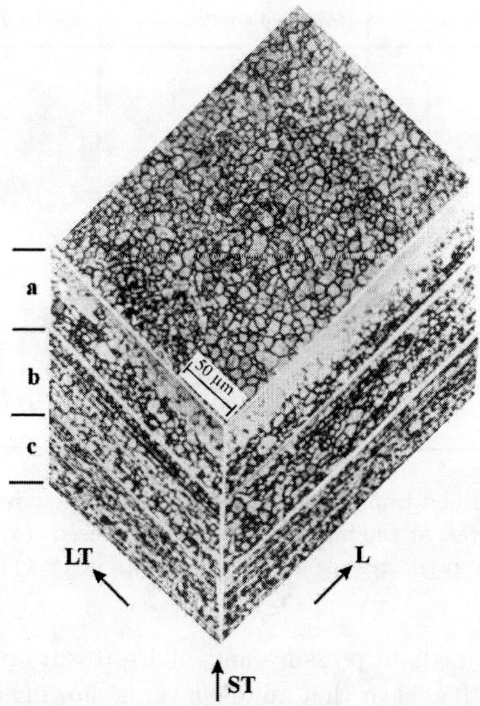

Fig. 1. Three-dimensional view of the microstructure after annealing at 540°C for 5 min: (a) Li depleted layer, (b) surface layer with recrystallized microstructure and (c) middle layer of unrecrystallized grains

face layer at low, intermediate and high pressure, respectively. Figs. 2 *d*, *e*, and *f*, show hemispherical bulges formed using middle layer at low, intermediate and high pressure, respectively.

Under optimum superplastic conditions and for material with high m value bulge should acquire spherical geometry. Deviation from the spherical geometry indicates poor formability in terms of bulge profile. The shape of the bulge can be characterized using the polynomial expression in terms of height (Z) and radial distance (X) of the point along the bulge given by Yang and Mukherjee [6]

$$X^2 = b_0 + b_1 Z + b_2 Z^2,$$

where, b_0, b_1 are constants and b_2 is the coefficient that determines the bulge profile. According to the equation, the bulge profile for $b_2 = -1$ will be circular, whereas, it will be prolate spheroid for b_2 between -1 and 0. Value of coefficient b_2 for surface layer at low pressure, intermediate pressure and high pressure of forming is 0.97, 0.82, and 0.68 respectively. On the other hand, value of b_2 for middle layer

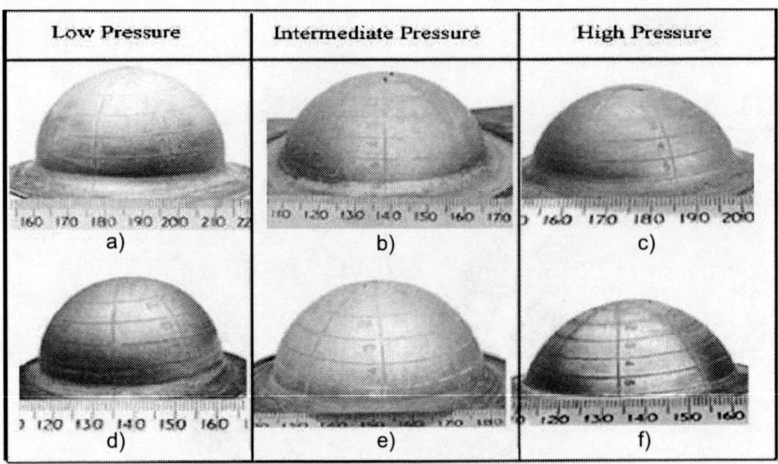

Fig. 2. The bulges formed from 1mm thick sheets at the low pressure used: (*a*) surface layer and (*d*) middle layer, at the intermediate pressure used: (*b*) surface layer, (*e*) middle layer, at the high pressure used: (*c*) surface layer, (*f*) middle layer

at low pressure, intermediate pressure and high pressure of forming is 0.93, 0.75, and 0.63 respectively. It is clear that middle layer is showing inferior properties at all pressures of forming, compared to surface layer, because of its starting microstructure which is not suitable for superplastic deformation. However, both the layers showed bulge profile very close to spherical geometry at low pressure of forming.

3.2. *Effect of starting microstructure on bulge shape on thickness variation*

Even during uniaxial tensile testing local strain variation is present along the gauge length. The local strain is substantially more halfway along the gauge length and very small close to the shoulder of the sample [10]. The local variation in strain is attributed to low m value of the material and to the constrained imposed by the deformation condition. Similarly thickness variation is present from apex to the edge of the superplastically formed bulge. This is usually characterized by the thinning factor $s/\bar{s}$, which is the ratio of actual thickness (s) to the ideal thickness ($\bar{s}$). Ideal thickness is assumed as the uniform thickness at all locations for selected overall strain. Here $\bar{s} = s_0/(1 + H_r^2)$ is the ideal thickness of the sheet [26], if it has been deformed into a hemisphere with uniform thickness, s and s_0 are the actual and initial thickness, respectively, and H_r is relative dome height ($H_r = h_p/R_0$, where h_p is height of the dome and R_0 is the die radius). Thinning factor of the bulge formed at low, intermediate and high pressures from surface and middle layers are plotted in Fig. 3. Slope of the straight line fitted to the data was taken as mea-

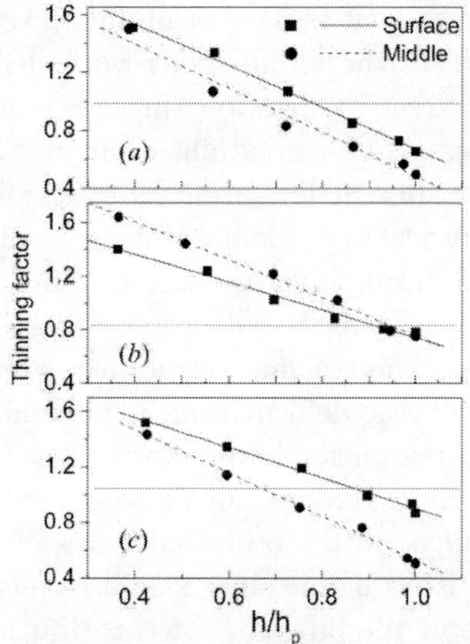

Fig. 3. Thinning factor plotted against fractional height (h/h_p). (a) low pressure, (b) intermediate pressure, and (c) high pressure of forming

sure of deviation from uniform thickness i.e. higher slope indicates higher thickness variation. Slope of the line fitted to thinning factor data for surface layer at low, intermediate and high pressure is 1.44, 1.03, and 1.09 respectively and the same data for middle layer is 1.52, 1.41, and 1.56. It is clear that the thickness variation is present in all cases; however, surface layer is showing much lower thickness variation than the middle layer for all pressures of forming. It has been reported in the literature that high value of strain rate sensitivity promotes forming with uniform thickness. Electron backscattered diffraction (EBSD) analysis showed equiaxed and well recrystallized microstructure in the surface layer with high percentage of high angle grain boundary, whereas in the middle layer the microstructure contain unrecrystallized, elongated grains with low fraction of high angle grain boundaries. Fan et.al. [9] reported high starting value of m for surface layer compared to the middle layer in the same alloy. Therefore, microstructure favorable to superplasticity with high m value in surface layer is playing an important role in reducing the thickness variation along the bulge.

3.3. *Microstructural evolution*

The starting microstructure of the two layers of the sheet of Al-Li 8090 alloys is different but microstructural evolution takes place with deformation at high temperature. This is more prominent in case of middle layer than surface layer which only exhibit grain growth. In the middle layer with deformation, unrecrystallized and elongated grains show recrystallization through deformation induced continuous recrystallization process. The recrystallization fraction is increasing from 0.56 at the edge to 0.61 at the apex at low pressure of forming, similarly from 0.32 at the edge to 0.62 at the apex at intermediate pressure of forming and more or less remaining constant at 0.25 during high pressure forming. In the middle layer of this alloy with deformation induced recrystallization, strain rate sensitivity index also improves substantially [9]. Though microstructure evolves in the middle layer to the one suitable for superplastic deformation, it does not affect formability of the layer to that extent. The reason for lower formability is that the microstructural evolution is taking place only in those regions where layer is experiencing substantial strain i.e. centre of the bulge, on the other hand close to the edge of the bulge the microstructural evolution is negligible since strain is very small. That implies that material close to the edge of the bulge is not contributing to the deformation and the material around centre of the sheet is deforming freely and hence increasing the thickness variation in the middle layer over and above the variation due to forming constraints.

CONCLUSIONS

Superplastic forming of layer with different starting microstructure at three different pressures of forming was attempted in the present work.

1. Surface layer showed better formability than the middle layer because; the starting microstructure of surface layer was suitable for superplastic deformation.

2. Microstructural evolution in the middle layer increases the recrystallization fraction with strain in the middle layer, but these changes did not improve the formability of the middle layer.

3. Formability of both the layers was best at low pressure of forming.

REFERENCES

1. F. Jovane, "An Approximate Analysis of the Superplastic Forming of a Thin Circular Diaphragm: Theory and Experiments," Int. J. Mech. Sci. **10**, 403–427 (1968).
2. Z.X. Guo and N. Ridley, "Testing Models for Superplastic Bulge Forming of Domes," Mater. Sci. Technol. **6**, 510–515 (1990).

3. D.L. Holt, "An Analysis of the Bulging of a Superplastic Sheet by Lateral Pressure," Int. J. Mech. Sci. **12**, 491–497 (1970).
4. Z.X. Guo and N. Ridley, "Modeling of Superplastic Bulge Forming of Domes," Mater. Sci. Eng. A **114**, 97–104 (1989).
5. F.U. Enikeev and A.A. Kruglov, "An Analysis of the Superplastic Forming of a Thin Circular Diaphragm," Int. J. Mech. Sci. **37**, 473–483 (1995).
6. H.S. Yang and A.K. Mukherjee, "An Analysis of the Superplastic Forming of a Circular Sheet Diaphragm," Int. J. Mech. Sci. **34**, 283–297 (1992).
7. L. Qing, H. Xiaoxu, Y. Mei, and Y. Jinfeng, "On Deformation Induced Continuous Recrystallization in a Superplastic Al–Li–Cu–Mg–Zr Alloy," Acta Metall. Mater. **40**, 1753–1762 (1992).
8. M. Eddahbi, C.B. Thomson, F. Carreno, and O.A. Ruano, "Grain Structure and Microtexture after High Temperature Deformation of an Al–Li (8090) Alloy," Mater. Sci. Eng. A **284**, 292–300 (2000).
9. W. Fan, B.P. Kashyap, and M.C. Chaturvedi, "Effect of Layered Microstructure and Its Evolution on Superplastic Behaviour of AA8090 Al–Li Alloy," Mater. Sci. Technol. **17**, 439–445 (2001).
10. V. Pancholi and B.P. Kashyap, "Effect of Local Strain Distribution on Concurrent Microstructural Evolution during Superplastic Deformation of Al–Li 8090 Alloy," Mater. Sci. Eng. A **351**, 174–182 (2003).

AN EXPERIMENTAL AND NUMERICAL STUDY OF EDGE IMPACT OF 7.62 MM AP PROJECTILE ON PERFORATED HIGH HARDNESS ARMOUR STEEL PLATES

B. Ramakrishna[1], B. Mishra[1], V. Madhu[1*], T.B. Bhat[1], and N.K. Gupta[2]**

ABSTRACT

Armour steels have typically been used in the construction of light armoured vehicles. Improvement in performance of armour is usually achieved by increasing the strength of the steel. In this paper, an experimental study has been carried out to investigate the performance improvement in steel through incorporation of perforations in the plate. Circular holes of diameter equal to that of the projectile were drilled in high hardness armour steel plates. These plates were subjected to ballistic impact of 7.62 mm AP projectiles by backing them with thick plate of 7017 Aluminium alloy. Improvement in efficiency of such perforated plates was evaluated by comparing the DOP under different conditions of impact. Results show that there is an improvement of about 37% in thickness efficiency and about 70% in mass efficiency of high hardness armour steel only due to perforations. Computational simulations were also carried out to correlate the effects of perforated armour. Results of experiments and simulations showed good correlation. Photographs of the target and backing plate after the experiments along with analysis of the ballistic efficiency of the perforated armour have been presented in this paper.

Key words: impact, penetration, high hardness steel, perforated armour, simulation, projectile, ballistic

[1]Defence Metallurgical Research Laboratory, Kanchanbagh — 500058, Hyderabad, India

[2]Department of Applied Mechanics, Indian Institute of Technology Delhi, New Delhi — 110016, Hauz Khas, India

[*]E-mail: *madhu_vemuri@hotmail.com*. Phone: +914024342252

[**]E-mail: *nkgupta@am.iitd.ernet.in*

INTRODUCTION

Response of armour to high velocity impact and penetration of projectiles depends on angle of impact, projectile material and geometry and the arrangement, geometry and mechanical properties of target plates. Typical rolled homogenous armour steels having tempered martensitic structure have hardness in the range of 300 VHN [1]. The hardness of the armour is much less in comparison to the hardness of armour piercing hard steel projectile like 7.62 AP bullet which is typically about 750 to 900 VHN. The penetration of the projectile can be reduced by either making the material harder or by designing the armour in such a way as to deflect, crack or break the projectile. Some high hardness steels having more than 550 BHN have shown a higher performance and can be used as add-on armour on light armoured vehicles. Such high hardness steels are therefore going to play an important role in the design of fighting vehicles for giving required protection against the kinetic energy threats with improved performance.

Attempts have been made in the past to decrease the weight of armour steel by making them perforated which has also been found to improve the ballistic performance. Chocron et al. [2] have demonstrated the defeat of the projectile and breakage of its core upon edge impact against a 3 mm thick RHA plate. Perforated plates have been used in shock tubes to disrupt shock waves by Medvedev [3] and Chaoa et al. [4]. Perforated designs of steel armour using circular holes [5], triangular holes [6] and other types of perforations [7] have been employed as front facing panel of the armour assembly to defeat the projectile. Langdon, et al. [8] have shown perforated plates to be promising passive mitigation systems against explosive blast. However, there is no information available in the open literature on the quantitative improvement of ballistic efficiency of perforated steel armour.

In the present study, an attempt has been made to understand the projectile-armour interaction and to evaluate the ballistic efficiency of perforated high hardness steel armour having circular holes by experimental as well as by computer simulations. Results of experiments as well as simulations have been presented.

1. EXPERIMENTAL

1.1. *Materials*

Rolled homogeneous armour steel plates with hardness of about 300 VHN of the size of 150 mm×150 mm and thickness of 5 mm were used in the present experiments. In metals, the strength of material and its toughness can be improved by refining the grain size, by micro-alloying and by heat treatment. The aim of the present work is to study the improvement in performance of steel target when they are perforated

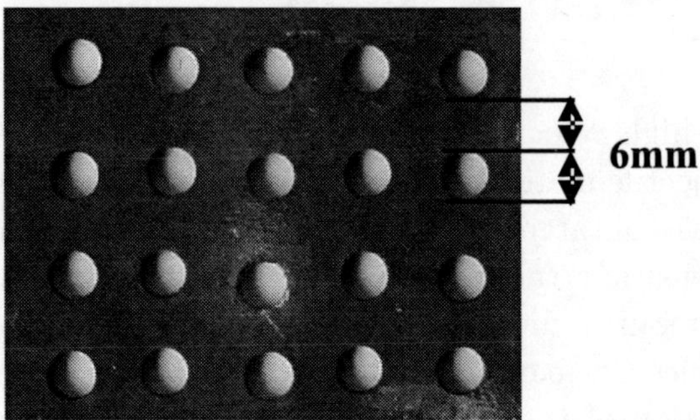

Fig. 1. Perforated high hardness steel plate design

so that during an impact event, the projectile encounters an edge impact when hit near the hole, thereby, creating lateral forces that would deflect the shot as well as create more lateral and shear stresses and reduce its penetrating capability.

A perforated plate design was formulated, as shown in Fig. 1. Only one geometry was considered in this study with 36 equiv-spaced holes arranged in 6 rows and columns with an inter-se spacing of 6mm on a plate of size 72 mm×72 mm. Since drilling of holes in high hardness steels is difficult, circular holes of 6 mm diameter were drilled before carrying out heat treatment on these plates.

This arrangement of holes gives a weight saving of about 20% over a complete plate of same size and thickness. This in effect would mean that for same area, the density of the plate is reduced by 20% i.e a steel plate of density 7.85 g/cc would in effect behave like a plate with density of 6.30 g/cc. These plates were then heat-treated to attain a yield strength of about 1400 MPa. In order to carry out efficiency evaluation, standard DOP test method [9] has been used which is more clearly described in the following section. The standard backing material used in the present set of experiments is the 7017 aluminium alloy plates.

1.2. *Testing*

The schematic diagram of the ballistic test is given in Fig. 2. Plates of heat treated high hardness steel (both non-perforated as well as perforated) were backed by 65 mm thick plates of 7017 aluminium alloy. These plates were impacted with 7.62 mm armour piercing projectiles in a small arms range using standard service rifle at velocities of about 820 ± 10 ms^{-1}. The hardened steel core has a length of 28 mm, diameter of 6.1 mm and weighs about 5 gm. It has a hardness of about 900 VHN at the tip.

Backing plate was fixed along with the target steel plate without any gap be-

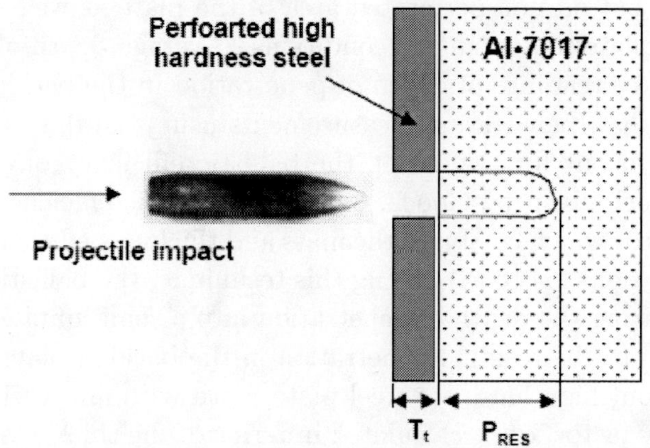

Fig. 2. Ballistic test configuration for evaluation of armour

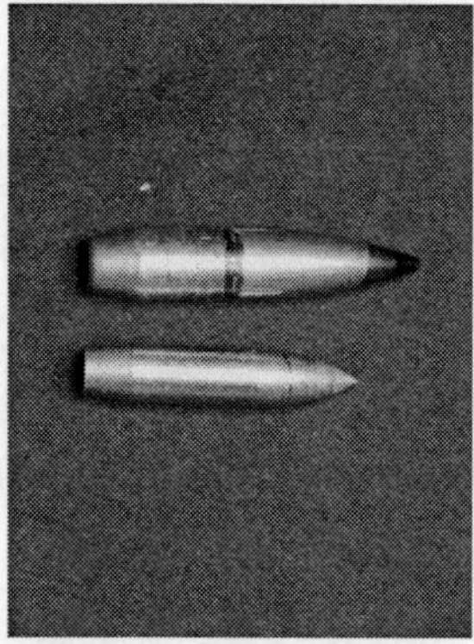

Fig. 3. Complete shot (upper) and core (lower) of 7.62 mm AP projectile

tween them. The whole configuration was then subjected to the impact of 7.62 AP projectiles. The angle of attack was normal to the target plate. The velocity of the projectile was measured using infrared light emitting diode photovoltaic cell by measuring the time interval between the interceptions caused by the projectile running across two transverse beams placed 2 meters apart (placed at 6 m and 8 m distance from the muzzle of the gun). The projectile was fired from a rifled gun from a distance of 10 meters. Fig. 3 shows photographs of the original shot and the core of the projectile.

The residual depth of penetration in backing aluminium plate was measured after each experiment by cutting the plate at the center of the crater and measuring the

DOP. To compare the ballistic efficiency of the perforated steel armour, experiments were also carried out on non-perforated area of the plate as well as on the backing aluminium alloy plate under similar conditions. Average depth of penetration was obtained from three readings of depth of penetration in the backing plate.

Standard depth of penetration measurements using the thick backing technique [10–11] was carried out for evaluating the ballistic efficiency of each system. Following the approach given by Gooch, et al [12], ballistic efficiency is assessed by a dimensionless factor which combines the mass and thickness efficiency of the material as defined in equations (1) to (3). Using this technique, the ballistic efficiency is calculated by comparing the depth of penetration into a semi-infinite thick aluminium alloy plate [P_{REF}] to the residual penetration in the backing plate [P_{RES}] after perforation in the front high hardness steel plate, both with and without perforations. Ballistic efficiency factor, q^2, is calculated in terms of the thickness efficiency E_t and mass efficiency E_m as follows:

$$E_t = \frac{P_{\mathrm{REF}} - P_{\mathrm{RES}}}{T_t}, \tag{1}$$

$$E_m = E_t \frac{\rho_{\mathrm{REF}}}{\rho_t}, \tag{2}$$

$$q^2 = E_t x E_m. \tag{3}$$

The ballistic efficiency factor has significance for armour designers as this factor connects both the mass and thickness efficiencies. A value of more than 1.0 indicates that the material under evaluation is more efficient than the equivalent reference backing material (7017 aluminium alloy in this case) for a specified threat.

2. NUMERICAL SIMULATIONS

Numerical simulation of impact and penetration problem was performed using Autodyn-3D^{TM} code. The nominal dimensions of the core were used to model the projectile, although the exact details of the ogive nose were approximated as per the standard dimensions of 7.62 mm AP projectile. The core was modeled as linearly elastic since the strain to failure of the core is as low as 2% at a stress of 2.3 GPa as evaluated by Chocron, et al. [2]. In the actual test, the projectile makes an edge impact when it hits the perforated portion of the plate.

The projectile was modeled with a density of 7850 kg/m^3, bulk modulus of 159 GPa, shear modulus of 81.7 GPa and yield strength of 2300 MPa. The aluminium target was modeled as elastic, perfectly plastic, with the following properties: density 2910 kg/m^3, shear modulus of 27.3 GPa, yield strength of 425 MPa and failure strain of 150%. Similarly the high hardness steel target was also modeled as elastic, perfectly plastic, with the following properties: density 7850 kg/m^3, shear modulus of 64 GPa, yield strength of 1400 MPa and failure strain of 150%.

Lagrangian mesh with an erosion strain was used in the present numerical study. When a computational element exceeds the failure strain, the element is no longer able to support tensile or shear stresses; however it can still support compressive stresses. Erosion strain is an important parameter to be specified for projectile as well as target material. The erosion strain is given to prevent excessive mesh distortion and entanglement. It is based on the criteria that the elements do not significantly contribute to the penetration process, if the effective plastic strain reaches a critical value i.e. the erosion strain [3]. The element is discarded on exceeding the erosion strain limit; the sliding interface between projectile and target are dynamically redefined, thereby allowing the computation to be carried out without rezoning the distorted region. In the present simulations, the high hardness steel plate and the backing aluminium alloy plate were modeled together using join command. The projectile was unconstrained and free to rotate as dictated by the dynamics of the problem. The projectile was modeled using 700 elements and the target was modeled with 7600 elements with aspect ratio of 1.0. The edges of the target plate and backing were fixed. Perforation in the target plate was modeled by creating a void and filling with air. A velocity of 820 m/s was imparted to the projectile as the initial boundary condition. The material properties employed were modified in the AUTODYN material library and used in the simulations. Johnson-Cook strength model was employed for modeling the problem.

Three different simulations were performed. The first being the impact of the projectile on a 65 mm thick aluminium alloy plate to obtain the depth of penetration of the projectile. The second one was the impact of the projectile on a 5 mm thick high hardness armour steel when backed by the aluminium alloy plate. The third was the impact of projectile on a 5 mm thick perforated high hardness (PS) plate when backed by Al-alloy plate. The target plate is modeled with 5 mm thick high hardness steel in the front (both perforated as well as non-perforated) and Al-7017 of 65 mm thick block as the backing. Fig. 4 shows the initial configuration of the model used for simulation of the impact of projectile on perforated steel plate backed by aluminium alloy. The depth of penetration and the velocity of the projectile and stresses and strains in the target and projectile were monitored in all the three cases.

3. RESULTS AND DISCUSSION

3.1. *Experimental results*

Results of the three sets of experiments are summarized in Table 1. From the results, it is seen that the thickness and mass efficiency of high hardness steel armour

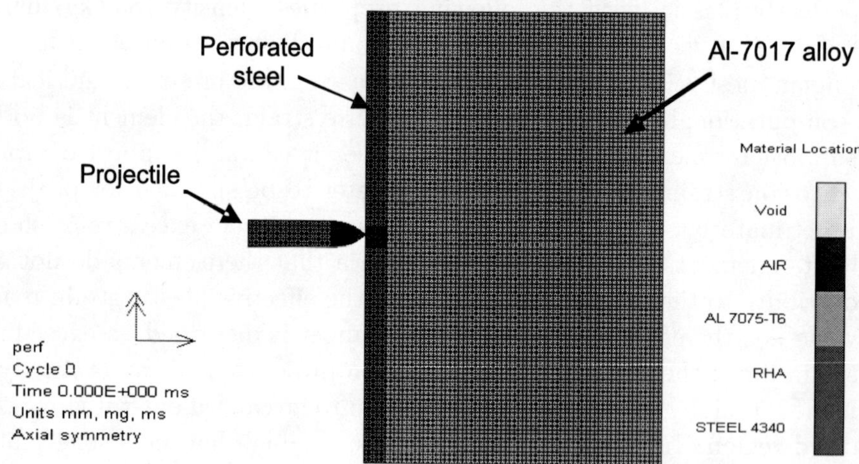

Fig. 4. Initial setup of the numerical model of impact on PS backed by Al-alloy

Table 1. Ballistic efficiency of perforated high hardness steel

Target configuration	Projectile velocity (m/s)	Depth of penetration (mm)	Thickness efficiency (E_t)	Mass efficiency (E_m)	Ballistic efficiency factor (q^2)
Al-alloy	823.7	39	1.0	1.0	1.0
HHS+Al-alloy	827.2	20	3.8	1.4	5.3
PS+Al-alloy	819.8	13	5.2	2.4	12.5

is about 3.8 and 1.4 times better than the standard armour grade Al-alloy. The thickness and mass efficiency of the same high hardness steel when used in perforated form is further enhanced to 5.2 and 2.4 respectively. This indicates that there is an improvement of about 37% in thickness efficiency and about 70% improvement in mass efficiency of high hardness armour steel only due to perforation.

The results indicate that there is an overall increase of about 2.4 times in the ballistic efficiency of high hardness steel against 7.62 mm AP projectiles when the steel plate is used in the form of perforated armour.

These improvements of ballistic performance of perforated armour can be attributed to the following mechanisms: (i) deflection of the incoming projectile, thus causing formation of an angle between the longitudinal axis of the projectile and its new, deflected trajectory; (ii) increase in frictional force and (ii) creation of bending moment on the projectile. Perforated armour plate effectively breaks the incoming projectile or at least diverts it from its incidence trajectory, thus substantially reduc-

Fig. 5. Photographs of broken projectiles after impact on PS plate

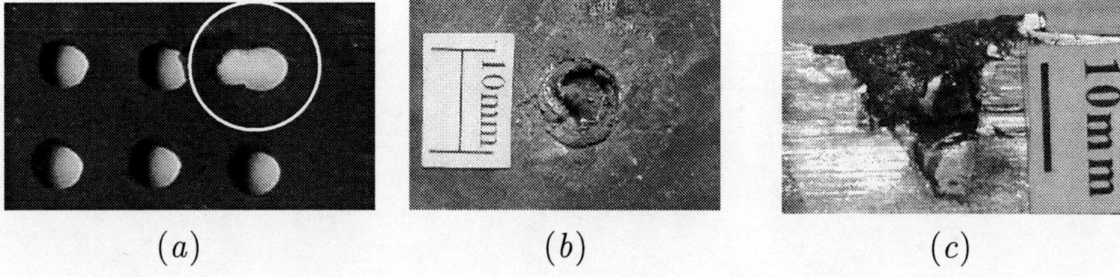

(*a*) (*b*) (*c*)

Fig. 6. Photographs of front of (*a*) perforated plate, (*b*) backing plate, and (*c*) DOP in backing

ing the residual penetration capability of the projectile into the backing material. Usually the holes or slits are designed to interact with the characteristic diameter of the projectile. The performance of such armour is dominated by the plate material and thickness. It is found that the projectile is fractured after perforation through the high hardness steel plates. A picture of a broken core recovered after the experiment is shown in Fig. 5.

Fig. 6 shows pictures of the front portion of the perforated target, front of the backing plate and cross section of backing plate after projectile impact. Fig. 7 shows the front portion of the high hardness steel target without perforation, front of the backing plate and cross section of backing plate after projectile impact.

3.2. *Numerical results*

Numerical simulations give an in depth understanding of the phenomena of penetration. Data on the velocity of the projectile, depth of penetration, stresses and many other parameters can be monitored at very close time intervals, which is not possible in experiments. Fig. 8 shows the depth of penetration of the projectile in all the three configurations extracted from the data monitored during simulations. In perforated steel plate backed by Al-alloy, the penetration during the initial stages

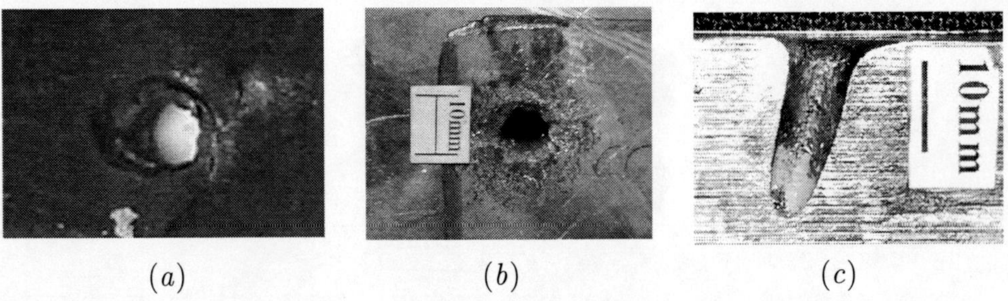

Fig. 7. Photographs of front of (*a*) high hardness plate, (*b*) backing plate, and (*c*) DOP in backing after impact of HHS+Al configuration

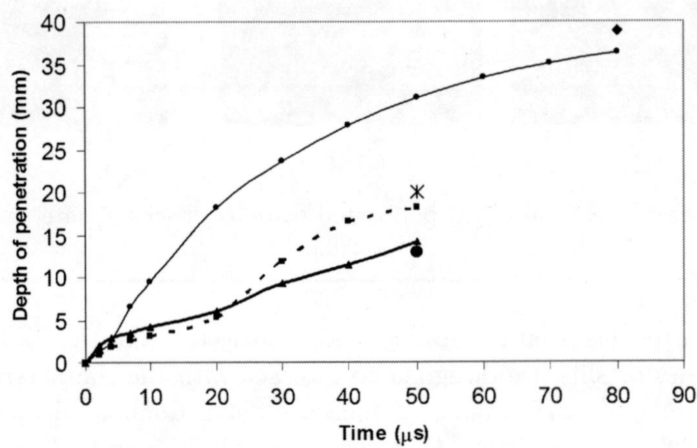

Fig. 8. Depth of penetration of projectile with respect to time

is seen to be higher than that in the HHS plate backed by Al-alloy. The reason for this is that in the perforated plate, the projectile first encounters a hole. The initial tip of the projectile near the ogive portion does not have interaction with the plate till the tip and the lateral surface of the projectile comes in contact with the target material. From the figure it can also be seen that the DOP matches closely with the experiments in all the cases.

The velocity of the projectile during the penetration process was monitored. Fig. 9 shows the profile of the projectile velocity under different conditions of impact. From the figure it is seen that initiation of the process of slowing down of the projectile is delayed when impacted on perforated steel plate.

This is indicative of the fact that when the projectile tip encounters a hole

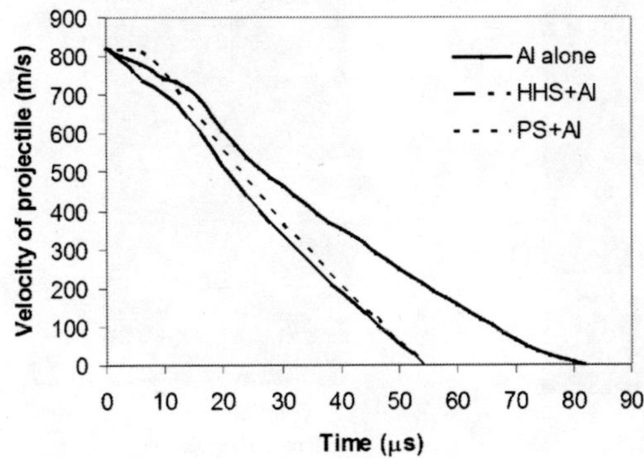

Fig. 9. Velocity of projectile during penetration obtained from simulations

initially, the velocity remains constant. Slowing down of the projectile is initiated at about 8 μs when the tip of the projectile and the lateral surface encounter the resistance of the target plate which is also evident from Fig. 8. Upon impact on both the high hardness steel plate as well as the perforated plate, the velocity of projectile comes to zero in much shorter time than in the case of aluminium plate.

Fig. 10 shows step-by-step simulation pictures of the deformation contours during impact of projectile on the three different target configurations.

In the case of impact on perforated steel plate, it is clearly seen that the projectile tends to bend upon impact. There is a deflection of the path of the projectile. The strain contours show that the strains are maximum at 1/3rd of the length of the projectile measured from nose tip. The corresponding stress values are also seen to exceed the fracture strength of the projectile which causes it to fracture, thereby reducing the overall depth of penetration.

CONCLUSIONS

An experimental as well as numerical study has been made to evaluate the ballistic efficiency of perforated high hardness steel armour having circular holes. Perforated plate gives a weight saving of about 20% over a complete plate of same size and thickness which in effect means that for same area, the density of the plate is reduced by 20%. Experiments were carried out using reference aluminium 7017 alloy plates alone as well as high hardness steel armour and perforated steel armour, when backed by the reference Al-alloy plates. Standard depth of penetration measurements using the thick backing technique were carried out for evaluating the ballistic efficiency of each target configuration. Results indicate that the thickness and mass efficiency of high hardness steel armour is about 3.8 and 1.4 times better than the reference 7017

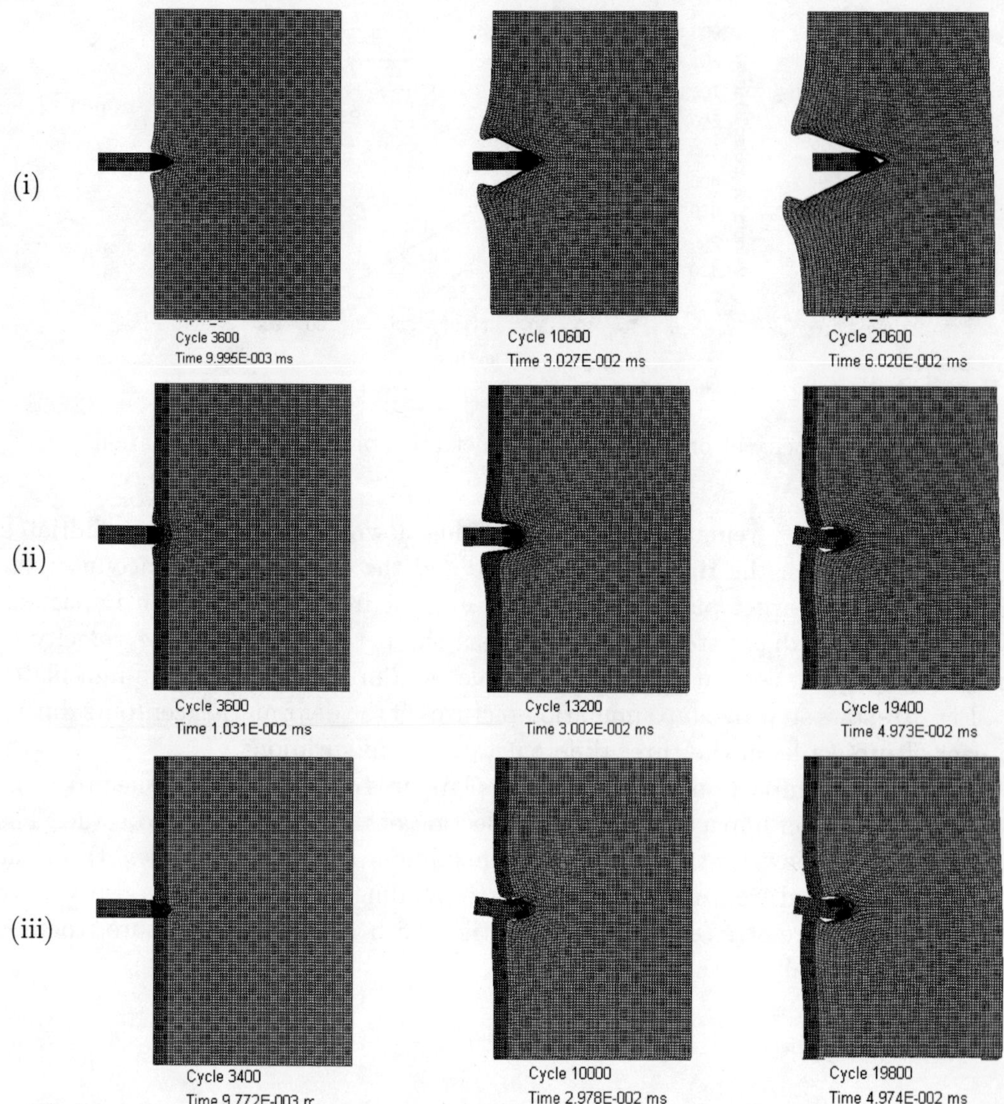

Fig. 10. Deformation contours from simulation of impact of 7.62 AP projectile on (i) Al 7017 alloy, (ii) high hardness steel backed by Al-alloy, and (iii) perforated steel backed by Al-alloy

Al-alloy. This is further enhanced to 5.2 and 2.4 respectively when the high hardness steel is used in perforated form. This indicates that there is an improvement of about 37% in thickness efficiency and about 70% improvement in mass efficiency of high hardness armour steel only due to perforation. Numerical simulations using

Autodyn 3D code were carried out to obtain an understanding of the penetration phenomena. Results of the simulations matched well with the experiments.

ACKNOWLEDGMENTS

The authors thank Director, DMRL for giving permission to publish this work. The authors also wish to acknowledge the support of staff of small arms range of DMRL for carrying out the ballistic tests.

REFERENCES

1. T.B. Bhat, "Principles of Armour Design," Trans. IIM **37** (4), 313–335 (1984).
2. S. Chocron, C.E. Anderson Jr., D.J. Grosch, and C.H. Popelar, "Impact of 7.62 mm APM2 Projectile Against the Edge of a Metallic Target," Int. J. Impact Engng **25**, 423–437 (2001).
3. S.P. Medvedev, S.V. Khomik, and H. Oliver, "Experimental Setup, Measurement Technique and Test Conditions for Explosions with Active/Passive Additives," in *EXPRO Deliverable Report D5, Contract EVG1-CT-2001-00042* (March 2003).
4. J. Chaoa, T. Otsukab, and J.H.S. Leea, "An Experimental Investigation of the Onset of Detonation," Proc. Comb. Inst. **30** (2), 1889–1897 (2005).
5. Y. Trasi, D. Ben-Moshe, and G. Rosenberg, "An Armour Assembly for Armoured Vehicles," European Patent No. EP 0 209 221 A1 (1987).
6. R.A. Auyer, R.J. Buccellato, A.J. Gidynski, et al., "Perforated Plate Armor," International Patent No. WO 89/08233 (1989).
7. R. Moshe and Y. Hirschberg, "Perforated armor plates," European Patent No. EP 1 705 452 A1 (2006).
8. G.S. Langdon, G.N. Nurick, V.H. Balden, and R.B. Timmis, "Perforated Plates as Passive Mitigation Systems," Defence Sci. J. **58** (2), 238–247 (2008).
9. B. James, "Depth of Penetration Testing," Ceramic Trans. **134**, 165–172 (2001).
10. Z. Rosenberg, Y. Yeshurun, and J. Tsaliah, "More on the Thick Backing Technique for Ceramic Tile against AP Projectiles," in *12th International Symposium on Ballistics, San Antonio, Texas, USA* (ADPA, 1990), pp. 197–201.
11. V. Madhu, K. Ramanjaneyulu, T.B. Bhat, and N.K. Gupta, "An Experimental Study of Penetration Resistance of Ceramic Armour Subjected to Projectile Impact," Int. J. Impact Engng **32**, 337–350 (2005).
12. W.A. Gooch and M.S. Burkins, "Ballistic Development of US High Density Tungsten Carbide Ceramics," in *Proceedings of Pac Rim IV international Conference on Advanced Ceramics and Glass, Hawaii* (2001), pp. 53–61.

3D NUMERICAL SIMULATIONS OF DUCTILE TARGETS SUBJECTED TO OBLIQUE IMPACT BY SHARP NOSED PROJECTILES

M.A. Iqbal[1], G. Gupta[1], and N.K. Gupta[2*]

ABSTRACT

The present study deals with the 3D numerical simulations of ductile targets subjected to oblique impact by sharp nosed projectiles. 1100-H12 aluminium target plates of 1 mm thickness were impacted by 19 mm diameter ogive-nosed steel projectile and Weldox 460E steel plates of 12 mm thickness were impacted by 20 mm diameter conical-nosed steel projectile. The obliquity of the target was increased until ricochet occurred. The angles of obliquity of the target were taken as 0°, 15°, 30°, 45°, 60°, and 75°. Numerical simulations were carried out with ABAQUS/Explicit finite element code. Johnson–Cook elasto-viscoplastic material model was used to carry out the simulations. Ballistic limit of the target was obtained for each angle of impact. The ballistic resistance of the target was found to increase with an increase in its obliquity. 1 mm thick aluminium plates failed through petal formation when impacted by ogive-nosed projectile. However, no petal formation was observed in the case of 12 mm thick steel plates impacted by conical-nosed projectile.

Key words: 3D simulations, oblique impact, ricochet, ballistic limit

INTRODUCTION

Oblique impact is a very important subject of terminal ballistic interaction. Since, in most of the cases of impact the projectile strikes the target at some obliquity. The angle of obliquity is an important parameter which affects the ballistic resistance of a target. It is defined as the angle subtended by the velocity vector and the normal

[1]Department of Civil Engineering, Indian Institute of Technology Roorkee, Roorkee — 247667, Uttarakhand, India

[2]Department of Applied Mechanics, Indian Institute of Technology Delhi, New Delhi — 110016, Hauz Khas, India

[*]E-mail: *nkgupta@am.iitd.ernet.in*

to the surface of the target [1]. It is revealed from the experimental investigations [2] that the behavior of target at low obliquity (¡ 30) is more or less same to that of the normal impact. However, at high obliquity (¿ 45) there is significant change in the resistance of the target. It has also been witnessed from the experiments [1]-[3] that at very high obliquity, a stage comes when the projectile strikes the target and rebounds back from the front side itself with or without penetrating the target. This phenomenon is called projectile ricochet, and the angle of incidence, the critical angle of ricochet.

Goldsmith and Finnegan [4] carried out an experimental program of normal and oblique impact of cylindro-conical and blunt cylindrical projectiles of 12.7 mm diameter, on aluminum and steel plates in the velocity range of 20 to 1025 m/s. Hard steel and soft aluminum Projectiles were impacted at 0°-50° obliquity. The exit obliquity angle increased more rapidly with initial angle of incidence for 3.175 mm thick aluminum plates. However, it varied linearly for 6.35 mm thick aluminum plates as well as steel plates.

Virostek et al. [5] carried out an experimental investigation of the forces produced by the penetration and perforation of thin aluminium and steel plates by cylindro-conical and hemispherical-tipped projectiles at 0°, 15°, 30° and 45° angles of incidence. An analytical model was also developed to predict peak forces for hemispherical and conical projectiles. For cylindro-conical projectiles, the observed peak forces were found to increase with the angle of incidence while an opposite trend was observed for hemispherical projectiles.

Gupta and Madhu [6–7] carried out experiments wherein spinning armor piercing projectiles of core diameter 6.2 mm were fired on mild steel, RHA steel and aluminum of thicknesses 4.7–40 mm, in a velocity range of 800–880 m/sec. The angle of obliquity was increased from 0 ° until ricochet occurred. The incident angle of a given plate thickness t, for which the incident velocity was ballistic limit, was determined. In oblique impact, the angle of exit was smaller or greater than the angle of incidence depending on the thickness of the plate and hardness of its material. The difference between the angle of obliquity for ballistic limit and critical ricochet increased with the increase in plate thickness.

Johnson and Cook [8] presented the Lagrangian EPIC code computations for oblique, yawed-rod impact on thin single as well as spaced steel plates at different velocities. The results were found in good agreement with experiments as well as previously published computational results with Eularian MESA code. The computation time required in the case of EPIC code was found comparatively lesser than that of the MESA code.

Scheffler [9] used the Eulerian CTH hydrocode to model the perforation behavior of 30° oblique aluminum targets impacted by hemispherical nosed steel projectiles. The simulations were mainly focused on testing the capability of the Eulerian hydrocode to predict correctly the residual velocity and projectile-target deformations. It was concluded that pure Eulerian codes cannot model the experiments and the need for a hybrid code was felt which couples Lagrangian and Eulerian methods. Yoo and Lee [10] studied the protective effectiveness of an oblique metallic plate against long rod projectile through a three dimensional dynamic finite element computer program NET3D. Tungsten alloy hemispherical projectiles were impacted on 4030 steel plates of 3.5, 7 and 10.5 mm thickness at 45°, 60°, and 75° obliquity. It was found that protection performance of an oblique plate was highest in case that the ratio of (line of sight) plate thickness to projectile diameter was around two.

Zhou and Stronge [11] studied the ballistic resistance of monolithic, double-layered and sandwiched steel plates with the help of experiments and numerical simulations. Flat and hemispherical nosed projectiles were impacted at 0°–45° obliquity. In the case of oblique impact by flat nosed projectile, layered plates were found to have higher ballistic limit than monolithic plates. In the case of oblique impact by hemispherical nosed projectile, monolithic plates and sandwich panels conceived almost same ballistic limit. For obtaining better results in the numerical simulations they suggested the refinement of material constitutive relations.

The studies available in literature revealed that oblique impact of target plates has been studied by varying angle of incidence mostly through experiments. Numerical simulation of the problem has also been carried out in few studies. However, such studies are rare, and, at low angle of obliquities, providing little information about the phenomenon. However, to ascertain the actual behavior of target during oblique impact, it is important to hit the target at high obliquity i.e. till ricochet occurred. In the present investigation therefore, 3D numerical simulations have been performed to study the ballistic resistance of ductile target plates at angles of incidence 0°–75°. 12 mm thick Weldox 460E steel plates were impacted by 20 mm diameter conical nosed steel projectiles and 1 mm thick 1100-H12 aluminum target plates were impacted by 19 mm diameter ogive nosed steel projectiles. Ballistic limit of the target was obtained for each angle of incidence. To investigate the effect of thickness on the failure mode of target, one plate was considered of high thickness (12 mm) and the other of comparatively low thickness (1 mm).

Three dimensional numerical models were prepared and analyzed with ABAQUS/Explicit finite element code. In both the cases the plates were circular. The diameter of 1100-H12 aluminum plate was 255 mm [12] and that of Weldox steel plate was 500 mm [13]. The impact velocities of conical nosed projectiles were kept same as obtained from the experimental investigation carried out by Borvik et al [13] and that of the ogive nosed projectile were kept identical to those obtained from the experimental investigation carried out by Gupta et al. [12]. The target plates were hit at same velocities at each angle of incidence and the ballistic limit was obtained. Effect of critical parameters such as element size as well as its aspect ratio on the numerical results has also been studied. Aspect ratio of the elements was kept unity in the influenced region. The influenced region is the region of target where the deformation is significant. In the other portion of the target plate also the aspect ratio of the elements was kept close to unity.

At low obliquity ($< 30°$) the deformation behavior as well as the ballistic resistance was found to be almost same. However, at thigh obliquity ($> 45°$) significant increase in the ballistic resistance was observed in both the target plates. Both the target plates failed through ductile hole enlargement. However, 1 mm thick plates failed thorough petal formation. On the other hand, 12 mm thick plates failed by formation of a circular hole in the target plate instead of petals. At normal impact the number of petals formed in 1 mm thick target plate was four. However, as the obliquity increased the number of petals decreased to two. On the other hand the circular hole formed in 12 mm thick target plates took the form of an ellipse at high obliquity.

1. NUMERICAL INVESTIGATION

The present numerical study deals with the normal and oblique behavior of 1 mm thick 1100-H12 aluminum and 12 mm thick Weldox 460E steel plates subjected to impact by ogival and conical nosed cylindrical steel projectiles respectively. Three dimensional finite element model of the projectile and the target plate was made using ABAQUS/CAE. For the case of oblique impact the target was rotated about its in-plane horizontal axis to get inclination at the desired angles. Projectile was modelled as analytical rigid and the target plate as a deformable body. Fig. 1 shows the numerical model of both the cases investigated in the present study. Eight noded brick elements were used in all the simulations carried out in this study. Contact was assigned using the kinematic contact algorithm of ABAQUS/CAE. In the case of normal impact outer surface of the projectile was modelled as the first surface and the contact region of the target as node based second surface. In the case of oblique impact a larger region of target comes in contact with that of the projectile. Therefore the contact region of the target during oblique impact was calculated in advance on the basis of angle of obliquity. In the case of 1100-H12 aluminum target plates the effect of friction was considered negligible between the projectile and target. In the case of Weldox steel plates however, a coefficient of friction of 0.05 was considered between the projectile and target in the same fashion as done by [14]. The value of coefficient of friction was considered identical for the case of normal as well as oblique impact. In both the cases the plates were modelled of circular shape with diameters equivalent to those of the actual plates used during the experiments carried out by Borvik et al. [13] and Gupta et al. [12]. The diameter of 1100-H12 aluminum plate was 255 mm and that of the Weldox steel plate was 500 mm. The

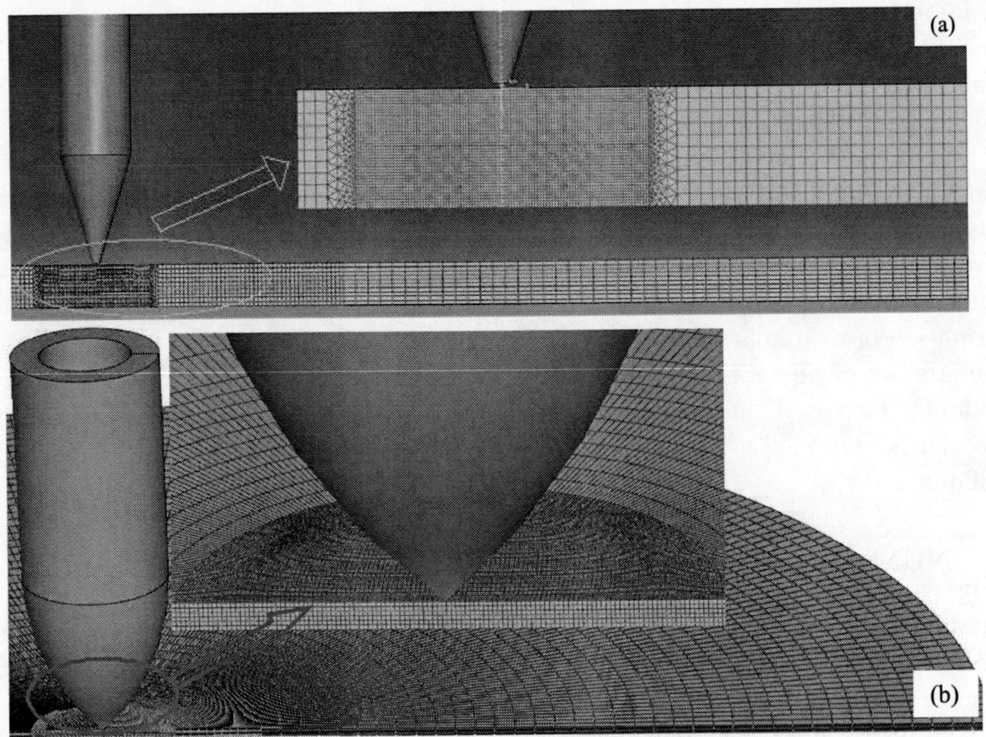

Fig. 1. Finite element model; (a) 12 mm thick Weldox steel plate; (b) 1 mm thick 1100-H12 aluminum plate

geometry of the conical and ogival projectiles was modelled identical to those of the projectiles used by Borvik et al. [13–14] and Gupta et al. [12] respectively. Tip of the conical nosed projectile was also truncated in the present investigation in the same fashion as done by Borvik et al [14]. The target plate was restrained at its periphery with respect to all degrees of freedom to model the periphery of the target fixed in accordance with those of the experiments.

2. MESHING STRATEGY

Extensive mesh study was performed before preparing the final mesh of the target plate. Optimal size of the element was selected. It was found that the aspect ratio of the element close to unity gives best results. However, keeping aspect ratio unity in the whole plate causes enormous increase in the total number of elements, due to which the computational cost of the analysis increases. Further, if the aspect ratio of the element is close to unity the size of increment (time increment) during the analysis remains appropriate. However, as the aspect ratio of element increases, the size of increment decreases and hence the total computational time of the problem increases. The aspect ratio of the element was kept unity in the influenced region. However, it was allowed to increase slightly beyond this region. It is pertinent to mention here that influenced region of the target is not the only region which comes in contact with that of the projectile. However, it is the region which is affected significantly due to the impact. In other words it is the region of the target where the deformation is significant. The size of element is another important factor which affects the quality of computational results. In the present investigation convergence of results was obtained by taking 6 elements over the thickness of 1 mm thick plate, giving a total number of 516104 elements in the target plate. However, at high obliquity, since the influenced region of target was comparatively larger. Hence, if the aspect ratio was kept unity in this region, the total number of elements in target was exceeding beyond the computational efficiency. To restrict the total number of elements under computational limit and keep unity aspect ratio in the influenced region, the number of elements at the thickness of 1 mm thick plate was restricted to 5 in a few cases.

A final mesh for 12 mm thick steel plates was prepared after taking sufficient number of trials for getting mesh independent results. The meshing of target plate was done in such a fashion that the results can be obtained accurately within the efficiency of available computational facility. In the central influenced region of the target 40 elements were taken over the thickness and aspect ratio of these elements was kept unity. This configuration produced total number of 651804 elements in the whole plate. However, at high obliquity the influenced region of target was increased. Therefore to keep the aspect ratio unity in this region again the total number of element was tremendously increased. Therefore to keep aspect ratio unity in the influenced region and keep total number of elements in the target within the efficiency of machine, the number of elements at the thickness of plate was reduced to 25 in some cases.

3. CONSTITUTIVE MODELING

The material behavior of the target for the numerical simulations of 1100-H12 aluminum as well as Weldox 460E steel plates was incorporated using Johnson-Cook elasto-viscoplastic model [15, 16] that is capable to predict the flow and fracture behavior of the target. Johnson and Cook developed a constitutive model to describe material behavior under impact loading. The equivalent stress of the Johnson-Cook model is expressed in the following form;

$$\bar{\sigma}(\bar{\varepsilon}^{\text{pl}}, \dot{\bar{\varepsilon}}^{\text{pl}}, T) = [A + B(\bar{\varepsilon}^{\text{pl}})^n]\left[1 + C \ln\left(\frac{\dot{\bar{\varepsilon}}^{\text{pl}}}{\dot{\varepsilon}_0}\right)\right](1 - \hat{T}^m), \tag{1}$$

$$\hat{T} = \frac{T - T_0}{T_{\text{melt}} - T_0}, \quad \text{where } T_0 \le T \le T_{\text{melt}}, \tag{2}$$

where $\bar{\sigma}$ is the von Mises stress, $\bar{\varepsilon}^{\text{pl}}$ is the effective plastic strain, A is quasi-static yield stress, B is a hardening constant, n is the hardening exponent, C is the strain-rate sensitivity parameter, m is the temperature sensitivity parameter, and $\dot{\varepsilon}_0$ is reference strain rate. The model accounts for isotropic strain hardening, strain rate hardening and temperature softening. The equivalent fracture strain of Johnson-Cook fracture models is expressed in the following form:

$$\bar{\varepsilon}_f^{\text{pl}}(\sigma_m/\bar{\sigma}, \dot{\bar{\varepsilon}}^{\text{pl}}, T) = \left[D_1 + D_2 \exp\left(D_3 \frac{\sigma_m}{\bar{\sigma}}\right)\right]\left[1 + D_4 \ln\left(\frac{\dot{\bar{\varepsilon}}^{\text{pl}}}{\dot{\varepsilon}_0}\right)\right](1 + D_5 \hat{T}), \tag{3}$$

in which the equivalent plastic strain $\bar{\varepsilon}_f^{\text{pl}}$ is expressed as a function of stress triaxiality $\sigma_m/\bar{\sigma}$, equivalent plastic strain rate $\dot{\bar{\varepsilon}}^{\text{pl}}$, and temperature $\hat{T}$. All the material parameters used in the present investigation are given in Tables 1–2 for 1100-H12 aluminum and Weldox steel plates respectively.

4. RESULTS AND DISCUSSION

The results obtained in the present investigation are shown in Table 3 and Table 4 for conical nosed projectile impacted on 12 mm thick steel plate and ogival nosed projectile impacted on 1 mm thick aluminium plate respectively. The results are presented in the form of impact and residual velocities of projectiles subjected to impact on the target plates at 0°, 15°, 30°, 45°, 60°, and 75° obliquity. Fig. 2 compares the results of present investigation with the previous studies. A close correlation was found between the previous experimental results and 3D numerical results of the present investigation for 12 mm thick Weldox steel plates impacted by conical projectiles and 1 mm thick 1100-H12 aluminum plates impacted by ogive nosed projectiles. Both the targets failed thorough ductile hole enlargement. However, 1 mm thick aluminium plates failed through petal formation (Fig. 3*a*). However, 12 mm thick steel plates failed through the formation of a circular hole and a bulge in the target plate around the projectile (Fig. 3*b*). Four petals were formed in 1 mm thick aluminium target plate when impacted normally by the projectile. However, as the obliquity of the target increased the size of upper two petals reduced and that of the lower two petals increased. At 45° obliquity, the upper two petals vanished and

Table 1. Material parameters for 1100-H12 aluminum plates

Parameter	Value
Modulus of Elasticity, E (N/mm^2)	65762
Poison's Ratio, ν	0.3
Density, ρ (kg/m^3)	2700
Yield Stress, A (N/mm^2)	148.361
B (N/mm^2)	345.513
n	0.183
Reference Strain Rate, $\dot{\varepsilon}_0$ (s^{-1})	1.0
C	0.001
m	0.859
T_{melt} (K)	893
T_0 (K)	293
Specific heat, C_p (J/(kg·K))	920
Inelastic heat fraction, α	0.9
D_1	0.071
D_2	1.248
D_3	−1.142
D_4	0.0097
D_5	0.0

Table 2. Material parameters for Weldox 460E steel plates

Parameter	Value
Modulus of Elasticity, E (N/mm^2)	2×10^5
Poison's Ratio, ν	0.33
Density, ρ (kg/m^3)	7850
Yield Stress, A (N/mm^2)	490
B (N/mm^2)	807
n	0.73
Reference Strain Rate, $\dot{\varepsilon}_0$ (s^{-1})	5×10^{-4}
C	0.0114
m	0.94
T_{melt} (K)	1800
T_0 (K)	293
Specific heat, C_p (J/(kg·K))	452
Inelastic heat fraction, α	0.9
D_1	0.705
D_2	1.732
D_3	−0.54
D_4	−0.01
D_5	0.0

the lower two petals got merged together giving an elliptical shape to the opening formed in the target plate. The failure mode of 1 mm thick aluminium target plate at each angle of obliquity is shown in Fig. 4. The circular hole formed in 12 mm thick steel plate also took the shape of an ellipse at high obliquity.

In general it was observed that the ballistic resistance of both the target plates increased with an increase in their obliquity. However, at low obliquity (up to 30°) the ballistic limit of 12 mm tick steel plate was found to be almost same to that of the normal impact. The impact and residual velocity curves of conical nosed projectile at 0°, 15°, and 30° obliquity were found to overlap each other (Fig. 5 *a*). This revealed that there was no significant change in the behavior of target at low obliquity. At 45° and 60° obliquity however, the figure revealed that there is change

Table 3. Experimental and numerical results of conical nosed projectile impact on 12 mm thick Weldox steel plates

Conical nosed projectile (diameter= 20 mm)								
Experimental results [13]		Axi-symmetric numerical results with adaptive meshing [14]	3D numerical results of present investigation					
Monolithic plate of 12 mm thickness			Monolithic plate of 12 mm thickness					
Normal impact			0° obliquity (normal impact)	15° obliquity	30° obliquity	45° obliquity	60° obliquity	75° obliquity
V_i (m/s)	V_r (m/s)	V_r (m/s)	V_r (m/s)	V_r (m/s)	V_r (m/s)	V_r (m/s)	V_r (m/s)	V_r (m/s)
600.0		523.0	539.58	540.77	537.09	476.3	352.1	Critical projectile ricochet
405.70	312.0	304.0	306.77	303.1	301.06	277.68	192.25	
355.60	232.3	228.9	229.16	229.88	227.55	196.21	42.35	
330.6	—	—	—	—	—	—	0.0	
317.90	155.8	161.3	157.39	161.19	159.53	104.72	—	
300.30	110.3	120.6	115.38	120.93	120.83	27.91	—	
280.90	0.0	46.1	50.35	58.79	50.0	0.0	—	
277.0	—	—	—	—	0.0	—	—	
276.20	—	—	—	0.0	—	—	—	
275.0	—	—	0.0	—	—	—	—	

in the behavior of target due to which significant increase in the ballistic resistance was found. There was an increase of 2.1% and 16.6% in the ballistic limit of 12 mm thick Weldox steel plate at 45° and 60° obliquity respectively as compared to that of the normal impact. At 75° obliquity, the conical nosed projectile ricocheted at all impact velocities when impacted on 12 mm thick steel plate.

The ballistic behavior of 1 mm thick aluminium target plate was somewhat different to that of the 12 mm thick steel plate. In this case (1 mm thick aluminium plate), there was a continuous increase in the ballistic limit of target with an increase in obliquity (Fig. 5 *b*). An increase of 5.95%, 8.58%, 8.87%, and 13% was found in the ballistic limit of 1 mm thick 1100-H12 aluminum target plate at 15°, 30°, 45°, and 60° obliquity respectively as compared to that of the normal impact. At 75° obliquity, the ogival projectile ricocheted as a result of impact on 1 mm thick

Table 4. Experimental and numerical results of ogive nosed projectile impact on 1 mm thick 1100-H12 aluminum plates

Ogive nosed projectile (diameter= 19 mm)								
Experimental results [12]		Axi-symmetric numerical results [12]	3D numerical results of present investigation					
Monolithic plate of 1 mm thickness			Monolithic plate of 1 mm thickness					
Normal impact			0° obliquity	15° obliquity	30° obliquity	45° obliquity	60° obliquity	75° obliquity
V_i (m/s)	V_r (m/s)	V_r (m/s)	V_r (m/s)	V_r (m/s)	V_r (m/s)	V_r (m/s)	V_r (m/s)	V_r (m/s)
150.0			141.26	139.23	135.25	129.2	121.7	
112.72	99.11	95.64	99.98	97.49	95.59	91.82	87.94	
97.23	78.26	73.25	81.35	78.68	76.52	71.98	67.67	
82.97	61.62	55.71	63.54	59.7	56.7	52.77	45.84	
81.91	58.19	53.27	62.04	57.57	55.02	49.8	41.6	Critical projectile ricochet
73.30	44.38	38.67	49.29	43.27	40.8	33.82	30.41	
65.80	29.68	26.04	35.10	28.49	26.06	21.71	17.48	
60.0	—	—	—	—	—	—	0.0	
57.28	17.86	15.93	15.42	6.79	2.3	0.0	—	
57.1	—	—	—	—	0.0	—	—	
55.5	—	—	—	0.0	—	—	—	
52.2	—	—	0.0	—	—	—	—	
51.28	8.72	0.0	—	—	—	—	—	
49.8	—	—	—	—	—	—	—	
45.3	0.0	0.0	—	—	—	—	—	

aluminium plate at all the impact velocities. Fig. 6 shows the variation of ballistic limit with angle of obliquity of target plates. It can be noticed that in the case of 12 mm thick steel plate there is nominal increase in the ballistic limit up to 30° obliquity. However, thereafter the ballistic limit increased sharply (Fig. 6 *a*). On the other hand, in the case of 1 mm thick aluminium plate a continuous increase in the ballistic limit was found with increasing obliquity (Fig. 6 *b*). Figs. 7 describes the deformation behavior of 1 mm thick aluminium plate impacted by ogive nosed projectile and 12 mm thick steel plate impacted by conical nosed projectile at different obliquities. With an increase in obliquity it was observed that the deformation of the target became more localized. The global deformation was highest at normal impact and thereafter it decreased with an increase in target obliquity.

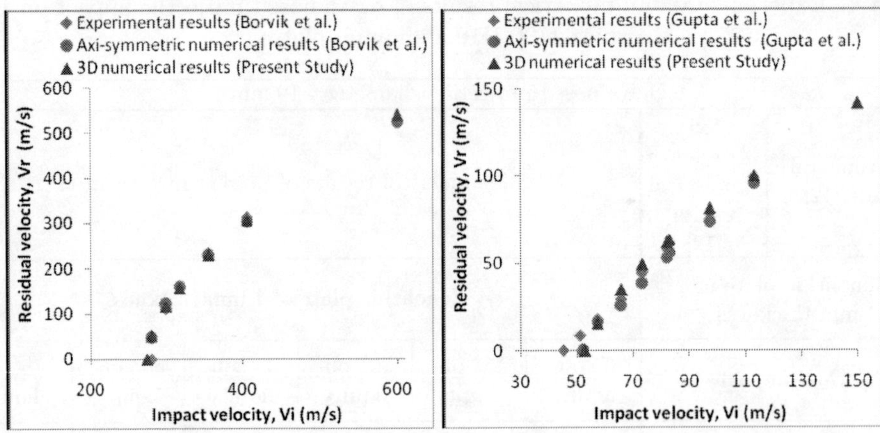

Fig. 2. Normal impact of projectiles on 12 mm thick Weldox steel Plates and 1 mm thick 1100-H12 aluminum plates; a comparison between the present study with the previous experimental and numerical results

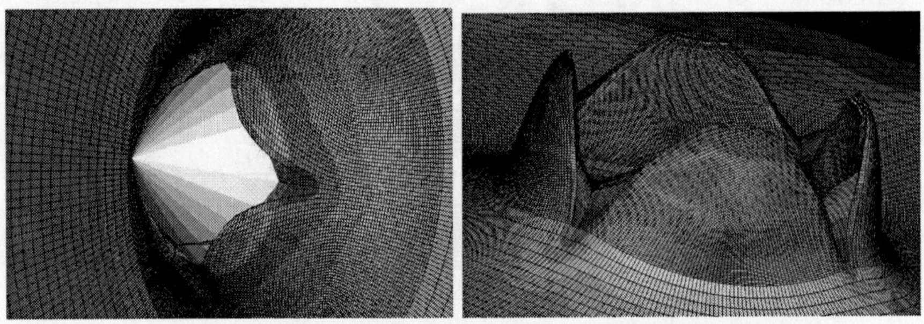

(a) 1 mm thick 1100-H12 aluminum plate impacted normally by ogive nosed projectile

(b) 12 mm thick Weldox steel plate impacted normally by conical nosed projectile

Fig. 3. Typical failure modes of target plates

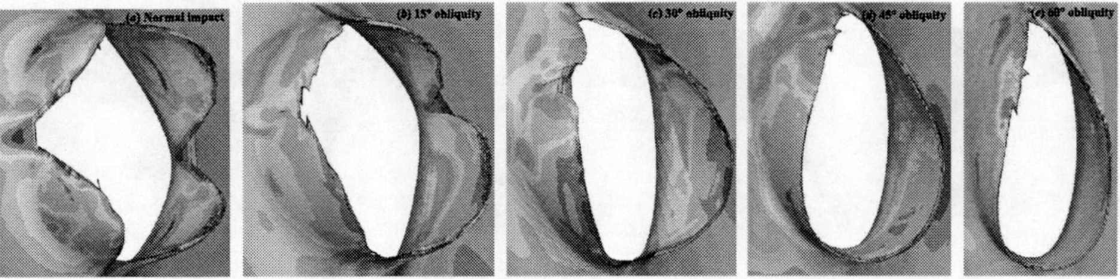

Fig. 4. Failure mode of 1100-H12 aluminum target plates at different obliquity

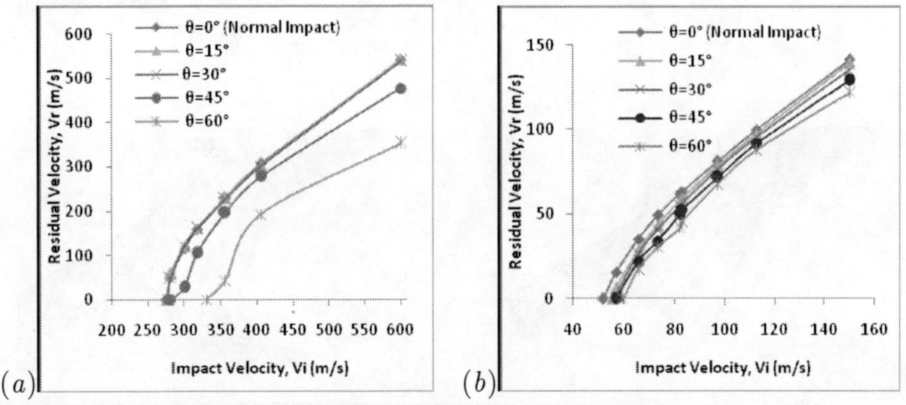

Fig. 5. Impact and residual velocity curves of projectiles; (*a*) Conical nosed; (*b*) Ogive nosed

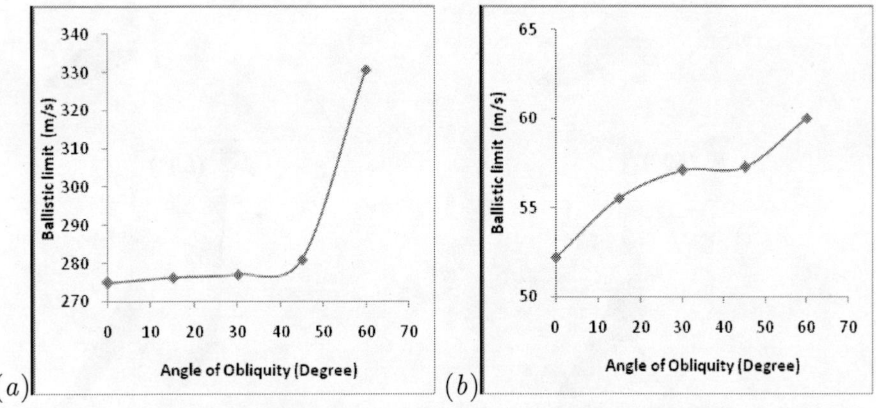

Fig. 6. Variation of ballistic limit with angle of obliquity; (*a*) 12 mm thick Weldox steel plate; (*b*) 1 mm thick 1100-H12 aluminum plate

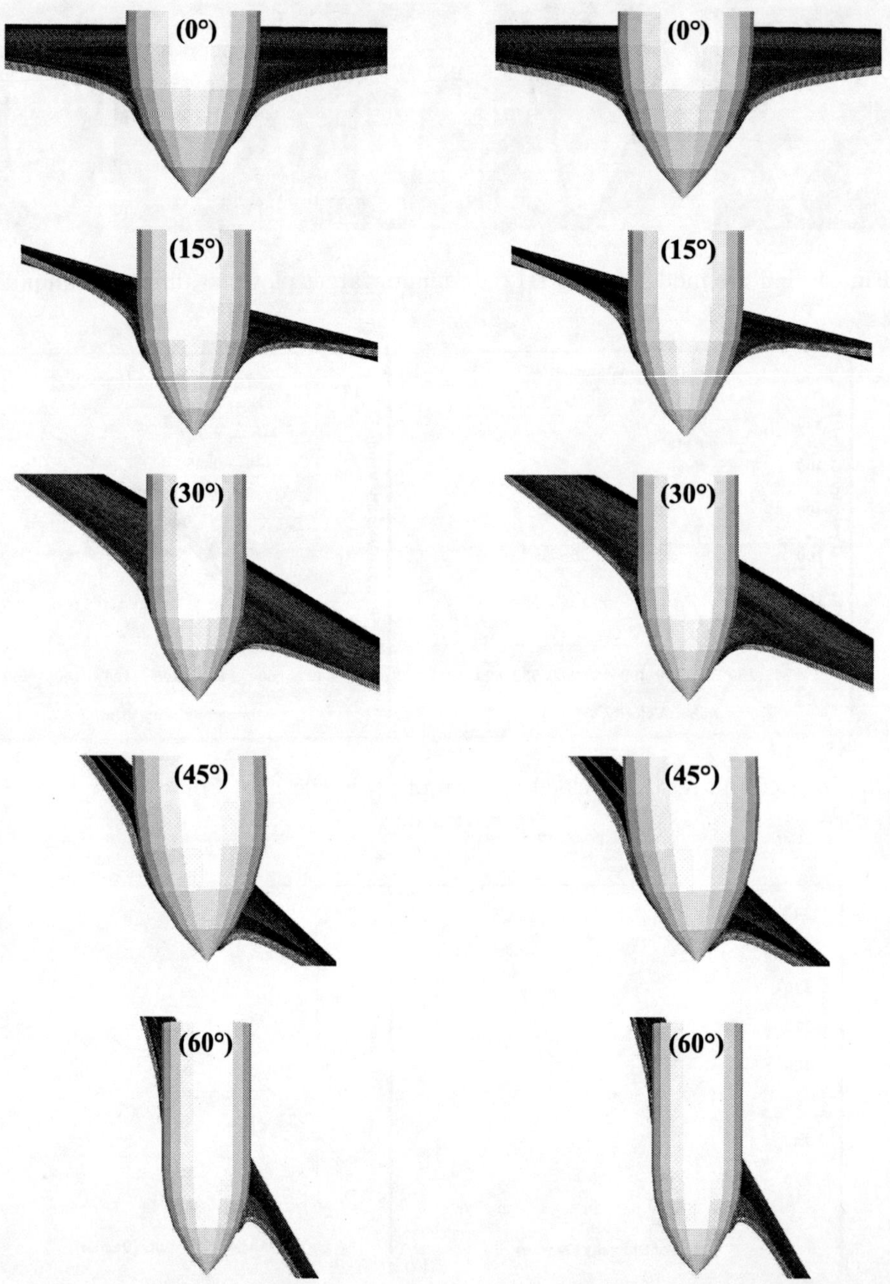

Fig. 7. 1 mm thick 1100-H12 aluminum plates impacted by ogive nosed projectile

Fig. 8. 12 mm thick Weldox steel plates impacted by conical nosed projectile

5. CONCLUSIONS

3D numerical simulations were performed with sharp nosed projectiles impacted on ductile target plates at 0°, 15°, 30°, 45°, 60°, and 75° obliquity. 12 mm thick Weldox 460E steel plates were impacted by conical nosed projectiles and 1 mm thick 1100-H12 aluminum plates were impacted by ogive nosed projectiles. Ballistic limit of target was obtained at each angle of incidence. The following conclusions can be drawn.

1. Ballistic limit of target increased with an increase in obliquity. However, the ballistic limit of 12 mm thick steel plate was almost same up to 30° obliquity. Thereafter it increased sharply. On the other hand, in the case of 1 mm thick target plate a continuous increase in the ballistic limit with increasing obliquity was observed.
2. When compared to the normal impact an increase of 0.44%, 0.72%, 2.1%, and 16.81% was found in 12 mm thick steel plate at 15°, 30°, 45°, and 60° obliquity respectively. While in the case of 1 mm thick aluminium plate the same was found to be 5.95%, 8.58%, 8.87%, and 13% respectively at 15°, 30°, 45°, and 60° obliquity.
3. Both the targets failed through ductile hole enlargement. However, 12 mm thick Weldox steel plate failed through the formation of a bulge around the projectile at the rare surface of target. On the other hand 1 mm thick aluminum target plate failed through petal formation.
4. Four petals were formed at each angle of impact in 1 mm thick 1100-H12 aluminum target plate. The size of upper two petals was found to decrease and lower two petals was found to increase with an increased in obliquity. At 45° obliquity the upper two petals were vanished.
5. At 75° obliquity, critical projectile ricochet was witnessed in both the target plates.

REFERENCES

1. M.E. Backman and W. Goldsmith, "The Mechanics of Penetration of Projectiles Into Targets," Int. J. Engng Sci. **16**, 1–99 (1978).
2. G.G. Corbet, S.R. Reid, and W. Johnson, "Impact Loading of Plates and Shells by Free-Flying Projectiles: a Review," Int. J. Impact Engng **18**, 141–230 (1996).
3. W. Johnson, A.K. Sengupta, and S.K. Ghosh, "High Velocity Oblique Impact and Ricochet Mainly of Long Rod Projectiles: an Overview," Int. J. Mech. Sci. **24**, 425–436 (1982).

4. W. Goldsmith and S.A. Finnegan, "Normal and Oblique Impact of Cylindro-Conical and Cylindrical Projectiles on Metallic Plates," Int. J. Impact Engng **4**, 83–105 (1986).
5. S.P. Virostek, J. Dual, and W. Goldsmith, "Direct Force Measurement in Normal and Oblique Impact of Plates by Projectiles," Int. J. Impact Engng **6**, 247–269 (1987).
6. N.K. Gupta and V. Madhu, "Normal and Oblique Impact of Kinetic Energy Projectile on Mild Steel Plates," Int. J. Impact Engng **12** 395–414 (1992).
7. N.K. Gupta and V. Madhu, "An Experimental Study of Normal and Oblique Impact of Hard-Core Projectile on Single and Layered Plates," Int. J. Impact Engng **19**, 395–414 (1997).
8. G.R. Johnson and W.H. Cook, "Lagrangian Epic Code Computations for Oblique, Yawned-Rod Impacts onto Thin-Plate and Spaced-Plate Targets at Various Velocities," Int. J. Impact Engng **14**, 373–383 (1993).
9. D.R. Scheffler, "Modeling Non-Eroding Perforation of an Oblique Aluminum Target Using the Eulerian CTH Hydrocode," Int. J. Impact Engng **32**, 461–472 (2005).
10. Y.H. Yoo and M. Lee, "Protection Effectiveness of an Oblique Plate Against a Long Rod," Int. J. Impact Engng **33**, 872–879 (2006).
11. D.W. Zhou and W.J. Stronge, "Ballistic Limit for Oblique Impact of Thin Sandwich Panels and Spaced Plates," Int. J. Impact Engng **35**, 1339–1354 (2008).
12. N.K. Gupta, M.A. Iqbal, and G.S. Sekhon, "Effect of Projectile Nose Shape, Impact Velocity and Target Thickness on Deformation Behavior of Aluminum Plates," Int. J. Solids Struct. **44**, 3411–3439 (2007).
13. T. Borvik, M. Langseth, O.S. Hopperstad, K.A. Malo, "Perforation of 12 mm Thick Steel Plates by 20 mm Diameter Projectiles with Flat, Hemispherical and Conical Noses Part I: Experimental Study," Int. J. Impact Engng **27**, 19–35 (2002).
14. T. Borvik, O.S. Hopperstad, T. Berstad, and M.Langseth, "Perforation of 12 mm Thick Steel Plates by 20 mm Diameter Projectiles with Flat, Hemispherical and Conical Noses Part II: Numerical Simulations," Int. J. Impact Engng **27**, 37–64 (v).
15. G.R. Johnson and W.H. Cook, "A Constitutive Model and Data for Metals Subjected to Large Strains, High Strain Rates and High Temperatures," in *Proceedings of the Seventh International symposium on Ballistics. The Hague* (1983).
16. G.R. Johnson and W.H. Cook, "Fracture Characteristics of Three Metals Subjected to Various Strains, Strain Rates, Temperatures and Pressures," Engng Fract. Mech. **21** (1), 31–48 (1985).

IMPACT STUDIES ON CIRCULAR COMPOSITE AND STEEL TUBES

R. Muralikannan[1], R. Velmurugan[2*], and N.K. Gupta[3]**

ABSTRACT

Thin walled tubes are used as energy absorbing elements in automobile applications. The circular tube proves to be a popular energy absorber because it provides a reasonably constant operating force which is the prime characteristics of the energy absorber. Composite and advanced high strength steel tubes can be utilized to absorb crash energy and minimize intrusion in the occupant zone. In this work, circular tubes made up of composites and dual phase steel are subjected to axial compression in static and dynamic loading. The composite tubes are made up of epoxy resin and glass fiber in the form of woven roving mat of 610 gsm as reinforcement. The hand lay-up process is employed for making the composite tubes. The aspect ratio of 2 is maintained for all the tubes. Dual phase steels have a microstructure of hard martensite dispersed throughout the soft ferrite matrix. The strength level of these grades is related to the amount of martensite in the microstructure. As a result of the soft ferrite matrix they can maintain their formability and the low yield to tensile strength ratio enhances the crash performance. The results are compared with those of annealed tubes of same cross section. The effect of impact velocity is also studied by keeping the striking mass constant and varying the drop height. The energy absorption rate, which is the important factor in assessing the crashworthiness of structures, is also studied for the above tubes. The experimental results are compared with the theoretical results and good agreement is found.

Key words: epoxy/glass fiber composite tubes, dual phase tubes, dynamic loading, plastic deformation, specific energy absorption

---

[1]Department of Mechanical Engineering, Indian Institute of Technology Madras, Chennai — 600036, Tamil Nadu, India

[2]Department of Aerospace Engineering, Indian Institute of Technology Madras, Chennai — 600036, Tamil Nadu, India

[3]Department of Applied Mechanics, Indian Institute of Technology Delhi, New Delhi — 110016, Hauz Khas, India

[*]E-mail: *ramanv@iitm.ac.in*

[**]E-mail: *nkgupta@am.iitd.ernet.in*

INTRODUCTION

The introduction of stronger safety legislation and increased fuel prices, automotive manufacturers must respond with higher car body stiffness for safety and lower body weight for fuel efficiency. Fiber reinforced composites and high strength steel components are now used in automobiles to reduce the weight of the structure and improve crashworthiness. Fiber reinforced composites provide significant functional and economic benefits such as enhanced strength, durability, weight reduction and lower fuel consumption [1]. Moreover they have been found to ensure enhanced level of structural vehicle crashworthiness. Composite tubes are capable of collapsing progressively in a controlled manner that ensures high crash energy absorption in the event of a sudden collision. On the contrary to the response of metals and polymers however, progressive crushing of composite collapsible energy absorbers is dominated by extensive micro-fractures instead of plastic deformation. The failure modes depend on geometry, material characteristics and testing parameters.

In all cases, it is observed that composite tubes, which collapse in a stable, progressive and controlled manner, can dissipate large amount of energy. From the wide experimental results of axisymmetric tubes made from fibre-reinforced polymer matrix composite materials, the following main modes of failure were observed during their axial compression [2].

(1) Progressive crushing with micro fragmentation of the composite material associated with large amounts of crush energy (Mode I).

(2) Brittle fracture of the component resulting in catastrophic failure with little energy absorption (Mode II and III; depended on the crack form).

(3) Progressive folding and hinging similar to the crushing behavior of thin-walled metal and plastic tubes, showing a medium energy absorbing capacity (Mode IV).

Although significant experimental work has been done on static loading, studies on dynamic loading are limited. Circular tubes are proved to be a popular energy absorber over other cross sections because of its geometry. In this work circular composite tube made up of glass fiber and epoxy are subjected to static and dynamic loading. The response of composite structures mainly depends on the velocity of impact. The effect of impact velocity is studied by keeping the striking mass constant and varying the impact velocity from 7.00 m/s to 9.40 m/s.

Dual phase steel is one of the advanced high strength steel which offers high strength and good formability. Dual phase steels have a microstructure of hard martensite dispersed in soft ferrite matrix. The strength level of these grades is related to the amount of martensite in the microstructure. As a result of the soft ferrite matrix they can maintain their formability and its low yield to tensile strength ratio enhances the crash performance. Circular tubes made up of dual phase steel and annealed mild steel are subjected to axial impact and the characteristics are compared with the composite tubes. The specific energy absorption and energy absorption rate of composite and dual phase steel are studied to assess the crashworthy behavior of these components.

1. MATERIALS

Composite tubes of circular cross section are made up of glass fibers and epoxy resin. Hand lay up technique is used for the fabrication. Glass fiber woven roving mat of 610 gsm is used as reinforcement. Structural grade epoxy resin LY 556 is used as matrix. The Araldite hardener HY 951 used for curing. The resin to hardener ratio is taken as 10:1. The tube consists of six layers. The outer diameter of the tube is 75 mm and the thickness is 3.6 mm. The dual phase steel tube is made from the mild steel by suitable heat treatment. The circular tube made up of mild steel with the dimensions of (75×1.6 mm) is heated to 760°C±10°C followed by quenching using water jet to generate dual phase microstructure. The dual phase tube has 70% ferrite and 30% martensite in its microstructure. Ferrite content accounts for ductility and the martensite content accounts for the increased strength of dual phase steel tubes

2. STATIC TESTS

Circular composite tubes are subjected to axial loading in Universal Testing Machine. L/D ratio of 2 is maintained for all the tubes. The specimens undergo progressive crushing with micro fragmentation. The specimens are shown in Fig. 1. The load displacement curves are shown in Fig. 2. The mean load of crushing is almost in the same range after the initial peak. The initial peak load is dependent on the material and geometric imperfections. The thickness of dual phase tubes is 1.6 mm. The dimension and weight of each specimen is measured. The average length of the circular specimen is 150 mm and the mass is 0.447 kg. The circular tubes exhibit mixed mode in static compression. The folding process starts with axisymmetric mode and then has shifted to diamond mode. The outward fold length is higher than inward fold length. The fold length in diamond mode is higher than the axisymmetric mode. The average fold length of first fold is 19.5 mm. The fold length of subsequent folds has increased up to 24 mm. After the formation of diamond mode there is no revert back to axisymmetric mode.

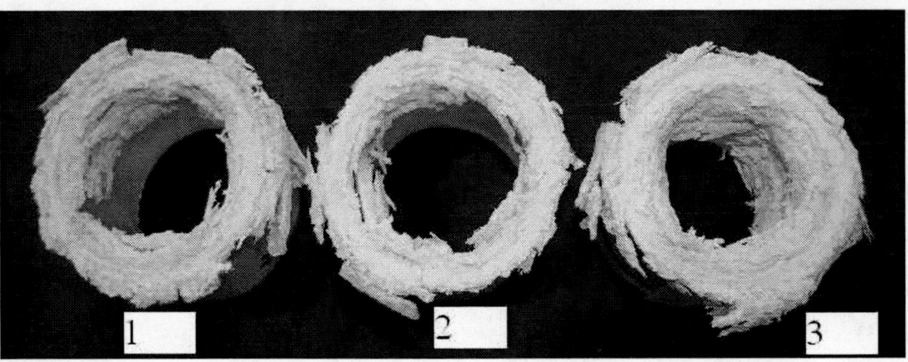

Fig. 1. Composite tubes subjected to quasi static loading

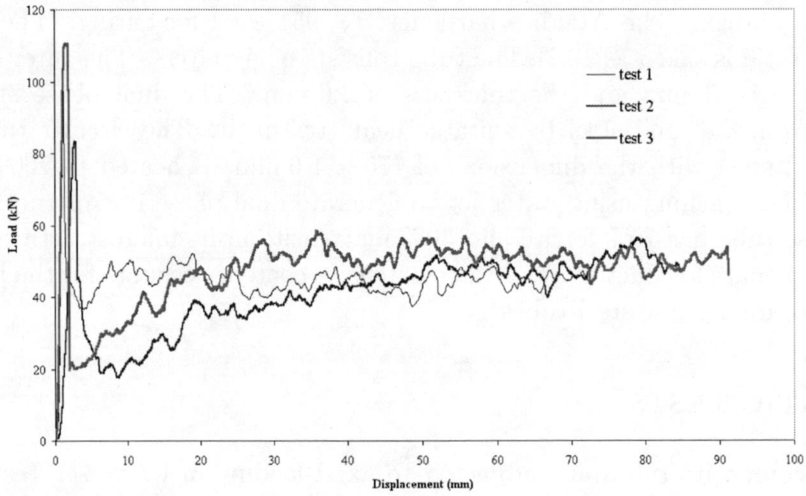

Fig. 2. Load displacement curves of the composite tubes subjected to quasi static loading

Table 1. Quasi static test results of composite tubes

S. No	Crush distance (mm)	Crushed mass (g)	Work done (J)	Mean load (kN)	Specific energy absorption (kJ/kg)
1	72.40	94.98	3278.90	45.29	34.52
2	84.50	107.72	3455.40	40.89	32.08
3	91.20	119.15	4364.20	47.85	36.63

Fig. 3. Instrumented drop mass facility

3. DYNAMIC TESTS

Dynamic tests are performed on instrumented drop mass setup. Fig. 3 show the experimental facility used for the impact studies. It has the flexibility to vary the impact mass and drop height, by which we can provide various input energy levels to the specimens. The striking mass is an assembly of circular discs of 7 kg and 10 kg mass. Depending on the requirement we can add the discs to the assembly so the striking mass can be varied from 10 to 150 kg. The assembly is lifted by an electro magnet through the guide pipe. After reaching the desired height the electromagnet is de-magnetized and the striking mass is dropped through the guide pipe. The guide pipe ensures the perpendicularity of the striking mass. In this setup, the maximum velocity that can be achieved is 10 m/s. The velocity of the striking mass just before the impact is measured using the laser — photo diode setup. Two photo diodes are separated by a known distance of 80 mm and they sense the laser light and are placed in parallel. When the striking mass cuts the first beam the counter starts and when it cuts the second beam the counter stops.

The velocity of the striking mass is calculated from the time taken by the mass to cross a particular known distance The load-time history is recorded by a high sensitivity, strain gage type load cell (HBM make) of 400 kN capacity. The frequency response of the load cell is 15 kHz and it is force washer type. One hardened washer is placed below the load cell and another one is placed above the load cell. The load cell with washers are placed in a fixture. The fixture consists of two parts. The lower part accommodates the load cell with hardened washers. The upper part of the fixture is placed over the top washer. The force is transmitted to the load cell through upper part of the fixture and hardened washer. The load cell with fixture is placed directly under the specimen. The stress wave created by impact passes through the specimen and is measured by the load cell. This signal is filtered by using a 1 kHz low pass filter. The load cell is calibrated using a Universal Testing Machine for various cross head speeds. The relationship between applied load and measured load is found to be linear. The calibration coefficients are almost same for all strain rates, which shows the consistency of measurement by the load cell. The striking mass is a rigid assembly of steel plates. An accelerometer with a frequency response of 10 kHz is mounted on the striking mass at the top. It is assumed that the displacement of the specimen is equal to the displacement of the striking mass. The signal obtained from the accelerometer is filtered by a low pass filter of 1 kHz and integrated twice to get the displacement of the specimen. The data is captured by National Instruments Data Acquisition Card and Labview® software. The load displacement curve for impact test is derived from the load time and displacement time curves. The drop mass used is 77.5 kg. The impact tests are performed for the impact velocities of 7.0 m/s, 7.67 m/s, 8.28 m/s, 8.858 m/s, and 9.4 m/s. The tubes are placed over the load cell without any support. The load time history and displacement time history are cross plotted to get the load displacement curve. From the load displacement curves obtained from the experiments, the work done is calculated. The mean load is calculated from load displacement curve.

3.1. *Composite tubes*

The composite tubes are subjected to dynamic loading by using the drop mass setup. The striking mass dropped on the specimen for different impact velocities ranges from 7.0 m/s to 9.4 m/s. The specimens impacted at different velocities are shown in Fig. 4. Unlike static tests the impacted specimens show mode I and mode IV type of failure modes. Some of the tubes exhibit progressive crushing and hinging similar to the crushing behavior of thin walled metal tubes in impact loading.

Fig. 5 shows the tubes impacted at the velocity of 7.67 m/s. The results of the tests are given in Table 2. The load displacement curves are shown in Fig. 6. At the early stage of loading the shell initially behaved elastically, and as soon as the load attains a peak value, cracks formed on the circumference and propagated

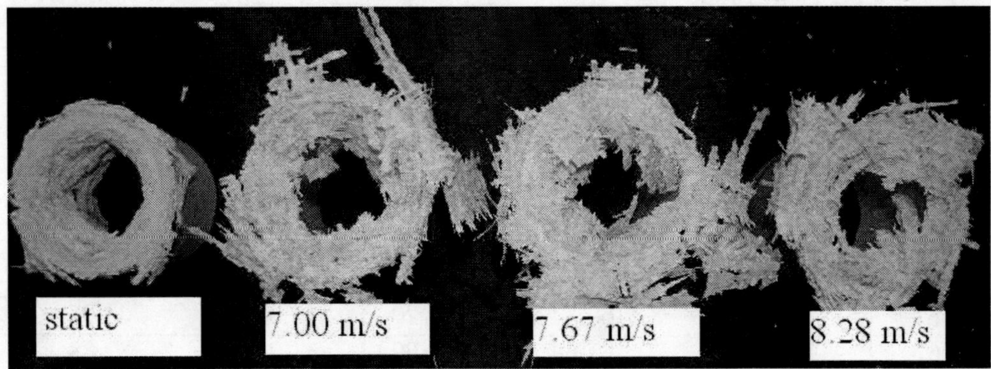

Fig. 4. Composite tubes subjected to impact loading

Fig. 5. Composite tubes impacted at 7.67 m/s velocity

downwards along the tube axis. As deformation progresses the shape of the load displacement curve, dependent heavily on the mode of collapse. The mean load of some specimens which undergo spiral and longitudinal cracks propagating along the shell circumference is less than other specimens

In static tests all the specimens undergo mode I type of crushing, but in impact loading mode IV type of crushing occur in most of the specimens. Mode IV type of failure absorb less energy but the mean load of crushing in dynamic loading is on far with static loading in most cases. This is mainly attributed to the dynamic friction between the striking mass and tube surface. The dynamic friction coefficient and mean load increased with the increase in impact velocity. After the initial elastic response the load displacement curve don't show much variation like steel tubes. The curve comprise of small peaks and valleys. The peaks indicate the formation of hinge

Table 2. Dynamic test results of composite tubes

Impact velocity (ms^{-1})	Crush distance (mm)	Crushed mass (g)	Work done (J)	Mean load (kN)	Specific energy absorption (kJ/kg)
7.00	40.50	52.37	1635.80	40.39	31.24
7.00	45.00	57.19	1657.70	36.84	28.99
7.00	47.50	61.44	1786.40	37.61	29.08
7.67	54.30	71.62	2027.00	37.33	28.30
7.67	46.50	60.34	2110.20	45.38	34.97
7.67	47.70	61.64	1995.60	41.84	32.38
8.28	62.60	78.15	2358.60	37.68	30.18
8.28	60.20	78.41	2399.70	39.86	30.60
8.28	61.40	80.45	2412.50	39.29	29.99
8.86	73.00	91.61	2751.50	37.69	30.03
8.86	70.00	91.13	2778.60	39.69	30.49
8.86	76.00	92.10	2867.20	37.73	31.13
9.40	78.50	88.04	3040.30	38.73	34.53
9.40	78.00	98.39	3088.00	39.59	31.39
9.40	76.00	97.14	3120.70	41.06	32.13

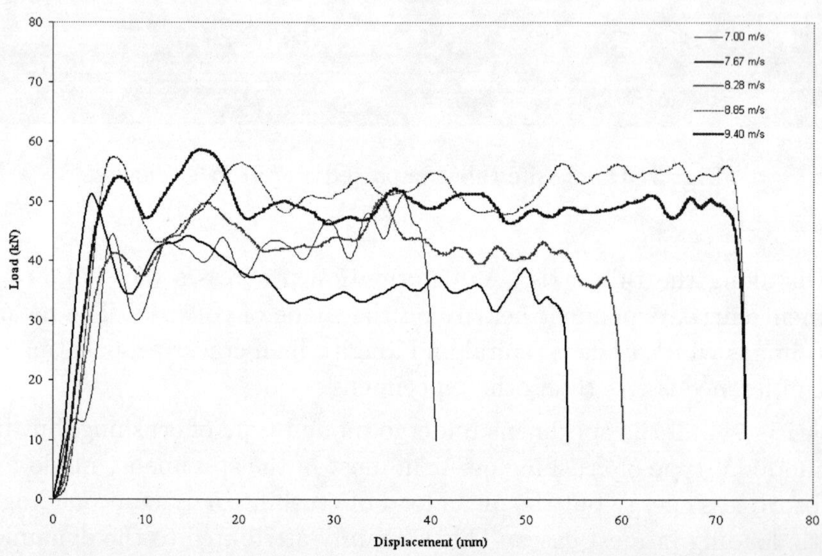

Fig. 6. Load displacement curves of the composite tubes subjected to impact loading

Table 3. Dynamic test results of dual phase steel tubes

Impact velocity (ms^{-1})	Crush distance (mm)	Peak load (kN)	Work done (kJ)	Mean load (kN)	Specific energy absorption (kJ/kg)
7.00	26	202.4	2038.9	78.42	26.76
7.00	29	146.4	2043.5	70.46	24.05
7.00	26	153.7	2038.1	78.38	26.75
7.67	35	144.5	2752.9	78.65	26.84
7.67	25	130.7	2575.1	73.57	25.11
7.67	34	172.9	2622.5	77.13	26.32
8.28	38.5	184.6	3378.1	87.74	29.94
8.28	36	193.3	3360.4	93.34	31.85
8.28	39	131.4	3077.1	78.9	26.92

points and valleys indicate the crushing of the tubes. The steady load displacement curve shows the suitability of composite tubes for crashworthiness application.

3.2. *Dual phase steel tubes*

The circular tubes undergo diamond mode of crushing in dynamic tests. The deformed circular specimens are shown in Fig. 7. The results of the tests are given in Table 3. The load displacement curves for different impact velocities are given in Fig. 8. The load value increases up to the elastic limit and drops after the formation of plastic hinges. The tube first deforms elastically when impacted by the striking mass. Once the plastic hinges are formed it deforms plastically. So the load value drops after the hinge formation. The first peak depends on the local stress up to elastic limit and also on the geometric imperfection. The first peak load which causes the first fold is clearly seen in figure and subsequent peaks are also visible. The first fold absorbs certain amount of impact energy and rest of the energy creates subsequent folds.

There is a substantial difference between the mean load of initial fold and mean load of subsequent folds. This causes the uneven energy absorption characteristics. Annealed mild steel tubes of same cross sections are also subjected to impact loading for comparison with dual phase steel tubes. The impacted annealed tubes are shown in Fig. 9.

The tubes undergo axisymmetric mode of collapse for the velocity range up to 7.67 m/s. The specimens undergo mixed mode of collapse for the impact velocity of 8.28 m/s. The load displacement curves are shown in Fig. 10. The results of the tests are given in Table 4. The mean load of crushing for annealed tube is less than dual phase steel tube.

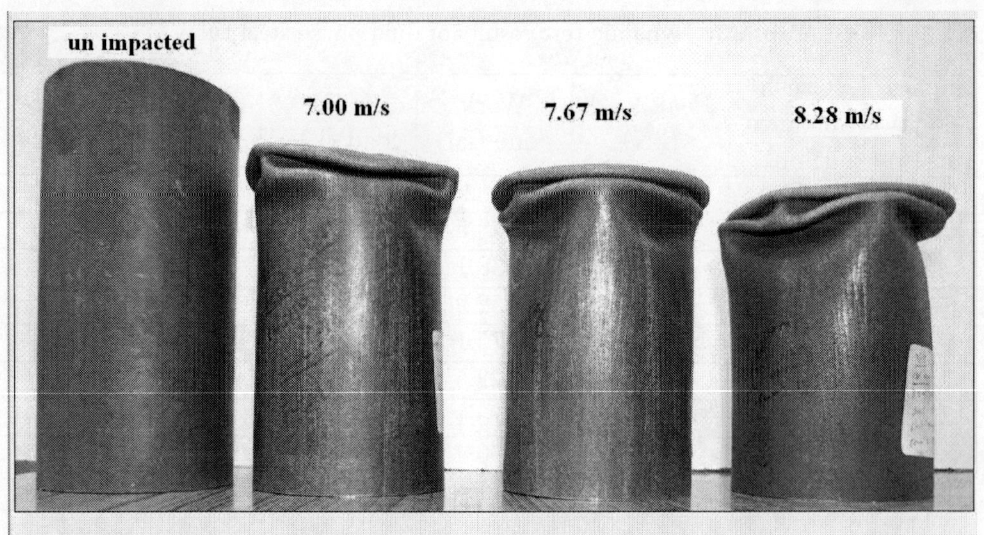

Fig. 7. Dual phase steel tubes subjected to impact loading

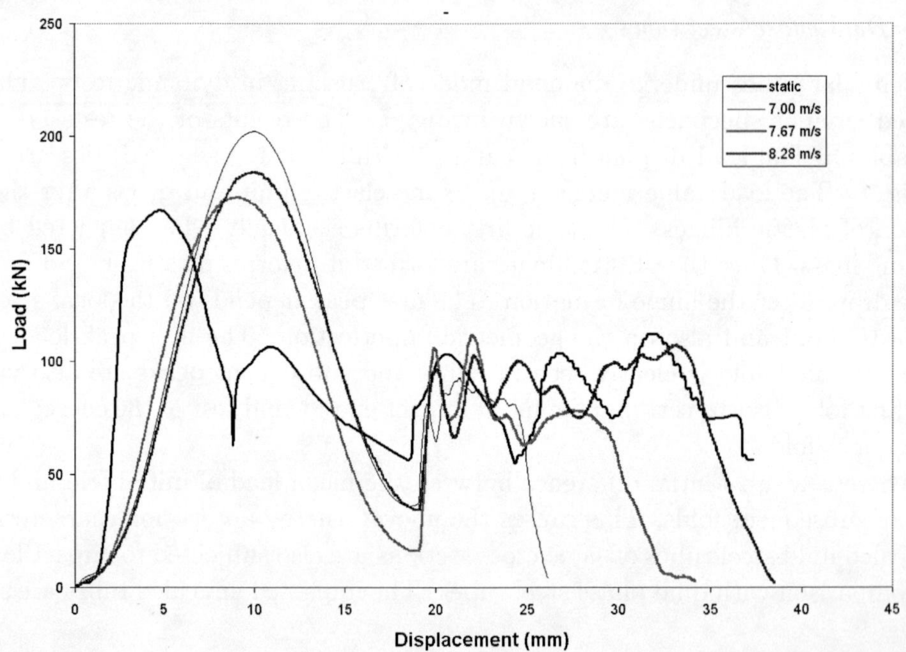

Fig. 8. Load displacement curves of Dual phase steel tubes subjected to impact loading

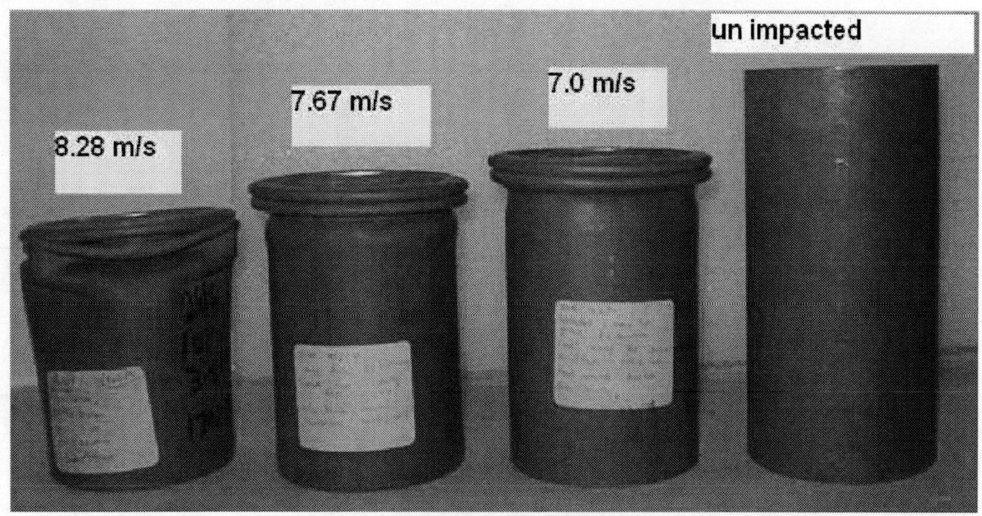

Fig. 9. Annealed steel tubes subjected to impact loading

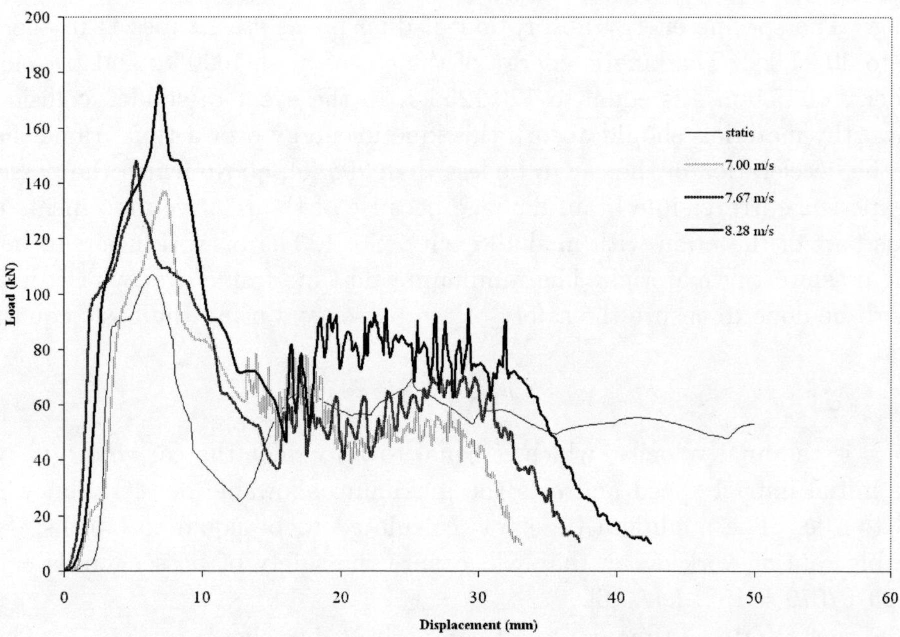

Fig. 10. Load displacement curves of Annealed steel tubes subjected to impact loading

4. COMPARISON BETWEEN COMPOSITE AND DUAL PHASE STEEL TUBES

Specific energy absorption is one of the important parameter in crashworthy applications. The specific energy absorption of the structure plays an important role

Table 4. Dynamic test results of annealed steel tubes

Impact velocity (ms^{-1})	Crush distance (mm)	Peak load (kN)	Work done (kJ)	Mean load (kN)	Specific energy absorption (kJ/kg)
7.00	33.1	135.42	1950.30	58.92	19.77
7.00	33.8	132.50	2048.20	60.60	20.33
7.00	31.7	140.00	1902.10	60.00	20.14
7.67	36.5	141.50	2238.30	61.32	20.58
7.67	37.2	147.35	2395.30	64.39	21.61
7.67	34.8	146.20	2264.80	65.08	21.84
8.28	39.2	152.30	2652.20	67.66	22.70
8.28	42.4	175.24	3028.30	71.42	23.97
8.28	45.1	169.80	3193.20	70.80	23.76

in reducing the weight of the total structure. The composite tubes have the specific energy absorption of 34 kJ/kg in quasi static loading and 31 kJ/kg in impact loading. The specific energy absorption of dual phase steel tubes is in the range of 25 to 30 kJ/kg. The kinetic energy of the car of mass 1000 kg and travelling at a velocity of 15.5 m/s is equal to 120.125 kJ. In the event of sudden collision, the crashworthy materials should absorb this kinetic energy over a time frame that ensures the deceleration of the car to be less than $20g$ [6], above which the passengers will experience irreversible brain damage because of the relative movements of the various part of the brain with in skull cavity. So, 120 kJ of work needs to be done on the crashworthy material. The minimum safe time frame over which this work needs to be done to ensure the safety of the passengers using the basic equation of motion:

$$v = u - at, \tag{1}$$

where v is the final velocity, which is equal to zero since the car comes to rest; u is the initial impact speed and a is the maximum allowable deceleration which is equal to $20g$. The minimum time was calculated to be equal to 0.079 s. So the allowable rate of work decay that will ensure the safety of passengers is equal to 120.125/0/079 = 1521 kJ/s [6].

The energy rate vs time curves of composite tubes shown in Fig. 11. The rate of energy absorption is high at the initial stages of impact and as the crushing progresses it reduces gradually. Even though the energy absorption rate is high in the initial stages of impact

The rate of energy absorption is less than 500 kJ/s through out the crushing process. There is not much difference in rate of energy absorption between 7.00 m/s

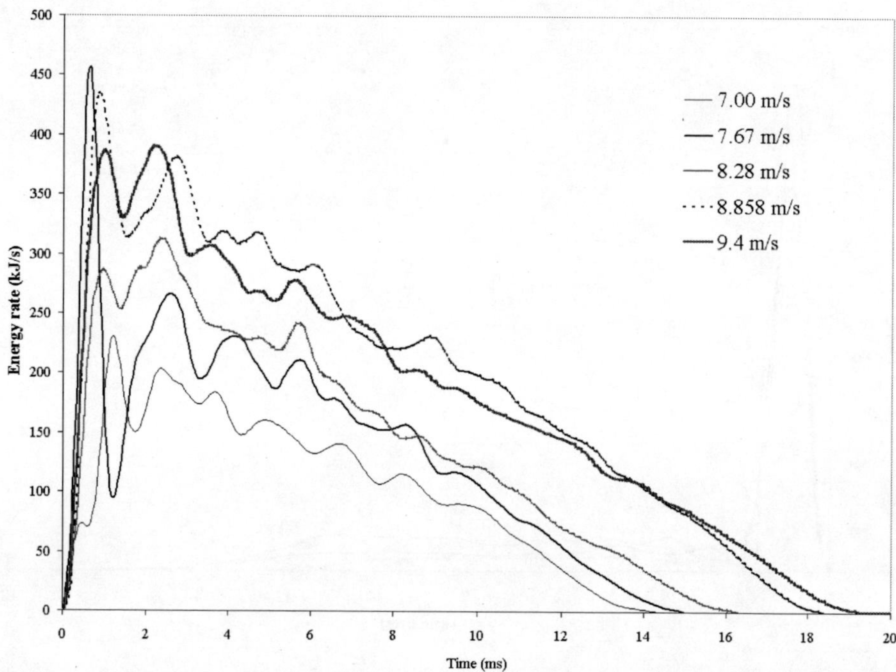

Fig. 11. Energy rate vs time curves of composite tubes subjected to impact loading

and 9.4 m/s. The time taken to complete the entire process is increased with the increase in impact velocity. This is shown in Fig. 11. The rate of energy absorption of dual phase steel tubes is shown in Fig. 12. The energy absorption rate is very high for the initial fold than the subsequent folds. The rate is higher than allowable energy rate, but this can be reduced by suitable trigger mechanism. If one end of the tube chamfered, the initial peak value is reduced to allowable level.

CONCLUSION

Composite tubes and dual phase steel tubes are subjected to axial loading in static and dynamic conditions. From the results, it is observed that the composite tubes absorb more energy in quasi static loading than impact. The tubes exhibit different types of failure modes in impact when compared to quasi static. The specific energy absorption for composite tubes is higher than those of dual phase and annealed steel tubes. The experiments show that the specific energy absorption is higher for dual phase tubes than annealed tubes. The rate of energy absorption is the main criteria in the design of energy absorption systems. The composite tubes energy absorption rate is with in the allowable limit. The dual phase steel tubes energy absorption

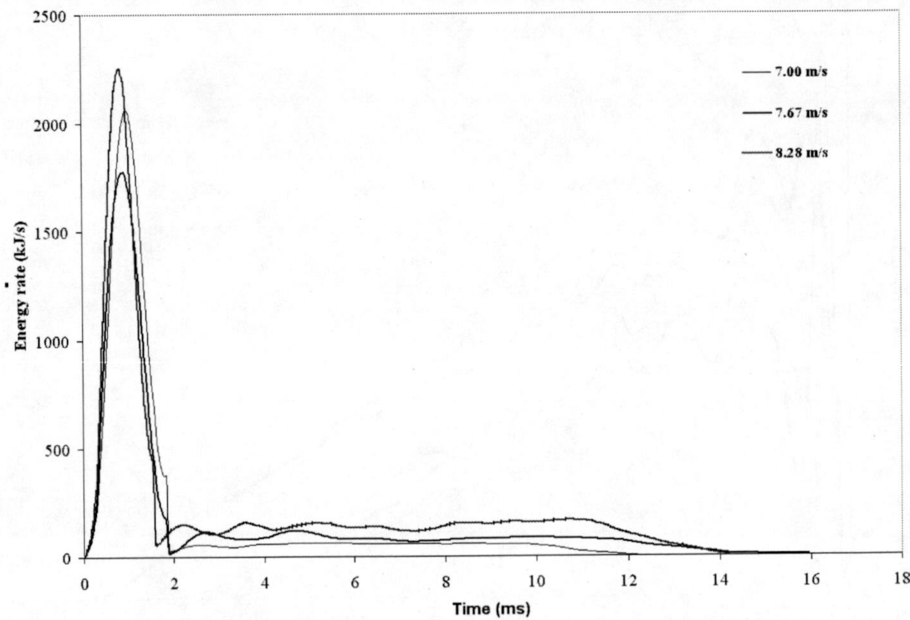

Fig. 12. Energy rate vs time curves of dual phase steel tubes subjected to impact loading

rate has very high variation when compared to composite tubes. In the initial stage of impact the energy absorption rate of dual phase steel tubes is higher than the allowable limit but it can be rectified by suitable triggering mechanisms.

REFERENCES

1. N.K. Gupta, R. Velmurugan, and S.K. Gupta, "An Analysis of Axial Crushing of Composite Tube," J. Compos. Mat. **33** (13), 1262–1286 (1997).
2. A.G. Mamalis, D.E. Manolakos, G.A. Demosthenous, and M.B. Loannidis, "The Static and Dynamic Axial Crumbling of Thin-Walled Fiberglass Composite Square Tubes," Composites. Part B **28B**, 439–441 (1997).
3. N.K. Gupta and R. Velmurugan, "Axial Compression of Empty and Foam Filled Composite Conical Shells," J. Compos. Mat. **33** (6), 567–591 (1999).
4. A.G. Mamalis, D.E. Manolakos, M.B. Ioannidis, and Papapostolou, "On the Response of Thin-Walled CFRP Composite Tubular Components Subjected to Static and Dynamic Axial Compressive Loading: Experimental," Compos. Struct. **69**, 407–420 (2005).
5. D. Hull, "Energy Absorbing Composite Structures," Sci. Tech. Rev., Univ. of Wales **3**, 23–30 (1998).
6. G.C. Jacob, J.F. Fellers, S. Simunovic, J.M. Starbuck, "Energy Absorption in Polymer Composites for Automotive Crashworthiness," J. Compos. Mat. **36** (7), 813–849 (2002).

A MODEL OF DEEP PENETRATING ANCHORS

I. Sharma[1*] and H.E. Huppert[2]

ABSTRACT

We present a simple model for the penetration of deep penetrating anchors and demonstrate how to obtain numerical results for the depth of penetration as a function of the anchor's incidental velocity. Such anchors are potentially important candidate solutions for the problem of supporting floating offshore facilities in deep waters.

INTRODUCTION

Deep penetrating anchors offer a simple solution to anchoring floating offshore facilities and provide an economically viable alternative to fixed offshore platforms. First proposed by Lieng et al. (1999), this anchoring system involves a torpedo shaped anchor (see Fig. 1) that when released from a height above the seabed penetrates into the seabed due its inertia. A major advantage of this anchoring method is the absence of any external energy source. A more detailed description is contained in the above reference.

In order to make use of this anchoring system it is important to be able to predict the embedment depth and also estimate the subsequent holding capacity. Several experimental studies have been carried out in the past, with the most recent ones (O'Loughlin et al., 2003) carried out using scaled down models of the deep penetrating anchor in centrifuge tests (Murff 1996, Schofield 1980). O'Loughlin et al. (2003) also summarize past research in this area, and further indicate the close parallel of this problem with the problem of nuclear waste disposal in seabeds. Concurrently, a finite element approach was employed by Einav et al. (2003) to estimate the embedment depth of such anchors. Here we present a very simple

[1]Department of Mechanical Engineering, Indian Institute of Technology Kanpur, Kanpur — 208016, India

[2]Institute of Theoretical Geophysics, DAMTP, Cambridge University, UK

[*]E-mail: *ishans@iitk.ac.in*

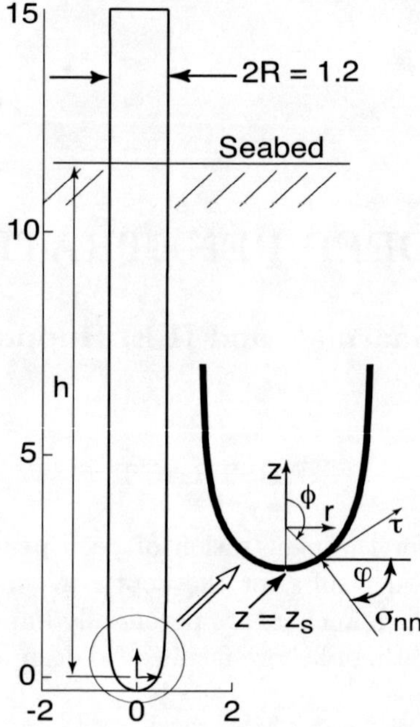

Fig. 1. Schematics of the anchor. Also shown is the coordinate system employed, and the relevant dimensions in SI units

analytical model of the deep penetrating anchor that yields an estimate close to the ones seen by O'Loughlin et al. (2003) in their centrifuge studies and by Einav et al. (2003) in their numerical study.

In passing we mention the analogous problem of deep-sea coring (Skinner et al. 2003). Despite superficial similarity with deep-anchoring problems, the two are quite different. The depth of penetration in deep-coring problems is fixed by the length of the core, and the important question is the effect of the coring process on the seabed. In contrast, the penetration depth is unknown in anchoring problems.

1. MODEL

1.1. *The sea-bed*

It is necessary to model the response of the seabed. O'Loughlin et al. (2003) used Kaolin clay in their experiments. We consider the seabed as water saturated clay, an assumption justified by the fact that most frontiers of offshore hydrocarbon development are on soft saturated clay.

Assuming that the penetration process is fast enough, so that the seabed responds in an undrained manner, we model the seabed as an incompressible rigid-perfectly plastic material. We provide further details below.

1.2. *The anchor shape*

In general, the typical deep penetrating anchor is characterized by a cylindrical shaft capped by a tapered tip. In addition, flukes may be present at the tail of the anchor. Rather than modelling the anchor in detail, we take the view that the most important geometric factor is the anchor's slender shape. Further, to aid simplicity, we disregard the possible presence of flukes. We assume that the exact shape of the tip is not of much importance, as long as the shank of the anchor is approximately cylindrical and slender. Thus, we will characterize the anchor's shape by the radius of the (nearly) cylindrical shank, capped by a curved tip. The exact shape is determined as a part of the calculation below.

For specific evaluation, the diameter ($2R$) and length (L) of the anchor are taken to be 1.2 m and 15 m respectively, as shown in Fig. 1. The density of the anchor's material is 8000 kg/m^3, while its mass is approximately 100 metric tons. These dimensions are typical of deep-penetrating anchors (Ehlers 2004, Lieng et al. 1999 and O'Loughlin et al. 2003). The solution to different values of these parameters can be easily determined.

2. SOLUTION STRATEGY

The deceleration a of the anchor is obtained by a straightforward force balance,

$$a = \dot{U} = g - N/m, \tag{1}$$

where N is the total force exerted by the seabed on the anchor, g the acceleration due to gravity, m the mass, and U the penetration speed.

To successfully use the above equation it is necessary to estimate N. This is usually done via cone-factors that capture the seabed's resistance to penetration (Teh et al. 1991). These cone-factors are in turn obtained by separately solving a cavity expansion problem in a plastic material (Yu 1998). We will instead follow an inverse procedure based on the "Strain Path Method" due to Baligh (1985). The advantages of this alternative approach lie in its relative simplicity because we avoid having to solve a plasticity problem, and also in its potential to be extended to study anchors with different tip shapes and a wider variety of seabeds. Here we limit ourselves to blunt tipped anchors in water-saturated clay seabeds for reasons explained above.

In order to approximate the force N we will require an estimate for stresses induced in the seabed due to the anchor's penetration. These stresses are in turn related to the soil's deformation. Because we subsequently model the seabed as a rigid-perfectly plastic material, the first step is to develop the appropriate constitutive law relating the stresses $\boldsymbol{\sigma}$ during plastic flow to the deformation rates in such a material.

2.1. *Constitutive law*

We model water saturated clay seabeds as an incompressible rigid-perfectly plastic material with a von Mises failure criterion and a non-associated flow rule derived from an appropriate plastic potential. Further details about the plasticity theory employed here may be found in Chen and Han (1988).

We begin with the von Mises yield surface that is given by

$$|\boldsymbol{\sigma}'|^2 = k^2, \tag{2}$$

where $\boldsymbol{\sigma}'$ is the deviatoric part of the stress tensor given by

$$\boldsymbol{\sigma}' = \boldsymbol{\sigma} - \frac{1}{3}(\mathrm{tr}\,\boldsymbol{\sigma})\mathbf{1}, \tag{3}$$

with $\mathbf{1}$ being the identity, while $|\boldsymbol{\sigma}'|^2 = \sigma'_{ij}\sigma'_{ij}$ using the summation convention, and k is a parameter related to the maximum tensile strength Y and the maximum shear stress K by $k = \sqrt{2/3}Y = \sqrt{2}K$. In their experiments, O'Loughlin (2003) measured a value for K that varied approximately linearly with depth. This linear variation with depth $h - z$ (see Fig. 1) in the seabed's maximum shear strength K can be modelled by redefining

$$K = \overline{K}\frac{h-z}{L}, \tag{4}$$

where $\overline{K}$ has the same dimensions as K and is given by

$$\overline{K} = 0.96L,$$

with 0.96 being the average gradient of the shear strength with depth, as reported by O'Loughlin et al. (2003). The above gives a value of 14.4 kPa for $\overline{K}$. The parameter k now becomes a function of z given by

$$k = \sqrt{2}K = \sqrt{2}\,\overline{K}\frac{h-z}{L}. \tag{5}$$

Next, we employ the plastic potential (see Chen and Han 1988)

$$g = \frac{1}{3}\mathrm{I}_\sigma^2 - \mathrm{II}_\sigma,$$

where

$$\mathrm{I}_\sigma = \mathrm{tr}\ \boldsymbol{\sigma} \quad \text{and}$$
$$\mathrm{II}_\sigma = \frac{1}{2}\left(\mathrm{I}_\sigma^2 - \mathrm{I}_{\sigma^2}\right),$$

are the first and second stress invariants. The above plastic potential is employed because it preserves volume, which is a good approximation for water saturated clays. From it we obtain the flow rule

$$d\varepsilon = \boldsymbol{\sigma}' dq,$$

where $d\varepsilon$ is the incremental strain. This is equivalent to, after converting to a rate form,

$$\boldsymbol{D} = \dot{q}\boldsymbol{\sigma}'. \tag{6}$$

where $\boldsymbol{D}$ is the symmetric part of the strain rate tensor that captures the stretching rates, and $\dot{q}$ is a constant. Combining (6) with the failure criterion (2), we obtain

$$\dot{q} = \frac{1}{k}\,|\boldsymbol{D}|,$$

where $|\boldsymbol{D}|^2 = D_{ij}D_{ij}$ as before. Using the above in (6), we obtain the plastic shear stress in terms of the strain rate

$$\boldsymbol{\sigma}' = k\frac{\boldsymbol{D}}{|\boldsymbol{D}|}, \tag{7}$$

which when combined with (3) yields the plastic stresses

$$\boldsymbol{\sigma} = -p\boldsymbol{1} + k\frac{\boldsymbol{D}}{|\boldsymbol{D}|}, \tag{8}$$

where p is the pressure required to maintain a constant volume, and is given by

$$p = -\frac{1}{3}\mathrm{tr}\ \boldsymbol{\sigma}. \tag{9}$$

Due to axi-symmetric conditions

$$[\boldsymbol{D}] = \begin{pmatrix} D_{rr} & 0 & D_{rz} \\ 0 & D_{\theta\theta} & 0 \\ D_{rz} & 0 & D_{zz} \end{pmatrix}, \tag{10}$$

where, because volume is preserved

$$D_{rr} + D_{\theta\theta} + D_{zz} = 0. \tag{11}$$

Utilizing (10) and (11) in (8), we obtain the following explicit relations for the plastic stress in axi-symmetric conditions

$$\sigma_{rr} = -p + k\frac{D_{rr}}{|\boldsymbol{D}|}, \tag{12}$$

$$\sigma_{\theta\theta} = -p + k\frac{D_{\theta\theta}}{|\boldsymbol{D}|}, \tag{13}$$

$$\sigma_{zz} = -p + k\frac{D_{zz}}{|\boldsymbol{D}|}, \tag{14}$$

$$\text{and} \quad \sigma_{rz} = k\frac{D_{rz}}{|\boldsymbol{D}|}, \tag{15}$$

with the remaining stresses being zero.

From the above we note that if we knew the plastic strains, the plastic stresses could be easily calculated, and a knowledge of the plastic stresses would immediately yield the forces acting on the anchor. In general, in order to calculate the plastic stresses we have to solve a plasticity problem, which is a rather complicated process, except in the simplest of shapes. We will, as mentioned earlier, follow an inverse approach, where we start by assuming a plastic flow pattern and recover the stresses from the above equations. This strategy was shown by Baligh (1985) to yield good results in problems involving deep penetration of slender bodies in soils.

2.2. *Plastic law*

Baligh (1985) showed that the addition of a constant velocity term to the velocity field of a growing spherical cavity gave rise to a velocity field that captured essential features of the material flow around a penetrating slender body. Using this velocity field it is a straightforward matter to derive the following plastic strain rates

$$D_{rr} = \frac{UR^2}{4\rho^3}\left(\cos^2\phi - 2\sin^2\phi\right), \tag{16}$$

$$D_{\theta\theta} = \frac{UR^2}{4\rho^3}, \tag{17}$$

$$D_{zz} = \frac{UR^2}{4\rho^3}\left(\sin^2\phi - 2\cos^2\phi\right), \tag{18}$$

and

$$D_{rz} = -\frac{3UR^2}{4\rho^3}\sin\phi\cos\phi, \tag{19}$$

where R is the radius of the shank far away from the tip (see Fig. 1), U is the penetration speed, $\phi = \tan^{-1}(r/z)$ and $\rho = \sqrt{r^2 + z^2}$.

As Baligh (1985) showed, the above plastic flow models well the incompressible flow around a penetrating slender body whose shape is given by the

$$z = G(r) = r\left[\frac{1}{(2r^2/R^2-1)^2} - 1\right]^{-1/2}, \tag{20}$$

corresponding to the zero velocity stream line, and is shown in Fig. 1. We denote F to be the inverse of G, i.e. $r = F(z)$, for future reference. We note that once R is chosen, the anchor's shape, including its tip, is completely specified. However, as mentioned earlier, the shape of the tip should not be very significant. It is also to be pointed out that because the velocity field assumed by Baligh (1985) to model the flow around a penetrating slender body differs from that due to a growing spherical cavity by a constant term, the above plastic strain rates for the two cases are, in fact, the same. Thus, the resulting plastic stresses will also be the same.

2.3. *Stresses*

Using (16)–(19) in (12)–(15), we obtain the stresses in terms of the pressure p

$$\sigma_{rr} = -p + k\sqrt{\frac{8}{3}}\frac{\cos^2\phi - 2\sin^2\phi}{\sqrt{13+3\cos 4\phi}}, \tag{21}$$

$$\sigma_{\theta\theta} = -p + k\sqrt{\frac{8}{3}}\frac{1}{\sqrt{13+3\cos 4\phi}}, \tag{22}$$

$$\sigma_{zz} = -p + k\sqrt{\frac{8}{3}}\frac{\sin^2\phi - 2\cos^2\phi}{\sqrt{13+3\cos 4\phi}} \tag{23}$$

and

$$\sigma_{rz} = -k\sqrt{\frac{8}{3}}\frac{3\sin\phi\cos\phi}{\sqrt{13+3\cos 4\phi}}. \tag{24}$$

Once the pressure p is known, so too are the stresses, and these may then be used to obtain the force acting on the anchor as we do below.

2.4. *Pressure*

Eqs. (21)–(23) express the stress in the seabed in terms of the pressure p, and it remains to obtain an equation for the latter. In addition to the pressure due to plastic flow, there will be stresses due to the soil's self weight. We assume these stresses are hydrostatic. There are no dynamical effect of the weight of the overlying waves, which is reflected in the fact that we set the pressure on the surface of the seabed to zero.

In order to obtain the pressure p we appeal to the balance of linear momentum in the r direction, while neglecting inertial effects,

$$\frac{\partial \sigma_{rr}}{\partial r} + \frac{\partial \sigma_{rz}}{\partial z} + \frac{\sigma_{rr} - \sigma_{\theta\theta}}{r} = 0, \tag{25}$$

and use the expressions for the stresses from (21)–(24), to obtain, after simplification, the partial differential equation

$$\frac{\partial p}{\partial r} = -\frac{2\sqrt{6}\sin\phi\cos\phi}{\sqrt{13+3\cos\phi}}\frac{\partial k}{\partial z} - \frac{k}{r}\frac{2\sqrt{6}(13 - 15\cos 2\phi + 3\cos 4\phi - \cos 6\phi)}{(13+3\cos 4\phi)^{3/2}}, \tag{26}$$

where the first term picks up the contribution due to linear variation in K. The above equation for p can be integrated along constant z-lines from $r = F(z)$ to $r = \infty$ to yield the pressure on the anchor's surface when the surface is at a height h (see Fig. 1)

$$\begin{aligned} p|_{r=F(z)} &= \rho_S g(h-z) - \overline{K}\frac{1}{L}\int_{F(z)}^{\infty} \frac{4\sqrt{3}\sin\phi\cos\phi}{\sqrt{13+3\cos\phi}}dr \\ &\quad + \overline{K}\frac{h-z}{L}\int_{F(z)}^{\infty} \frac{4\sqrt{3}(13 - 15\cos 2\phi + 3\cos 4\phi - \cos 6\phi)}{r(13+3\cos 4\phi)^{3/2}}dr, \end{aligned} \tag{27}$$

where the far-field pressure is taken to be due to the soil's own weight alone, so that $p(r = \infty) = \rho_S g(h - z)$, with ρ_S being the density of the soil. In the above equation the middle term is due to the variation of the shear strength, and we have substituted for k from (5). The above equation provides an estimate of the pressure in the seabed during the anchor's penetration.

It is seen that the pressure becomes negative at a certain length along the anchor's shank. This is an indication of the fact that because the seabed's strength decreases near the surface, it is not able to constrain the plastic flow into a narrow zone near the anchor's surface as is required by our assumed deformation field. We will continue by further assuming that because soils are usually unable to support negative pressures, the material above a certain length of the anchor cavitates, and we will neglect any force interaction with the soil above this length, designated by h^*.

A natural question to ask is whether the pressure obtained from the above satisfies linear momentum in the z direction. The answer to this is, unfortunately, no, because our assumed plastic strain rates are only approximate. However, it is possible to improve the pressure's estimate by an iterative strategy that is outlined in Baligh (1985).

In passing we note that changes in pore pressure are captured automatically by our employing a von Mises failure criterion to characterize the soil's strength, because the undrained strength includes dynamic pore pressure effects.

2.5. *Force*

From Fig. 1 we observe that the vertical traction T at any point on the anchor's surface may be expressed in terms of the normal (σ_{nn}) and tangential (τ) tractions at that point by the equation

$$T = \sigma_{nn} \sin \varphi + \tau \cos \varphi. \tag{28}$$

Baligh's (1985) approximation of the plastic strain rate was initially derived for a body with a smooth surface, yielding $\tau \equiv 0$. However, as he suggests, deep penetration problems tend to be strain controlled, so that it is plausible to assume that while that normal traction can be estimated using the plastic stresses derived above, the tangential traction on the anchor's surface is given in terms of the maximum shear strength K by

$$\tau = \alpha K, \tag{29}$$

where α, the adhesion factor, will initially be taken to be one following Einav et al. (2003). Thus, (28) becomes

$$T = \sigma_{nn} \sin \varphi + K \cos \varphi, \tag{30}$$

which, in polar coordinates is

$$T = \frac{F'(2\sigma_{rz}F' - \sigma_{rr} - \sigma_{zz}F'^2)}{(1+F'^2)^{3/2}} + K\frac{1}{\sqrt{1+F'^2}}. \tag{31}$$

The total vertical force on the anchor is obtained by integrating the traction given by (31) over the anchor's surface to obtain

$$\begin{aligned} N &= \int T dA \\ &= 2\pi \int_{z_s}^{h^*} \left\{ \overline{K}\frac{h-z}{L}\frac{1}{\sqrt{1+F'^2}} + \frac{F'(2\sigma_{rz}F' - \sigma_{rr} - \sigma_{zz}F'^2)}{(1+F'^2)^{3/2}} \right\} F\, dz, \end{aligned} \tag{32}$$

where $z = h^*$ is, as noted above, the distance along the anchor where the pressure first becomes negative, $z = z_s$ locates the anchor's tip (see Fig. 1), and we have substituted for K from (4). The above equation yields the total vertical force acting on the anchor at any moment of time as a function of the penetration depth h.

3. SOLUTION

With the total force N given by (32) we can integrate the deceleration equation (1) till the anchor comes to a stop, yielding the final depth of penetration. The results for different values of initial impact velocities U_0 are shown in Figs. 2 and 3.

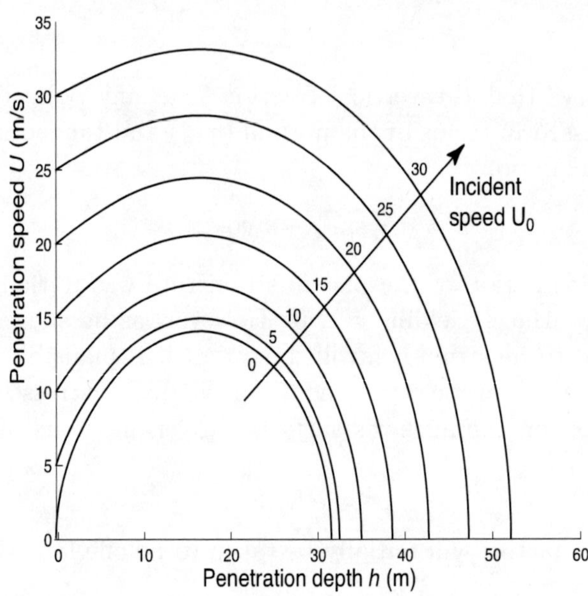

Fig. 2. Penetration speed as a function of the depth of penetration. The arrow indicates curves of increasing incident speed U_0, with the numbers next to it indicating U_0's magnitude in m/s.

4. DISCUSSION

Fig. 2 shows the variation of penetration speed U with the anchor's penetration depth h for different incident (impact) speeds U_0. They match the corresponding results of Einav et al. (2003) well qualitatively. Some quantitative difference is to be expected because their results were for anchors with flukes (fins), which tend to stop at lower penetration depths compared to anchors without flukes. From Fig. 2, we note that the anchor's penetration speed increases initially due to gravity and a relatively low seabed resistance. As the anchor penetrates further, the seabed's resistance increases both due to an increase in shear strength and in the pressure due to the soil's weight, leading to the anchor's speed reaching a maximum, after which the anchor decelerates to a stop at a penetration depth h_f. Note that the penetration achieved depends on the impact speed U_0.

Fig. 3 meanwhile plots the final penetration depth h_f achieved as a function of the incident speed U_0. As expected, the depth of penetration increases with increasing impact speed. Also shown in the figure is the experimental data of O'Loughlin et al. (2003). Comparing with the experimental data, we see that we have managed to capture the trend of the experimental data very well. However, we over-predict

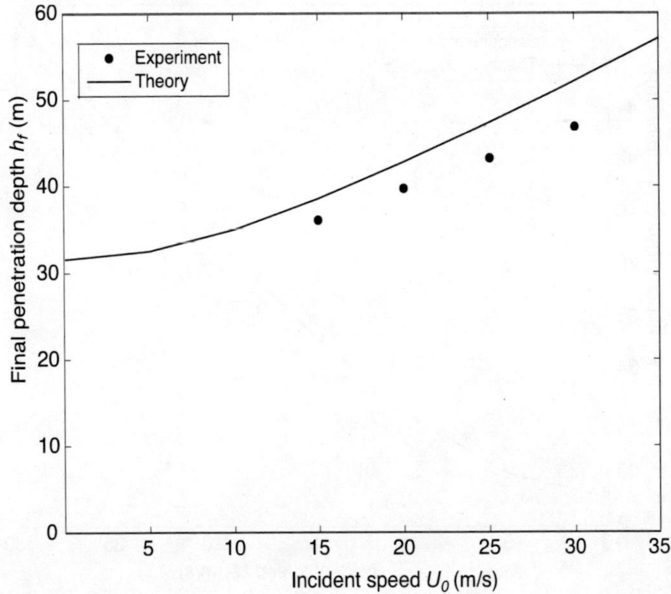

Fig. 3. Final penetration depth h_f as a function of the incident speed U_0.

the penetration depth which may be explained by reasons discussed in the following section. Nevertheless, the matching achieved is very encouraging considering the simplicity of the theory.

5. RATE DEPENDENCE

In the above we have used an adhesion factor α (see Eq. (29)) of one. O'Loughlin et al. (2004) suggest employing an adhesion factor of about 0.4, in accordance with their experiments. This would increase our predicted penetration depths even further. However, O'Loughlin et al. (2004) further note that the seabed's shear strength increases with the rate of straining (Mitchell and Soga 2005), and suggest capturing this dependence by augmenting the shear strength K by the factor

$$R_f = 1 + \lambda \log \frac{U}{U_s}, \tag{33}$$

where U_s is a reference penetration speed representative of undrained conditions taken by them to be 3 mm/s, and λ is a constant taken to be 0.24, indicating a 24% increase in the seabed's strength per log cycle. Our choice of λ is consistent with those of O'Loughlin et al. (2004).

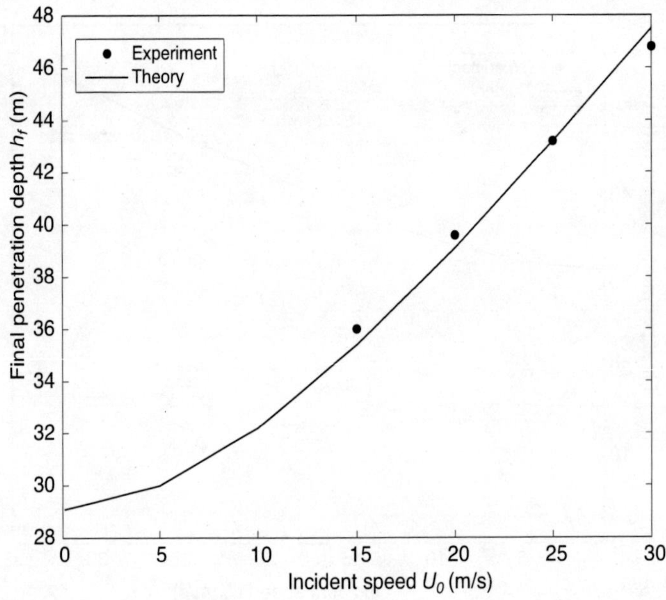

Fig. 4. Final penetration depth h_f as a function of the incident speed U_0. Effects of strain rate on the seabed's shear strength from (33) are included; cf. Fig. 3

By setting $\alpha = 0.4$ in (29), and employing $R_f K$ for K throughout the previous calculations, we recalculate the final penetration depth as a function of incident speed. The results are shown in Fig. 4. We see that the match between our theory and the experiment of O'Loughlinhas et al. (2004) been improved considerably, while also modelling the seabed's response better.

CONCLUSION

In order to obtain a more sophisticated theory, several avenues are available. First, this is a zero-order theory, in the sense that the stresses are the same as the ones due to the expansion of a spherical cavity. Thus, it may be worthwhile to explore the effects due to the anchor's slender shape on the stresses. It is also possible to explore the effects of flukes and different tip shapes under the present theory. This can be done by noting that flukes predominantly provide extra surfaces of frictional interaction. Different tip shapes may also be modelled by superposing several deformational fields, as is shown in the context of piles by Levedoux et al. (1980).

Next, we employed a deformation field that, while a good assumption for deep penetration in uniform solids, may not be a good approximation in solids with varying strength profiles. It is very possible that moving away from anchor's tip, the zone within which there are large deformations in the seabed increases, as the seabed becomes weaker towards the surface. This is not captured by the presently assumed deformation field that predicts a uniformly thin zone of large deformation along the anchor's length.

In the meantime, we have derived a new zero-order model which is amenable to simple interpretation and straighforward numerical evaluation. We envisage that the present approach will find further important applications in the future.

6. ACKNOWLEDGMENTS

We would like to thank and M.D. Bolton and D.J. White at the Schofield Centre, J.T. Jenkins at TAM, Cornell, and E.J. Hinch and J.R. Willis at DAMTP for helpful discussions.

REFERENCES

1. M.M. Baligh, "Strain Path Method," J. Geotech. Engng **111**, 1108–1136 (1985).
2. W. Chen and D. Han, *Plasticity for Structural Engineers* (Springer-Verlag, New York, 1988).
3. C.J. Ehlers, A.G. Young, and J.H. Chen, "Technology Assessment of Deepwater Anchors," in *Proceedings of 2004 Offshore Technology Conference, Houston, USA* (Houston, 2004), paper No. 16840.
4. I. Einav, A. Klar, C. O'Loughlin, and M. Randolph, "Numerical Modelling of Deep Penetrating Anchors," in *9th Australian and New Zealand Conference on Geomechanics, Auckland, New Zealand, University of Auckland*, Vol. 2 (2003), pp. 591–597.
5. J.N. Levadoux and M.M. Baligh, "Pore Pressure during Cone Penetration," in *Research Report R80-15 for Order No. 666* (Dept. of Civil Engineering MIT, Mass. USA, 1980).
6. J.T. Lieng, F. Hove, and T.I. Tjelta, "Deep Penetrating Anchor: Subseabed Deepwater Anchor Concept for Floaters and Other Installations," in *The Proceedings of the 9th (1999) International Offshore and Polar Engineering Conference* (Brest, France, 1999).
7. J.K. Mitchell and K. Soga. *Fundamentals of Soil Behavior.* 3rd ed. (John Wiley and Sons, New York, 1960).
8. J.D. Murff, "The Geotechnical Centrifuge in Offshore Engineering," in *Proc. Offshore Technology Conf., Houston* (Houston, 1996), OTC 8265.
9. C. O'Loughlin, M. Randolph, and I. Einav, "Physical Modelling of Deep Penetrating Anchors," in *9th ANZ Conference on Geomechanics, Auckland, 8–11 Feb 2004* (2003).
10. O'Loughlin, C., M. Randolph, and M. Richardson, "Experimental and Theoretical Studies of Deep Penetrating Anchors," in *Proceedings of 2004 Offshore Technology Conference, Houston, USA* (Houston, 2004), paper No. 16841.

11. A.N. Schofield, "Cambridge Geotechincal Centrifuge Operations," Geotechnique **30**, 227–268 (1980).
12. L.C. Skinner and I.N. McCave, "Analysis and Modelling of Gravity- and Piston Coring Based on Soil Mechanics," Marine Geology **199**, 181–204 (2003).
13. C.I. Teh and G.T. Houslby, "An Analytical Study of the Cone Penetrometer Test in Clay," Geotechnique **41**, 17–34 (1991).
14. H.S. Yu, *Cavity Expansion Methods in Geomechanics*, 2nd ed. (Springer, New York, 2000).

EFFECTS OF LOADING RATE ON THE DYNAMIC RESPONSE OF SHAPE MEMORY ALLOYS

D.R. Mahapatra[1*], K. Dasharathi[1], and V. Sutrakar[1]

ABSTRACT

A comprehensive multiscale framework is constructed in this paper to simulate the dynamic response of Shape Memory Alloy (SMA) materials due to various loading rates and for various sample dimensions. The approach employs the ideas of thermodynamic dissipation and phase transforming microstructure to formulate the evolution kinetics and estimate the associated scaling parameters. A detailed treatment is given to energy of surface and interfaces of crystalline grains. A macroscopic constitutive model is considered, which uses various material constants prescribed in a phase transformation diagram approach. Microscopic phases are expressed by volume average phase fraction. Based on this continuum description and first-order kinetics, we show the effect of stress rate on the hysteresis loop in NiTi wire under tensile loading-unloading.

Key words: shape memory alloy, free energy, finite element, phase transformation, strain rate, size effect

INTRODUCTION

Shape Memory Alloys (SMAs) are an interesting class of functional materials capable of recovering large inelastic deformations [1]. Quasi-static response of SMA is categorized into two, namely, Shape Memory effect (SME) and pseudo-elastic effect. For rate dependent loading, the adiabatic conditions lead to combined and more complicated effects, which are very little understood. Dynamic response of SMA is an important and relatively new areas of research, given the fact that SMAs have very good impact damping properties. The impact damping properties of the SMA arise

[1]Department of Aerospace Engineering, Indian Institute of Science, Bangalore — 560012, India
[*]E-mail: *droymahapatra@aero.iisc.ernet.in*

due the energy dissipation resulting from its hysteretic behavior. Previous studies have established that the hysteretic behavior of SMA is rate-dependent [2, 3]. Thus, a systematic study of the dynamic response of the SMAs under impact loads can facilitate the understanding required to successfully develop an efficient SMA based damper. Also, the effect of localized phase transformation under impact loading, the associated problems in training of SMA dampers for optimal damping, effect of loading rate on fatigue and post-impact re-cycling of SMA dampers are very little understood till date.

In this paper, we first review the constitutive modeling approaches and their applications in context of dynamic loading of SMA wires. Next we review the existing approaches to link the macroscopic dynamic response to microscopic and microstructural responses and the role of certain spatio-temporal scales of microstructure on the evolution of martensites due to rate-dependent loading. To this end, we formulate a Gibbs free energy based framework wherein the material constants are derived from atomistic system using embedded atom potential. Effect of strain rate loading on the stress-strain response of pure phase and mixed phases in SMA materials at nano-scale are reported. Effect of microstructural scales are quantified analytically based on the free energy formulation.

0.1. *Impact dynamic response of SMA*

Several researchers have reported experimental investigations of dynamic behavior of SMA wires and also there have been numerous attempts to predict the SMA behavior under impact or dynamic loads using phenomenology based models. Results based on lumped parameter model clearly indicate that the damping effect varies as a function of area of the hysteresis loop. Also, it was observed that the dynamic loading of SMA causes significant changes in its temperature, which in turn affects the SMA transformation behavior. Propagation of phase transformation front or detwinning shock waves through an SMA body subjected to a dynamic or impact loading has been analyzed in ref. [3]. In this study, a rate-independent constitutive model was employed, which was calibrated using quasi-static tests. With this, numerical simulations of shock loading were performed using finite element method.

Empirical findings reveal that the occurrence of stress and thermal gradients over the SMA wire severely affects the performance, reliability and life of the SMA based devices. In impact damping application, it is important to understand the ways one can utilize the hysteresis loop optimally and also how one can re-train the materials following a complete annealing. Another important aspect is that in SMA based vibration dampers in particular, the SMA element is subjected to cyclical loads. It is then imperative to examine the fatigue behavior of the SMA wire. Valuable experimental contribution towards understanding the fatigue behavior of the SMA components has been made in ref. [4]. The fatigue behavior of SMAs can be divided boradly into 'structural fatigue' and 'functional fatigue'. Structural fatigue refers to the degradation failure of the SMA components under cyclical loads due to the accumulation of the microstructural damages. Failure of SMA components due to structural fatigue is synonymous to the failure of any common engineering material. Functional fatigue refers to the degradation of the exploitable functional properties such as maximum recoverable strain and dissipated energy. Results reported in ref. [4] clearly show that the microstructure plays an important role in the hysteretic behavior. Also, fatigue studies reported in ref. [5] on polycrystalline SMA specimens establish the significance of crystallographic orientation (texture) on the functional fatigue of the SMA components.

0.2. *Modeling constitutive behavior of SMA*

SMA materials exhibit non-linear and non-smooth behavior, which is due to changes in the lattice structure. There are a number of mathematical models available in this context to study the behavior of SMA materials. Depending on the extent of physical details, these models can be categorized as (1) phenomenology based models using the generalized framework of plasticity [6–9], (2) models based on the theory of compatible microstructures [10–13] which deal with the quasi-static pattern of the austenite-martensite bands and (3) phase field models based on the theory of first-order kinetics [14–16]. In the phenomenological models of SMAs, a polycrystalline microstructure is represented by an effective volume fraction of phases. One of such phenomenological model was proposed by Brinson [17]. This model builds upon an earlier phenomenological model proposed by Liang [18]. In Liang's model, the internal variable for martensite volume fraction represents the percentage of material transformed into martensite, and does not distinguish between multiple variants generated during various types of thermo-mechanical loading. Based on the micromechanics of SMA, Brinson's model [17] extends the phase fraction definition by introducing a temperature induced martensite and a stress induced martensite. However, this model accommodates for the presence of multiple martensite variants only in a spatially averaged phase faction. Thus, there is currently a limitation in such a modeling approach based on phenomenology of pseudoelasticity and shape memory effect (SME), where the true phase inhomogeneity is not being considered. In order to analyze the response of a microstructure under dynamic loadings, it is essential to include the contribution of phase inhomogeneity and their evolution. From crystallographic standpoint, there may be three or more variants of martensite. In a quasi-static sense, a detailed description of a transformed texture within a polycrystalline domain has been reported by Siredey et al. [9] and later by Niclaeys et al. [19] and several others for multivariant interactions in three dimensions.

In the context of engineering application, it is desirous to have fast computation without compromising the important physics information. A coupled thermomechanical model for SMAs for an idealized single crystal is reported in ref. [20]. They propose to capture the inhomogeneities in the polycrystalline SMA using a 'representative single crystal' wherein the average strain within the control volume would be determined by statistical homogenization. However, the constitutive model they employ (see ref. [21]) relies on empirical parameters such as width of the hysteresis. Thus, it is effective in capturing the behavior of SMA rather than 'predicting' it. Also, the prediction of the variation in the hysteretic behavior during cyclical loading has not been reported till date in published literature. More intensive computation based on compatible microstructure and Gibbs free energy based variational framework are available (see ref. [22]). However, a device-scale simulation is not straightforward. Plasticity effects and degradation of the shape memory behavior during cyclical loading are very little understood. A one-dimensional model has been reported in ref. [23], which is capable of simulating the hysteretic behavior of the SMA wire under cyclical loading. It considers training of a single crystal SMA wire. In such an approach, the austenite-martensite transformation is modeled using exponential functions, which are based on the transformation stress thresholds of untrained and fully trained specimen. This model faithfully follows the experimental results. However, predicting the polycrystalline material behavior under different boundary conditions requires more substantial investigations. An attempt has been made in this paper to simulate the behavior of inhomogeneous SMA wires under dynamic loading.

The behavior of SMA with microstructural inhomogeneities under dynamic loading will be drastically different compared to the quasi-static behavior. In order to incorporate the microstructural inhomogeneity and to study related effects, a finite element model of SMA wire has been developed in this paper. In the past few years, there have been several efforts towards development of a finite element model [24–31]. However, in these models the microstructural inhomogeneity effects have not been incorporated and the impact loading aspects have not been addressed.

Another important reason for the difference in the dynamic and quasi-static behavior of the SMA is the rate-dependent phase evolution and its propagation. These aspects have not been examined in details in published literature. Classical rate-independent constitutive models can capture the SMA behavior under impact loads with limited accuracy. One way to resolve this problem is to employ kinetic relations which govern the evolution of phases as a function of time. Such refinements and physical details now appear to be the key towards understanding the defects in SMA and successfully designing the SMA integrated systems.

The evolution kinetics relates the rate of formation of martensite volume fraction to the various driving forces, which in turn depends upon the dissipation of free energy and non-local effects. Attempts have been made earlier by several researchers to define the kinetics of SMA in a phenomenological sense [17], however they have limited success and require tuning of several parameters in the kinetics. Since the evolution of martensite in the SMA depends on the interplay between the thermo-mechanical force and the interface driving force, it is essential to study the relative influence of these forces. The importance of the proposed formulation lies in the introduction of two scaling parameters in the kinetics and systematically identifying them for investigating the effect of microstructural variation on the driving forces and consequently on the evolution of martensite volume fraction. This paper reports a new formulation on the kinetics by considering microstructural inhomogeneity, which is based on Gibbs free energy formalism. It is discussed how such kinetics can be implemented while employing a phenomenological constitutive model. Before we proceed with the issues in kinetics, first the thermodynamics framework is discussed next.

1. GIBBS FREE ENERGY BASED THERMODYNAMIC FRAMEWORK FOR MICROSTRUCTURAL MOTION

In this section we review the Landau-Ginzburg theory in context of first-order phase evolution applicable to SMAs [14–16, 32]. The Gibbs free energy density G for multivariant phase transformation can be expressed as

$$G(\eta) = -\frac{1}{2}\boldsymbol{\sigma}^T\left[\mathbf{S}^{(0)} + \sum_{k=1}^{N}(\mathbf{S}^{(k)} - \mathbf{S}^{(0)})\Phi(\eta_k)\right]\boldsymbol{\sigma} - \boldsymbol{\sigma}^T\sum_{k=1}^{N}\boldsymbol{\varepsilon}^{t,(k)}\Phi(\eta_k) + \sum_{k=1}^{N}f(T,k)$$
$$- \boldsymbol{\sigma}^T\left[\boldsymbol{\varepsilon}_T^{(0)} + \sum_{k=1}^{N}(\boldsymbol{\varepsilon}_T^{(k)} - \boldsymbol{\varepsilon}_T^{(0)})\Phi(\eta_k)\right] + \sum_{i=1}^{N-1}\sum_{j=i+1}^{N}F_{ij}(\eta_i,\eta_j), \tag{1}$$

where $\boldsymbol{\sigma}$ is the stress, $\mathbf{S}$ is the second order fourth-rank compliance tensor, superscript (0) indicates Austenite (A) phase, superscript (k) indicates kth variants of martensite phase, $\boldsymbol{\varepsilon}_T^0 = \alpha_0(T - T_k)$, $\boldsymbol{\varepsilon}_T^k = \alpha_k(T - T_e)$, T is the temperature, T_e is the equilibrium temperature, α_0 and $\alpha_k$are the thermal expansion co-efficient tensors for A and M_k phases, f is the chemical part of the free energy. $\boldsymbol{\varepsilon}^{t,(k)}$ denotes the transformation stress tensor for $A \leftrightarrow M_k$ transformation and it can be obtained by relating the lattice constants before and after transformation. Φ is a function which introduces the contribution of the evolved phases to the material constants (see ref. [22] for details). For a variant-variant interaction, F_{ij} denotes an interaction potential required to preserve the frame-invariance of $G(\eta)$ with respect to the point group of symmetry and uniqueness of the multivariant phase transformation at a given material point. The transformation energy associated with $A \leftrightarrow M_k$

transformation is given by

$$G(\boldsymbol{\sigma}, T, \bar{0}) - G(\boldsymbol{\sigma}, T, \bar{1}) = \boldsymbol{\sigma}^T \boldsymbol{\varepsilon}^{t,(k)} - \Delta G^T, \tag{2}$$

where ΔG^T is the transformation enthalpy involving changes in the chemical part of the free energy alone. The consequence of this jump as well as the jump in the strain across the $A - M_j$ interface is a thermodynamic driving force, which has to be dissipated. The forcing term would eventually be balanced by the kinetic force causing evolution or self-accommodation of the variants. In the context of first-order kinetics, the variational formulation reported in ref. [33] shows that the non-equilibrium thermodynamics leads to a framework known as time-dependent Ginzburg-Landau phase kinetics. At this point, it is worth mentioning the consistency conditions that can be naturally derived following the Clausius-Duhem inequality [34] and employed in the light of dissipation in shape memory alloy phase transformation as shown in refs. [35].

For usual representation of the balance of energy, balance of mass, balance of momentum and entropy and the Clausius-Duhem inequality, the consistency conditions lead to the following:

$$\varepsilon = -\frac{\partial G}{\partial \sigma}, \quad E = -\frac{\partial G}{\partial T}, \quad \Gamma + \frac{\partial G}{\partial \eta_k}\frac{\partial \eta_k}{\partial t} = 0,$$

where E denotes the entropy and Γ denotes the rate of energy dissipation density. The first two conditions are automatically satisfied due to the present choice of Gibbs free energy and transformation energy (see Eqs. (6) and (2)). In the third condition, since $\Gamma \geq 0$ according to the Clausius-Duhem inequality, the terms $\partial G/\partial \eta_k$ and $\partial \eta_k/\partial t$ must be of opposite signs. Hence, by balancing the thermodynamic force with the kinetic force, one can write

$$C' \frac{\partial \eta_k}{\partial t} + \frac{\partial G'}{\partial \eta_k} = 0, \tag{3}$$

where C' is a constant and G' is the modified Gibbs free energy including the gradient terms to account for the non-local nature of the interface energy or the grain boundary energy and the free surface energy. The displacements are related to the strain $\boldsymbol{\varepsilon}$ via the standard inelastic strain displacement relation

$$\varepsilon = \varepsilon_{\text{el}} + \varepsilon_{\text{transformation}} + \varepsilon_{\text{thermal}}, \tag{4}$$

while the stress is given by $\boldsymbol{\sigma} = \mathbf{C}(\eta)\varepsilon_{\text{el}}$ with $\mathbf{C}(\eta)$ as the inhomogeneous stiffness tensor for the single grain. The true displacement in the SMA is obtained by integrating the total strain $\boldsymbol{\varepsilon}$. For the single grain with evolving phases, one can write the stiffness tensor as

$$C(\eta_k) = \left[S^{(0)} + \sum_{k=1}^{N} (S^{(k)} - S^{(0)}) \Phi(\eta_k) \right]^{-1}, \tag{5}$$

where the compliance tensors $\mathbf{S}$ are either assumed or estimated from experimental data and the structure of $\Phi(\eta_k)$ should be known (see ref. [22]). However, effects of temperature, rate of transformation and non-equilibrium nature of the thermodynamics during crystal transformation are difficult to accommodate analytically or by fitting experimental data. Most often, obtaining experimental data for such constructions is very difficult. As we will see later in this paper that one can resort to molecular dynamic simulations to obtain such details.

For polycrystalline microstructure with grain boundaries, one can rotate the crystallographic axis for defining the stress tensor in Eq. (5). For single crystalline grain, the stiffness tensor and other material constants are strongly dependent on the nucleating phases, temperature and strain rate. These can be obtained by minimizing a suitable atomic potential for a block of atoms undergoing displacive phase transformation. A detailed analysis will be done later. We first proceed with the analysis of various constants related to the kinetics of microstructure due to Eq. (3).

By careful observation of Eq. (6), we now express the evolution kinetics in a simpler form:

$$\dot{\xi} = A\frac{\partial G}{\partial \xi} + B\nabla^2 \xi, \tag{6}$$

where the overhead dot denotes the time derivative in a realistic time scale, ξ is the grain average of η_k distribution, G is the Gibbs free energy density in the context of the one-dimensional SMA wire, A and B are two scaling parameters which are introduced here in order to investigate the effect of microstructural on the driving forces and the competition between grain boundary/interface energy and the strain energy.

1.1. *Microscopically homogenized constitutive model*

We consider an analytical method to characterize the scaling parameters (A, B) in the kinetics in Eq. (6) although in a reduced one-dimensional form. To start with, we consider a one-dimensional constitutive model of SMA wire [18]. The differential form of the one-dimensional constitutive model is given by

$$d\sigma = D(\xi)\, d\varepsilon - \varepsilon_l D(\xi)\, d\xi_S + (\varepsilon - \varepsilon_l \xi_S)(D_m - D_a)\, d\xi + \Theta\, dT. \tag{7}$$

The integrated form of the same constitutive model is given by

$$\sigma - \sigma_0 = D(\xi)\varepsilon - D(\xi_0)\varepsilon_0 - \varepsilon_l D(\xi)\xi_S + \varepsilon_l D(\xi_0)\xi_{S0} + \Theta(T - T_0), \tag{8}$$

where D is the stiffness, ε_l is the maximum residual strain (grain average of $\varepsilon^{t,(k)}$ in Eq. (4), Θ is the effective thermal coefficient of expansion which is a homogenized function of (α_0, α_k) in Eq. (6), and the total martensite fraction $\xi = \xi_S + \xi_T$. Again, here also, ξ is essentially a homogenized function of η_k in Eq. (6). The components ξ_S (stress-induced martensite) and ξ_T (temperature induced martensite) are due to the applied stress and the applied temperature, respectively. The incremental form of the above constitutive relation is written as

$$\begin{aligned} \Delta\sigma = (D_m - D_a)\varepsilon_0\,\Delta\xi + (D_m - D_a)\,\Delta\xi\,\Delta\varepsilon + D_0\,\Delta\varepsilon + \Theta\,\Delta T \\ - \varepsilon_l[(D_m - D_a)\xi_{S0}\,\Delta\xi + D_0\,\Delta\xi_S + (D_m - D_a)\,\Delta\xi\,\Delta\xi_S], \end{aligned} \tag{9}$$

where $\Delta\sigma$ denotes the stress increment, $\Delta\varepsilon$ denotes the strain increment and $\Delta\xi$ denotes the phase. Next, the following conditions for phase transformation are applied.

1.2. *Transformation conditions in phase diagram based approach*

The conditions for austenite to detwinned martensite transformation are dependent on the instantaneous and critical values of stresses and temperatures. These critical values of stress and temperature, often termed as transformation stresses and temperatures, typically represent the transformation surfaces in the transformation diagram in the $\sigma - T$ space and divides it into various zones (see ref. [17] for details). These critical transformation stresses and temperatures mark the onset and completion of the transformation without referring to the thermodynamic force in a realistic time scale. This is after all a quasi-static approximation wherein the quantities of interest are the stress-strain or the stress-temperature at some location of the SMA wire. In such a model, the critical transformation temperatures are determined under a stress-free condition and denoted by T_{MS}, T_{Mf}, T_{AS}, and T_{Af}, where T_{MS} is the temperature at which martensite evolution starts, T_{Mf} is the temperature at which the martensite evolution ends, T_{AS} is the temperature at which austenite evolution starts and T_{Af} is the temperature at which austenite evolution ends. The critical transformation stresses are represented by σ_S^{cr} and σ_f^{cr} for temperatures below T_{MS}. Above the value of T_{MS}, the critical transformation stress σ^{cr} is a function of temperature. The slope of these transformation boundaries in the $\sigma - T$ space is given by C_M and C_A as shown in Fig. 1*a*. The conditions which determine the transformation of austenite to detwinned martensite can be found in ref. [17, 27].

We formulate the criteria for the onset and saturation of transformation by taking into account the effect of dissipative force on the transformation surface as

$\sigma^{\rm cr} \leftarrow \sigma^{\rm cr} \pm \dot{\xi}/A$ where A is the scaling parameter in Eq. (6) which needs to be estimated.

1.3. *Scaling parameters in kinetics*

Observation of phase transition in a SMA specimen consisting of a microstructure reveals that there exist several regions of pure phases (variants of martensite in a self accommodated manner). More regions of new phases nucleate, and they grow/decay during the phase transition. Therefore, the transition proceeds by nucleation and growth of new phases within the old one. We approximate the new phases with an accumulated phase fraction $\Delta\xi$ and assume that their growth/decay takes place according to the Weibull distribution over time. Then, Fourier transformation and a finite difference scheme are applied in a step-by-step procedure to arrive at the scaling parameters (see Eq. (6)).

According to the thermodynamic force balance law and in the light of the second law of thermodynamics, let us now consider Eq. (6). In a simplified form, the Gibbs free energy density is first written as $G = -(1/2)\sigma^T\varepsilon_{\rm el} + G_{\rm chem}(T,\xi)$, where $G_{\rm chem} = (T-T_e)(a_0+a_1\xi+a_2\xi^2+a_3\xi^3)$, with four transformation conditions enforced as $\partial G/\partial\xi = 0$ when $\xi = 0, 1$; $G_{\rm chem} = G_0$, $T = T_{A_f}$, $\xi = 0$; $G_{\rm chem} = G_0 + \Delta G$, $T = T_{M_f}$, $\xi = 1$. A schematic representation of the energy landscape is given in Fig. 1*b*. In Fig. 1*b*, the reference energy G_0 is the free-energy of the stress-free martensite and it can also be dependent on the grain boundary density and dislocation potentials. The four coefficients a_0, a_1, a_2, and a_3 in the cubic polynomial are determined uniquely with the help of the four conditions of transformation conditions. Eliminating these four coefficients, we finally get

$$G_{\rm chem} = (T-T_e)\left\{\frac{G_0+\Delta G}{T_{A_f}-T_e} - 3\left[G_0\frac{T_{M_f}-T_{A_f}}{(T_{M_f}-T_e)(T_{A_f}-T_e)} + \frac{\Delta G}{T_{A_f}-T_e}\right]\xi^2\right.$$
$$\left. + 2\left[G_0\frac{T_{M_f}-T_{A_f}}{(T_{M_f}-T_e)(T_{A_f}-T_e)} + \frac{\Delta G}{T_{A_f}-T_e}\right]\xi^3\right\}. \quad (10)$$

$$\dot{\xi} = A\frac{\partial}{\partial\xi}\left[-\frac{1}{2}D(\xi)\varepsilon_{\rm el}^2\right] + A\frac{\partial G_{\rm chem}}{\partial\xi} + B\nabla^2\xi. \quad (11)$$

Next, we derive the explicit form of the scaling parameters A and B using a finite difference scheme with assumed nucleation of phases across two opposite phase boundaries ξ_0 according to a Weibull distribution W such that $\xi(x,t) = \xi_0 + \Delta\xi(x,t)$, $\Delta\xi(x,t) = \hat{\xi}_0(x,t)W(t,k)$, $\hat{\xi}_0(x,t) = \sum\tilde{\xi}(x,\omega)e^{i\omega t}$. With this, we estimate B as function of A for a given characteristic distribution over phase boundary (see Fig. 2)

$$B = \frac{1}{128\{{\rm Re}[\tilde{\xi}_j(x,\omega)] - \xi_0'\}}\left(\left\{8\bar{C}\lambda_0^2\cos(\omega t) + 16\bar{\bar{A}}_2\lambda_0^2{\rm Re}[\tilde{\xi}_j(x,\omega)] - 8\xi_0'\bar{\bar{A}}_2\lambda_0^2\right\}\right.$$

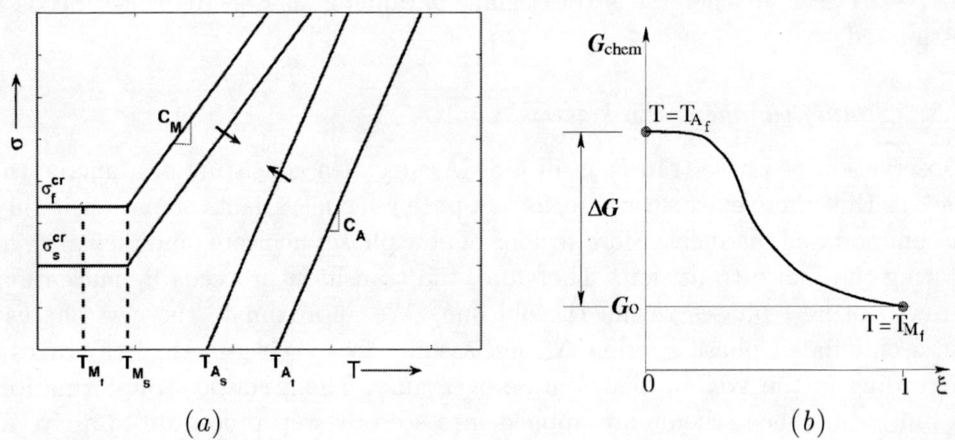

Fig. 1. (a) Phase diagram showing the transformation curves. The arrow marks shows region in which the existing model can produce spurious results. (b) Variation in chemical part of free energy with martensitic volume fraction ξ

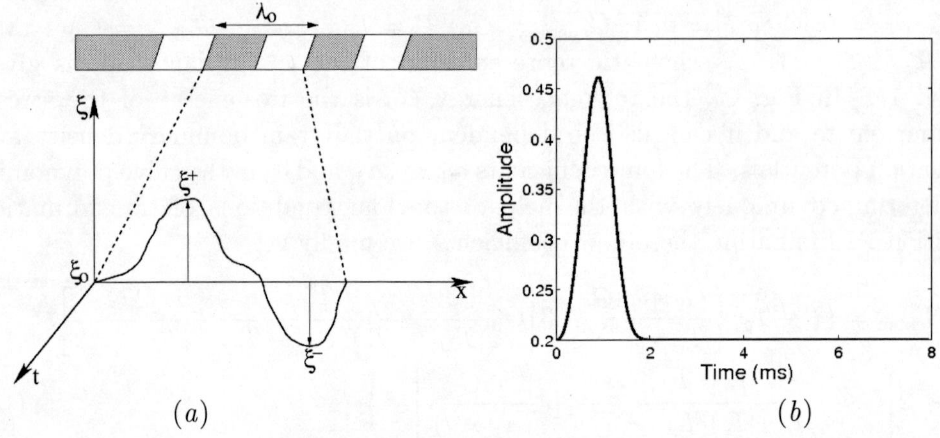

Fig. 2. (a) Schematic diagram showing SMA wire with alternate phases. (b) Evolution of $\Delta\xi$ with time

$$\pm \Big\{ \big(8\bar{C}\lambda_0^2\cos(\omega t) + 16\bar{\bar{A}}_2\lambda_0^2 \mathrm{Re}[\tilde{\xi}_j(x,\omega)] - 8\xi_0'\bar{\bar{A}}_2\lambda_0^2\big)^2 - 4\big(64\mathrm{Re}[\tilde{\xi}_j(x,\omega)] - 64\xi_0'\big)$$
$$\times \big(\bar{\bar{A}}_2\bar{C}\lambda_0^4\cos(\omega t) + \lambda_0^4\omega\bar{C}\sin(\omega t) + \lambda_0^4[\bar{\bar{A}}_2^2 + \omega^2]\mathrm{Re}[\tilde{\xi}_j(x,\omega)]\big)\Big\}^{1/2}\Big). \qquad (12)$$

Fig. 3 a shows a comparison between the stress-strain hysteresis with empirical fitting using trigonometric function and with the first-order kinetics model discussed above. It is observed from Fig. 3 a that the present kinetics gives rise to sharp

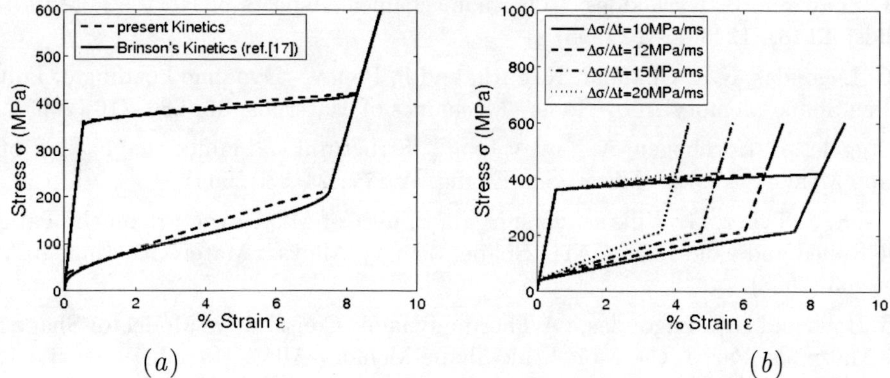

Fig. 3. (*a*) Comparison between the present kinetics and Brinson's kinetics [17]. (*b*) Hysteresis loop for various loading rates

hysteresis curves with straight segments. This is due to the fact that, the present kinetics, when used alone for the entire transformation path in the $\sigma - T$ space, does not have empirically fitted trigonometric functions to represent martensite evolution in several grains in an equivalent sense. Therefore, the phase ξ evolves naturally according to the first-order kinetics for a single type of interface in a single grain. Although this is the case obtained for single type of phase boundary and unidirectional interfaces, however, for multivariant interface problem, this is expected to produce more realistic stress-strain curves as smeared effect due to spatial homogenization as the grain size/number increases. Fig. 3 *b* shows the hysteresis loop for various loading (stress) rates. From Fig. 3 *b*, it can be observed that as the stress rate is increased from 10 MPa/ms to 20 MPa/ms, the hysteresis loop shrinks, indicating partial transformation without adequate time given for relaxation of the stresses at the phase boundaries.

CONCLUSIONS

A comprehensive multiscale physical and mathematical framework is constructed to simulate the rate dependent response of SMA materials. The approach employs the ideas of thermodynamic dissipation and phase transforming microstructure to arrive at the continuum scale constitutive model.

REFERENCES

1. K. Otsuka and C.M. Wayman, *Shape Memory Materials* (Cambridge University Press, 1998).

2. J.A. Shaw and S. Kyriakides, "Thermomechanical Aspects of NiTi," J. Mech. Phys. Solids, **43** (8), 1243–1281 (1995).
3. D.C. Lagoudas, K.R. Chandar, K. Sarh, and P. Popov, "Dynamic Loading of Polycrystalline Shape Memory Alloy Rods," Mechanics of Materials, **35**, 689–716 (2003).
4. G. Eggeler, E. Hornbogen, A. Yawny, et al., "Structural and Functional Fatigue of NiTi Shape Memory Alloys," Mater. Sci. Engng: A **371**, 24–33 (2004).
5. K. Gall, J. Tyber, G. Wilkesanders, et al., "Effect of Microstructure on the Fatigue of Hot-Rolled and Cold-Drawn NiTi Shape Memory Alloys," Mater. Sci. Engng: A **486**, 389–403 (2008).
6. J.G. Boyd and D.C. Lagoudas, "A Thermodynamic Constitutive Model for Shape Memory Materials. Part I. the Monolithic Shape Memory Alloy," Int. J. Plasticity, **12** (6), 805–842 (1996).
7. R.C. Goo and C. Lexcellent, "Micromechanics-Based Modeling of Two-Way Memory Effect of a Single Crystalline Shape Memory Alloy," Acta Mater. **45**, 727–737 (1997).
8. M. Brocca, L.C. Brinson, and Z.P. Bazant, "Three-Dimensional Constitutive Shape Memory Alloy Based on Microplane Model," J. Mech. Phys. Solids **50**, 1051–1077 (2002).
9. N. Siredey, E. Patoor, M. Bervellier, and E. Aberhardt, "Constitutive Equations for Pollycrystalline Thermoelastic Shape Memory Alloys, Part I. Intragrannular Interactions and Behavior of the Grain," Int. J. Solids Struct. **36**, 4289–4315 (1999).
10. J.M. Ball and C. Cartensen, "Compatibility Conditions for Microstructures and the Austenite-Martensite Transitions," Mater. Sci. Engng: A **273**, 231–236 (1999).
11. K. Bhattacharya, *Microstructure of Martensite* (Oxford University Press, 2003).
12. R.D. James and K.F. Hane, "Martensite Transformations and Shape Memory Materials," Acta Mater. **48**, 197–222 (2000).
13. J.J. Zhu and K.M. Lew, "Describing the Morphology of 2H Martensite Using Group Theory Part I: Theory," Mech. Advanced Mater. Struct. **11** (3), 197–225 (2004).
14. R. Abeyaratne, C. Chu, and R.D. James, "Kinetics of Materials with Wiggly Energies; The Evolution of Twinning Microstructure in a Cu-Al-Ni Shape Memory Alloys," Phil. Mag. **73A**, 457–496 (1996).
15. A. Artemev, Y. Wang, and A.G. Khachaturyan, "Three-Dimensional Phase Field Model and Simulation of Martensitic Transformation in Multilayer Systems under Applied Stress," Acta Mater. **48**, 2503–2518 (2000).
16. T. Ichitsubo, K. Tanaka, M. Koiwa, and T. Yamazaki, "Kinetics to Cubic Tetragonal Transformation under External Field by the Time Dependent Ginzburg-Landau Approach," Phys. Rev. B **62**, 5435 (2000).
17. L.C. Brinson, "One Dimensional Constitutive Behavior of Shape Memory Alloys: Thermomechanical Derivation with Non-Constant Material Functions," J. Intelligent Material Systems and Structures **4** (2), 229–242 (1993).
18. C. Liang, *The Constitutive Modeling of Shape Memory Alloy* Ph.D. thesis (Virginia Tech., USA, 1990).

19. C. Niclaeys, T.B. Zineb, S.A. Chirani, and E. Patoor, "Determination of Interaction Energy in the Martensitic State," Int. J. Plasticity **18**, 1619–1647 (2002).
20. O. Heintze and S. Seelecke, "A Coupled Thermomechanical Model for Shape Memory Alloys - From Single Crystal to Polycrystal," Mater. Sci. Engng: A, 481–482, 389–394 (2008).
21. R.C. Smith, S. Seelecke, M. Dapino, and Z. Ounaies, "A Unified Framework for Modeling Hysteresis in Ferroic Materials," J. Mech. Phys. Solids **54**, 46–85 (2006).
22. D.R. Mahapatra and R.V.N. Melnik, "Finite Element Analysis of Phase Transformation Dynamics in Shape Memory Alloys with a consistent Landau-Ginzburg Free Energy Model," Mech. Adv. Mater. Struct. **13**, 443–455 (2006).
23. F. Auricchio, S. Marfia, and E. Sacco, "Modelling of SMA Materials: Training and Two Way Shape Memory Effects," Computers and Structures **81**, 2301–2317 (2003).
24. J.G. Boyd and D.C. Lagoudas. "A Thermodynamical Constitutive Model for Shape Memory Materials. Part 1. The Monolithic Shape Memory Alloy," Int. J. Plasticity **12** (6), 805–842 (1996).
25. X. Gao, M. Huang, and L.C. Brinson, "A Multivariant Micromechanical Model for SMAs. Part 1. Crystallographic Issues for Single Crystal Model," Int. J. Plasticity **16**, 1345–1369 (2000).
26. M. Huang, X. Gao, and L.C. Brinson, "A Multivariant Micromechanical Model for SMAs. Part 2. Polycrystal Model," Int. J. Plasticity **16**, 1371–1390 (2000).
27. L.C. Brinson and R. Lammering, "Finite Element Analysis of the Behavior of Shape Memory Alloys and their Applications," Int. J. Solids Struct. **30** (23), 3261–3280 (1993).
28. M.Y. Serry, D.W. Raboud, and W.A. Moussa, "Finite-Element Modeling of Shape Memory Alloy Components in Smart Structures, Part I: Non-linear Coupled Field Finite Element Procedure" in *Proceedings of the International Conference on MEMS, NANO and Smart Systems. IEEE Computer Society, Alberta* (Alberta, 2003).
29. X. Gao, O. Rui, and L.C. Brinson, "Phase Diagram Kinetics for Shape Memory Alloys: A Robust Finite Element Implementation," Smart Mater. Struct. **16**, 2102–2115 (2007).
30. L.C. Brinson, A. Bekker, and S. Hwang, "Deformation of Shape Memory Alloys Due to Thermo-Induced Transformation," J. Intelligent Mater. Syst. Struct. **7**, 97–107 (1996).
31. J.J. Amalraj, A. Bhattacharyya, and M.G. Faulkner, "Finite-Element Modeling of Phase Transformation in Shape Memory Alloy Wires with Variable Material Properties," Smart Mater. Struct. **9**, 622–631 (2000).
32. F. Falk and P. Kanopka, "Three-Dimensional Landau Theory Describing the Martensitic Phase Transformation of Shape Memory Alloys," J. Phys. Condens Matter **2**, 61–77 (1990).
33. P. Van, "The Ginzburg-Landau equation as a consequence of the second law," Continuum Mech. Thermodyn. **17**, 165–169 (2005).
34. K.R. Rajagopal and A.R. Srinivasa, "On the Thermomechanics of Shape Memory Wires," ZAMP **50**, 459–496 (1999).
35. D. Brenardini, "Model of Hysteresis in the Framework of Thermomechanics with Internal Variables," Physica B **306**, 132–136 (2001).

CONTROL OF FRICTION-DRIVEN OSCILLATIONS BY TIME-DELAYED FEEDBACK

A. Saha[1] and P. Wahi[1*]

ABSTRACT

In this paper, we discuss time delayed feedback control of friction-driven oscillations. Efficacy of the control force in quenching these oscillations are compared for two different kinds of linear control forces derived from delayed position feedback. We have also considered two different mathematical models for representing the nonlinear frictional forces. Linear stability analysis of the corresponding mathematical model gives stability boundaries separating regions in the plane of the control parameters with stable and unstable equilibria. The equilibrium is found to lose stability via. a Hopf bifurcation. The criticality of the bifurcation is obtained using two different methods viz. the averaging method and the method of multiple scales. It is shown that the results obtained using the method of multiple scales (MMS) corresponds much better with the numerical solution as compared to the results obtained using the averaging method. The bifurcation is found to be supercritical for one of the frictional force model while subcritical for the other model raising questions on the true nature of the bifurcation for friction-driven oscillations which will be answered in our future analysis.

INTRODUCTION

In this paper, we use linear time-delayed position feedback for the control of friction-driven vibrations.

Friction-induced vibration is a very serious problem in many engineering systems [1, 2] (e.g., brake-squeal, train curve squeal, clutch chatter, machine-tool chatter, vibrations in controlled positioning systems. The drooping characteristic of the friction force (decrease in the friction force with an increase in the relative velocity) in the low relative velocity regime has been established as one of the major causes

[1]Department of Mechanical Engineering, Indian Institute of Technology Kanpur, Kanpur — 208016, India

[*]E-mail: *wahi@iitk.ac.in*. Phone: +915122596092. Fax: +915122597408

of such violent vibrations [1, 2]. With an increase in the relative velocity past some critical value, the friction force either increases with an increase in the relative velocity or attains a constant value. As a result, the system undergoes bounded periodic motions, the so called limit cycle oscillations. Since these self-excited vibrations are highly undesirable in most engineering applications, it is desirable to obtain a better understanding of the phenomena and to develop efficient methods for controlling such oscillations.

Heckl and Abrahams [3] propose an active control scheme based on a phase-shifted velocity feedback for quenching friction-induced oscillations. High frequency tangential excitation (the so-called dither) was proposed by Thomsen [1] in controlling these self-excited oscillations and the effectiveness of tangential dither in controlling stick-slip chaos in a system driven by frictional forces was experimentally studied by Feeny and Moon [4]. Chatterjee [5] has analyzed normal force modulation as a means to control these vibrations.

In recent times, time-delayed feedback control of vibrations has become very popular among researchers. Atay [6] and Macarri [7, 8] investigated the use of time-delayed state feedback method in controlling free, forced and parametric vibrations in the Van der Pol oscillator. Hu [9] studied the primary resonance and 1/3 subharmonic resonance of a harmonically forced Duffing oscillator under state feedback control with a time delay.

Very limited studies have, however, concentrated on the use of time-delayed feedback control of friction-induced oscillations. Das and Mallik [10] considered time-delayed PD feedback control of self-excited and forced vibration of a friction-driven system. Chatterjee [2] has investigated time-delayed feedback control of different types of friction-induced instabilities viz. the Stribeck instability, the mode-coupling instability, and the sprag-slip instability. In our earlier analysis [11], we have considered linear time-delayed feedback control of the frictional model suggested by Thomsen [1] and obtained that the bifurcation to instability is a supercritical Hopf bifurcation for the control law chosen therein. In this paper, we have also considered the exponential friction force model [2] for which the bifurcation is subcritical in nature. This observation guides us towards performing a detailed experimental study to ascertain the true nature of the bifurcation relevant to friction-driven oscillations which we would be undertaking in the future.

1. THE MATHEMATICAL MODEL

We consider a simplified model for the study of friction-induced vibrations consisting of a spring- mass-damper system with the mass in contact with a friction belt moving with a constant velocity as shown in Fig. 1.

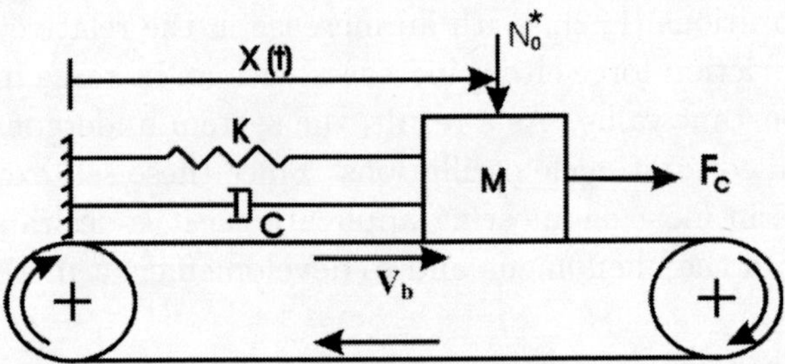

Fig. 1. Harmonic oscillator on a moving belt as a model of friction-driven vibration

The equation of motion of the system is given as

$$M\frac{d^2X}{dt^2} + C\frac{dX}{dt} + KX = F(V_r) + F_c, \tag{1}$$

where $F(V_r) = N_0^*\mu(V_r)$ is the friction force (N_0^* is the normal load and $\mu(V_r)$ is the friction function) and F_c is the control force. Two different control forces viz.

$$(F_c)_1 = -K_{c1}^*[X - X(t - T_1^*)], \tag{2a}$$

$$(F_c)_2 = K_{c2}^*X(t - T_2^*) \tag{2b}$$

will be considered for analysis in this paper. K_{c1}^*, K_{c2}^* are the control gains and T_1^*, T_2^* are the time-delays.

The non-dimensionalized form of the equations of motion with the control forces 1 and 2 are

$$x'' + 2\xi x' + x = \mu(v_r) - K_{c1}[x - x(\tau - T_1)], \tag{3a}$$

$$x'' + 2\xi x' + x = \mu(v_r) + K_{c2}x(\tau - T_2), \tag{3b}$$

respectively, where the non-dimensional quantities are

$$\tau = \omega_0 t, \quad T_1 = \omega_0 T_1^*, \quad T_2 = \omega_0 T_2^*, \quad \omega_0 = \sqrt{\frac{K}{M}},$$

$$x_0 = \frac{N_0^*}{M\omega_0^2}, \quad \xi = \frac{C}{2M\omega_0}, \quad x = \frac{X}{x_0}, \quad v_b = \frac{V_b}{\omega_0 x_0},$$

$$K_{c1} = \frac{K_{c1}^*}{M\omega_0^2}, \quad K_{c2} = \frac{K_{c2}^*}{M\omega_0^2}, \quad v_r = \frac{V_r}{\omega_0 x_0} = \frac{V_b - \dot{X}}{\omega_0 x_0} = v_b - x'.$$

We will also consider two different friction functions. The first one is a polynomial type representation (as used by Thomsen [2])

$$\mu(v_r) = \mu_s \mathrm{sign}(v_r) - \frac{3}{2}(\mu_s - \mu_m)\left[\frac{v_r}{v_m} - \frac{1}{3}\left(\frac{v_r}{v_m}\right)^3\right], \tag{4}$$

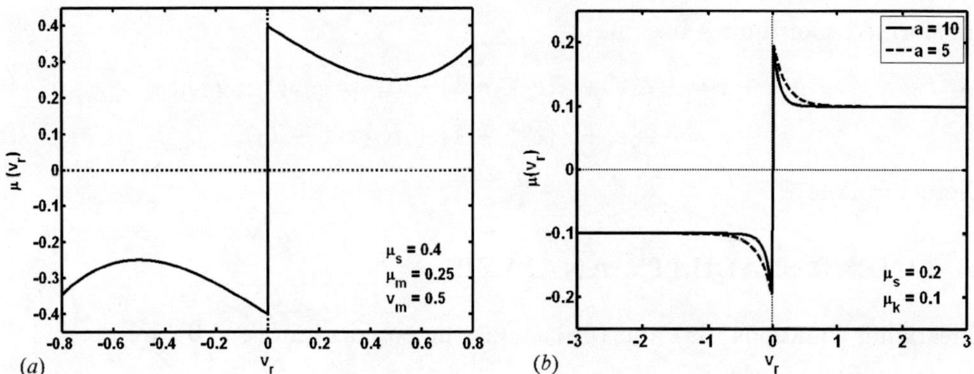

Fig. 2. (*a*) Friction function as given by equation (4), (*b*) Exponential friction function as given by equation (5)

where $v_r = v_b - x'$ is the (non-dimensional) relative velocity between the mass and the belt, μ_s is the coefficient of static friction, and v_m is the velocity corresponding to the minimum coefficient μ_m of kinetic friction, $\mu_m \leq \mu_s$. The friction characteristic corresponding to equation (4) for some typical parameter values is shown in Fig. 2.

The second one we will use is the exponential friction model as given by the following friction function

$$\mu(v_r) = (\mu_k + \Delta\mu e^{-a|v_r|})\text{sign}(v_r), \tag{5}$$

where $\Delta\mu = \mu_s - \mu_k$, μ_s is the coefficient of static friction, μ_k is the minimum coefficient of kinetic friction and a is the slope parameter.

Now consider the equation of motion (3a). The steady state ($x'' = x' = 0$) solution is $x_{\text{eq}} = \mu(v_b)$. Similarly, the steady state solution for the equation of motion (3b) is $x_{\text{eq}} = \mu(v_b)/(1 - K_{c2})$. Shifting the origin to the equilibrium solution by introducing the coordinate transformation $z = x - x_{\text{eq}}$ (i.e. $z' = x'$, $z'' = x''$) and after some straightforward manipulations, the equation of motion (3a) and (3b) reduce to

$$z'' + z = h_1 z' + h_2 z'^2 + h_3 z'^3 - K_{c1}[z - z(\tau - T_1)], \tag{6a}$$

$$z'' + z = h_1 z' + h_2 z'^2 + h_3 z'^3 + K_{c2} z(\tau - T_2), \tag{6b}$$

where

$$h_1 = -2\xi + \alpha_1\left[1 - \left(\frac{v_b}{v_m}\right)^2\right], \quad h_2 = \frac{\alpha_1}{v_m^2} v_b, \quad h_3 = -\frac{\mu_s - \mu_m}{2v_m^3}, \quad \alpha_1 = \frac{3}{2}\frac{\mu_s - \mu_m}{v_m} > 0 \tag{7}$$

for the friction force represented by function (4). When the friction force is represented by the function (5), the non-dimensionalized equation of motion with the transformed coordinates become

$$z'' + z = -2\xi z' + \gamma(e^{az'} - 1) - K_{c1}[z - z(\tau - T_1)], \tag{8a}$$

$$z'' + z = -2\xi z' + \gamma(e^{az'} - 1) + K_{c2} z(\tau - T_2), \tag{8b}$$

where $\gamma = \Delta\mu e^{-av_b}$.

2. LINEAR STABILITY ANALYSIS

Linearizing equations (6a) and (6b) about the equilibrium ($z = 0$), we get

$$z'' + (1 + K_{c1})z - h_1 z' = K_{c1} z(\tau - T_1), \tag{9a}$$

$$z'' + z - h_1 z' = K_{c2} z(\tau - T_2). \tag{9b}$$

If we linearize equations (8a) and (8b), the linearized form will be exactly similar to the equations (9a) and (9b) with $h_1 = \gamma a - 2\xi$.

The corresponding characteristic equations are

$$s^2 - h_1 s + (1 + K_{c1}) - K_{c1} e^{-sT_1} = 0, \tag{10a}$$

$$s^2 - h_1 s + 1 - K_{c2} e^{-sT_2} = 0. \tag{10b}$$

On the stability boundary, the most dominant pair of roots becomes purely imaginary. Therefore, we substitute $s = j\omega$ in equations (10a) and (10b), where ω determines the frequency of the ensuing periodic solution. Next, a separation of the real and imaginary parts of the equations gives us two real equations that can be solved for two unknowns ($K_{c1,c2}$ and $T_{1,2}$) with the other unknown (ω) as the parameter. The parametric dependence of $K_{c1,c2}$ and $T_{1,2}$ on ω for the equations (10a) and (10b) is given by

$$\begin{aligned} &\omega^4 + [h_1^2 - 2(1 + K_{c1})]\omega^2 + 1 + 2K_{c1} = 0, \\ &T_1 = \frac{1}{\omega_c}\left[\tan^{-1}\left(\frac{h_1\omega_c}{1 + K_{c1} - \omega_c^2}\right) + 2i\pi\right] \quad \forall i = 0, 1, 2, \ldots, \infty, \end{aligned} \tag{11a}$$

$$\begin{aligned} &\omega^4 + (h_1^2 - 2)\omega^2 + 1 - K_{c2}^2 = 0, \\ &T_2 = \frac{1}{\omega_c}\left[\tan^{-1}\left(\frac{h_1\omega_c}{1 + K_{c2} - \omega_c^2}\right) + 2i\pi\right] \quad \forall i = 0, 1, 2, \ldots, \infty. \end{aligned} \tag{11b}$$

For the stability of the equilibrium to switch at the critical values of the bifurcation parameter under consideration (i.e., a Hopf bifurcation to occur), the eigenvalues must cross the imaginary axis transversally. Therefore, the sign of the following quantities must also be checked:

$$R_1(\omega_c, K_{c1o}) = \text{sign}\left[\text{Re}\left(\frac{ds}{dK_{c1}}\bigg|_{j\omega_c, K_{c1o}}\right)\right]$$

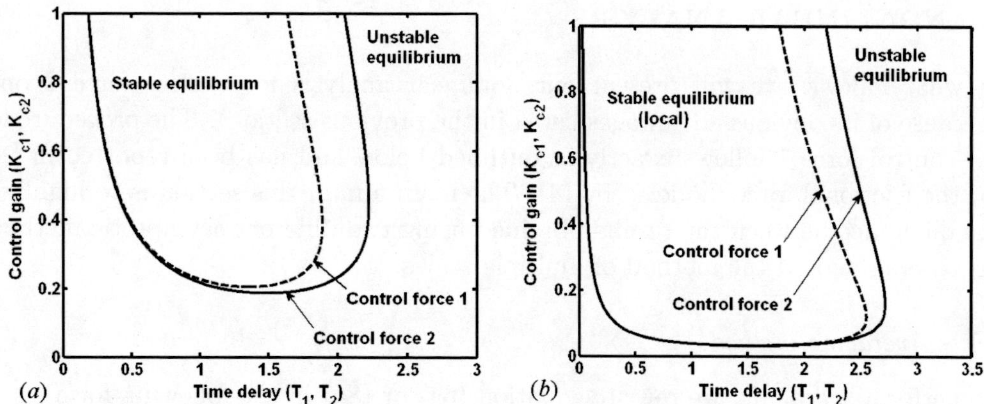

Fig. 3. Comparison of stability boundaries for control forces 1 and 2. (*a*) For friction function (4) with $v_b = 0.3$, $\xi = 0.05$, $\mu_s = 0.4$, $\mu_m = 0.25$, $v_m = 0.5$. (*b*) For friction function (5) with $v_b = 0.25$, $\xi = 0.025$, $\mu_k = 0.1$, $\Delta\mu = 0.1$, $a = 10$

$$= \text{sign}\{(h_1 + K_{c1o}T_1)[1 - \cos(\omega_c T_1)] - 2\omega_c \sin(\omega_c T_1)\}, \tag{12a}$$

$$R_2(\omega_c, K_{c2o}) = \text{sign}\left[\text{Re}\left(\frac{ds}{dK_{c2}}\bigg|_{j\omega_c, K_{c2o}}\right)\right]$$
$$= \text{sign}\left[K_{c2o}T_2 - h_1 \cos(\omega_c T_2) - 2\omega_c \sin(\omega_c T_2)\right], \tag{12b}$$

The equilibrium becomes stable if $R_{1,2} < 0$ and unstable if $R_{1,2} > 0$ when $K_{c1,c2}$ increases past the critical value.

Using equations (11a) and (12a), the stability regions of the equilibrium for control force 1 can be plotted in the plane of controlled parameters (K_{c1} vs. T_1). Similarly, using equations (11b) and (12b), the stability regions of the equilibrium for control force 2 can be plotted in the plane of controlled parameters (K_{c2} vs. T_2).

Figure 3 compares the stability boundaries for control forces 1 and 2 for the two different models for frictional forces. It is observed that the stability region for control force 2 is much larger than that for the control force 1 in both the cases. This implies that for the same control gain, the static equilibrium is stable for a larger range of time delay for the control force 2. Normally, the time delay in the feedback loop cannot be controlled very accurately and hence, for robustness of the control, it is desirable to have a larger margin of error for the time-delay which is provided by control force 2. Also $K_{c2,\min} < K_{c1,\min}$ for a given belt velocity for the frictional force modeled by (4). Therefore, the control force 2 can achieve a similar effect (stable equilibrium) as the control force 1 with a smaller control force.

3. NONLINEAR ANALYSIS

In what follows, we will present our nonlinear analysis for control force 2 only (because of its obvious advantages listed in the previous section). The procedure for the control force 1 follows exactly as outlined below and has been reported in [11] for the frictional force modeled by (4). The main aim of this section is to highlight the difference between the results obtained using two different asymptotic methods viz. averaging and the method of multiple scales.

3.1. *Averaging method*

To perform averaging, we rewrite equation (6a) or (8a) in the following form

$$z'' + z = \varepsilon f(z(\tau), z'(\tau), z(\tau - T_2)), \tag{13}$$

where,

$$\varepsilon f(z(\tau), z'(\tau), z(\tau - T_)) = h_1 z' + h_2 z'^2 + h_3 z'^3 + K_{c2} z(\tau - T_2), \quad 0 < \varepsilon \ll 1 \tag{14}$$

for equation (6a) and

$$\varepsilon f(z(\tau) z'(\tau), z(\tau - T_2)) = -2\xi z' + \gamma(e^{az'} - 1) + K_{c2} z(\tau - T_2), \quad 0 < \varepsilon \ll 1. \tag{15}$$

for equation (8a).

In essence, we are treating the friction force and the control force as small and the unperturbed system is a simple harmonic oscillator. The slowly varying solution of equation (13) is assumed (as in [2, 12]) to be

$$z = A_1(\tau)\sin(\psi_1(\tau)), \quad z' = \omega_0 A_1(\tau)\cos(\psi_1(\tau)), \quad \psi_1(\tau) = \omega_0\tau + \varphi_1(\tau), \tag{16}$$

$A_1(\tau)$ and $\varphi_1(\tau)$ are slowly varying functions of time τ. So it is appropriate to assume $A_1(\tau - T_2) \approx A_1(\tau)$, $\varphi_1(\tau - T_2) \approx \varphi_1(\tau)$ for $T_1 < 2\pi$. Thus, following the standard averaging method, we obtain

$$A_1' = \frac{1}{2}[h_1 - K_{c2}\sin(T_2)]A_1 + \frac{3}{8}h_3 A_1^3, \tag{17}$$

for friction function (4), where h_1 and h_3 are given by equation (7) and

$$A_1' = -\frac{1}{2}[2\xi + K_{c2}\sin(T_2)]A_1 + \gamma I_1(aA_1), \tag{18}$$

for friction function (5). $I_1(aA_1)$ is the modified Bessel's function of order one. We expand $I_1(aA_1)$ in Taylor series around $A_1 = 0$ and retain up to the third order terms to get

$$A_1' = -\frac{1}{2}[K_{c2}\sin(T_1) - h_1]A_1 + \frac{\gamma a^3}{16}A_1^3, \tag{19}$$

where $h_1 = \gamma a - 2\xi$.

The coefficient of the linear term contains the bifurcation parameter and changes sign at some critical value of the bifurcation parameter (the stability boundary) which is given by the zero level set of the functions

$$L(T_2, K_{c2}) = h_1 - K_{c2} \sin(T_2), \tag{20}$$

and

$$L(T_2, K_{c2}) = -h_1 + K_{c2} \sin(T_2), \tag{21}$$

for equations (18) and (19) respectively.

We observe from equation (17) that the linear term becomes positive past the critical value and since the coefficient of the cubic term is negative ($h_3 < 0$), we get a stable limit cycle signifying a supercritical Hopf bifurcation. In case of equation (19), the linear term again becomes positive past the critical value, but the coefficient of the cubic term is positive signifying a subcritical Hopf bifurcation. Hence, unstable limit cycles exist around the linearly stable equilibrium and there are no stable small amplitude limit cycles around the unstable equilibrium. Thus we observe that the two different mathematical models representing the same physical phenomenon viz. frictional force between two bodies in relative motion behave differently. While the polynomial friction function (4) shows supercritical Hopf bifurcation, subcritical Hopf bifurcation is observed for the exponential friction model. This raises the question about the true nature of the bifurcation related to nonlinear frictional forces in friction-induced vibrations and detailed experiments need to be conducted to resolve the issue.

We can also get the critical belt velocity (belt velocity corresponding to the Hopf bifurcation point) from the zero level set of the function $L(T_2, K_{c2})$. The critical belt velocities for friction functions (4) and (5) are

$$(v_b)_c = v_m \sqrt{1 - \frac{1}{\alpha_1}[2\xi + K_{c2} \sin(T_2)]}, \tag{22}$$

and

$$(v_b)_c = \frac{1}{a} \ln \left(\frac{a \Delta\mu}{2\xi + K_{c2} \sin(T_2)} \right). \tag{23}$$

For operating belt velocities larger than this critical value, due to supercriticality of the Hopf bifurcation, the static equilibrium is globally stable for the friction model (4). However, since the type of Hopf bifurcation is subcritical for the friction model (5), the static equilibrium is only locally stable above the critical belt velocity [2].

Figure 4 compares the stability boundary obtained using the averaging method with the exact stability boundary obtained using linear stability analysis for the

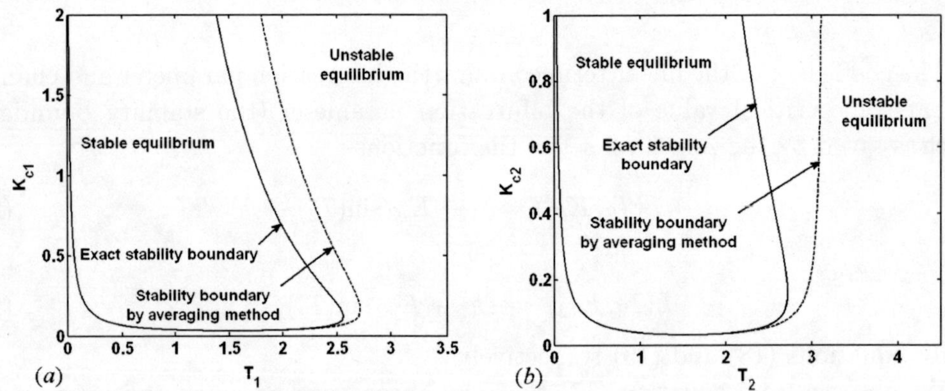

Fig. 4. Comparison of the exact stability boundary with the stability boundary obtained from averaging method for friction model (5). (a) for control force 1, (b) for control force 2. $v_b = 0.25$, $\xi = 0.025$, $\mu_k = 0.1$, $\Delta\mu = 0.1$, $a = 10$

friction force modeled by equation (5). The corresponding figures for the friction force model (4) have already been reported in [11]. It is observed that the stability boundaries obtained by the averaging method does not match the exact stability boundaries very well. In what follows, we apply the method of multiple scales (MMS) [13, 14] wherein no assumption is made on the magnitude of the friction force or the control force, and the unperturbed equation corresponds exactly to the linearized DDE studied in section 2.

3.2. *Method of multiple scales (MMS)*

To cast equation (6b) in the standard form for applying the method of multiple scales, we require a small parameter say $0 < \varepsilon \ll 1$and we further require the nonlinear terms to be small. In order to achieve this goal, we transform coordinate as $z = \varepsilon y$ which after some straightforward manipulation leads to

$$y'' + y - h_1 y' - \varepsilon h_2 y'^2 - \varepsilon^2 h_3 y'^3 - K_{c2} y(\tau - T_2) = 0. \tag{24}$$

Next, we identify the Hopf point (i.e., the critical value of the bifurcation parameter K_{c2o} at $T_2 = T_{2o}$ obtained from linear stability analysis) and perturb it as follows: $K_{c2} = K_{c2o} + \varepsilon^2\Delta$. Equation (24), then, becomes

$$y'' + y - h_1 y' - \varepsilon h_2 y'^2 - \varepsilon^2 h_3 y'^3 - (K_{c2o} + \varepsilon^2\Delta) y(\tau - T_2) = 0. \tag{25}$$

For $\varepsilon = 0$, equation (25) reduces exactly to the linearized equation (9a) whose solution is known to us.

We next define multiple time scales as: $t_0(\tau) = \tau$, the original time scale; $t_1(\tau) = \varepsilon\tau$, $t_2(\tau) = \varepsilon^2\tau$, and so on. The solution to equation (25) is assumed to be $y(\tau) = Y(t_0, t_1, t_2)$, which can be expanded in powers of ε, as

$$y(\tau) \equiv Y(t_0, t_1, t_2) = Y_0(t_0, t_1, t_2) + \varepsilon Y_1(t_0, t_1, t_2) + \varepsilon^2 Y_2(t_0, t_1, t_2) + \ldots \quad (26)$$

We substitute the above expression in equation (25) and collect various powers of ε. The equation at the first order, i.e., ε^0 is:

$$\frac{\partial^2 Y_0}{\partial t_0^2} + Y_0 - h_1 \frac{\partial Y_0}{\partial t_0} = K_{c2o} Y_0(t_0 - T_{2o}). \quad (27)$$

The steady-state solution to equation (27) (from linear analysis) is:

$$Y_0 = C_1 \sin(\omega t_0) + C_2 \cos(\omega t_0), \quad (28)$$

where, $C_1 = C_1(t_1, t_2)$ and $C_2 = C_2(t_1, t_2)$. The equation at order ε^1 after substituting equation (28) for Y_0 is:

$$\frac{\partial^2 Y_1}{\partial t_0^2} - h_1 \frac{\partial Y_1}{\partial t_0} + Y_1 - K_{c2o} Y_1(t_0 - T_{20})$$
$$+ P_1 \sin(\omega t_0) + P_2 \cos(\omega t_0) + P_3 \sin(2\omega t_0) + P_4 \cos(2\omega t_0) + P_5 = 0 \quad (29)$$

where,

$$P_2 = [-h_1 + K_{c2o} T_{2o} \cos(\omega T_{2o})] \frac{\partial C_1}{\partial t_1} - [2\omega - K_{c2o} T_{2o} \sin(\omega T_{2o})] \frac{\partial C_2}{\partial t_1},$$
$$P_5 = -\frac{h_2 \omega^2}{2}(C_1^2 + C_2^2),$$
$$P_2 = [2\omega - K_{c2o} T_{2o} \sin(\omega T_{2o})] \frac{\partial C_1}{\partial t_1} + [-h_1 + K_{c2o} T_{2o} \cos(\omega T_{2o})] \frac{\partial C_2}{\partial t_1},$$
$$P_3 = h_2 \omega^2 C_1 C_2, \quad P_4 = -\frac{h_2 \omega^2}{2}(C_1^2 - C_2^2).$$

To avoid secular terms, the coefficients of $\sin(\omega t_0)$ and $\cos(\omega t_0)$ in equation (29) are set to zero. This gives us $\partial C_1/\partial t_1 = 0$ and $\partial C_2/\partial t_1 = 0$. Equation (29) can then be solved for Y_1in the following form

$$Y_1 = C_3 + C_4 \sin(2\omega t_0) + C_5 \cos(2\omega t_0), \quad (30)$$

by considering the particular integral solution only. The coefficients C_3, C_4, and C_5 can be obtained in terms of the parameters in equation (29). The results obtained so far, including the expressions $\partial C_1/\partial t_1$ and $\partial C_2/\partial t_1$, are now substituted into the equation at order ε^2. Again there are terms containing $\sin(\omega t_0)$ and $\cos(\omega t_0)$ which can give rise to secular terms. Accordingly, and as before, the coefficients of $\sin(\omega t_0)$ and $\cos(\omega t_0)$ in the order ε^2 equation are set to zero which can then be solved to obtain $\partial C_1/\partial t_2$ and $\partial C_2/\partial t_2$. The above procedure can in principle be continued indefinitely (to any order). Finally, the rate of change of C_1 and C_2 in the original time τ is given by

$$C_1' = \frac{\partial C_1}{\partial \tau} = \varepsilon \frac{\partial C_1}{\partial t_1} + \varepsilon^2 \frac{\partial C_1}{\partial t_2} + O(\varepsilon^3), \quad C_2' = \frac{\partial C_2}{\partial \tau} = \varepsilon \frac{\partial C_2}{\partial t_1} + \varepsilon^2 \frac{\partial C_2}{\partial t_2} + O(\varepsilon^3).$$

The actual expressions involved, obtained here using the symbolic algebra package MAPLE, are long and not displayed here for lack of space. We further change variables to polar coordinates as: $C_1(\tau) = R_1(\tau)\cos(\varphi_1(\tau))$ and $C_2(\tau) = R_1(\tau)\sin(\varphi_1(\tau))$. The solution now becomes $y(\tau) = R_1(\tau)\sin(\tau + \varphi_1(\tau))$. The amplitude equation (parameter values as shown in Figure 5 *a*) and the limit cycle amplitude are found as

$$\begin{gathered} \frac{dR_1}{d\tau} = [(-0.1998170859\Delta)R_1 + (-0.186404544)R_1^3]\varepsilon^2, \\ A_1 = \varepsilon R_1 = \varepsilon\sqrt{\frac{-0.1998170859\Delta}{0.186404544}}. \end{gathered} \tag{31}$$

The previous analysis can also be performed by considering the belt velocity $v_b$as the bifurcation parameter. The relevant equation for that case can be obtained by substituting $h_1 = h_{1o} + \varepsilon^2\Delta_1$ and $h_2 = h_{2o} + \varepsilon^2\Delta_2$ in equation (24). Following the procedure outlined above, we find the amplitude equation (parameter values as shown in Figure 5 *b*) and the limit cycle amplitude as

$$\begin{gathered} \frac{dR_1}{d\tau} = [(-0.5079501564\Delta_1)R_1 + (-0.186404544)R_1^3]\varepsilon^2, \\ A_1 = \varepsilon R_1 = \varepsilon\sqrt{\frac{-0.5079501564\Delta_1}{0.186404544}}. \end{gathered} \tag{32}$$

As observed from Figures 5 *a* and *b*, the MMS solution approximates the amplitudes obtained numerically much better as compared to the results obtained using the averaging method. The agreement between analytical prediction by MMS and numerical simulation improves with a decrease in the distance from the Hopf bifurcation point.

Expanding the exponential term up to the third order, we can transform the equation (8a) in the form of the equation (6a) with $h_1 = \gamma a - 2\xi$, $h_2 = \gamma a^2/2$, $h_3 = \gamma a^3/6$, and $\gamma = \Delta\mu e^{-av_b}$. We can then perform the similar analysis as before

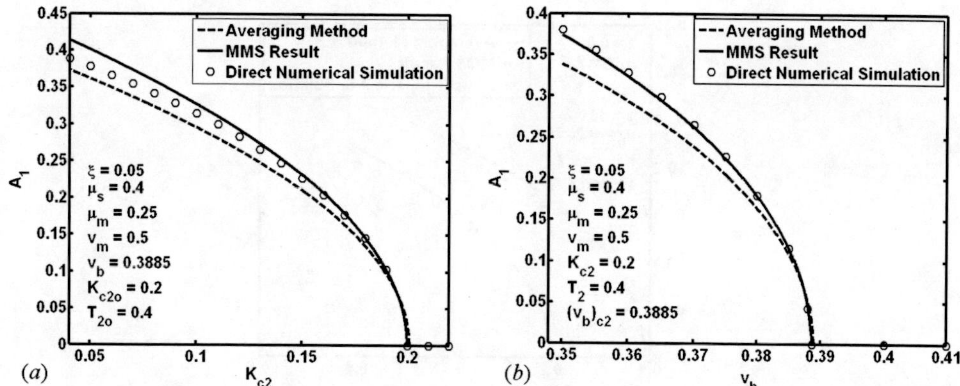

Fig. 5. Amplitude of pure slipping motion of the mass as a function of (*a*) control gain, (*b*) belt velocity. Circles correspond to numerical integration of equation (6b). Solid and dashed lines correspond to MMS results and results obtained from averaging method respectively. $(v_b)_{c1}$ denotes the critical belt velocity for control force 1

and get the amplitude equation (parameter values are as shown in Figure 6) and the limit cycle amplitude (with control gain K_{c1} as the bifurcation parameter) as

$$\begin{aligned} \frac{dR_1}{d\tau} &= [(-0.1976153565\Delta)R_1 + (0.8242027821)R_1^3]\varepsilon^2, \\ A_1 &= \varepsilon R_1 = \varepsilon\sqrt{\frac{0.1976153565\Delta}{0.8242027821}}. \end{aligned} \tag{33}$$

As expected, the MMS result matches closely with the direct numerical simulation result. It has to be noted at this point that due to the subcritical nature of the bifurcation in this case, the limit cycles are unstable and coexist with the stable equilibrium solution. As a result, getting the limit cycles by direct numerical simulation is very difficult. Moreover, there will be large amplitude limit cycles existing in the unstable regime but they involve stick-slip motion as opposed to the pure slip motion considered in this paper. The stick-slip motion will involve special analytical as well as numerical techniques which are not explored in this paper. The differing nature of the bifurcation for the different frictional models draws our attention to a detailed study required to unravel the true nature of the bifurcation for friction-driven vibrations. This would be undertaken in our subsequent study.

4. CONCLUSIONS

We have compared two types of control forces derived from time-delayed feedback in their effectiveness in controlling friction driven oscillations. We have also considered

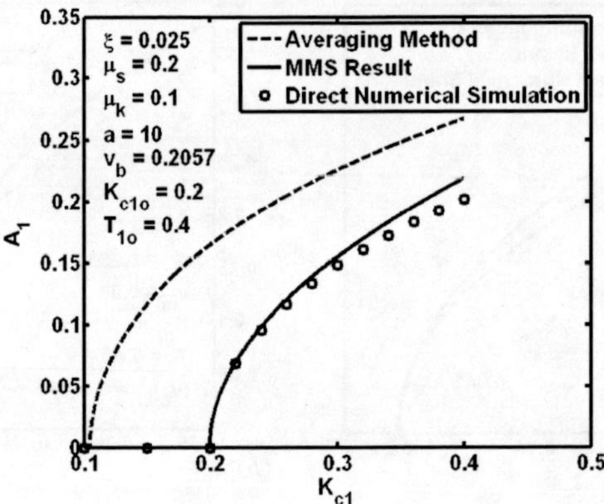

Fig. 6. Amplitude of pure slipping motion of the mass as a function of control gain. Circles correspond to numerical integration of equation (8a). Solid and dashed lines correspond to MMS results and results obtained from averaging method respectively

two different models for the frictional forces which are popular in the literature. Both linear and nonlinear stability analysis is performed. Stability boundaries separating regions with stable and unstable equilibria are plotted in the plane of the control parameters (control gain and delay). Stability is found to be lost via. a Hopf bifurcation. We have also compared two different perturbation methods viz. the method of averaging and the method of multiple scales (MMS) for the nonlinear analysis. It has been observed that the method of multiple scales is superior to the method of averaging in the sense that the results obtained from MMS corresponds much better with the numerical solution than the method of averaging. It has been further observed that the Hopf bifurcation is supercritical for one of the friction models while subcritical for another model. A detailed study to investigate the true nature of the bifurcation for the vibrations excited due to the nonlinear frictional forces would be carried out in the future.

ACKNOWLEDGMENTS

We thank Prof. A.K. Mallik and Prof. B. Bhattacharya for useful discussions.

REFERENCES

1. J.J. Thomsen, "Using Fast Vibrations to Quench Friction Induced Oscillations," J. Sound Vibr. **228** (5), 1079–1102 (1999).
2. S. Chatterjee, "Time-Delayed Feedback Control of Friction-Induced Instability," Int. J. Non-Lin. Mech. **42**, 1127–1143 (2007).

3. M.A. Heckl and I.D. Abrahams, "Active Control of Friction Driven Oscillations," J. Sound Vibr. **193** (1), 417–426 (1996).
4. B.F. Feeny and F.C. Moon, "Quenching Stick–Slip Chaos with Dither," J. Sound Vibr. **237** (1), 173–180 (2000).
5. S. Chatterjee, "Non-Linear Control of Friction-Induced Self-Excited Vibration," Int. J. Non-Lin. Mech. **42**, 459–469 (2007).
6. F.M. Atay,"Van der Pol Oscillator under Delayed Feedback," J. Sound Vibr. **218** (2), 333–339 (1998).
7. A. Maccari, "Vibration Control for the Primary Resonance of the Van der Pol Oscillator by a Time Delay State Feedback," Int. J. Non-Lin. Mech. **38**, 123–131 (2003).
8. A. Maccari, "The Response of a Parametrically Excited Van der Pol Oscillator to a Time Delay State Feedback," Nonlin. Dyn. **26**, 105–119 (2001).
9. H. Hu, E.H. Dowell, and L.N. Virgin, "Resonances of a Harmonically Forced Duffing Oscillator with Time Delay State Feedback," Nonlin. Dyn. **15**, 311–327 (1998).
10. J. Das and A.K. Mallik, "Control of Friction Driven Oscillation by Time-Delayed State Feedback," J. Sound Vibr. **297** (3–5), 578–594 (2006).
11. A. Saha and P. Wahi, "Control of Friction-Induced Vibration by Time-Delayed Position Feedback," in *9th International Conference on Vibration Problems, Kharagpur, India, January 19–22, 2009* (2009).
12. P. Wahi and A. Chatterjee, "Averaging Oscillations with Small Fractional Damping and Delayed Terms," Nonlin. Dyn. **38**, 3–22 (2004).
13. S.L. Das and A. Chatterjee, "Multiple Scales without Center Manifold Reductions for Delay Differential Equations near Hopf Bifurcations," Nonlin. Dyn. **30**, 323–335 (2002).
14. P. Wahi and A. Chatterjee, "Regenerative Tool Chatter near a Codimension 2 Hopf Point Using Multiple Scales," Nonlin. Dyn. **40**, 323–338 (2005).

WAVELET BASIS FINITE ELEMENT SOLUTION FOR TRANSIENT PROBLEMS

K. Gopikrishna[1*] and M. Shrikhande[1]**

ABSTRACT

A new approach for computing the dynamic response of linear structural systems is presented. The novelty of this approach lies in utilizing Daubechies' wavelets as basis functions in finite element method for temporal approximation of the resulting semi-discretized equations (in finite element sense) with time as the independent variable. The time-frequency localization properties of wavelets and the inherent multiresolution capability helps in representing the response at different resolutions, thereby providing a means for developing hierarchical form of solution. The present approach using wavelets as basis functions provides a better control over the error in approximation as the equilibrium is satisfied for the entire duration in the weighted integral sense. This approach reduces the semi-discrete system of equations in time to be solved to a single algebraic problem, in contrast to step-by-step-time integration methods, where a sequence of algebraic problems is to be solved to compute the solution. The present paper is restricted to the analysis of linear problems that arise in structural dynamics. The proposed formulation has been compared against the respective analytical solutions and the Newmark-β method. It is found to compare favorably well in terms of accuracy and computation time. The stability and accuracy characteristics of the proposed formulation has been examined and is found to be energy conserving. Numerical experiments are performed to demonstrate the formulation.

Key words: hierarchical formulation, finite element, multi-resolution, time integration, wavelet basis

[1]Department of Earthquake Engineering, Indian Institute of Technology Roorkee, Roorkee — Uttarakhand, 247667, India

[*]E-mail: *kgopi.iitr@gmail.com*

[**]E-mail: *mshrifeq@iitr.ernet.in*

INTRODUCTION

Transient problems in structural dynamics are characterized by the coupled second order ordinary differential equations (ODE) in time, obtained through a finite element approximation in spatial domain for the governing partial differential equations, as:

$$\mathbf{M}\ddot{\mathbf{u}} + \mathbf{C}\dot{\mathbf{u}} + \mathbf{K} = \mathbf{f} \quad \text{with} \quad \mathbf{u}(t=0) = \mathbf{u}_0 \text{ and } \dot{\mathbf{u}}(t=0) = \dot{\mathbf{u}}_0, \tag{1}$$

where, $\mathbf{M}$, $\mathbf{C}$, and $\mathbf{K}$ respectively denote the system mass, damping and stiffness matrices, $\mathbf{f}$ represents the vector of external forces, $\mathbf{u}$, is the vector of nodal displacements. The single field formulation represented by Eq. (1), can be transformed into an equivalent set of first order ODEs by letting $\dot{\mathbf{u}} = \mathbf{v}$ and $\ddot{\mathbf{u}} = \dot{\mathbf{v}}$ and rearranging as:

$$\dot{\mathbf{d}} + \mathbf{A}\mathbf{d} = \mathbf{g}, \tag{2}$$

where, $\mathbf{A} = \begin{bmatrix} 0 & -\mathbf{I} \\ \mathbf{M}^{-1}\mathbf{K} & \mathbf{M}^{-1}\mathbf{C} \end{bmatrix}$, $\mathbf{d} = \begin{Bmatrix} \mathbf{u} \\ \mathbf{v} \end{Bmatrix}$, and $\mathbf{g} = \begin{Bmatrix} 0 \\ \mathbf{M}^{-1}\mathbf{f} \end{Bmatrix}$. This treatment leads to a mixed two-field formulation for its solution, as displacement and velocity can be treated independently. These problems are generally solved by using a suitable step-by-step time-marching procedure, wherein the equilibrium is established at closely spaced discrete time instants by assuming a suitable variation for system kinematics (displacement, velocity and acceleration) within the interval between the two time instants. This implies that a continuous solution is not sought over the entire duration and the solution is obtained by solving a sequence of simultaneous linear algebraic equations at each time instant. This requires massive computational resources for large scale problems which often require very small time steps for acceptable accuracy. Further, higher frequencies (or small timescale variations) are not accurately resolved by finite element mesh thereby causing spurious oscillations to creep into the computed solution for certain wave-propagation problems whenever the time step of integration approaches the Courant-Friedrich-Lewy (CFL) limit [3]. Further, the step-by-step procedures are vulnerable to the accumulation of errors at each time step due to finite precision arithmetic. To provide a control over the total error in the computed solution over the entire response duration we explore the possibility of constructing a wavelet basis approximation in time by using an hierarchical approach for solving transient problems in structural dynamics. The proposed procedure utilizes Daubechies' wavelets as basis functions in the hierarchical finite element formulation in time dimension. Daubechies' wavelets possess several desirable qualities like orthogonality, compact support and exact representation of polynomials of a fixed degree, which makes them suitable for representing the solution of partial differential equations (PDE) [2, 4, 10]. The multiresolution capability of wavelet approximations further aids in the analysis of transient problems.

1. DAUBECHIES' WAVELETS

The term wavelet refers to a spatially localized function with finite energy and a class of such functions with compact supports was proposed by Daubechies [5]. These wavelets have been studied extensively and are known to possess several features like orthogonality, compact support, and exact representation of polynomials of a fixed degree—known as regularity—which provide efficient and stable representation of regions with strong gradients or oscillations [10]. Daubechies' wavelet consists of scaling and wavelet functions defined by the following refinement relations and the values of these functions are calculated by recursively from these refinement relations.

$$\phi(t) = \sum_{i=0}^{N-1} h_i \phi(2t - i), \tag{3}$$

$$\text{and } \varphi(t) = \sum_{i=0}^{N-1} (-1)^i h_{N-1-i} \phi(2t - i), \tag{4}$$

where, h_i and h_{N-1-i} are scaling and wavelet coefficients (also known as the filter coefficients). The supports of scaling and wavelet functions are given by $\text{supp}(\varphi) = [0, N-1]$ and $\text{supp}(\phi) = [1 - N/2, N/2]$. The scaling and wavelet functions are defined with the following properties:

1. The area under the scaling function is normalized to one, i.e.,

$$\int_{-\infty}^{\infty} \phi(t)\, dx = 1.$$

2. The scaling function and its translates are orthonormal, i.e.,

$$\int_{-\infty}^{\infty} \phi(t-j)\phi(t-m)\, dx = \delta_{j,m}.$$

3. The wavelet function has $N/2$ vanishing moments, i.e.,

$$\int_{-\infty}^{\infty} t^k \varphi(t)\, dx = 0, \quad k = 0, 1, \ldots, N/2 - 1.$$

4. The scaling and wavelet functions are orthogonal, i.e.,

$$\int_{-\infty}^{\infty} \phi(t)\varphi(t-m)\,dx = 0, \quad m \in Z,$$

   where, Z is a set of integers.

5. The translates of scaling and wavelet functions on each fixed scale j span the orthogonal subspaces, i.e.,

$$V_j = 2^{j/2}\phi(2^j t - k), \quad k = \ldots, -1, 0, 1, \ldots,$$
$$W_j = 2^{j/2}\varphi(2^j t - k), \quad k = \ldots, -1, 0, 1, \ldots$$

   such that V_j form a sequence of embedded subspaces

$$V_{j+1} = V_j \oplus w_j, \quad V_0 \subset V_1 \subset \ldots \subset V_{j+1},$$
$$\bigcap_{j\in Z} V_j = 0, \quad \bigcup_{j\in Z} V_j = L^2(R)$$
$$V_{j+1} = V_0 \oplus W_0 \oplus W_1 \oplus \ldots \oplus W_j,$$

   where, Z is a set of integers and $L^2(R)$ denotes the space of square integrable functions.

An approximation of $f(t)$in $L^2(R)$ using scaling function as basis at a certain level (resolution) j is then represented as:

$$P_j(f(t)) = \sum_k c_{j,k}\phi_{j,k}(t), \quad k \in Z,$$

where, $P_j(f(t))$ be the approximation of function, k represents translation parameter and $c_{j,k}$ represent expansion (approximation) coefficients. Similarly, let $Q_j(f(t))$ be the approximation of function $f(t)$ in $L^2(R)$ using wavelet function as basis at a certain level (resolution) j, and is represented as:

$$Q_j(f(t)) = \sum_k d_{j,k}\varphi_{j,k}(t), \quad k \in Z.$$

The approximation at next higher level of resolution is then given by:

$$P_{j+1}(f(t)) = P_j(f(t)) + Q_j(f(t)),$$

which, forms the basis for multi-resolution analysis of wavelet approximation [5].

2. WAVELET FINITE ELEMENT METHOD

The orthogonality, compact support and multiresolution capability of Daubechies' wavelets make them good candidates for choice as basis functions in finite element approximations by using Galerkin weighted residual approach. Considering the governing equation of motion for a single degree of freedom system given by:

$$m\ddot{u} + c\dot{u} + ku = f.$$

The wavelet basis approximation for displacement, velocity and forcing function at scale (resolution) j, for the entire duration may be given by:

$$u(t) = \sum_k u_k \phi(2^j t - k), \quad k \in Z, \tag{5}$$

$$v(t) = \sum_k v_k \phi(2^j t - k), \quad k \in Z, \tag{6}$$

$$f(t) = \sum_k f_k \phi(2^j t - k), \quad k \in Z, \tag{7}$$

where, u_k, v_k and f_k denote the expansion coefficients in wavelet space for displacement, velocity, and forcing function respectively at location defined by translation parameter k, and Z represents the set of integers. Substituting the wavelet basis approximations of displacement, velocity and forcing function in the equivalent first order form of the equation of motion (Eq. (2)), the residual of approximation may be given by:

$$R = 2^j \begin{Bmatrix} \sum_k u_k \dot{\phi}(2^j t - k) \\ \sum_k v_k \dot{\phi}(2^j t - k) \end{Bmatrix} + \begin{bmatrix} 0 & -1 \\ m^{-1}k & m^{-1}c \end{bmatrix} \begin{Bmatrix} \sum_k u_k \phi(2^j t - k) \\ \sum_k v_k \phi(2^j t - k) \end{Bmatrix} - \begin{Bmatrix} 0 \\ m^{-1} \sum_k f_k \phi(2^j t - k) \end{Bmatrix}. \tag{8}$$

We consider the strong form of weighted residual approach and adopt the translates (in time) of the scaling function belonging to the same space and given by $\phi(2^j t - l)$ (where, $l \in Z$ represents the translation parameter) as the weighting function for constructing the weighted residual statement as:

$$\int_{\Omega_T} W R \, d\Omega_T = 0,$$

where, Ω_T denotes the total duration ($[0, t_f]$), and W represents a weighting function. Considering vanishing of the inner product of the residual with the weighting function and considering the orthogonality property of the scaling function, the final set of algebraic equations can be arranged in matrix form as:

$$\mathbf{B}\hat{\mathbf{d}} = \hat{\mathbf{f}} \tag{9}$$

where,

$$\mathbf{B} = \begin{bmatrix} \mathbf{\Omega}^{0,1}_{k-l} & \delta_{k,l}\mathbf{I} \\ m^{-1}k \times \delta_{k,l}\mathbf{I} & \mathbf{\Omega}^{0,1}_{k-l} + m^{-1}c \times \delta_{k,l}\mathbf{I} \end{bmatrix}, \quad \hat{\mathbf{d}} = \begin{Bmatrix} \mathbf{u}_k \\ \mathbf{v}_k \end{Bmatrix}, \quad \hat{\mathbf{f}} = \begin{Bmatrix} 0 \\ \mathbf{f}_k \end{Bmatrix},$$

and $\mathbf{\Omega}^{0,1}_{k-l}$ are known as the connection coefficients as defined by Latto [8], and are nonzero only in the interval $k-l = -DN+2$ to $k-l = DN-2$ where, DN refers to the order of the wavelet used for approximation and $\delta_{k,l}$ is the Kronecker delta. In this formulation the domain $[0, t_f]$ is discretised by say $N_x = 2^j + 1$ points and expanded on either side by $DN-1$ points therefore, the indices k and l range from $(-DN+1)$ to (N_x+DN-1) thereby resulting in a linear system of $N_x + 2(DN-1)$ equations and unknowns. Since two-field formulation has been employed here it contains $2(N_x + 2(DN-1))$ number of equations and unknowns (displacement and velocity components). The evaluation of the connection coefficients that arises in the wavelet-Galerkin formulation of the ODE forms a crucial part of the solution procedure. The connection coefficients depend on the basis function used for approximation and on the order of the derivatives that exists in the weighted residual formulation of the governing ODE given by Eq. (9). These are independent of the problem domain, therefore can be computed *a priori* and used as an input in the solution procedure for populating the wavelet-Galerkin system matrix. The computation scheme for evaluation of the connection coefficients is suggested by Restrepo and Leaf [11] based on a procedure devised by Latto, et al. [8]. Subsequent to the computation of connection coefficients the initial conditions $u(0) = u_0$ and $v(0) = v_0$ are incorporated in wavelet space by augmenting the set of algebraic defined by Eq. (9) with the following constraints:

$$\sum_{k=-1+DN}^{N_x+DN-1} u_k \phi(-k) = u_0, \tag{10}$$

$$\sum_{k=-1+DN}^{N_x+DN-1} v_k \phi(-k) = v_0 \tag{11}$$

by replacing the appropriate coefficients as suggested by Lu [9] for solution of boundary value problems (BVPs). This procedure preserves the original size of the system of algebraic equations which can be subsequently solved using a linear equation solver like Gauss elimination to compute the unknown coefficients (i.e., expansion coefficients of displacement and velocity in wavelet space). The solution in physical space (displacement and velocity response) is computed by transforming the coefficients from wavelet space to physical space by using the wavelet approximation (Eqs. (5) and (6)).

3. NUMERICAL EXAMPLES

The suggested computational scheme for solving equation of motion is illustrated by means of the following numerical examples:

Free Vibration of a SDOF System. Computation of the free vibration response of an undamped SDOF system with $m = 1$ kg, $k = 100$ N/m and a damped SDOF system with similar characteristics and with damping ratio of $\zeta = 0.05$, with initial conditions $u(0) = 1$ m and $v(0) = 0$ m/s.

Forced Vibration of a SDOF System. Computation of the forced vibration response of an un damped SDOF system, starting from rest, with $m = 1$ kg, $k = 10$ N/m excited by an harmonic force $f = \sin t$.

The free vibration response of the SDOF oscillators computed by the proposed wavelet basis hierarchical finite element (WFE) approximation is compared with the analytical and numerical solutions (obtained by using Newmark constant average acceleration scheme) in Figs. 1 and 2. Similar comparison for the case of forced vibration of SDOF system is shown in Fig. 3.

It can be observed that WFE solutions compared favorably with corresponding analytical solutions as well as with Newmark's solutions for the test problems. The computation time for WFE solution is similar to that required for Newmark scheme.

For the general case of multidegree of freedom system (MDOF), the coupled governing differential equations of motion can be decoupled using mode decomposition method. The WFE approximations can be developed for each modal equation independently by the procedure outlined above for the response of SDOF oscillator. The computed solution in modal space is then transformed back to the physical space to get the desired response of MDOF system. The advantage of the proposed formulation in using compactly supported orthogonal Daubechies' wavelets for MDOF systems lies in utilizing its inherent multi-resolution capability of representing the solution. The response of each modal equation can be approximated at a certain level of resolution determined by the frequency of the oscillator for a desired level of accuracy defined by the error threshold value. Further, the solution of lower modes can then be interpolated suitably to the resolution adopted for the high frequency oscillators without any loss of accuracy and with little extra cost as described by Gordan and Shampine [6]. Hence, this solution procedure exploits the multiresolu-

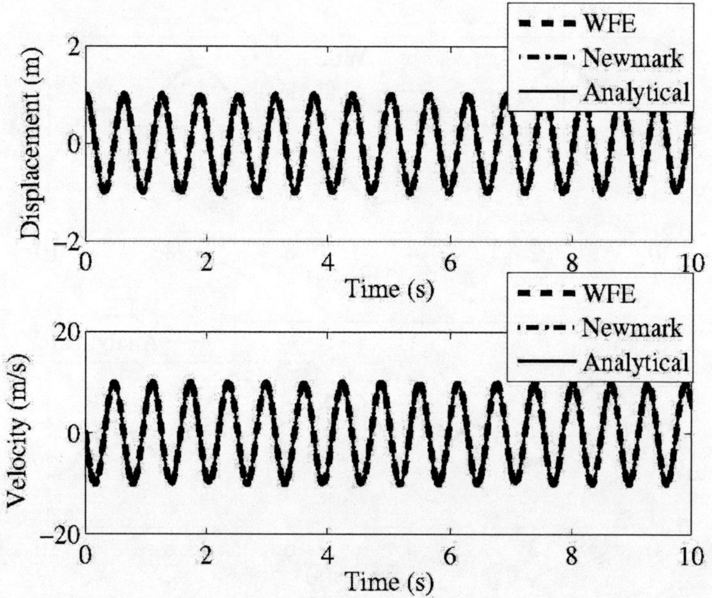

Fig. 1. Undamped free vibration of SDOF system

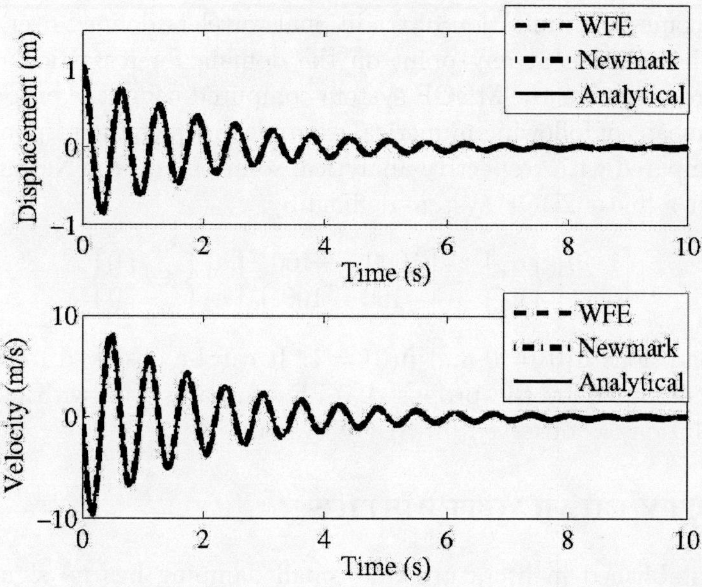

Fig. 2. Damped free vibration of SDOF system with $\zeta = 0.05$

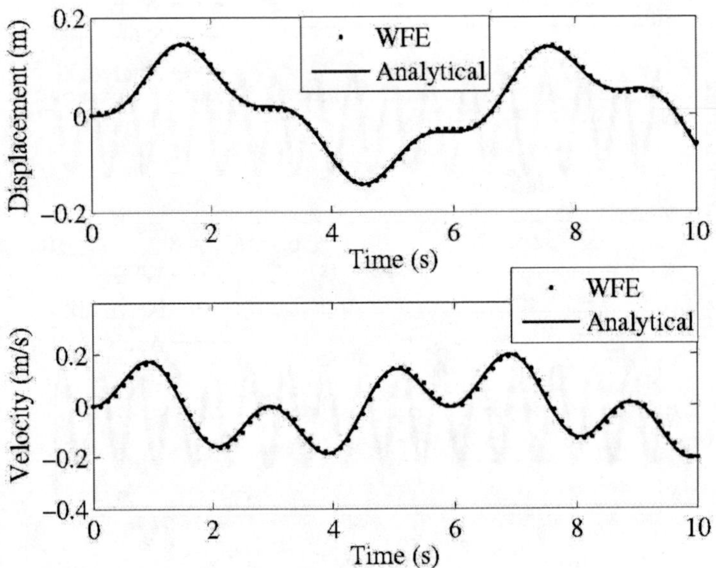

Fig. 3. Undamped forced vibration of SDOF system

tion features of wavelets to provide a hierarchical form of solution. Moreover, this approach overcomes the limitation of traditional wavelets defined over dyadic grids in obtaining the solution at any point on the domain for a particular resolution. The dynamic response for the MDOF system computed using the proposed WFE is illustrated by means of following numerical example in which the response computed by WFE is compared with respective analytical solution and the Newmark method. Let us consider a linear 2DOF system defined by:

$$\begin{bmatrix} 1 & 0 \\ 0 & 1 \end{bmatrix} \begin{Bmatrix} \ddot{u}_1 \\ \ddot{u}_2 \end{Bmatrix} + \begin{bmatrix} 1000 & -100 \\ -100 & 100 \end{bmatrix} \begin{Bmatrix} u_1 \\ u_2 \end{Bmatrix} = \begin{Bmatrix} 0 \\ 0 \end{Bmatrix}$$

with initial conditions $\mathbf{u}(0) = 0$ and $\dot{\mathbf{u}}(0) = 1$. It can be observed from Fig. 4 that the response computed by the proposed WFE compares well with analytical and Newmark's solutions as observed for SDOF system.

4. STABILITY CHARACTERISTICS

It has been established in literature that small damping has no significant effect on stability and accuracy of algorithms [13]. Hence the algorithmic analysis and evaluation is generally restricted to free and undamped vibration problems. Also MDOF system can be uncoupled using modesuperposition method and it has been

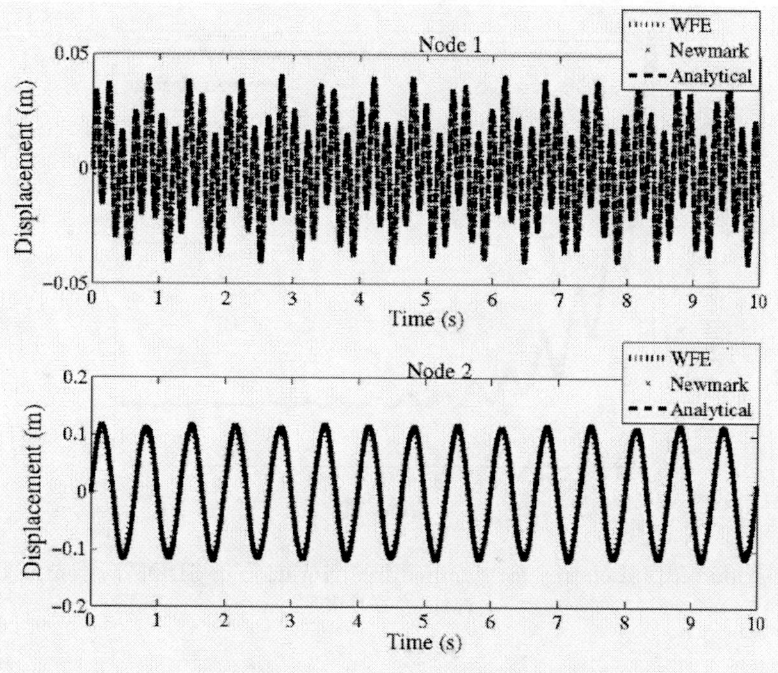

Fig. 4. Undamped free vibration of MDOF system

well established that integration of uncoupled equations is equivalent to integration of original system. Further, the conclusions on stability and accuracy of linear Single degree of freedom (SDOF) systems are valid for MDOF systems also [13]. In this study the stability characteristics are evaluated using energy method for linear SDOF systems, as energy balance provides an excellent way of checking the stability of explicit calculation [7]. Energy balance is defined as

$$|W_{\text{int}} + W_{\text{kin}} - W_{\text{ext}}| \leq \delta \|W\|,$$

where, W_{int} refers to internal energy or Potential energy, W_{ext} refers to work performed by external nodal forces and W_{kin} corresponds to work done by kinetic energy, $\|W\|$ refers to energy norm and δ refers to tolerance, which is problem dependent. For convenience, the energy norm is taken as $\|W\| = |W_{\text{ext}}| + |W_{\text{int}}| + |W_{\text{kin}}|$, as this quantity is indicative of total energy in the system. For examining the present formulation the energy parameters are defined as: $W_{\text{ext}} = F_{ext}(t) * u(t)$, $W_{\text{kin}} = 0.5 * m * v(t)^2$, $W_{\text{int}} = 0.5 * k * u(t)^2$, and $W_{\text{damp}} = c * v(t)$ where $u(t)$ and $v(t)$ are the displacement response and velocity response computed using proposed approach and c is the dissipative (damping) force. It can be observed from energy balance equation that in case of free undamped vibration the tolerance parameter is

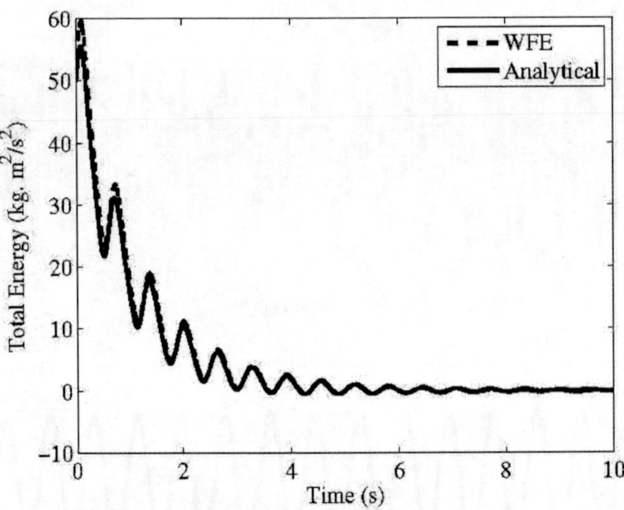

Fig. 5. Variation of total energy for damped free vibration of SDOF system with damping ratio $\zeta = 0.05$

always unity. This implies that it is energy conserving for free undamped SDOF system. The proposed formulation has been examined for the stability characteristics of various free undamped and damped SDOF systems and found to be energy conserving for undamped free vibration systems and in case of damped free vibration, it is found that the total energy approaches zero with time. Hence, it is L-stable as shown in phenomenon can be observed in the results depicted in Fig. 5. Further, it can be observed that the total energy of damped system follows the total energy of analytical solution. Hence the proposed methodology mimics the true behavior of analytical or exact solution.

5. ACCURACY CHARACTERISTICS

The accuracy of the computed response (displacement and velocity)by WFE depends on the resolution adopted for approximation as given by Eqs. (3) and (4) which is a function of error threshold value. Therefore the resolution to be adopted for computing the response of a particular oscillator can be chosen based on the frequency content of the oscillator for a given error threshold value. In the present formulation an L^2 norm of relative error has been adopted for fixing the error threshold value and is written as:

$$\|\text{Error}\|_2 = \int_0^{Td} \left(\frac{u_{\text{WFE}} - u_{\text{Analytical}}}{u_{\text{Analytical}}} \right)^2 dt. \tag{12}$$

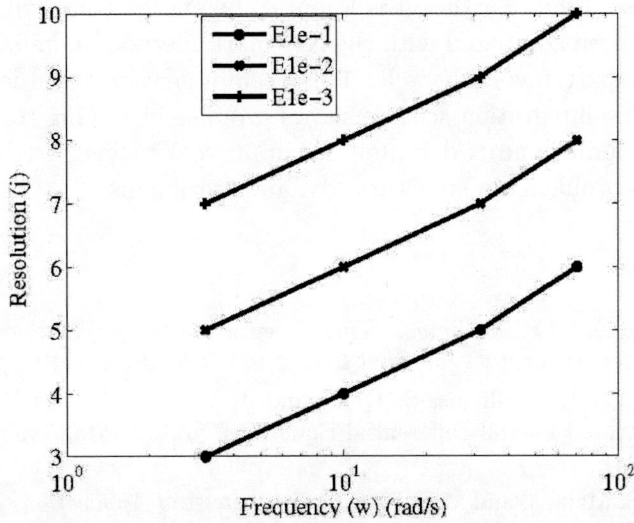

Fig. 6. Variation of required resolution with frequency

The variation of required resolution for approximating a harmonic wave within a frequency content (radians/sec) for different specified error norm is shown in Fig. 6 for various frequencies. This gives an idea of resolution to be adopted for approximating the response of SDOF oscillators of certain natural frequency and also specifies the global error involved in the approximation in terms of error threshold value. In case of forced vibration problems the resolution can be chosen on the basis of the highest frequency response to be represented in the solution. This feature facilitates in reducing the computational cost of the solution procedure in the analysis of MDOF systems using the mode superposition method and also aids in the development of hierarchical form of solution.

6. CLOSURE

A new time-integration scheme (WFE) has been proposed for temporal approximation of structural dynamic problems. This method features the ability to utilize traditional finite element method for spatial approximation and utilizes Daubechies' wavelets as basis functions in finite element method for further approximation of semi-discretised ODE in time. Two-field formulation has been considered, as displacement and velocity can be treated independently. Further, fictitious domain approach is adopted for incorporating initial conditions. The proposed formulation is examined for algorithmic characteristics and found that it mimics the true behavior of analytical solution. Further it is found to be energy conserving and L-Stable. Moreover, it has been compared with the Newmark method and analytical solutions and found to compare favorably well. This preliminary investigation suggests that wavelet based time integration scheme shows promise in solving transient problems in structural dynamics and is definitely an improvement over traditional FEM for solving transient problems in structural dynamic problems.

REFERENCES

1. K. Amaratunga and J.R. Williams, "Time Integration Using Wavelets," in *Proceedings of SPIE, Wavelet Application for Dual Use, 2491, Orlando, FL* (1995), pp. 894–902.
2. K. Amaratunga, J.R. Williams, S. Qian, and J. Wiess, "Wavelet-Galerkin Solutions for One-Dimensional Partial Differential Equations," Int. J. Num. Meth. Engng **37** (1), 2703–2716 (1994).
3. R.D. Cook, D.S. Malkus, and M.E. Plesha, *Concepts And Applications of Finite Element Analysis*, Third edition (John Wiley and Sons, Singapore, 1989).
4. W. Dahmen, "Wavelet Methods for PDEs Some Recent Developments," J. Computat. Appl. Math. **128** (1/2), 133–185 (2001).
5. I. Daubechies, "Orthonormal Bases of Compactly Supported Wavelets," Comm. Pure Appl. Math. **41**, 906–966 (1988).
6. M.K. Gordon and L.F. Shampine, "Interpolating Numerical Solutions of Ordinary Differential Equations," in *ACM 74: Proceedings of the 1974 Annual Conference*, Vol. 1 (ACM, New York, 1974), pp. 46–53 [http://doi.acm.org/10.1145/800182.810378].
7. T.J.R. Hughes, "Analysis of Transient Algorithms with Particular Reference to Stability Behavior," in *Computational Methods for Transient Analysis*, Ed. by T. Belytschko and T.J.R. Hughes, Part 2 (North-Holland, Amsterdam, 1983), pp. 67–155.
8. A. Latto, H.L. Resnikoff, and E. Tanenbaum, "The Evaluation of Connection Coefficients of Compactly Supported Wavelets," in *Proceedings of the FrenchUSA Workshop on Wavelets and Turbulence, Princeton university* (Springer-Verlag, New York, 1991).
9. D. Lu, T. Ohyoshi, and L. Zhu, "Treatment of Boundary Conditions in the Application of Wavelet-Galerkin Method to a SH Wave Problem," citeseer.ist.psu.edu/84953.html.
10. S. Qian and J. Weiss, "Wavelets and the Numerical Solution of Partial Differential Equations," J. Comput. Phys. **106** (1), 155–175 (1993).
11. J.M. Restrepo and G.K. Leaf, "Wavelet-Galerkin Discretization of Hyperbolic Equations," J. Comput. Phys. **122**, 118–128 (1995).
12. C.H. Romine and B.W. Peyton, "Computing Connection Coefficients of Compactly Supported Wavelets on Bounded Intervals," citeseer.ist.psu.edu/romine97computing.html.
13. K.K. Tamma, X. Zhou, and D. Sha, "The Time Dimension: A Theory Towards the Evolution, Classification, Characterization and Design of Computational Algorithms for Transient/Dynamic Applications," Arch. Comput. Meth. Engng **7** (1), 67–286 (2000).

MATHEMATICAL MODELLING OF HUMAN CAROTID IN HEALTHY, AFFECTED OR POST-CORRECTIVE SURGERY CONDITIONS

L.Yu. Kossovich[1*], I.V. Kirillova[1], Yu.P. Gulaev[1], D.V. Ivanov[1], A.V. Kamenskiy[2], V.O. Polyaev[2], N.V. Ostrovskiy[3], and K.M. Morozov[4]**

ABSTRACT

The following investigation of carotid artery bifurcation(CAB) behavior in healthy, affected or post-corrective surgery conditions (Russian Foundation for Basic Research Grant No. 06-01-00564) is aimed at solving the medical and social problems relevant to optimization of the cerebral circulation disorders surgical treatment.

Key words: atherosclerosis, carotid endarterectomy, patch material, clinical follow-up, FSI coupled simulation, stress strain state

1. STATEMENT OF THE PROBLEM

Over a long period, cerebrovascular diseases have been taking one of the leading positions in morbidity and mortality rate structures in the world [1]. According to Prof. A.V. Pokrovskiy's [2] statistics, stroke is the second mortality reason in modern Russia after heart attack. The subsequent formidable socio-economic damage is obvious, as the speed of the stroke-caused mortality rate growth is at its highest

[1]Saratov State University, Saratov, Russia
[2]University of Nebraska-Lincoln, USA
[3]Saratov State Medical University, Saratov, Russia
[4]Bakoulev Scientific Center of Cardiovascular Surgery of the Russian Academy of Medical Sciences, Moscow, Russia
[*]E-mail: *rector@sgu.ru*
[**]E-mail: *mcm07@sgu.ru*

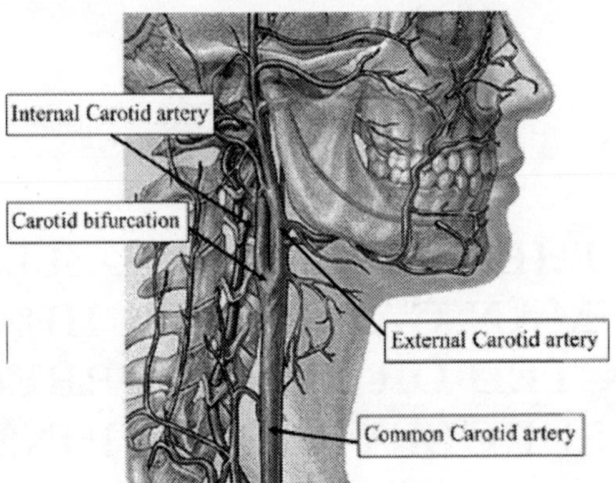

Fig. 1. Carotid artery

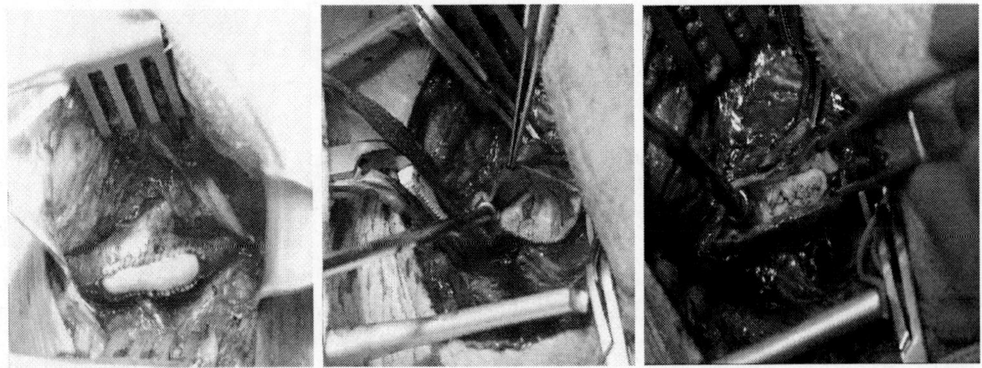

Fig. 2. Carotid endarterectomy surgery phases

in people aged 30–50. Moreover, 50% of the survived patients go through occasional strokes in the course of the next five years. It has been detected, that 80% of strokes are of ischemic geneses. The reasons for ischemic strokes are both, ischemia, associated with the vessel stenosis or occlusion, and blood flow disorders caused by the cerebral vessels embolism (arterio-arteriolar embolism type). Most researchers presume that atherosclerosis is the reason of stenosis and occlusion of the extracranial parts of cerebral arteries in 84–90% of ischemic damage cases. In 70% of the cases the carotid zone (Fig. 1) is the source of arterio-arteriolar embolism. Unfortunately, carotid artery disease used to be treated only by means of surgery. Carotid endarterectomy (CEA) is the standard surgical procedure. The latest minimally invasive endovascular intervention is called carotid artery angioplasty with stenting. A lot of research has been done on stent modeling recently [2].

In carotid endarterectomy, an atherosclerotic plaque that has formed on the inside of the carotid artery wall is removed surgically (Fig. 2). Most of the authors have historically agreed on the need of making patching operation after endarterectomy [4–7]. Many of them believe patching the operated zone of the vessel to reduce the risk of restenosis development. At present, therefore, CEA with the following arteriotomic passage patching is considered to be the "golden standard" of such surgical correction.

The choice of patching materials is rather wide, but since there are no indications for using a definite type of patches for a specific patient, the surgeon's preferences are based on his/her experience and intuition rather than on objective individual peculiarities of the patient's carotid artery. That may contribute to the number of restenoses that occur in the reconstructed zones. According to various authors, this number varies substantially [8], but generally ranges around 10–15%. Therefore, the question of choosing the most adequate patching material for closing the arteriotomic hole of the external carotid artery (ECA) outfall after endarterectomy that would have minimum negative effects on the patient is still open.

Developing indications for choosing patching materials is closely associated with the recent theory of hemodynamic atherogenesis. The essence of the theory lies in determination of hemodynamic and mechanical factors that would signal of atherosclerosis formation in the carotid bifurcation zone. A lot of attempts were made over years to reveal this sort of parameters [9–12]. Among these factors are the low or oscillating wall shear stress (WSS), high cyclic strain (CS) and high effective stress (ES) [13–15].

It was determined, that one may observe monocytes adherence to the endothelium in the low value WSS zones (<1.5 Pa), which is considered to be an early stage of atherogenesis.

Even though a good correlation between the low WSS regions and zones of the atherosclerosis formation was noted by many authors, the low WSS theory fails to explain some zones of atherosclerotic thickening in the inner wall of ECA and flow divider, where the WSS admits high values [16].

It was also determined, that endothelial tissue damage, which might be caused by mechanical intervention (such as high CS and high ES) leads to the increase of its flow capacity and is the reason of the lipids accumulation by the unstriated musclescells [17–19]. As the process progresses, the cells die to form an atherosclerotic plaque.

Thus, the major goal of the present study consists in working out the objective indications for carotids patching. This requires building a maximally precise mathematical model and studying the vessel mechanics from the point of view of the hemodynamic theory of atherogenesis.

Development of the model makes it possible to conduct a "virtual operation" at the pre-operational stage of the patient examination.

Research was carried out in the following directions:

- series of experiments were made to determine the physical and mechanical parameters of the carotids wall (common, external and internal carotid artery (ICA)), circle of Willis' arteries [20, 21], as well as patches made of synthetic and biological materials used in modern patching;
- comparative analysis of the common carotid artery (CCA) and its branches minute structures were made;
- geometrical and morphometrical parameters database of the CCA, ICA and ECA of the human was created (age, sex, and body types were taken into account) based on the morphometrical investigation;
- parameter selection method for the vessel wall and patch models was worked out on the basis of experimental data;
- 3-D reconstructions of velocity profiles, based on the *in vivo* data of the ultrasound Doppler sonography of various carotids zones were performed;
- methods for constructing precise 3-D geometries of arteries at health and different stages of medical condition as well as post-surgical intervention were worked out;
- a numerical and analytical 3-D model was constructed of the blood flow dynamics and stress strain state (SSS) for various models of human vessel walls (linear model, nonlinear isotropic model, nonlinear orthotropic model) in the carotids in healthy, affected and post-reconstructive surgery conditions;
- types of pathological tortuosities of carotid artery (kinking, coiling (or looping)) (Figs. 3 and 4) were examined;
- data on blood movement and artery walls was obtained from the numerical experiment results. The results were analyzed and compared with the clinical data.

2. METHODS

2.1. *Mechanical properties*

Generally, the carotid wall behavior is anisotropic [22]; specially developed biaxial experiments are required [23] to capture its response. Those were not available, and uniaxial tensile tests on TiraTest 28005 (registration number 23512-02 in the State Register of the Russian Federation) with 100N loadcell were performed instead. The values of the applied force (H) and stretch (mm) in the application of force direction were recorded in the course of the experiment.

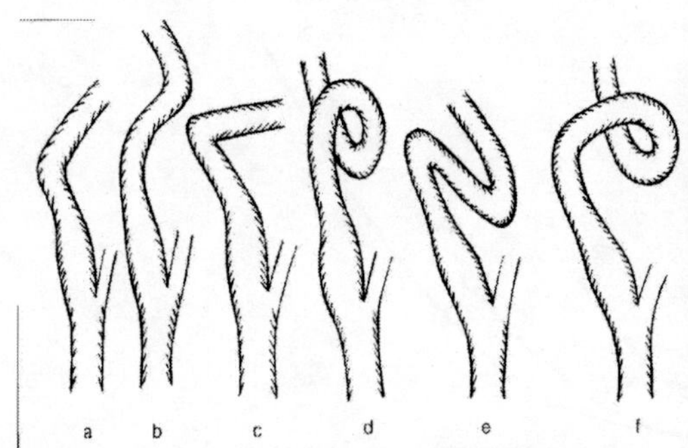

Fig. 3. Different types of carotid artery tortuosities (a, b, c, e — kinking, d, f — coiling)

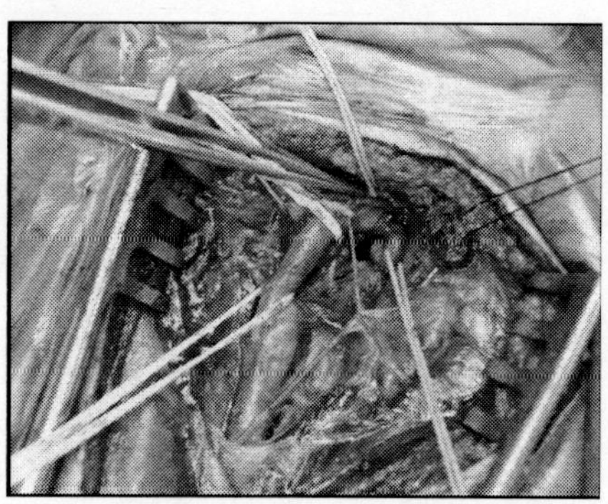

Fig. 4. Perioperative image showing the coiling lesion of the left extracranial ICA

Table 1. Cadaveric data

No	Sex	Years of age	Cause of death	Identification No
1	M	1967–2005	emetic mass aspiration	3411
2	F	1942–2005	acute coronary deficiency (ACD)	3449
3	M	1987–2005	blunt head trauma	3448
4	M	1956–2005	lungs cancer	3615
5	M	1923–2005	acute coronary deficiency	3625
6	F	1924–2005	peritonitis	3627
7	M	1974–2005	alcohol poisoning	3631
8	M	1972–2005	pneumonitis	3681
9	F	1971–2005	fall from the height	3676
10	M	1957–2005	alcohol poisoning	3767
11	M	1953–2005	foreign body aspiration	3843
12	M	1931–2005	acute coronary deficiency	3900
13	M	1957–2005	acute coronary deficiency	3923
14	M	1957–2005	destructive pancreonecrosis	3920
15	M	1940–2005	drown	3981

The total of 15 cadaveric human common carotids was tested for mechanical properties. Samples were harvested no later than 16–18 hours after death and stretched no later than two hours after autopsy in order to avoid any changes in the tissue mechanical properties. Table 1 presents the cadaveric data on the tested carotids, inclusive of sex, age, cause of death and identification number.

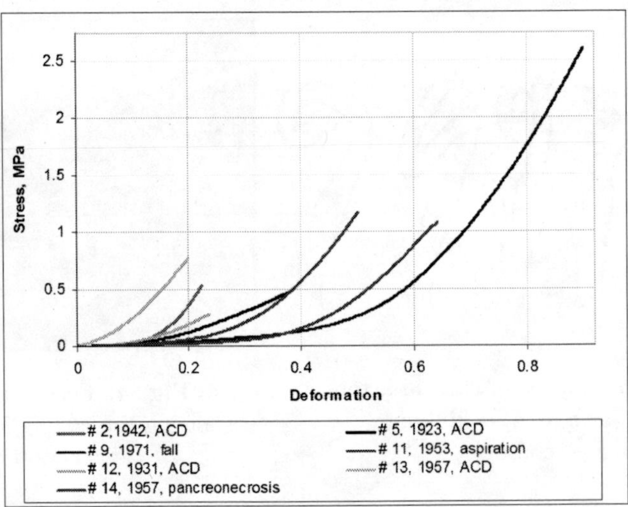

Fig. 5. Strain–Stress (kPa) experimental relations on seven carotids under uniaxial stretching

Alongside with the load and stretch in the principal direction, the changes in the sample thickness were recorded optically during the experiment. Stretch in the transverse direction was computed from the tissue incompressibility condition [24]. It turned out, however, that upon the load application, the sample thicknesses changed insignificantly.

The stress-strain curves for all the tested carotids were computed after 10 cycles of preconditioning. Seven stress-strain curves are presented in Fig 5.

Mechanical properties of all the samples differed substantially; nevertheless, common non-linear behavior all of them may be observed.

We also note that carotids highly affected with atherosclerosis turned out to be more stiff compared to less affected ones, even though samples did not contain any signs of fibrous plaque.

We have tested synthetic and biological types of patches, such as polytetrafluorinethylene (PTFE), knitted polyester, bovine pericardium used by the surgeons in Bakoulev's Scientific Center of Cardiovascular Surgery of the Russian Academy of Medical Sciences (see table 2).

Mechanical properties were determined with similar uniaxial extension technique in either of the two directions in order to determine the necessity of distinguishing the patch orientation during its placement. Behavior of the sample under uniaxial loading was almost linear on the applied load range. Bovine pericardium patches stretched along the fibers demonstrated the same quality results as arteries (Fig. 6).

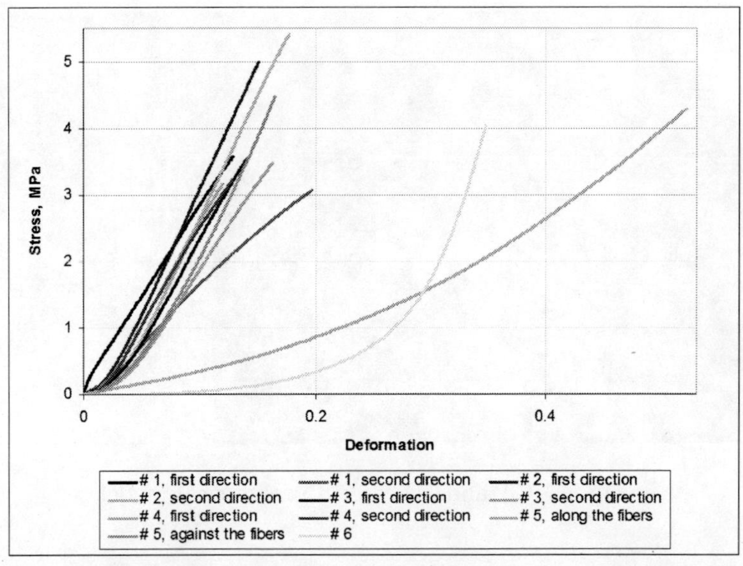

Fig. 6. Strain - Stress (MPa) experimental relations for patching materials stretched separately in two different directions

Table 2. Patch materials descriptions

Patch #	Description	Thickness, mm	Density, gram/cm^3
1	Polytetrafluorinethylene PS 442 93 0101	0.4	0.5979
2	Polytetrafluorinethylene PS 433 15 0101	0.4	0.369
3	Polytetrafluorinethylene PS 442 19 0107	0.6	0.3548
4	Polytetrafluorinethylene PS 442 94 0101	0.6	0.6296
5	Textile Belorussian Medical patch for carotid operations	0.45	0.5081
6	Xenopericardium	0.45	1.6179

2.2. *3D and Finite element modeling*

Since the 3D geometrical reconstruction of the carotids is difficult, we have developed a method for model construction from the *in vitro* data, the computer tomography

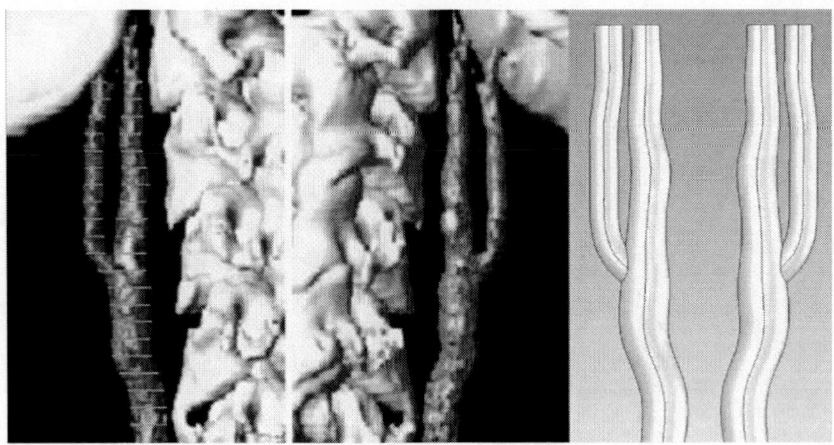

Fig. 7. CT and 3D carotid geometrical model

(CT) and magnetic resonance imaging (MRI) *in vivo*; this allows to restore the vessel geometry which is maximally close to the original one. MRI is an ideal imaging modality for providing geometric models for fluid-solid interaction analysis. The special program package SolidWorks 2007 (SolidWorks Corporation) was used to create the precise 3D carotid geometrical model (Fig. 7).

Construction of a 3D model consists of three base stages: data acquisition, image segmentation and transformation of the 2D model into the 3D structure.

The obtained geometrical model is imported into the finite element model software ADINA 8.4.4. (ADINA R& D, Inc., USA). ADINA performs finite element analysis of the models of healthy, affected and reconstructed vessels, fluids, and fluid flows with structural interactions. Solid bodies of the vessel walls, the blood flow and the patch were imported into the ADINA-Structures and ADINA CFD modules as an assemble through Parasolid format.

Blood was modeled as a laminar flow of an incompressible, Newtonian fluid with 0.004 Pa·s viscosity and 1100 kg·m^3 density. Due to geometry complexity, unstructured meshes were used for both, fluid and solid domains. We have used 29767 4-node linear elements to model the fluid and 35185 11-node quadratic elements to model the vessel wall. 11-node elements for solid bodies were used for stability considerations while solving the problems involving hyperelastic materials.

We have considered three models for vessel wall material: linear elastic isotropic, hyperelastic isotropic and hyperelastic orthotropic with large displacements and large strains, while the patch was assumed to be linear elastic, with the model parameters, determined from uniaxial experiments. Simple rigid contact through the bounding surfaces between the patch and the wall was used to force the synchronous

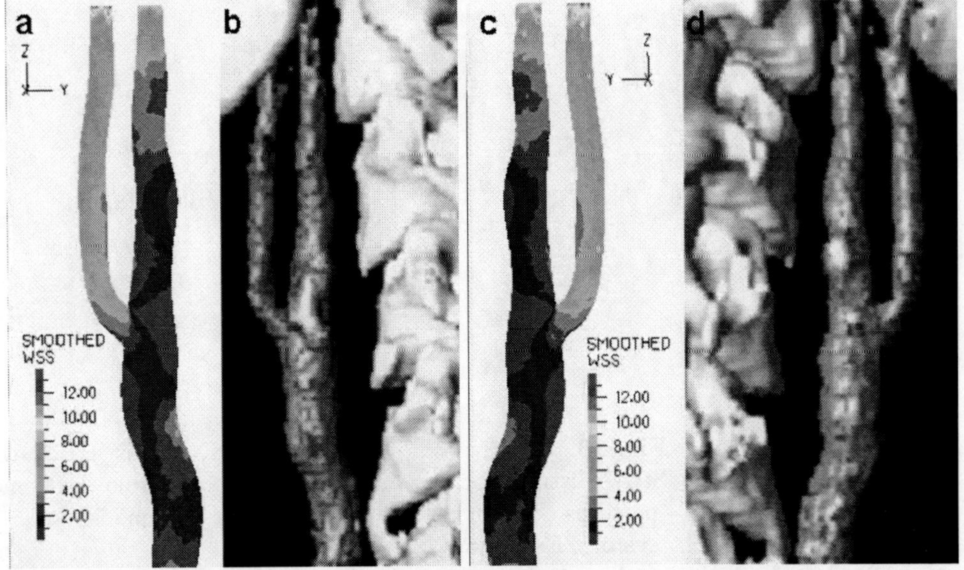

Fig. 8. Wall Shear Stress distribution patterns (Pa) on both sides of the carotid at systole, smoothed patterns; angiogram of the carotid, both sides

deformation of these two bodies. The inlet of the common carotid artery was loaded with the normal pressure, while the velocity profiles were applied to the nodes of the ECA and ICA outlet cross sections. No-slip boundary conditions on the walls were used for the blood flow.

3. RESULTS AND DISCUSSION

3.1. *No-patch case*

The SSS of the vessel was investigated in terms of the hemodynamic parameters relevant to atherogenesis. The calculations were carried out for the cases of linear isotropic, nonlinear isotropic and nonlinear orthotropic materials. Realization of the orthotropic model with by means of FEM presented certain difficulties during the process of its separation into simple volumes, where the orthotropy orientation may be univalently determined in any point. Comparative analysis (based on the problem solutions for the blood flow through the healthy carotid bifurcation with rigid and elastic walls) was carried out in order to analyze the vessel wall influence on the blood flow. The resulting solution of the problem showed a correlation between the low WSS zones (< 1.5 Pa) [10] in the CA bulb and the atherosclerosis-affected zones, visible on the CT (see Fig. 8).

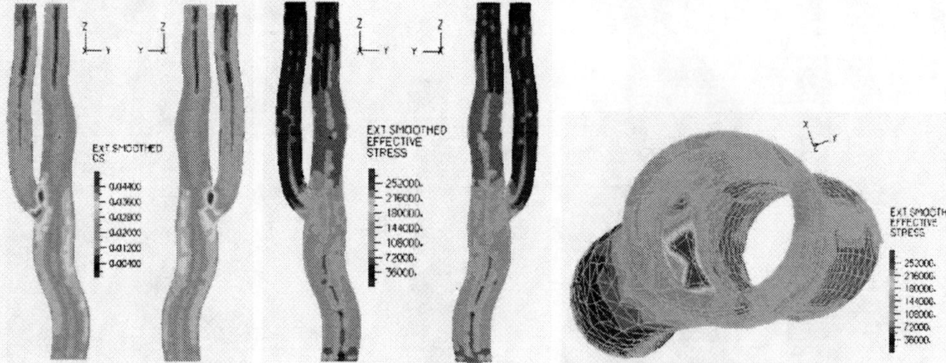

Fig. 9. Von Mises Cyclic Strain patterns for both sides of the carotid. Cyclic strain is computed as the difference between corresponding systolic and diastolic values. Extreme smoothing technique is used

Fig. 10. Effective Stress distribution patterns (Pa) for systole. Extreme smoothing technique is used

Fig. 11. Effective stress (Pa) in the apex cut zone for systole. Extreme smoothing technique is used

The comparison of WSS distributions for one linear and two non-linear models showed a non-significant difference of results. The low values were observed in the CA bulb region, while in the apex the WSS value was high. A slightly broader zone of low WSS was observed in the nonlinear isotropic case.

The fact that the zones of high WSS were in the apex zone (stream division zone) where the atherosclerotic plaques can be clearly seen on the angiogram, proves, that it is necessary to consider additional mechanic criteria, such as CS [13, 16, 25] and ES distribution [26]. CS extreme smoothing distribution is presented in Fig. 9.

The maximum value of CS is reached at the apex from the side wall of the ICA. The second largest value observed is located at the external-common adjoining wall. This is especially evident in the left side pattern. These two areas correspond to the atherosclerotic plaque formation seen on the angiogram.

Besides, the high CS areas are located at the CCA wall, opposite to the apex. The upper part of this area also corresponds to the atherosclerosis formation on the angiogram. It is also worth noticing that the maximum CS values are observed at the inner wall of the vessel, closer to the flow. The maximum value registered is 0.04901, it corresponds to the apex zone inner wall.

CS distribution comparison for the three models showed a qualitative and rather significant difference. The high CS values were most localized in the case of the non-linear orthotropic model. The closest correspondence of the high CS zones and atherosclerosis-affected zones on CT was observed in this model, as well.

ES distribution patters (Pa) for systole are presented in Fig. 10. The maximum ES values are observed at the apex (see Fig. 11) zone. This result is quite obvious and predictable, since it is reasonable to expect high stress values at the point where the flow crashes into the divider wall at high speed, possessing significant force at a small area of the apex.

ES distribution is relatively uniform, except for the zones near the divider at the external-common adjoining wall. This zone is similar to the high CS zones obtained above. The ES values are higher at the inner wall of the vessel than at the outer wall in the zones that correlate with atherosclerosis affection seen on the angiogram [26].

The most non-homogeneous ES distribution was observed in the linear isotropic model. The non-linear isotropic model showed a bit more homogeneous distribution. The high ES values in this case were more concentrated. The highest ES localization was observed in the non-linear orthotropic model.

3.2. *Patch case*

"Virtual" operation was performed based of the surgeon's recommendation. A plaque as well as a fraction of the adhering media layer was eliminated. The detailed comparison of patches inter se as well as with the healthy vessel case was carried out.

Information of all six patching materials was utilized in the finite element model. The focus was made on the following quantities: WSS, CS and ES. We assumed the similarity of the flow conditions for all types of patches used with one geometry.

WSS distributions for all patches turned out to be very similar.

CS distributions, calculated with the use of logarithmic strains, for carotids reconstructed with patch No 1 and No 6 are given on Fig. 12. These patches were chosen to outline the differences between the two cases. As seen from the figure, high patch rigidity (patch No 1) on one hand inhibits the motion of the patch, but on the other excites the oscillation of the wall in the bulb area which may contribute to cellular proliferation and lipid uptake in this region [15]. We note, that high CS values for all cases were also observed on the wall-patch boarder at the inner side of the carotid wall, closer to the flow.

In order to compare the ES distributions for all the cases considered we have selected six different locations on the patch, where high ES values were registered. The highest ES values were observed in the apex and patch boarders at the side closer to the flow. ES is the most suitable criteria for patch comparison, because it gives the direct characterization of the patch material response. Fig. 13 contains information on ES values, calculated at systole in six locations. On this diagram we

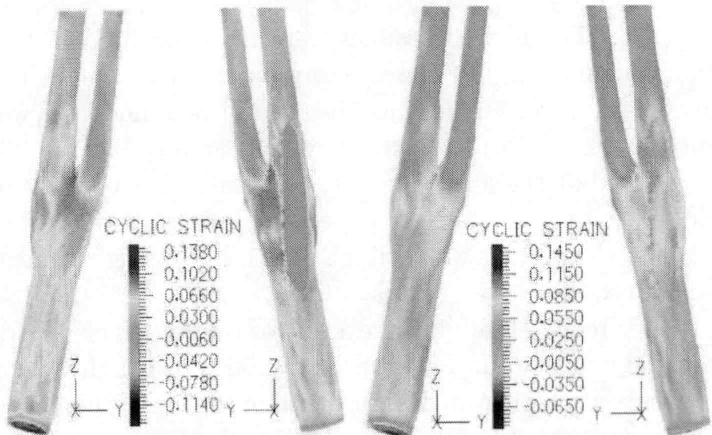

Fig. 12. CS distribution obtained for carotids reconstructed with a) patch No 1 (PTFE) and b) patch No 6 (bovine pericardium)

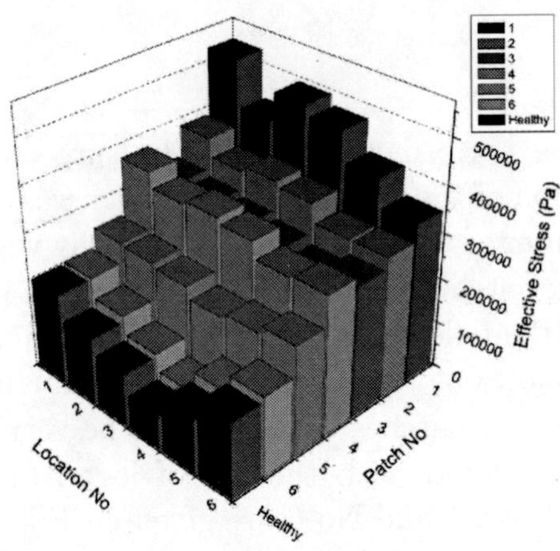

Fig. 13. Comparison of the ES (Pa) values at selected locations for all six patches and healthy carotid

can see that the largest difference from the case of healthy carotid is observed for PTFE patch No 1 and the smallest – for bovine pericardium patch No 6.

Thus, analysis showed that the most obvious differences from the healthy carotid artery case belonged to the vessels reconstructed with polytetrafluorethylene patches. The patch considered to be the closest to the normal case was made of pericardium. Prof. L.A. Bokeria and Prof. Z.K. Pirtshalaishvily clinically obtained similar results for groups of patients with synthetic and biological patches.

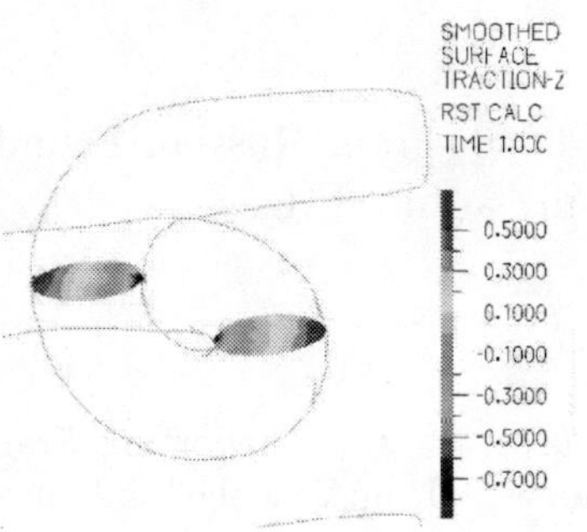

Fig. 14. Local bloodflow pressure in transverse cut

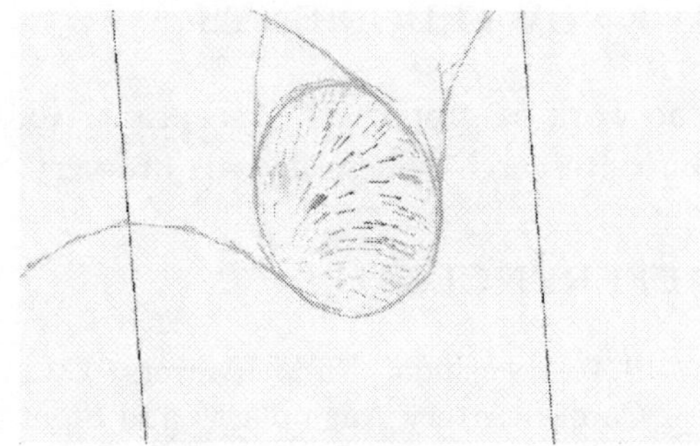

Fig. 15. Transverse blood flow circulation streams

3.3. *Carotid tortuosities*

Knowledge on the blood flow and vessel walls movement of pathological tortuosities of carotids was obtained by means of numerical simulation. The result analysis and correlations with the clinical data [27, 40, 41] showed that in the transverse cut of the vessel arch, local blood pressure was minimal at the outer wall, grew while approaching to the inner wall and took its maximum value immediately at the latter structure (see Fig. 14). There was an inverse relationship for the blood flow speed: it reached its maximum at the outer wall of the vessel arch. The transverse circulation streams of the blood flow appeared due to the blood pressure disparity between the inner and the outer radii of the vessel arch (Fig. 15). More detailed description will be published later.

3.4. *Conclusions*

For almost half a century carotid endarterectomy with patching has been the most widespread way of treating the artery stenosis. A large variety of patching materials are recently available on the market, most of them are widely used by vascular surgeons all over the world. Unfortunately, lack of indications for the use of a specific material for a specific patient leaves the surgeon making decisions based on his own experience and intuition which may contribute to the number of restenoses of the reconstructed vessels. Development of such indications is connected with a thorough study of the patched vessel hemodynamics. In this paper we generalized of the our investigation results of human carotid in healthy, affected or post-corrective surgery conditions reconstructed carotid artery [42–47].

ACKNOWLEDGMENTS

The work is supported by a grant No. 06-01-00564 from Russian Foundation for Basic Research. These awards are very gratefully acknowledged.

REFERENCES

1. R.T. Higashida, Ph.M. Meyers, C.C. Phatouros, et al., "Reporting Standards for Carotid Artery Angioplasty and Stent Placement," J. of Vascular and Interventional Radiology **15** (5), E1–E24 (2004).
2. A.V. Pokrovsky, "Vascular Surgery Possibilities in Ischemic Stroke Prevention," Vestnik Ross. Akad. Nauk **11**, 34–38 (2003) [Herald Russ. Acad. Sci. (Engl. Transl.)].
3. D.A. Healdy, R.E. Zierler, S.C. Nicholls, et al. "Long-Term Follow-Up and Clinical Outcome of Carotid Restenosis," J. Vasc. Surg. **10** (6), 662–669 (1989), discussion pp. 668–669.
4. J.P. Archie, "Prevention of Early Restenosis and Thrombosis-Occlusion after Carotid Endarterectomy by Saphenous Vein Patch Angioplasty," Stroke **17** (5), 901–905 (1986).
5. N.R. Hertzer and E.G. Beven, "A prospective study of vein patch angioplasty during carotid endarterectomy," Ann. Surg. **206**, 628–635 (1987).
6. M.M. Katz, G.T. Jones, J. Degenhardt, et al., "The Use of Patch Angioplasty to Alter the Incidence of Carotid Restenosis Following Tromboendarterectomy," J. Cardiovasc. Surg. (Torino) **28** (1), 2–8 (1987).
7. G. Deriu and E. Ballotta, "The Rationale for Patch-Graft Angioplasty after Carotid Endarterectomy," Stroke **15** (6), 972–979 (1984).
8. T. Thom, N. Haase, W. Rosamond, et al., "Heart Disease and Stroke Statistics – 2006 update. A Report From the American Heart Association Statistics Committee and Stroke Statistics Subcommittee," Circulation **113** (6), e85–e151 (2006).
9. T.G. Papaioannou and C. Stefanadis, "Vascular Wall Shear Stress: Basic Principles and Methods," Hellenic J. Cardiol **46** (1), 9–15 (2005).
10. M.R. Kaazempur-Mofrad, A.G. Isasi, H.F. Younis, et al., "Characterization of the Atherosclerotic Carotid Bifurcation Using MRI, Finite Element Modeling, and Histology," Ann. Biomed Eng **32** (7), 932–946 (2004).
11. A.M. Malek, S.L. Alper, and S. Izumo, "Hemodynamics Shear Stress and Its Role in Atherosclerosis," JAMA **282** (21), 2035–2042 (1999).
12. J.N. Oshinski, J.L. Curtin, and F. Loth, "Mean-Average Wall Shear Stress Measurements in the Common Carotid Artery," J. Cardiovasc Magnetic Resonance **8** (5), 717–722 (2006).
13. T.J. Reape and P.H. Groot, "Chemokines and atherosclerosis," Atherosclerosis **147** (2), 213–225 (1999).
14. S. Weinbaum, G. Tzeghai, P. Ganatos, et al., "Effect of Cell Turnover and Leaky Junctions on Arterial Macromolecular Transport," Am. J. Physiol. Heart Circ. Physiol. **248** (6), H945–H960 (1985).

15. B.I. Tropea, S.P. Schwarzacher, A. Chang, et al., "Reduction of Aortic Wall Motion Inhibits Hypertension-Mediated Experimental Atherosclerosis," Artherioscler. Thromb. Vasc. Biol. **20** (9), 2127–2133 (2000).
16. M.R. Kaazempur-Mofrad, H.F. Younis, S. Patel, et al., "Cyclic Strain in Human Carotid Bifurcation and Its Potential Correlation to Atherogenesis: Idealized and Anatomically-Realistic Models," J. Eng. Math. **47** (3–4), 299–314 (2003).
17. B.V. Howard, E.I. Macarak, D. Gunson, and N.A. Kefalides, "Characterization of the Collagen Synthesized by Endothelial Cells in Culture," Proc. Nat. Acad. Sci. USA **73** (7), 2361–2364 (1976).
18. K. Perktold, M. Resch, and H. Florian, "Pulsatile Non-Newtonian Flow Characteristics in a Three-Dimensional Human Carotid Bifurcation Model," J. Biomech. Eng. **113** (4), 464–475 (1991).
19. O. Heinzlef, A. Cohen, and P. Amarenco, "An Update on Aortic Causes of Ischemic Stroke," Curr. Opinion Neurol. **10** (1), 64–72 (1997).
20. B. Hindze, *Arterial System of Human and Animal Brain.* Vol. 1 (Moscow, 1946).
21. www.brainport.ru — Federal Centre of pain surgery.
22. H. Frericks, J. Kievit, J.M. van Baalen, and J.H. van Bockel, "Carotid Recurrent Stenosis and Risk of Ipsilateral Stroke: A Systematic Review of the Literature," Stroke **29** (1), 244–250 (1998).
23. J.L. Glover, P.J. Bendick, R.S. Dilley, et al., "Restenosis Following Carotid Endarterectomy. Evaluation by Duplex Ultrasonography," Arch. Surgery **120** (6), 678–684 (1985).
24. V. Zbornikova, C. Lassvik, and A. Alm, "One Year of Prospective Follow-Up after Carotid Thrombendarterectomy — a Clinical and Duplex Study," Acta Neurol. Scand. **98** (4), 248–253 (1998).
25. H.F. Younis, M.R. Kaazempur-Mofrad, R.C. Chan, et al., "Hemodynamics and Wall Mechanics in Human Carotid Bifurcation and Its Consequences for Atherosclerosis: Investigation of Inter-Individual Variation," Biomechan. Model. Mechanobiol. **3** (1), 17–32 (2004).
26. A. Delfino, N. Stergiopulos, J.E. Moore, et al., "Residual Strain Effects on the Stress Field in a Thick Wall Finite Element Model of the Human Carotid Bifurcation," J. Biomech. **30** (8), 777–786 (1997).
27. www.venart-swiss.ru —vascular clinic Venart of Swiss Medical Center.
28. A. Abizaid, A.D. Richard, G.S. Mints, et al., "Acute and Long-Term Results of an IVUS-Guided PTCA/Provisinal Stent Implantation Strategy," Am. J. Cardiol. **84**, 1381–1384 (1999).
29. R. Mehran, G. Dangas, A.S. Abizaid, et al., "Angiographic Pattern of In-Stent Restenosis. Classification and Implication for Long-Term Outcome," Circulation **100** (18), 1872–1878 (1999).
30. M.D. Haust, in *Arterial endothelium and its potentials.* Ed. by G. Manning and M.D. Haust (Plenum Press, New York, 1977), pp. 34–51.
31. H. Bouissou, M-Th. Pieraggi, and M. Julian, M. Bull. Acad. Nat. Med. **163**, 515–522 (1979).

32. S.M. Schwartz and E.P. Benditt, "Studies on Aortic Intima. I. Structure and Permeability of Rat Thoracic Aortic Intima," Amer. J. Pathol. **66** (2), 241–264 (1972).
33. P. Constantinides, Excerpta med. Atherogenesis **II**, 51–65 (1973).
34. R. Ross and J.A. Glomset, "The Pathogenesis of Atherosclerosis (First of Two Parts)," New Engl. J. Mod. **295** (7), 369–377 (1976).
35. R. Ross and L. Harker, "Hyperlipidemia and Atherosclerosis," Science **193** (4258), 1094–1100 (1976).
36. R. Ross, J. Glomset, and L. Barker, *Atherosclerosis Reviews.* Ed. by R. Paoletti, A. Gotto (Raven Press, New York, 1978), pp. 69–76.
37. J.C.F. Poole, A.G. Sanders, and H.W. Flore, "Regeneration of Aortic Endothelium," J. Pathol Bacteriol **75**, 133–143 (1958).
38. W.G. Stehbens, "Reaction of Ìenous Endothelium to Injury," Lab. Invest **14**, 449–459 (1965).
39. E.Z. Hirsch and A.L. Robertson, "Selective Acute Arterial Endothelial Injury and Repair. I. Methodology and Surface Characteristics," Atherosclerosis, **28** (3), 271–287 (1977).
40. C. Özbek, U. Yetkin, A. Özelçi, et al., "Coiling Of Extracranial Internal Carotid Artery As A Cause Of Neurologic Deficits," The Internet Journal of Thoracic and Cardiovascular Surgery, **10** (1), 2007.
41. N.A. Ovchinnikov, R.T. Rao, and S.R. Rao, "Unilateral Congenital Elongation of the Cervical Part of the Internal Carotid Artery with Kinking and Looping: Two Case Reports and Review of the Literature," Head & Face Medicine, **3** (29), 2007.
42. N.V. Ostrovsky, I.V. Kirillova, A.S. Desyatova, et al., "Application of Computer Technology to Comparative Assessment of Patch Material Used Carotid Endarterectomy," Voprosy Rekonstrukt. Plast. Khirurg. **2** (17), 42–45 (2006).
43. L.A. Bokeria, Yu.P. Gulyaev, K.M. Morozov, et al., "Mathematical Modeling of Bifurcation of Carotid Artery (to the Question of Load Distribution in Asymmetric Bifurcations)," Region. Krovoobr. Mikrotsirkul. **1** (17), 5–12 (2006).
44. L.A. Bokeria, K.M. Morozov, L.Yu. Kossovich, et al., "Application of Different Patch Types for Carotid Endarterectomy," Biomed. Tekhnol. Radioelectronika **12**, 33–41 (2006).
45. L.Yu. Kossovich, I.V. Kirillova, Yu.P. Guljaev, et al., "Patches of Different Types for Carotid Patch Endarterectomy," Saratovsk. Nauch.-Med. Zh. **2** (12), 23–34 (2006).
46. L.Yu. Kossovich, I.V. Kirillova, Yu.P. Gulyaev, et al., "Mathematical Modelling of Arteries Behavior," in *Methods of Computer-Assisted Exclusion in Biomedicine. Study Guide for Students.* Ed. by D.A. Usanov (Izd-vo Sarat. Univ., Saratov, 2007), pp. 74–96 [in Russian].
47. I.V. Kirillova, Yu.P. Gulyaev, D.V. Ivanov, et al., "Mathematical Modeling of Human Carotids in Healthy, with Pathology and Post-Corrective Surgery Condition," in *Proceedings of the IX Russian Conference "Biomechanics 2008", N. Novgorod, 20-24 May 2008* (IPPh RAN, N. Novgorod, 2008), pp. 23–26 [in Russian].

ERECTION OF A HEAVY SEMICIRCULAR ARCH STRUCTURE

A.V. Manzhirov[1*] and D.A. Parshin[1]**

ABSTRACT

We study the influence of gravity forces on objects gradually formed in their presence. We consider the cases of viscoelastic aging and purely elastic materials. Within the framework of linear mechanics of accreted solids we solve the mathematical two-dimensional problem on the erection of a heavy semicircular arch structure on a smooth rigid foundation by the method of layer-by-layer thickening of the initially installed structure on the side of its internal surface. We show that, to estimate the strength and the bearing capacity of a heavy structure, it is extremely important to take into account the gravity forces acting on this structure during the entire erection process. We also demonstrate certain possibilities of a very efficient control of the stressed state of an accreted solid.

Key words: arch structure, erection, accretion, gravity, elasticity, viscoelasticity, rate of process, prestressed elements, local force support, strength, bearing capacity

INTRODUCTION

In many cases, the operating stresses and strains in structures bearing a given surface load are much greater in value than their components formed under the action of its own mass forces. Therefore, these forces are usually neglected in strength analysis of such structures. But the gravity forces play an extremely important role in the case of gradual erection of objects of considerable dimensions. Each time, the mass of new elements added to the structure inevitably leads to additional strains in the structure. As a result, the actual stress-strain state of the completed formed object placed in operation may cardinally differ from the state obtained by computations

[1]Ishlinsky Institute for Problems in Mechanics of the Russian Academy of Sciences, Moscow, Russia

[*]E-mail: *manzh@ipmnet.ru*

[**]E-mail: *parshin@ipmnet.ru*

without gravity forces taken into account. And under certain accretion conditions, the technological stresses in some parts of the structure may be comparable with or even significantly greater than the stresses caused by the assumed operating load.

For numerous structures, the gravity forces are the main forces acting on these structures, and, obviously, the analysis of their state with mechanical characteristics of the formation process taken into account can also lead to results that are principally different from those obtained by the classical scheme without taking these forces into account.

Very often, at the initial stage of erection, the structure has significant dimensions but rather small strength and hence can endure considerable stresses under the action of its own mass. The subsequent strengthening of such a structure by additional mass elements (and hence the subsequent increase in mass) can not only fail to eliminate but even increase the original stresses in the structure and thus result in situation in which the complete finished object has high stress regions that cannot be predicted by the analysis of the final configuration. In this case, it is necessary to take special measures to control the stress-strain state of the erected object. For example, as such measures, it is proposed to create preliminary stresses in the added construction elements or to organize a temporary local support of the structure in the process of its erection. It may happen that, as a result of such a control, the structure may be brought to a state even more profitable than the state obtained in the (possibly, hypothetical) version of its erection in which the gravity forces begin to exhibit significant influence on the structure only in the final configuration.

1. STATEMENT OF THE PROBLEM

We consider the problem of gradual erection of a circular arch structure (semicircular arch) on a smooth horizontal foundation. Note that a similar problem of strengthening such a structure without taking into account the gravity forces was considered in [1] (for general theory and solved problems of accreted solids mechanics see also [2–13]).

We assume that, at time $t = t_0$, a semicircular cylindrical arch (a blank) is installed on a smooth rigid horizontal foundation and fixed on it by sliding fixation so that the separation of the arch base from the foundation is forbidden but free sliding along the foundation is allowed. This arch was made without residual stresses form a homogeneous isotropic viscoelastic aging material, which is assumed to the prepared at the zero time instant. Further, starting from time $t_1 \geqslant t_0$, the installed structure is gradually thickened in N stages of continuous addition of uniform-in-thickness elementary layers of some additional material to its internal cylindrical surface. In the gaps between these stages and before and after the process of thickening, no

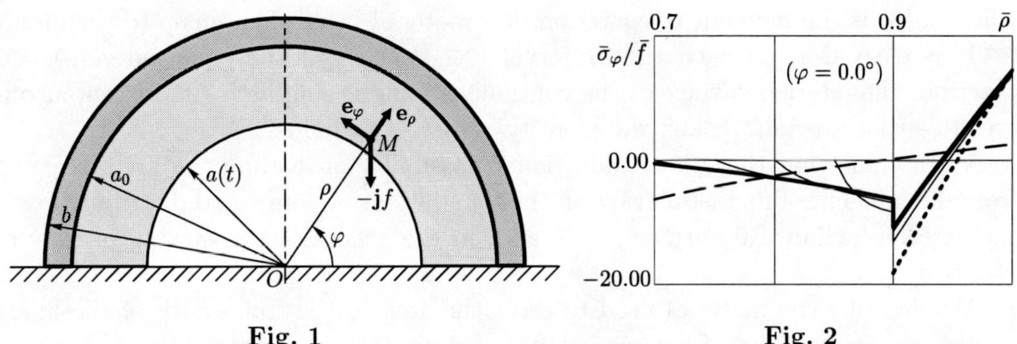

Fig. 1 **Fig. 2**

loads are applied to the internal surface of the already formed structure. The action of sliding fixation involves the entire current area of the base of the arch to be erected.

We study the process of formation of the stress-strain state of this structure, which arises under the action of gravity forces on the structure during and after the erection procedure described above. This process will be called piecewise continuous accretion. We also consider and analyze some possible versions of technological control of this process.

We assume that the added material is prepared simultaneously with the original material and has the same mechanical properties. We perform our consideration in small strains in the case of a plane strain state and neglect dynamical effects.

Let b be the external radius of the arch, let a_0 be its internal radius before the beginning of thickening, and let $a(t)$ be its internal radius at time $t \geqslant t_0$ (Fig. 1). We assume that $a(t)$ is a given continuous function of time strictly monotone decreasing on the intervals $t \in (t_{2k-1}, t_{2k})$ ($k = 1, \dots, N$) corresponding to the stages of arch continuous accretion and constant outside these intervals. The arch radius at the end of the kth stage is denoted by $a_k = a(t_{2k})$.

We associate the arch transverse cross-section plane with the circular cylindrical coordinate system (ρ, φ, z) with the right orthonormal local frame $\{\mathbf{e}_\rho(\varphi), \mathbf{e}_\varphi(\varphi), \mathbf{k}\}$, where ρ is the polar radius counted from the arch central axis, φ is the polar angle counted upwards from the foundation (Fig. 1), and z is the longitudinal coordinate.

We take into account the fact that some prestressed state may arise in the construction elements added to the arch. But in this case we consider only processes in which the load is applied to the layers of additional material only just at the moment of their accretion to the body. Under this assumption, the stresses at the arch points at the distance ρ from its axis arise at the time instant

$$\tau_0(\rho) = \begin{cases} t_0, & \rho > a_0, \\ \tau_*(\rho), & \rho < a_0, \end{cases} \tag{1}$$

where $\tau_*(\rho)$ is the moment of accretion of a material layer of radius ρ to the arch.

It is clear that on each time interval $t \in (t_{2k-1}, t_{2k})$ the equation $\tau_*(\rho) = t$ describes the internal surface of the continuously increasing arch (its instantaneous growth surface $\rho = a(t)$); i.e., the identity $\tau_*(a(t)) \equiv t$ holds.

We assume that there is an additional possibility of loading the arch, starting from the moment of its installation on the foundation, by some load distributed over the external cylindrical surface $\rho = b$ and, in the general case, varying in time as $\mathbf{t}(\varphi, t)$, $t \geqslant t_0$.

We describe the material used to erect the arch in the framework of the linear theory of viscoelasticity of homogeneously aging isotropic media [3,14]; i.e., we start from the equation of state

$$\mathbf{T}(\mathbf{r}, t) = G(t)\big(\mathcal{I} + \mathcal{N}_{\tau_0(\mathbf{r})}\big)\big[2\,\mathbf{E}(\mathbf{r}, t) + (\varkappa - 1)\,\mathbf{1}\,\mathrm{tr}\,\mathbf{E}(\mathbf{r}, t)\big], \tag{2}$$

where $\tau_0(\mathbf{r})$ is the time at which the stresses arise at the point $\mathbf{r}$ of the body (in our case, this time is equal to $\tau_0(\rho)$), $\mathbf{T}$ and $\mathbf{E}$ are the stress and small strain tensors, $\mathbf{1}$ is the unit tensor of rank 2; $G(t)$ is the elastic shear modulus, $\varkappa = (1 - 2\nu)^{-1}$, and $\nu = \mathrm{const}$ is Poisson's ratio. Here the viscoelasticity operator is determined by the relation

$$\begin{aligned}
&\mathcal{I} + \mathcal{N}_s = (\mathcal{I} - \mathcal{L}_s)^{-1},\\
&\mathcal{L}_s f(t) = \int_s^t f(\tau)K(t, \tau)\,d\tau, \quad \mathcal{N}_s f(t) = \int_s^t f(\tau)R(t, \tau)\,d\tau,\\
&K(t, \tau) = G(\tau)\,\partial\Delta(t, \tau)/\partial\tau, \quad \Delta(t, \tau) = 1/G(\tau) + \omega(t, \tau),
\end{aligned}$$

where $\mathcal{I}$ is the identity operator, $\mathcal{L}_s$ and $\mathcal{N}_s$ are the integral Volterra operators with parameter s, $K(t, \tau)$ and $R(t, \tau)$ are the creep and relaxation kernels, and $\Delta(t, \tau)$ and $\omega(t, \tau)$ ($t \geqslant \tau \geqslant 0$) are the specific strain function and the creep measure in the case of pure shear. By definition, the latter satisfies the identity

$$\omega(\tau, \tau) \equiv 0, \quad \tau \geqslant 0. \tag{3}$$

With the identity (3) taken into account, the specific strain function can be represented as

$$\Delta(t, \tau) = (\mathcal{I} - \mathcal{L}_\tau)G(t)^{-1}. \tag{4}$$

If we define the linear integral operator $\mathcal{H}_s f(t) = (\mathcal{I} - \mathcal{L}_s)[f(t)/G(t)]$ with parameter s and introduce the notation

$$g^\circ(\mathbf{r}, t) = \mathcal{H}_{\tau_0(\mathbf{r})}\,g(\mathbf{r}, t) \tag{5}$$

for an arbitrary function $g(\mathbf{r}, t)$ of the point $\mathbf{r}$ of the accreted body and time t, then the constitutive relation (2) can be rewritten as [2]

$$\mathbf{T}^\circ(\mathbf{r}, t) = 2\,\mathbf{E}(\mathbf{r}, t) + (\varkappa - 1)\,\mathbf{1}\,\mathrm{tr}\,\mathbf{E}(\mathbf{r}, t). \tag{6}$$

2. BOUNDARY VALUE PROBLEM FOR THE ORIGINAL BODY

All elements of the originally manufactured arch are loaded simultaneously at time $t = t_0$ of its installation on the foundation under the action of the homogeneous field of gravity forces of vector intensity $\mathbf{f} = -\mathbf{j}f$, where $\mathbf{j}$ is the unit normal to the foundation and f is the material specific weight. Because of this, the parameter of the integral operation in the constitutive relation (2) and in definition (5) is independent of the point of the original body and is equal to t_0 (see (1)).

Thus, the stress-strain state of the arch under study on the time interval from the moment of its installation on the foundation to the moment of the beginning of thickening is described by the classical boundary value problem of the linear theory of viscoelasticity:

$$\begin{aligned}
&\nabla\cdot\mathbf{T}+\mathbf{f}=\mathbf{0}, \quad \rho\in(a_0,b), \quad \varphi\in(0,\pi), \quad t\in[t_0,t_1];\\
&\mathbf{T}=G(\mathcal{I}+\mathcal{N}_{t_0})[2\,\mathbf{E}+(\varkappa-1)\,\mathbf{1}\,\mathrm{tr}\,\mathbf{E}]; \quad \mathbf{E}=(\nabla\mathbf{u}^{\mathrm{T}}+\nabla\mathbf{u})/2;\\
&\mathbf{e}_\rho\cdot\mathbf{T}=\mathbf{0}, \quad \rho=a_0; \qquad \mathbf{e}_\rho\cdot\mathbf{T}=\mathbf{t}(\varphi,t), \quad \rho=b;\\
&\mathbf{e}_\varphi\cdot\mathbf{T}\cdot\mathbf{e}_\rho=0, \quad \mathbf{e}_\varphi\cdot\mathbf{u}=0, \quad \varphi=0,\pi.
\end{aligned} \tag{7}$$

Here $\mathbf{u}(\mathbf{r},t) = \mathbf{e}_\rho u_\rho + \mathbf{e}_\varphi u_\varphi + \mathbf{k}\,u_z$ is the vector field of displacements of the arch points. In the case of plane strain under study, we have $u_z \equiv 0$ and $\partial u_\rho/\partial z \equiv \partial u_\varphi/\partial z \equiv 0$.

We apply the linear operator $\mathcal{H}_{t_0}$ to all relations of the boundary value problem (7) containing the stress tensor $\mathbf{T}$, use representation (4), and obtain the following equivalent statement of this problem [2]:

$$\begin{aligned}
&\nabla\cdot\mathbf{T}^\circ=\mathbf{j}\,f\Delta(t,t_0), \quad \rho\in(a_0,b), \quad \varphi\in(0,\pi), \quad t\in[t_0,t_1];\\
&\mathbf{T}^\circ=2\,\mathbf{E}+(\varkappa-1)\,\mathbf{1}\,\mathrm{tr}\,\mathbf{E}; \quad \mathbf{E}=(\nabla\mathbf{u}^{\mathrm{T}}+\nabla\mathbf{u})/2;\\
&\mathbf{e}_\rho\cdot\mathbf{T}^\circ=\mathbf{0}, \quad \rho=a_0; \qquad \mathbf{e}_\rho\cdot\mathbf{T}^\circ=\mathbf{t}^\circ, \quad \rho=b;\\
&\mathbf{e}_\varphi\cdot\mathbf{T}^\circ\cdot\mathbf{e}_\rho=0, \quad \mathbf{e}_\varphi\cdot\mathbf{u}=0, \quad \varphi=0,\pi,
\end{aligned} \tag{8}$$

which contains the time t already as a real parameter. We note that since the creep measure $\omega(t,\tau)$ is nonnegative (e.g., see [15]), it follows that the inequality $\Delta(t,t_0) > 0$ holds for any time instant $t \geqslant t_0$.

Integrating by parts and using identity (3), we calculate the function

$$\mathbf{t}^\circ(\varphi,t)=\mathbf{t}(\varphi,t_0)\,\Delta(t,t_0)+\int_{t_0}^{t}\frac{\partial\mathbf{t}(\varphi,\tau)}{\partial\tau}\,\Delta(t,\tau)\,d\tau \tag{9}$$

contained in the condition on the boundary $\rho = b$.

3. BOUNDARY VALUE PROBLEM FOR THE ACCRETED ARCH

In the process of accretion, the elements of additional material are added to the growing solid in the course of its strain motion in space. It is clear that the entire solid formed in such a way cannot in general have the original unstrained configuration. It is this fact that is decisive in the strain process for any accreted body and essentially distinguishes the mechanical behavior of such bodies from the behavior of bodies of constant composition (classical bodies in continuum mechanics) and of bodies whose boundary is variable because of the removal of the material. Owing to this characteristic property, it is impossible to define the strain measure of the accreted body by the method usually adopted in continuum mechanics; therefore, the Cauchy formulas do not hold for the total strain tensor components, and hence the Saint-Venant conditions of their compatibility are not satisfied.

Note, however, that the particles of additional material after its adhesion to the surface of growth continue their motion further as part of the solid. This means that, in the space region occupied by the entire accreted body at a given time, a sufficiently smooth velocity field of its particles is uniquely determined. Therefore, we can expect that the problem of deformation of such a body can be well defined for the *displacement velocities* (as in problems of fluid mechanics). Starting from this, we write out the velocity analog $\mathsf{S}(\mathbf{r},t) = 2\,\mathsf{D}(\mathbf{r},t) + (\varkappa - 1)\,\mathbf{1}\,\mathrm{tr}\,\mathsf{D}(\mathbf{r},t)$ of the constitutive relation (6) by introducing the tensor $\mathsf{S} = \partial \mathsf{T}^\circ/\partial t$ [2] and the strain rate tensor $\mathsf{D} = (\nabla \mathbf{v}^{\mathrm{T}} + \nabla \mathbf{v})/2$, where $\mathbf{v}(\mathbf{r},t) = \mathbf{e}_\rho v_\rho + \mathbf{e}_\varphi v_\varphi + \mathbf{k}\, v_z$ is the instantaneous velocity field of the accreted body particles. In the case of plane strain, we have $v_z \equiv 0$ and $\partial v_\rho/\partial z \equiv \partial v_\varphi/\partial z \equiv 0$.

The differential equation for the above-introduced tensor S in the region occupied at the current time by the accreted body can be obtained by applying the linear operator $\mathcal{H}_{\tau_0(\mathbf{r})}$ to the equilibrium equation $\nabla \cdot \mathsf{T} + \mathbf{f} = \mathbf{0}$ and by differentiating the result with respect to time t. In this case, at the stage of the body deformation after the beginning of accretion, this integral operator does not already commute with the divergence operator, as it was at the stage before the beginning of accretion (see the preceding section). This is related to the fact that the lower limit in the integral $\mathcal{L}_{\tau_0(\mathbf{r})}$ depends on a point of the body. But one can show [2] that in the accretion processes studied in the present paper, under the condition that each future and current surfaces of growth (before the beginning and during each of the stages of the continuous increase) does not receive any load, the stress tensor can be written as $(\nabla \cdot \mathsf{T})^\circ = \nabla \cdot \mathsf{T}^\circ$. In this case, for $t > t_1$, the following equations are satisfied in the entire accreted body:

$$\nabla \cdot \mathsf{T}^\circ + \mathbf{f}^\circ = \mathbf{0}, \qquad \nabla \cdot \mathsf{S} + \partial \mathbf{f}^\circ/\partial t = \mathbf{0}. \tag{10}$$

Since it is assumed in our problem that the additional material is under load just at the moment of its adhesion to the body and can acquire arbitrary preliminary stresses at this moment, it is necessary, at each time interval of the continuous increase on the growth surface, to introduce the stress tensor

$$\mathbf{T} = \mathbf{T}_*(\mathbf{r}), \quad \rho = a(t), \quad t \in (t_{2k-1}, t_{2k}), \tag{11}$$

which is consistent with the external load [16].

It follows from the above that the tensor function $\mathbf{T}_*(\mathbf{r})$ should satisfy the condition $\mathbf{e}_\rho \cdot \mathbf{T}_*(\mathbf{r}) = \mathbf{0}$, which means that there is no load on the growth surface. This condition agrees with the above statement of the problem and, as was already mentioned, is required for the further use of Eqs. (10).

We note that the set of conditions (11) for all moments at the kth stage of the continuous increase is equivalent to the following initial condition on the stress tensor at the points of the part of the body formed at this stage:

$$\mathbf{T}\big(\mathbf{r}, \tau_*(\rho)\big) = \mathbf{T}_*(\mathbf{r}), \quad \rho \in (a_k, a_{k-1}). \tag{12}$$

It is important that condition (11), which is nontraditional in mechanics, can be transformed [2,7,9] to the boundary condition

$$\mathbf{e}_\rho \cdot \mathbf{S} = \big[\nabla \cdot \mathbf{T}_*(\mathbf{r}) + \mathbf{f}\big] a'(t)/G(t), \quad \rho = a(t), \quad t \in (t_{2k-1}, t_{2k}), \tag{13}$$

on the components of the tensor $\mathbf{S}$ on the instantaneous growth surface, which is similar in form to the classical force condition.

Since, as was specified in the statement of the problem, the internal arch surface does not receive any load outside the accretion intervals (we recall that this is one of the conditions under which Eqs. (10) can be used to solve the accretion problem), the usual condition of the absence of load is posed on this surface after each stop in growth. Its obvious analog for the tensor $\mathbf{S}$ is given by the boundary condition $\mathbf{e}_\rho \cdot \mathbf{S} = \mathbf{0}$, $\rho = a_k$, $t \in (t_{2k}, t_{2k+1})$.

As we see, it follows from the definition of the function $a(t)$ that the condition in the form (13) on the internal arch surface formally remains valid also beyond the limits of the time intervals of growth; namely, it can be used for any $t > t_1$. But it is important to understand that the nature of this condition at the stages of arch continuous accretion and after a temporary or final termination of this process is substantially different.

As for the conditions on the other surfaces of the piecewise continuously accreted arch, in this case, the former relations remain valid (see the preceding section), which are obviously modified for the statement of the problem in velocity form.

Thus, after the beginning of the piecewise continuous thickening of the arch, the arch strain process is described by the following boundary value problem:

$$
\begin{aligned}
&\nabla\cdot\mathbf{S}=\mathbf{j}\,f\,\partial\omega\big(t,\tau_0(\rho)\big)/\partial t,\quad \rho\in\big(a(t),b\big),\quad \varphi\in(0,\pi),\quad t>t_1;\\
&\mathbf{S}=2\,\mathbf{D}+(\varkappa-1)\,\mathbf{1}\,\mathrm{tr}\,\mathbf{D};\quad \mathbf{D}=(\nabla\mathbf{v}^{\mathrm{T}}+\nabla\mathbf{v})/2;\\
&\mathbf{e}_\rho\cdot\mathbf{S}=\mathbf{p},\quad \rho=a(t);\qquad \mathbf{e}_\rho\cdot\mathbf{S}=\mathbf{q},\quad \rho=b;\\
&\mathbf{e}_\varphi\cdot\mathbf{S}\cdot\mathbf{e}_\rho=0,\quad \mathbf{e}_\varphi\cdot\mathbf{v}=0,\quad \varphi=0,\pi.
\end{aligned}
\tag{14}
$$

In this problem, just as in problem (8), time t is a parameter. We introduce the notation

$$
\mathbf{p}(\varphi,t)=\big\{[\nabla\cdot\mathbf{T}_*(\mathbf{r})]\big|_{\rho=a(t)}+\mathbf{f}\big\}\,a'(t)/G(t),
$$

$$
\mathbf{q}(\varphi,t)=\frac{\partial\mathbf{t}^\circ(\varphi,t)}{\partial t}=\mathbf{t}(\varphi,t_0)\,\frac{\partial\omega(t,t_0)}{\partial t}+\frac{\partial\mathbf{t}(\varphi,t)}{\partial t}\,\frac{1}{G(t)}+\int_{t_0}^{t}\frac{\partial\mathbf{t}(\varphi,\tau)}{\partial\tau}\,\frac{\partial\omega(t,\tau)}{\partial t}\,d\tau.
$$

We note that the last relation is the result of differentiation of the expression (9) with respect to t with the identity (3) taken into account, and the right-hand side of the differential equation in (14) is written using (4). It follows from the required properties of the creep measure (see, e.g., [15]) that $\partial\omega\big(t,\tau_0(\rho)\big)/\partial t\geqslant 0$ for $t>\tau_0(\rho)$.

After problem (14) has been solved, at each point $\mathbf{r}$ it is necessary to reconstruct the evolution of the tensor $\mathbf{T}^\circ$ for $t\geqslant\tau_1(\rho)$ from the obtained rate $\mathbf{S}$ of its variation using the corresponding initial condition for $\mathbf{T}^\circ$: $\mathbf{T}^\circ(\mathbf{r},t)=\mathbf{T}^\circ\big(\mathbf{r},\tau_1(\rho)\big)+\int_{\tau_1(\rho)}^{t}\mathbf{S}(\mathbf{r},\tau)\,d\tau$, $t\geqslant\tau_1(\rho)$. Here, by definition, $\tau_1(\rho)=t_1$ for $\rho>a_0$ and $\tau_1(\rho)=\tau_*(\rho)$ for $\rho<a_0$. The initial condition for points of the originally existing part of the body can be found from the solution of the classical problem (8) for $t=t_1$, and that for elements of the additional part can be found from their known stress state (12) at the time of adhesion to the accreted body by the formula $\mathbf{T}^\circ\big(\mathbf{r},\tau_1(\rho)\big)=\mathbf{T}_*(\mathbf{r})/G\big(\tau_*(\rho)\big)$, $\rho\in(a_k,a_{k-1})$, which holds by (5) and (1).

After problems (8) and (14) have been solved and the evolution of the tensor $\mathbf{T}^\circ$ has been determined for each element of the body from the moment of its loading to any arbitrary remote time moment, the total evolution of the stress tensor $\mathbf{T}$ at each point of the final body can be determined from definition (5) by the formula $\mathbf{T}(\mathbf{r},t)=G(t)\big(\mathcal{I}+\mathcal{N}_{\tau_0(\rho)}\big)\mathbf{T}^\circ(\mathbf{r},t)$, $t\geqslant\tau_0(\rho)$. This is in fact the resolvent representation of the solution of the integral Volterra equation of the second kind $\mathbf{g}(\mathbf{r},t)-\int_{\tau_0(\rho)}^{t}\mathbf{g}(\mathbf{r},\tau)\,K(t,\tau)\,d\tau=\mathbf{T}^\circ(\mathbf{r},t)$ for the unknown tensor $\mathbf{g}(\mathbf{r},t)=\mathbf{T}(\mathbf{r},t)/G(t)$ considered at an arbitrary point $\mathbf{r}$ of the body as a function of time t. If the closed analytical expression for the resolvent $R(t,\tau)$ of the kernel $K(t,\tau)$ is unknown or its calculation is too cumbersome, then this equation can be solved numerically, for example, by the quadrature method based on the trapezoid formula [17], as this is done in the present paper.

4. ON THE SOLUTION OF THE ABOVE-POSED BOUNDARY VALUE PROBLEMS

The boundary value problems (8) and (14) obtained at the stages of arch deformation before and after the beginning of its thickening are stated for different mechanical variables but, as we see, have the same mathematical structure. Either of the problems, up to the replacement of $\mathbf{T}^{\circ}$ and $\mathbf{S}$ by the stress tensor divided by the shear modulus and, in (14), also up to the replacement of the tensor $\mathbf{D}$ by a small strain tensor and of the vector $\mathbf{v}$ by the displacement vector, is the classical boundary value problem of mechanics of deformable rigid body about the equilibrium of a linearly elastic semicircular arch with variable (in the radius) specific weight on a smooth rigid horizontal foundation under the action of gravity forces and prescribed surface loads. The specific weight and the surface loads, as well as the arch internal radius in (14), depend on a real parameter t.

We note that it follows from the above condition of the arch plane strain that the stress field prescribed on its external (nonaccreted) surface must have zero projection onto the longitudinal direction. To preserve the symmetry of this problem, we assume that, for all $t \geqslant t_0$, the vector field $\mathbf{t}(\varphi, t)$ is symmetric with respect to the vertical longitudinal half-plane $\varphi = \pi/2$ passing through the arch axis.

Problems (8) and (14) can be solved analytically using the expansions of the radial components of the desired displacement vector $\mathbf{u}$ and the velocity vector $\mathbf{v}$ in cosine Fourier series and the expansions of the circular components in sines of angles that are multiples of $\varphi \in [0, \pi]$ with the symmetry of the strain picture taken into account. We do not present the solutions thus obtained because they are extremely cumbersome.

We only note that the stresses in the originally installed arch before the beginning of its thickening are independent of the time t and of the time t_0 of its erection on the foundation. It follows from the well-known correspondence principle in the linear theory of viscoelasticity [3] that these stresses also simultaneously correspond to the solution of a similar classical problem about the strain of a pure elastic arch on a smooth rigid foundation.

5. ELASTIC CASE

The purely elastic behavior of the material used to produce the arch under study is obviously a specific case of the originally accepted constitutive relation (2). In this case, it is necessary to assume that $\omega(t, \tau) \equiv 0$ and $G(t) \equiv \text{const}$ in this relation. Then we have $\Delta(t, \tau) \equiv 1/G$, and hence $\mathcal{I} - \mathcal{L}_s = \mathcal{I} + \mathcal{N}_s = \mathcal{I}$, $g^{\circ}(\mathbf{r}, t) \equiv g(\mathbf{r}, t)/G$. It is clear that all the above arguments and transformations remain valid in this case.

After the tensor $\mathbf{S}$ is calculated, the stress tensor $\mathbf{T}$ in the elastic case can be found simply by integration: $\mathbf{T}(\mathbf{r},t) = \mathbf{T}(\mathbf{r},\tau_1(\rho)) + \int_{\tau_1(\rho)}^{t} G\mathbf{S}(\mathbf{r},\tau)\,d\tau$.

We also note that the displacements of the elastic arch particles before the beginning of accretion and the velocities of their motion after the beginning of accretion are inversely proportional to the shear modulus G, while the stresses during the entire strain process are independent of G at all. In this case, the rate of stress variation with time after the beginning of accretion is independent of the Poisson ratio ν.

6. MATERIAL CHARACTERISTICS AND SOME CONVENTIONS

To analyze the behavior of the accreted viscoelastic aging arch numerically, we introduce the creep measure of the material under shear in the form [14] $\omega(t,\tau) = A(\tau)\big[1 - e^{-\gamma(t-\tau)}\big]$. The coefficient $\gamma > 0$ determines the creep rate. The factor $A(\tau)$ is usually called the aging function.

We pass to dimensionless variables as follows. We multiply all quantities with the dimension of time by γ and divide all quantities with the dimension of stress by the shear modulus $G_\infty = G(+\infty)$ of a very old material. We normalize linear variables to the value of the radius b of the arch external surface (which does not vary in the course of accretion). The other dimensional physical variables are reduced to dimensionless form by dividing them by the corresponding combination of the parameters γ, G_∞, and b. All the variables normalized in this way will be denoted by a bar over the letter.

The approximating expressions for the shear modulus and the aging function are taken in the form [15] $G(t) = G_\infty\big(1 - \delta G\, e^{-\alpha t}\big)$ and $A(\tau) = A_\infty + \Delta A\, e^{-\beta\tau}$, where in calculations we assume [3] that $\bar{\alpha} = 2$, $\delta G = 1/2$, $\bar{\beta} = 31/60$, $\bar{A}_\infty = 0.5522$, and $\Delta\bar{A} = 4$. The Poisson ratio is set to be $\nu = 0.1$. Note that these approximations to the mechanical material characteristics and the above values of their parameters agree well with experimental data obtained for the creep of several kinds of concrete [3, 14, 15, 18, 19].

Hereafter, to illustrate the results of numerical calculations, we agree about the following.

In all figures presenting the stress distributions over the arch thickness, the *dotted* lines correspond to the stress diagrams in the arch originally installed on the foundation (and playing the role of a blank in the technological process under study) before the beginning of its accretion. The graphs with *thin solid* lines present the evolution of these stress diagrams in the process of the subsequent thickening of the arch and after the completion of this process. The *thick solid* lines present the final (steady-state as $t \to +\infty$) stress distributions in the final structure. The *dashed* lines show the stresses in the arch of final dimensions (attained before the end of the process of its erection) but produced without the influence of the gravity forces and only after this installed on the horizontal foundation, i.e., in the traditional case.

In the case where several single-type dependencies for the chosen set of values of some numerical parameter are illustrated on the same coordinate field, the arrow shows the curves in ascending order of the corresponding values of this parameter.

In the present paper, we only discuss the results of studies of the processes of sufficiently slow (at a constant rate of variation in the internal radius $\overline{a'(t)} \equiv (\bar{a}_1 - \bar{a}_0)/(\bar{t}_2 - \bar{t}_1) = {}^{-1}/_{25}$) continuous erection of a thin-walled arch (with the initial and final radius $\bar{a}_0 = 0.9$ and $\bar{a}_1 = 0.7$). In this case, throughout the paper, we assume that the arch blank is installed on the foundation at a sufficiently small age of the material ($\bar{t}_0 = 0.1$) and its thickening begins rather soon (at the age $\bar{t}_1 = 0.3$).

We point out that the case of a thin-walled structure calculations is chosen here only because it is most interesting from the standpoint of applications. In the present paper, the smallness of the wall thickness of the erected arch is not used to simplify the model constructed above: the computations are performed according to the above general relations, which hold for arches with arbitrary thickness of the wall.

7. DEFORMATION OF THE ERECTED ARCH UNDER THE ACTION OF GRAVITY FORCES

In this section, we study the influence of the gravity forces on the arch structure erected under their action. To this end, we assume that, in the boundary value problems (8) and (14), the surface load and the preliminary stress tensor are zero: $\mathbf{t}(\varphi, t) \equiv \mathbf{0}$ and $\mathbf{T}_*(\mathbf{r}) \equiv \mathbf{0}$.

We focus our attention on the contact stresses $\sigma_\varphi\big|_{\varphi=0}$ (the contact pressure with opposite sign) acting on the base of the erected arch from the foundation.

Fig. 2 shows the evolution of these stresses in the erected arch. We see that, in the case under study, there is a negative pressure zone in the peripheral regions (near the arch external boundary). This means that, to realize the obtained strain picture, it is necessary to have constantly acting *bilateral constraints* on the foundation. The dimension of this zone remains approximately the same during the entire process of the arch erection and its completion. In this case, the maximal separation stress always acts near the external arch boundary. The largest (positive) pressure arises on the interface between the original and additional parts of the body and is exerted by the former.

Note that the potential separation from the smooth rigid foundation is typical not only of thin-walled arch structures. Even rather thick-walled arches installed on such a foundation (after they have been produced without residual stresses) tend to separate from it under the action of their own mass. For example, for the above value $\nu = 0.1$ of the Poisson ratio (typical of some sorts of concrete), the base separation does not occur only for the circular arches whose wall thickness is approximately equal to at least 68% of the external radius. For such (very thick) arches, the specific condition of sliding fixation on the foundation is obviously equivalent to the usual condition of smooth contact.

We note that if the thin-walled arch under study were installed on the foundation already after its manufacturing, then the contact pressure diagram would be practically linear (which, obviously, follows from the fact that the arch is thin-walled), would have a (positive) maximum near the body internal surface, and would have an almost twice larger negative pressure zone. In this case, the level of separation stresses would be by a factor of $5^1/_2$ less than that after the manufacturing under the operating load in the operation mode under study. All this allows us to conclude immediately that it is in principle necessary to take into account the gravity forces during the entire process of erection of heavy objects. Here we have no possibility to discuss the specific features of their influence on the viscoelastic aging and purely elastic circular arches erected under different operation modes.

We note also that the influence of the accretion rate on the evolution of the stress state of a thin walled arch structure was discussed in [8].

8. ARCH ACCRETION BY PRESTRESSED STRUCTURAL ELEMENTS

Thus, we have found that there are negative contact pressure zones in the peripheral regions of the support points of a heavy circular arch produced without residual stresses at the moment of its fixation on a smooth rigid foundation by means of sliding fixation if only the thickness of its wall is not too large compared with the external radius. We note that the thinner this arch, the larger part of its support points is occupied by these zones and the higher the level of the separation stresses is there. In the process of further accretion of such a arch, even after a significant thickening of the wall, the negative pressure zones on the bases are not only preserved but even become somewhat larger in dimensions.

Technologically, this fact means that, during the entire process of erection of an arch structure, it is under the action of bilateral constraints from the foundation, which prevent it from separation. But, from the operating standpoint, it is much more important that these constraints must also be preserved infinitely long after the structure is completely erected. But there are situations in which this is not desirable (for example, if the structure must be moved after the manufacturing to a different place, where the above constraints cannot be created). Therefore, it is meaningful to pose the problem of finding a modification of the modeled technological process, which would permit finishing with separation stresses on the bases after the final stage of the process but preserving the condition of the base sliding fixation in the process of the structure erection.

We again point out that this specific characteristic takes place even if the final thickness of the arch wall is so large that should this arch be manufactured without the straining factors and only then installed on the foundation, the separation of its bases under the action of the gravity forces would not be observed.

To attained the planned goal, we first try to perform the erection of the structure under study by its accretion by prestressed elements. This way of solution is suggested by the intuitive understanding of the fact that after a layer of an originally extended material has been added to the internal arch surface, this layer tends to compress thus contracting the bases of the arch sinking under the action of its own mass. Finally, this must decrease the pressure on the internal parts of the bases and increase the pressure on their external parts. As a result, we can expect that the separation stresses initially arising in the peripheral regions of the support points will be suppressed.

To simplify the calculations, we assume that each additional layer of the added material is uniformly extended in the circular direction. Then the preliminary stress tensor has the form

$$\mathbf{T}_*(\mathbf{r}) = \mathbf{e}_\varphi(\varphi)\mathbf{e}_\varphi(\varphi)\,\sigma_\varphi^*(\rho), \tag{15}$$

where the prescribed value of tensile stress ("preload") σ_φ^* depends only on the radius ρ of the corresponding material layer and is independent of the angular coordinate φ. As before, we assume that the surface load $\mathbf{t}$ on the arch external surface $\rho = b$ is identically zero both before and after the beginning of accretion.

We study the evolution of contact stresses on the base of the arch erected by the method modeled here for various intensities of the preliminary tension of the added elements. We define the initial tensile stress in the new additional layers, depending on their radius ρ, by the function

$$\sigma_\varphi^*(\rho) = f\,\alpha_\sigma \rho, \tag{16}$$

where the coefficient $\alpha_\sigma = \text{const}$ characterizes the extension intensity.

Fig. 3, *a* demonstrates the final distributions of contact stresses over the base of the finally manufactured arch for $\alpha_\sigma = 12$, 24, 60, and also for $\alpha_\sigma = 0$. (The last graph corresponds to the arch erection by using unstressed elements and is taken from Fig. 2.) In Fig. 3, *b*, as an example, we show the evolution of contact stresses for one of the above values: $\alpha_\sigma = 24$.

As we see, even for relatively small intensity of the constructive elements preload α_σ, the zone of negative values on the final contact pressure diagram already appears

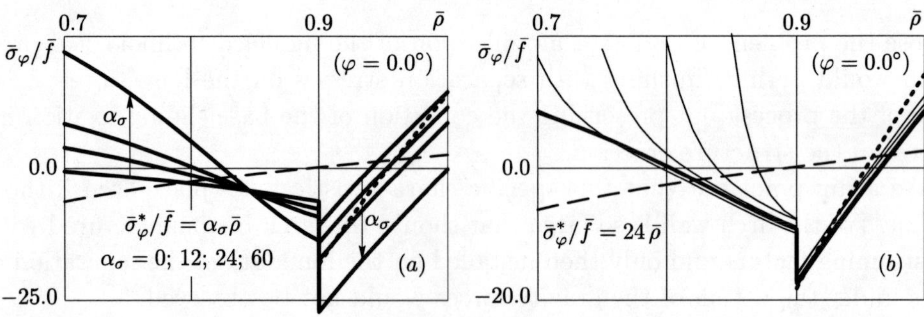

Fig. 3

also near the internal surface of the erected arch, although the prescribed initial tension of the added layers decreases as the structure becomes thicker. As α_σ increases, this region evolves. Its dimension is already significant, but the situations near the arch external surface does not vary significantly. Thus, the tendency to separation is now observed on either (internal and external) side of the contact region. An increase in the preload intensity leads to a gradual decrease in the final dimension of the external region of potential separation and to a decrease in the corresponding maximal stresses. But, at the same time, there is an increase in the internal region of separation stresses and their maximum in this region. For $\alpha_\sigma \approx 24$, the maxima attained, as a result, in the external and internal regions coincide in value and are approximately equal to 61% of the maximal separation stress on the base of the initially installed arch. Obviously, in the case under study, this is the minimal possible limit value of the maximal separation stress (we note that in the case of accretion with zero preload, the final maximum value practically coincides with the initial value). For $\alpha_\sigma = 24$, the internal region of potential separation occupies nearly half the entire part of the base created by the accretion, and its dimension is approximately three times larger than the dimension of the external region. In this case, the maximal (positive) pressure on the foundation at all time instants remains close to its original value (in the case of accretion without preload, the maximal pressure significantly decreases with time). Finally, for $\alpha_\sigma \approx 60$, the region of negative values on the steady-state contact pressure diagram in the external part of the base disappears completely, and the pressure near the external edge of the base becomes zero. But the value of the negative pressure near the internal edge is already more than $1^1/_2$ times greater than the maximal in absolute value negative pressure acting on the base of the arch accreted without tension. The zone of potential separation occupies approximately $^5/_8$ of the entire part of the supporting surface created in the course of accretion. The maximal pressure on the foundation during the erection increases approximately by a factor of $1^1/_2$.

As a result of the above analysis, we conclude that, in the case under study, there does not exists a value of the coefficient α_σ in the chosen law (16) of the preliminary tension of the added layers for which the final contact pressure on the foundation is positive everywhere.

But if the existence of bilateral constraints on the support points is admissible in the process of the structure operation, then, as the performed computations show, by varying the coefficient of the preload intensity, one can minimize the maximal level of contact separation stresses.

9. LOCAL FORCE SUPPORT OF THE ARCH IN THE PROCESS OF ERECTION

Obviously, the tendency of external parts of the bases of a not too thick-walled heavy arch to separate from the smooth foundation is due to the fact that the main part of the mass of such a structure is projected not on its support points (just as in the case of a very thick arch) but on the part of the foundation between them and there are no friction forces, which can counterbalance the bending moment arising in this case. It is clear that this situation can be corrected by providing a support of the arch near its vertex, for example, by using a suspension with a controlled tension force. In this case, the mass of the material between the supporting points is partially taken by the suspension, and the material located over the bases can press them completely to the foundation. For a sufficiently large constraining force, there is a situation opposite to the original: already the internal parts of the bases tend to separate. Of course, one can increase the force so that, in the case of bilateral constraints, the separation stresses begin to act in the entire contact and, in the case without such constraints, the arch completely separates from the foundation (hangs on the suspension).

Everything said above seems to be absolutely obvious for the arch completely prepared before its installation on the foundation. The main question is whether it is possible, using a local force support of the arch erected directly on a smooth foundation, to form a stress field in this arch such that, after the erection process is finished and the suspension is disconnected, the finished structure can stand without separation on the smooth foundation without any retaining constraints. This is just the problem studied in the present section.

We assume that, on the external surface $\rho = b$ of the arch originally installed on the foundation, in a neighborhood $\varphi \in (\pi/2 - \Delta\varphi, \pi/2 + \Delta\varphi)$ of the line connecting its vertices, where $\Delta\varphi$ is a small angle, the stresses distributed according to some known law $\mathbf{t}(\varphi, t)$ act from the moment of the arch installation $t = t_0$. Beyond this neighborhood, we set $\mathbf{t}(\varphi, t) \equiv \mathbf{0}$; i.e., we assume that there is no load on this part of the surface. Because of the assumption that the nonzero stresses act on a small part of the external boundary of the body, it follows from the Saint Venant principle that the specific form of their distribution on this segment affects the stress-strain state of the body only in its small part near this segment. But for the main (with respect to dimensions) part of the body, only the integral characteristics of this distribution are significant.

We assume that a given load $\mathbf{t}(\varphi, t)$ is statically equivalent to the force acting vertically upwards, which is uniformly distributed over the line of the arch vertices in the plane passing through the arch axis. We denote the value of this force per unit length of the arch axis by $P(t)$; i.e., we set $b\int_{\pi/2-\Delta\varphi}^{\pi/2+\Delta\varphi} \mathbf{t}(\varphi, t)\, d\varphi = \mathbf{j}\, P(t)$, $t \geqslant t_0$.

We also assume that on the external arch surface there are only normal stresses $\sigma_\rho\big|_{\rho=b} = t_\rho$, whose distributions at all times are symmetric with respect to the half-plane $\varphi = \pi/2$ and are similar to each other.

It does not make sense to change the load t_ρ at the stage $t \in [t_0, t_1]$ before the beginning of the accretion; therefore, we assume that, at this stage, it is constant and corresponds to some initial suspension tension force P_0. To control the process of erection after the beginning of thickening of the initially installed structure, the surface load must be already variable in time and correspond to some supporting time-dependent force $P(t)$. We introduce this law as follows. We assume that, before the beginning of accretion of the arch installed on the foundation, the constraining force P_0 acting on it is equal to a certain prescribed part $\alpha_P = \text{const}$ of the structure original mass per unit length along the axis, $W_0 = f\pi(b^2 - a_0^2)/2$. After the beginning of accretion (for $t > t_1$), we increase the supporting force so that the ratio of α_P to the increasing linear weight of the arch $W(t) = f\pi\left[b^2 - a^2(t)\right]/2$ remains unchanged. After the accretion is finished, we smoothly leave the arch and gradually decrease the value of the constraining force $P(t)$ form the value attained at the end of accretion to zero. We assume that the constraining force decreases exponentially.

We also assume that, in the structural elements used to erect the arch, preliminary stresses characterized by the well-known tensor of the form (15), (16) can be created.

We study the process of erection of the structure under study in the presence of the constraining force $P(t)$. The graph of the time-dependence of this force, realized in computations, is shown in Fig. 4, *a*. We set $\Delta\bar{t}_\mathrm{r} = 5$ and $\delta_\mathrm{r} = 10^{-3}$. The corresponding local normal stress distribution σ_ρ on the external circular cylindrical surface $\rho = b$ of the arch is shown in Fig. 4, *b*. The angular measure of half the arc on which the nonzero stresses act is chosen to be $\Delta\varphi = 5°$.

First, in the course of accretion, we use elements not preliminary stressed ($\alpha_\sigma = 0$). The evolution of the contact stresses and their final distributions in this case is shown in Fig. 4, *c*, *d*. As we see (Fig. 4, *c*), if in the process of the arch erection the

$W(t) = \frac{\pi f}{2}\left[b^2 - a^2(t)\right],\ t \in [t_1, t_2]$
P
$\alpha_P\, W(t_2)\, \delta_r^{(t-t_2)/\Delta t_r}$
$\alpha_P\, W(t)$
(a)
0
t_1
t_2
$t_2 + \Delta t_r$
t
$b \int_{\pi/2-\Delta\varphi}^{\pi/2+\Delta\varphi} \sigma_\rho|_{\rho=b} \sin\varphi\, d\varphi = P$
$\sigma_\rho|_{\rho=b}$
(b)
0
$\pi/2 - \Delta\varphi$
$\pi/2$
$\pi/2 + \Delta\varphi$
φ
$\bar{\sigma}_\varphi/\bar{f}$
(e)
6
($\varphi = 0.0°$)
3
0
−3
$\alpha_P = 0.600$
$\bar{\sigma}^*_\varphi/\bar{f} = 12\,\bar{\rho}$
−6
0.7
0.8
0.9
$\bar{\rho}$
0.7
0.8
0.9
$\bar{\rho}$
$\bar{\sigma}_\varphi/\bar{f}$
($\varphi = 0.0°$)
0
−3
$\alpha_P = 0.542$
(c)
−6
$\bar{\sigma}_\varphi/\bar{f}$
($\varphi = 0.0°$)
0
−3
$\alpha_P = 0.600$
(d)
−6
0.8
0.9
$\bar{\rho}$

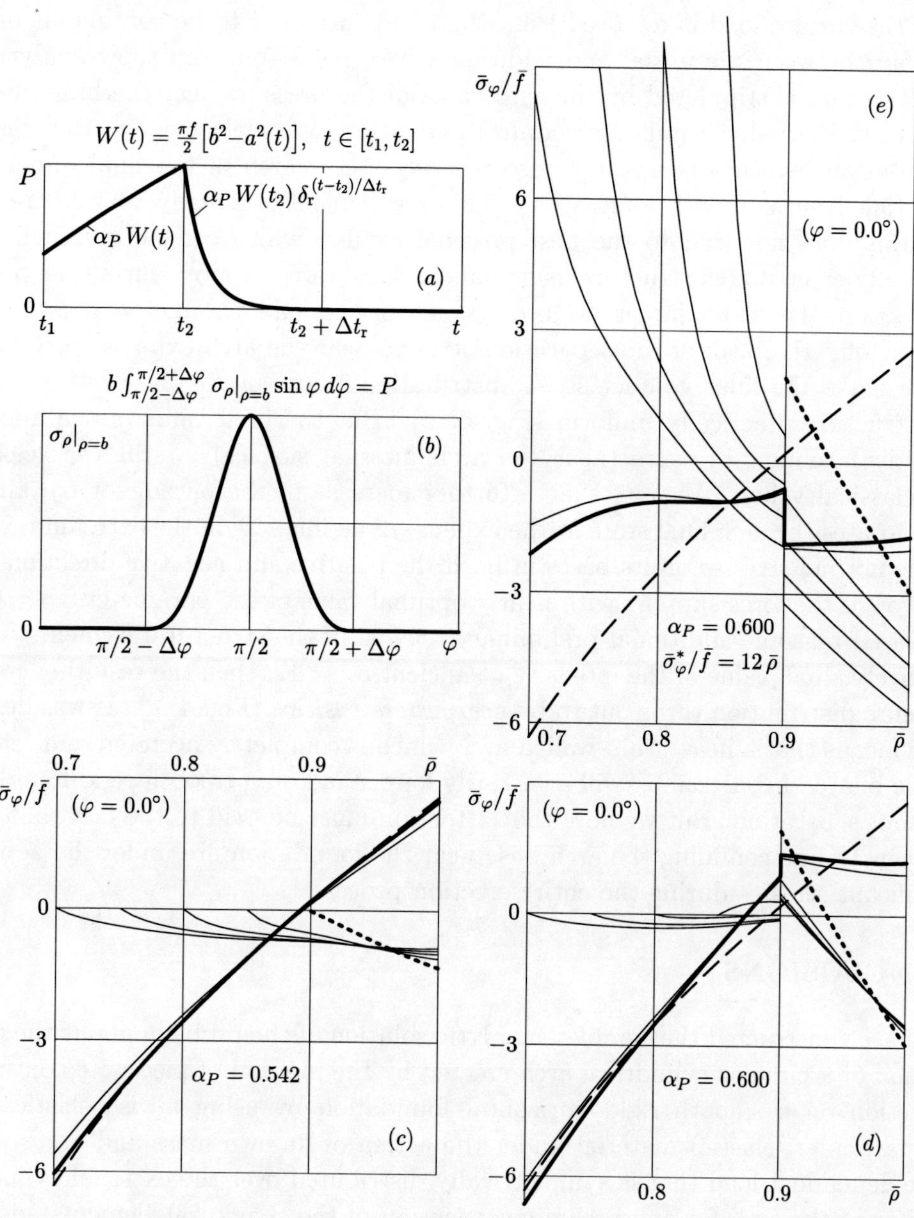

Fig. 4

constraining suspension takes a bit larger than half the arch mass ($\alpha_P \approx 0.542$), then the final contact pressure distribution (which is steady-state in a long time period after the surface load is removed from the arch) turns out to be continuous on the interface between the initial and additional base parts (one can show analytically that this fact is stipulated by the zero value of the pressure near the base internal edge of the initially installed structure) and to be extremely close to the classical distribution, which arises in the case where just the arch of the final dimensions is installed on a smooth foundation. This result itself is already very interesting. But this does not lead to the best possible results, which can be obtained for a gradually erected arch structure using only its local force support during the process of erection. For a bit larger value $\alpha_P = 0.6$ of the load parameter, it is possible to "cut off" the peak of the separation stresses near the arch external surface and hence make the limit contact stress distribution in the originally existing part of the structure practically uniform (Fig. 4, *d*). But the final value of the maximal (positive) contact pressure (near the arch internal surface) is still very close to the classical value. We note that a further increase in the parameter α_P already deteriorates the arch final state at the expense of an increase in the extremum values of the normal stresses on its bases in both the positive and negative directions.

Now if the force support with nearly optimal value of the parameter $\alpha_P = 0.6$ is increased by some additional preliminary tension of the structural elements with a relatively small value of the intensity coefficient $\alpha_\sigma = 12$, then the resulting contact pressure distribution turns out to be *everywhere positive* (Fig. 4, *e*), as was desired. This means that a heavy thin-walled arch which is completely accreted and released of the bilateral constraints, will arbitrarily long stand on a smooth rigid foundation without separation. But we note that attention must be paid to the fact that there the constraints confining the arch base near the foundation are under the action of significant stresses during the entire erection process.

CONCLUSIONS

We have constructed the complete analytic solution of the problem about the deformation of a circular cylindrical arch erected by the method of piecewise continuous accretion on a smooth rigid horizontal foundation by using a viscoelastic aging (in particular, elastic) material under the action of its own mass and an arbitrary time-dependent load that is symmetrically distributed over the external cylindrical surface of the arch for any preliminary tension of the structural elements added in the process of accretion. By analyzing the solutions thus obtained and performing numerical computations, we show that the neglect of the action of the gravity forces during the entire process of accretion of heavy objects may lead to radically wrong descriptions of their current and final state, including severe overestimation of the strength and the operating bearing capacity. The stress-strain state of such objects crucially depends on the technology and the operation mode of the erection process. We have demonstrated the possibility of a rather efficient control of this state by creating preliminary stresses in the additional material added to the accreted material and by using a temporal local load on the surface of this body.

ACKNOWLEDGMENTS

This work is supported by the Russian Foundation for Basic Research (under grants Nos. 08-01-91302-IND_a, 08-01-00553-a, and 06-01-00521-a), by the Department of Energetics, Mechanical Engineering, Mechanics and Control Processes of the Russian Academy of Sciences (Program No. 13 OE), and by the Russian Science Support Foundation.

REFERENCES

1. A.V. Manzhirov and V.A. Chernysh, "Strengthening of a Buried Arch by the Way of Accretion," Izv. Akad. Nauk. Mekh. Tverd. Tela No. 5, 25–37 (1992).

2. A.V. Manzhirov, "The General Non-Inertial Initial-Boundaryvalue Problem for a Viscoelastic Ageing Solid with Piecewise-Continuous Accretion," J. Appl. Math. Mech. **59** (5), 805–816 (1995).

3. N.Kh. Arutyunyan and A.V. Manzhirov *Contact Problems in the Theory of Creep* (Izd-vo NAN RA, Erevan, 1999) [in Russian].

4. A.V. Manzhirov and M.N. Mikhin "Plane Problem for an Accreted Solid," in *Modern Problems of Continuum Mechanics. Proceedings of the VI International Conference. Rostov-on-Don, June 19–23, 2000*, Vol. 2 (Izd-vo SKNTs VSh, Rostov-on-Don, 2001), pp. 106–109 [in Russian].

5. A.V. Manzhirov and M.N. Mikhin, "Methods of the Theory of Functions of a Complex Variable in Growing Solids Mechanics," Vestn. Samarsk. Gos. Univ. Estestv.-Nauch. Ser. No. 4 (34), 82–98 (2004).

6. A.V. Manzhirov and M.N. Mikhin, "Torsion of Accreted Solids," in *Modern Problems of Continuum Mechanics. Proceedings of the IX International Conference Devoted to the 85th Anniversary of Academician I.I. Vorovich, Rostov-on-Don, October 11–15, 2005.* Vol. 1 (Izd-vo "TSAR", Rostov-on-Don, 2005), pp. 131–136 [in Russian].

7. A.V. Manzhirov and D.A. Parshin, "Accretion of a Viscoelastic Ball in a Centrally Symmetric Force Field," Mech. Solids **41** (1), 51–64 (2006).

8. A.V. Manzhirov and D.A. Parshin, "Accretion of Solids under Mass Forces," in *Indo-Russian Workshop on Problems in Nonlinear Mechanics of Solids with Large Deformation. Proceedings. IIT Delhi, November 22–24, 2006* (IIT Delhi, New Delhi, 2006), pp. 71–79.

9. A.V. Manzhirov and D.A. Parshin, "Modeling the Accretion of Cylindrical Solids on a Rotating Mandrel with Centrifugal Forces Taken into Account," Mech. Solids **41** (6), 121–134 (2006).
10. A.V. Manzhirov and D.A. Parshin, "Modeling of the Deformation Process of Accreted Conic Solids," Vestn. Samarsk. Gos. Univ. Estestv.-Nauch. Ser. No. 4 (54), 290–303 (2007).
11. A.V. Manzhirov, "Mechanics of Accreted Solids: State-of-the-Art, Problems, and Perspectives," in: *Topical Problems of Continuum Mechanics. Proceedings of the International Conference Devoted to the 95th Anniversary of the Full Member of the National Academy of Sciences of the Republic of Armenia N.Kh. Arutyunyan, Taskmaster, 25–28 September 2007* (Erevan State Univ. Architecture and Civil Engineering, Erevan, 2007), pp. 243–246 [in Russian].
12. D.A. Parshin, "Accretion Problems for Elastic and Viscoelastic Solids under Mass Force Fields," in *Topical Problems of Continuum Mechanics. Proceedings of the International Conference Devoted to the 95th Anniversary of the Full Member of the National Academy of Sciences of the Republic of Armenia N.Kh. Arutyunyan. Taskmaster, 25–28 September 2007* (Erevan State Univ. Architecture and Civil Engineering, Erevan, 2007), pp. 318–321 [in Russian].
13. D.A. Parshin, "Strengthening of a Conic Column by the Way of Accretion," in *Modern Problems of Continuum Mechanics. Proceedings of the XI International Conference. Rostov-on-Don, 26–29 November 2007.* Vol. 1 (Izd. "TSAR", Rostov-on-Don, 2007), pp. 205–209 [in Russian].
14. N.Kh. Arutyunyan, *Some Problems in the Theory of Creep* (Paragon Press, Oxford, 1966).
15. N.Kh. Arutyunyan and V.B. Kolmanovskii, *Creep Theory of Inhomogeneous Bodies* (Nauka, Moscow, 1983) [in Russian].
16. N.Kh. Arutyunyan and V.V. Metlov, "Nonlinear Problem of the Creep Theory of Accreted Aging Bodies," Izv. Akad. Nauk SSSR. Mekh. Tverd. Tela No. 4, 142–152 (1983).
17. A.D. Polyanin and A.V. Manzhirov, *Handbook of Integral Equations*, 2nd ed. (Chapman & Hall/CRC Press, Boac Ratoon–London, 2008).
18. N.Kh. Arutyunyan "Creep of Aging Materials. Concrete Creep," in *Mechanics in USSR in 50 Years.* Vol. 3 (Nauka, Moscow, 1972), pp. 155–202 [in Russian].
19. I.E. Prokopovich and V.A. Zedgenidze *Applied Theory of Creep* (Sterilized, Moscow, 1980) [in Russian].

COUPLED THERMOMECHANICAL MODEL OF METAL CUTTING PROCESSES TAKING INTO ACCOUNT A FRACTURE AND A FRAGMENTATION

V.N. Kukudzhanov[1*] and A.L. Levitin[1]**

ABSTRACT

We use the finite element method to perform the three-dimensional modeling of unsteady process of cutting an elastoviscoplastic slab by an absolutely rigid cutting tool moving at a constant velocity V_0 at different inclinations α of the tool face (Fig. 1). The modeling was based on the coupled thermomechanical model of an elastoviscoplastic material. The adiabatic process of cutting was compared with the regime in which the slab material heat conduction is taken into account. The cutting process was parametrically studied for variations of the cutting tool and slab's thickness geometry, in the rate and depth of cutting, and in the properties of the processed material. The stressed state varied from the plane-stressed $\bar{H} = H/L \ll 1$ (thin plate) to the plane-strained $\bar{H} \gg 1$ (wide plate), where H is the slab thickness and L is the slab length. The problem was solved on a moving adaptive Lagrange–Euler mesh by the finite element method with splitting, by using the explicit-implicit integration schemes for constitutive equations [5]. It was shown that the numerical modeling of the problem in the three-dimensional statement permits studying the cutting processes with continuous chip formation and with chip destruction into separate pieces. The mechanism of this phenomenon in the case of orthogonal cutting ($\alpha = 0$) can be explained by the thermal softening with formation of adiabatic shear strips without using the damage models. In cutting by a sharper tool (the angle α is large), it is necessary to use the coupled model of thermal and structural softening. The dependencies of the force acting on the tool for different geometric and physical parameters of the problem was obtained. We also show that the quasimonotone and oscillating operation modes are possible and explain them from the physical standpoint.

[1]Ishlinsky Institute for Problems in Mechanics of the Russian Academy of Sciences, Moscow, Russia

[*]E-mail: *kukudz@ipmnet.ru*

[**]E-mail: *alex_lev@ipmnet.ru*

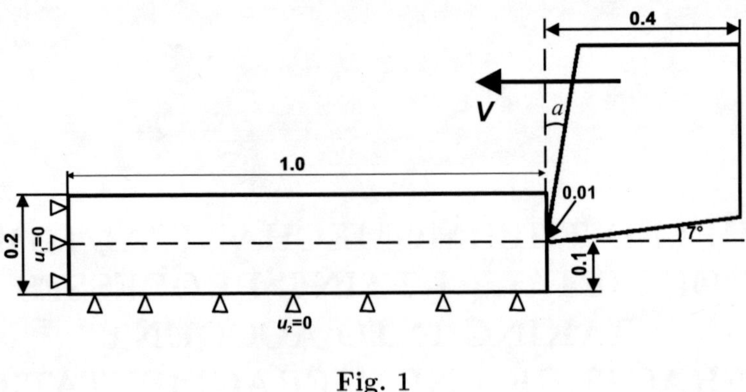

Fig. 1

Key words: plasticity, damage, metal cutting

INTRODUCTION

The cutting plays an important role in processing the hardly-deformed materials on turning and milling machines. The machining is the main price setting operation in manufacturing work pieces of complicated profile made of hardly-deformed materials such as titanium-aluminum and molybdenic alloys. The cuts formed in cutting such materials can split into separate pieces (chips), which results in a nonsmooth surface of the cut material and a strongly uniform pressure on the cutting tool. It is very difficult to determine the temperature and stress-strain states of the material under high-speed cutting experimentally. An alternative method is the numerical simulation of the process, which permits explaining the main specific characteristics of the process and to study the cutting mechanism in detail. The fundamental understanding of chips formation and destruction mechanisms is very important for efficient cutting. The mathematical simulation of the cutting process requires to take account of large strains and rates of strains and heating due to dissipation of the plastic strain, which lead to the high temperature of the material softening, fracture and fragmentation.

No exact solution of this process has yet been obtained, although the studies have been performed since the middle of the 20th century. The first works dealt with the simplest rigid-plastic computational scheme [2–5]. But the results obtained on the basis of the rigid-plastic analysis could not satisfy either practitioners or theoreticians, because this model did not answer the desired questions. In the literature, there is no solution of this problem in the 3D statement, where the nonlinear effects, the chips fracture, fragmentation, and the material thermomechanical softening were considered.

Recently, the numerical simulation made a certain progress in studying these processes. It was examined how the cutting angle, the thermomechanical properties of the work piece and the cutting tool affect the chips formation and destruction [6–8]. But in the majority of works, the cutting process was considered under significant restrictions: the statement of the problem was two-dimensional (plain strain); the influence of the initial stage of the unsteady process on the force acting on the tool was not considered; it was assumed that the fracture occurs on an a priori given interface. All these restrictions did not permit the complete study of the cutting process, in several cases, resulted in wrong understanding of the process itself.

Moreover, as experimental studies of the last years show [9, 10], at high rates of strain $\dot{\varepsilon} \geq 10^5 - 10^6\,\mathrm{s}^{-1}$, many materials exhibit an anomalous temperature dependence related to the reconstruction of the dislocation motion mechanism. The thermofluctuation mechanism is replaced by the background resistance mechanism and, as a result, the material resistance dependence on the temperature changes to the opposite, namely, the material strength increases with the temperature. Such effects may lead to great problems in the high-speed cutting. These problems have not been studied in the literature until recently. To model the high-speed cutting process, it is necessary to develop models taking account of the complicated dependencies of the material viscoplastic behavior and, first of all, the damage and fracture with crack formation and fragmentation of particles and pieces of the strained material. To take all these effects into account, one needs not only complicated thermophysical models but also modern computational methods for calculating large strains, which exclude the mesh limit distortions and take account of the material fracture and continuity violation [11–13]. These problems require a large amount of computations. It is necessary to develop high-speed algorithms for solving elastoviscoplastic equations with internal variables [1, 5, 14, 15].

1. STATEMENT OF THE PROBLEM

1.1. *Geometry*

We deal with the three-dimensional statement of the problem. Figure 1 shows the domain and the boundary conditions in the cutting plane. In the direction perpendicular to this plane, the slab has a finite thickness $\bar{H} = H/L$ (L is the slab length), which can vary in a wide region. The spatial statement admits a certain freedom in motion of the processed material away of the cutting plane and a more smooth outcome of chips, which ensures more favorable cutting conditions.

1.2. *Basic equations*

The complete couples system of thermoelastoviscoplastic equations consists of the momentum conservation law

$$\rho \frac{dv}{dt} = \sigma_{ij,j}, \tag{1.1}$$

the Hooke's law with temperature stresses

$$\frac{d\sigma_{ij}}{dt} = D_{ijkl}(\dot{\varepsilon}_{kl} - \dot{\varepsilon}^p_{kl} - \alpha\delta_{kl}\dot{\theta}), \tag{1.2}$$

and the heat influx equation

$$\rho C_e \frac{d\theta}{dt} = K\theta_{,ii} - (3\lambda + 2\mu)\alpha\theta_0\dot{\varepsilon}_{ii} + \kappa\sigma_{ij}\dot{\varepsilon}^p_{ij}, \tag{1.3}$$

where C_e is the heat capacity, K is the thermal conductivity, and κ is the Quinney–Taylor coefficient taking account of the material heating due to plastic dissipation.

We also have the associated law of plastic flow

$$\dot{\varepsilon}^p_{ij} = \dot{\lambda}\frac{dF}{d\sigma_{ij}} \tag{1.4}$$

and the plasticity condition

$$F(J_i, E^p_i, \chi_i, \theta) = J_2 - \sigma_Y(J_1, E^p_i, \chi_i, \theta) \le 0, \tag{1.5}$$

where J_i are the stress tensor invariants and E_i are the plastic strain tensor invariants, σ_Y is the yield limit. The evolution equations for the internal variables χ_i have the form

$$\frac{d\chi_i}{dt} = f_i(J_k, \chi_k, \theta). \tag{1.6}$$

1.3. *Model of the material*

In the present paper, we accept the Mises-type thermoelastoviscoplastic model, i.e., the model of plasticity with the yield strength in the form of multiplicative dependence (1.7) including the strain and viscoplastic strengthening and the thermal fracture [17]:

$$\sigma_Y(\bar{\varepsilon}^p, \dot{\bar{\varepsilon}}^p, \theta) = [A + B(\bar{\varepsilon}^p)^n]\left[1 + C\ln\left(\frac{\dot{\bar{\varepsilon}}^p}{\dot{\varepsilon}_0}\right)\right](1 - \hat{\theta}^m), \tag{1.7}$$

where σ_Y is the yield strength, $\bar{\varepsilon}^p$ is the plastic strain intensity, and $\hat{\theta}$ is the relative temperature referred to the melting temperature θ_m:

$$\hat{\theta} = \begin{cases} 0, & \theta < \theta_*, \\ \dfrac{\theta - \theta_*}{\theta_m - \theta_*}, & \theta_* \leq \theta \leq \theta_m, \\ 1, & \theta > \theta_m. \end{cases}$$

We assume that the work piece is made of a homogeneous material. We used rather soft material Al2024-T3 (the elastic constants: $E = 73\,\text{GPa}$, $\nu = 0.33$; the plastic constants: $A = 369\,\text{MPa}$, $B = 684\,\text{MPa}$, $n = 0.73$, $\dot{\varepsilon}_0 = 5.77 \cdot 10^{-4}$, $C = 0.0083$, $m = 1.7$, $\theta_* = 300\,\text{K}$, $\theta_m = 775\,\text{K}$, $\beta = 0.9$) and a more rigid material 42CrMo4 ($E = 202\,\text{GPa}$, $\nu = 0.3$, $A = 612\,\text{MPa}$, $B = 436\,\text{MPa}$, $n = 0.15$, $\dot{\varepsilon}_0 = 5.77 \cdot 10^{-4}$, $C = 0.008$, $m = 1.46$, $\theta_* = 300\,\text{K}$, $\theta_m = 600\,\text{K}$, $\beta = 0.9$). The adiabatic process of cutting was compared with the solution of the complete thermomechanical problem.

1.4. *Fracture*

The material fracture model is based on the continual Maenchen–Sack approach [11], where the fracture regions are modelled by discrete particles. For the fracture criterion we take the critical value of the plastic strain intensity ε_f^p:

$$\varepsilon_f^p = \left[d_1 + d_2 \exp\left(d_3 \frac{J_1}{J_2}\right)\right]\left[1 + d_4 \ln\left(\frac{\dot{\varepsilon}^p}{\dot{\varepsilon}_0}\right)\right](1 + d_5\hat{\theta}), \tag{1.8}$$

where d_i are the material constants determined experimentally.

If the fracture criterion is satisfied in a Lagrangian cell, then the bonds between the nodes in such cells become free and either the stresses relax to zero or only the resistance to compression is preserved. Under fracture, the Lagrangian nodal masses become independent particles carrying away the mass, momentum, and energy, moving as a rigid solids not interacting with the undestroyed particles. A detailed survey of these algorithms is given in [12, 13]. In the present paper, the fracture is determined by attaining the critical intensity of the plastic strain ε_f^p and no fracture surface is prescribed a priori. In our computations, $\varepsilon_f^p = 1.0$, and the tool velocity was $2\,\text{m/s}$ and $20\,\text{m/s}$.

1.5. *Method of integrating the equations*

To integrate the coupled system of thermoplasticity equations (1.1)–(1.8), it is expedient to use the splitting method developed in [5]. The scheme of splitting the elastoplastic equations consists in dividing the entire process into

1) a predictor, which is a thermoelastic process, where $\dot{\varepsilon}^p \equiv 0$ and all the operators related to the plastic strain are zero, and

2) a corrector, in which the total strain rate is $\dot{\varepsilon} \equiv 0$.

At the predictor stage, system (1.1)–(1.6) for the variables denoted by tilde takes the form

$$\begin{aligned} &\rho\frac{d\tilde{v}}{dt} = \tilde{\sigma}_{ij,j} \\ &\frac{d\tilde{\sigma}_{ij}}{dt} = D_{ijkl}(\dot{\tilde{\varepsilon}}_{kl} - \alpha\delta_{kl}\dot{\tilde{\theta}}), \\ &\rho C_e\frac{d\tilde{\theta}}{dt} = K\tilde{\theta}_{,ii} - (3\lambda + 2\mu)\alpha\theta_0\dot{\tilde{\varepsilon}}_{ii}, \\ &\dot{\varepsilon}^p_{ij} = 0, \quad \dot{\chi}_i = 0. \end{aligned} \tag{1.9}$$

System (1.9) can be integrated easily, because the last term in the heat influx equation, which is related to elastic volume deformation, is small even for high temperatures and can be neglected [18]. Then the heat conduction equation can be integrated separately from the others (according the explicit conditionally stable or by implicit absolutely stable scheme with splitting in different directions). Then we solve the isothermal dynamical problem following the explicit central-difference scheme for $\tilde{v}$ and $\tilde{\sigma}_{ij}$. The obtained values at predictor stage are used as the initial data at the corrector stage for the following system of equations for the variables denoted by "hat":

$$\rho\frac{d\hat{v}}{dt} = 0, \quad \dot{\hat{\varepsilon}}_{ij} = \dot{\hat{\varepsilon}}^e_{ij} + \dot{\hat{\varepsilon}}^p_{ij} = 0. \tag{1.10}$$

Hook's law and the associated law (1.4) lead to the relaxation equation

$$\frac{d\hat{\sigma}_{ij}}{dt} = -\frac{d\lambda}{t}D_{ijkl}\frac{\partial F}{\partial\hat{\sigma}_{kl}}. \tag{1.11}$$

The heat influx equation (1.3) becomes

$$\rho C_e\frac{d\hat{\theta}}{dt} = \kappa\sigma_{ij}\dot{\hat{\varepsilon}}^p_{ij}. \tag{1.12}$$

Equations (1.4)–(1.6) for the variables with "hat" remain unchanged.

2. COMPUTATIONAL RESULTS

Firstly, we consider the problem for $\bar{H} \gg 1$ (plane strain state). For a large plate thickness $\bar{H} \gg 1$, the material in strained state under the action of the moving tool, must move towards the free surface (Fig. 2 *a*, *b*). Figure 2 illustrates formation of shear localization strips and development of fracture surfaces in cutting a thick

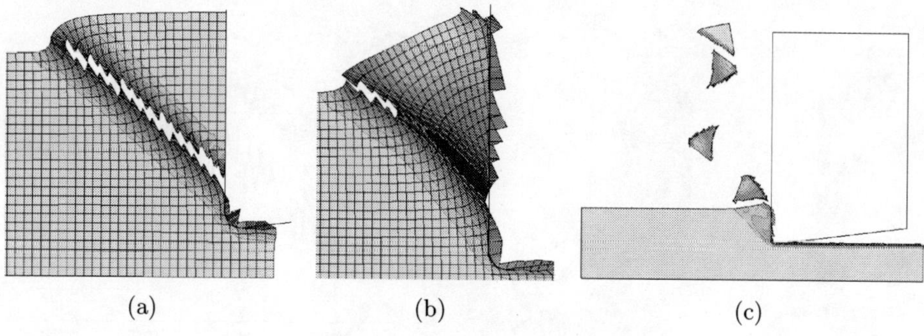

Fig. 2

plate made of alloy 42CrMo4: (a) — formation of the first chip, (b) — formation of the second fragment, (c) — separation of a series of fragments ($\alpha = 0°$, $V = 2\,\text{m/s}$, adiabatic process).

In the shear strip, the fracture occurs according to the following scheme. The extruded material forms a lump, which grows as the tool moves. In the case of orthogonal cutting, two shear strips originate, one near the tool tip and another on the lump surface (because of the loss of stability of the surface). As the tool moves, these two strips approach each other and the fracture occur as they merge (Fig. 2 b). A part of the material in chips form separates from the working piece. After this, a new lump is formed in the already heated and softened material. The process becomes nonmonotone and quasiperiodic and is determined by the moving tool velocity.

It is shown that, in the case of orthogonal cutting, fracture occurs along the shear localization strips and is related to the thermal softening due to dissipation of the plastic energy. After the first chip splits off, the process comes to the quasistationary operation mode. A continuous chip or chips as a sequence of separate fragments are formed (Fig. 2 c). This depends on the relation between the strength and thermomechanical properties of the working piece material. A more smooth cutting surface and continuous chips are formed in the case of small velocities and acute angles of cutting and if the heat conduction is taken into account.

Under the conditions of plane strain (the relative thickness is $\bar{H} \gg 1$), the material is strained in the lateral direction and the chip moves upwards in the tool plane. In the spatial statement for thin plates ($\bar{H} \ll 1$), the chips typically move sideways out of the tool motion plane. In this case, the formation of continuous chips occurs in a wider range of velocities and the material heat conduction (Fig. 3).

Figure 3 shows: (a) — the plastic strain intensity distribution and formation of spiral-shape continuous chips; (b) —variation in the total force acting on the tool in time. The slab material is 42CrMo4, $\alpha = 0°$, $V = 2\,\text{m/s}$, and the heat conduction

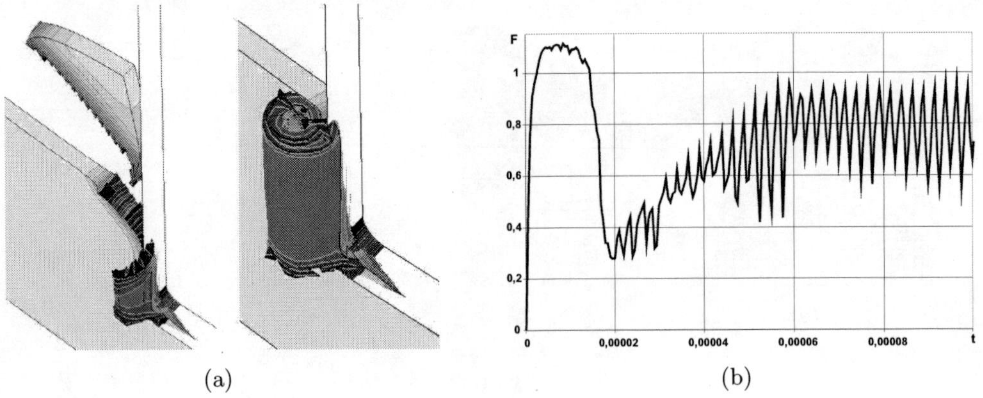

(a) (b)

Fig. 3

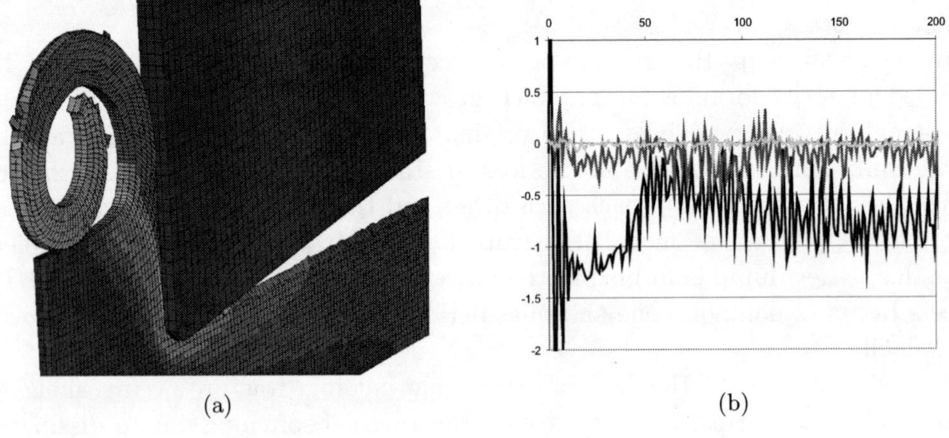

(a) (b)

Fig. 4

is taken into account in the cutting process.

Figure 4(a) shows cutting of the thin plate (plane stress state) and formation of continuous spiral chip. Zones of minimal temperature in the chip are shown by dark color and maximum by bright. The calculations are related to non steady thermoconduction process in elastoviscoplastic material with Johnson-Cook yield condition taking into account thermosoftening.

Figure 4(b) shows the corves of time dependence of integral forces acting on the tool in 3 different directions. At the beginning of the process there is a deep drops of the forces. After that, the forces grow and reach their steadystate values. The force acting in moving tool direction has a maximum drop.

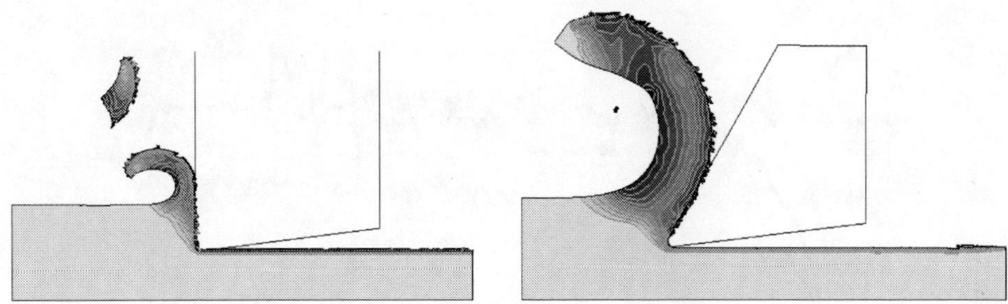

Fig. 5

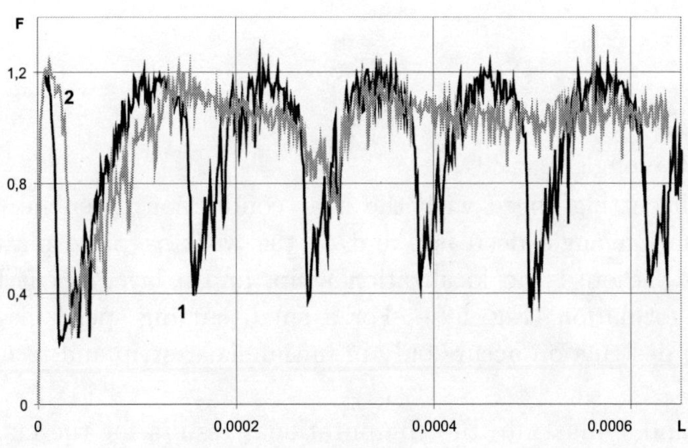

Fig. 6

The formation of continuous chips creates a monotone force of the tool reaction. As the chips tore in fragments, the reaction force has a saw-tooth form, which negatively affects the technological process of cutting (Fig. 6).

If the plate thickness is small, first, as the material is not yet heated, a shear strip is formed along which fracture occurs and the first chip separates from the basic material. After this, the tool moves along the slab on the already heated and softened material, flattening it and cutting the material that flows sideways and forms continuously curling spiral-shape chips (Fig. 3 *a*). The force acting on the tool after the initial unsteady part on which it is maximal decreases and becomes steady-state, and the cutting process evolves quasimonotonically. The high-frequency vibrations of this value are cause by the material destruction on the contact surface and by the fragmentation of small pieces (Fig. 3 *b*).

Figure 5 shows: (*a*) — chips fracture in the case of orthogonal cutting of the hardly-deformed material 42CrMo4; (*b*) — continuous cutting of the soft material

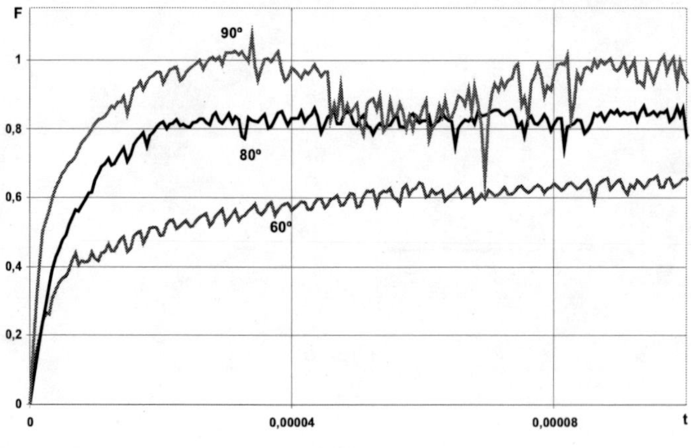

Fig. 7

Al2024-T3 at the angle $\alpha = 30°$.

For a small cutting speed with the heat conduction taken into account or as the tool inclination angle decreases and/or the working piece plasticity increases (duraluminium), there is no localization strip, and a layer is cut homogeneously without chips formation (Fig. 5 *b*). For a small cutting speed like to orthogonal ($\alpha \approx 0° - 10°$) destruction occurs only in initial shear strip and forms a single chip (Fig. 5 *a*).

Figures 6 and 7 illustrate the computational results for the cutting resistance force acting on tool in time. In Fig. 6, for a thick plate, one can see that there is an initial area of unsteady process related to formation of the primary shear strip, where the resistance force is maximal. In the case of orthogonal cutting of a thin plate (Fig. 3 *b*), after fracture and separation of a piece of the plate, there is a sharp decrease whose value depends on the cutting speed, after which the stabilization stage occurs and the resistance force comes to the quasisteady stationary value. A similar decrease occurs as the successive chips split off (Fig. 6, curve *1*). We note that the decrease in the reaction force is related to the fact that, after separation of the first piece, the tool presses the already heated and softened material weakened in the shear strip, which is easily fractured. With the material heat conduction in the shear strip taken into account, the softening is weaker and no fracture occurs in the second shear strip (Fig. 6, curve *2*).

For a more acute cutting angle and the soft material Al2024-T3 (Fig. 7), there is no chips fracture in the first shear strip and, as a consequence, no decrease in the reaction force. Typically, there is formation of continuous chips, which implies a monotonically increasing force of the tool reaction coming to the steady operation mode. In the case of piecewise chips, the reaction force has a saw-tooth form, which negatively affects the technological process of cutting (Fig. 6).

In Fig. 7, one can see that as the tool inclination angle decreases, the resistance force decrease and monotonically varies approaching the steady-state value. The mechanism of cutting and the chip fracture also varies. There is no fracture along the strips of adiabatic shear. The tool penetrates into the material, and the fracture is or wedging type. In front of the tool, there is a shear region with uniform tension, in which we must take account not only of the thermal but also of the structure softening. This region coincides with the zone of microdefects origination and growth.

CONCLUSIONS

In the present paper, it is shown that the cutting process with a constant speed has an unsteady character at the initial stage to which the maximal resistance force corresponds; after a decrease in this force, either oscillating or monotone operation mode is established for the resistance force depending on whether the cut-off layer (chips) splits into discrete fragments or remains continuous. To model the process, it is necessary to take the thermal softening, fracture, and fragmentation of the material into account.

It is shown that, depending on the geometry and the stress-strain state near the tool tip, the cutting process may evolve with formation of pore-type defects or without defects. In the case of nearly orthogonal cutting, no pores are formed, and the cause of the chips fracture is the thermal softening.

ACKNOWLEDGMENTS

The research was financially supported by the Russian Foundation for Basic Research (under grants Nos. 08-01-91302-IND_a and 06-01-00523-a).

REFERENCES

1. V.N. Kukudzhanov, "Decomposition Method for Elastoplastic Equations," Izv. Akad. Nauk. Mekh. Tverd. Tela, No. 1, 98–108 (2004) [Mech. Solids (Engl. Transl.) **39** (1), 73–80 (2004)].
2. M.E. Merchant, "Mechanics of the Metal Cutting Process. I. Orthogonal Cutting," J. Appl. Phys. **16**, 267–275 (1945); "II. Plasticity Conductions in Orthogonal Cutting," J. Appl. Phys. **16**, 318–324 (1945).
3. R. Hill, *The Mathematical Theory of Plasticity* (Oxford, Clarendon Press, 1950; Gostekhizdat, Moscow, 1956).

4. M.C. Shaw, "A Quantized Theory of Strain Hardening as Applied Cutting of Metals," J. Appl. Phys. **21** (6), 599–606 (1950).
5. M.C. Shaw, *Metal Cutting Principles* (Oxford Sci. Publ., New York, 1986).
6. K. Liu and S.N. Melkote, "Material Strengthening Mechanisms and Their Contribution to Size Effect in Micro-Cutting," Trans. ASME J. Manufact. Sci. Engng **128** (3), 730–738 (2006).
7. H. Miguelez, R. Zaera, A. Rusinek, et al., "A Numerical Modeling of Orthogonal Cutting: Influence of Cutting Conductions and Separation Criterion," J. Phys. IV France **34**, 417–422 (2006).
8. M. Baker, "Does Chip Formation Minimize the Energy?" Comput. Mat. Sci. **33**, 407–418 (2005).
9. E. Sakino, "Transition in Rate Controlling Mechanics of FFC Metal at Very High Strain Rates and High Temperatures," J. Phys. IV France **10**, 57–64 (2000).
10. G.I. Kanel, S.V. Razorenov, and V. E. Fortov, "Shock-Wave Compression and Tension of Solids at Elevated Temperatures: Superheated Crystal States, Pre-Melting, and Anomalous Growth of the Yield Strength," J. Phys. Condens. Matter. **16** (14), S1007–S1016 (2004).
11. G. Maenchen and S. Sack, "The "Tensor" Code," in *Methods Comput. Phys.*, Vol. 3 (Acad. Press, New York, 1964), pp. 188–210.
12. V.M. Fomin, A.I. Gulidov, G.A. Sapozhnikov, et al., *High-Speed Interaction of Bodies* (Izd-vo SO RAN, Novosibirsk, 1999) [in Russian].
13. N.G. Burago and V.N. Kukudzhanov, *Numerical Solution of Continual Fracture Problems.* Preprint No. 746, IPMekh RAN (Inst. for Problems in Mechanics RAS, Moscow, 2004) [in Russian].
14. V.N. Kukudzhanov, A.L. Levitin, and V.S. Sinyuk, "Numerical–Analytical Splitting Method for Modeling Quasi-Static Processes of Strain of Damaged Materials," in *Applied Problems of Strength and Plasticity. Mezh. VUZov. Sb.*, No. 68 (Nizhnii Novgorod, 2006), pp. 7–21 [in Russian].
15. V.N. Kukudzhanov, "Coupled Models of Elastoplasticity and Damage and Their Integration," Izv. Akad. Nauk. Mekh. Tverd. Tela, No. 6, 103–135 (2006) [Mech. Solids (Engl. Transl.) **37** (6), 83–109 (2006)].
16. N. Aravas, "On the Numerical Integration of a Class of Pressure-Dependent Plasticity Models," Intern. J. Numer. Methods in Eng-ing **24** (7), 1395–1416 (1987).
17. G.R. Johnson and W.H. Cook, "A Constitutive Model and Data for Metals Subjected to Large Strains, High Strain Rates and High Temperatures," in *Proc. 7th Intern. Symp. Ballistics* (1983), pp. 541–547.
18. B.A. Boly and J.H. Weiner, *Theory of Thermal Stresses* (Wiley, New York, 1960; Mir, Moscow, 1964).

ON THE STRENGTH AND ACCURACY OF SLS PROTOTYPES

P.K. Jain[1], K. Senthilkumaran[1], P.M. Pandey[1*], and P.V.M. Rao[1]

ABSTRACT

Inferior surface quality, poor accuracy and low strength are the major handicaps of rapid prototypes and many attempts have been made to improve these. With the increased scope of application of rapid prototypes as functional components, there is a need that the low strength and poor accuracy of RP parts which is due to deposition of layers and other factors like improper selection of process parameters etc. must be addressed. In recent years researchers have paid attention in developing new materials and combination of materials which can be processed using selective laser sintering (SLS) process. Attempts have also been made to improve accuracy of prototypes by optimizing hatch directions or selecting optimum process parameters.

In the present paper, description of an experimental investigation on strength aspect of SLS prototypes and statistical analysis to establish various significant process parameters has been given. Paper also presents details of effect of an important parameter namely 'delay time' on the part strength. The paper describes the feasibility study of laser sintering of nano-clay and polyamide blend.

The inaccuracy on X-Y plane in SLS prototypes is mainly due to improper shrinkage compensation. The shrinkage factors are also found to be dependent on process parameters and are non-uniform in nature. A major source of inaccuracy has been found due to various building strategies in case of SLS. The paper describes the various sources of inaccuracies and compensation strategies.

Key words: rapid prototyping, selective laser sintering, strength, accuracy

---

[1]Department of Mechanical Engineering, Indian Institute of Technology Delhi, New Delhi — 110016, Hauz Khas, India

[*]E-mail: *pmpandey@mech.iitd.ac*

INTRODUCTION

Customer driven market has resulted in tremendous reduction in time taken for new product development. One of the technologies which have made this possible is Rapid Prototyping (RP). The global acceptance and application of RP technology has greatly helped in reducing design and manufacturing cycle times and facilitated in building products faster, cheaper and better [1]. RP is a layer by layer material additive process capable of producing complex objects directly from the CAD model [2]. Stereolithography (SLA), 3D printing (3DP), Laminated Object Manufacturing (LOM), Fused Deposition Modeling (FDM), Selective Laser Sintering (SLS), Laser Engineered Net Shaping (LENS) etc. are popular RP systems commercially available today.

Initially RP was used to facilitate visualization of design concepts via prototypes produced from wax and other polymers. Nowadays, RP is emerging as a rapid manufacturing technique, which produces functional parts in small batches, particularly for aerospace and rapid tooling applications [3]. Therefore, prototypes should have sufficient strength and accuracy to ensure proper functional requirements. However the parts produced by SLS shows low strength, poor accuracy and short functional life than the parts produced by conventional processing techniques like injection moulding [1].

In the past, researchers studied surface finish and accuracy of prototypes which were considered as major handicaps for realizing parts for direct use [4–8]. With the increased scope of application of rapid prototypes as functional components there is a need to address the issue of low strength and accuracy of RP parts. The present paper describes some important process parameters and their influence on part strength and accuracy. An important parameter namely 'delay time' (Fig. 1) has been found to affect the part strength significantly. A computational algorithm has been proposed and implemented for improving part strength by varying delay time at the process planning stage. In recent years researchers have paid attention in developing new materials and combination of materials which can be processed using SLS [9]. In the present paper details of laser sintering of mixture of virgin polyamide powder with nano-clay has been given.

The sources of inaccuracy in SLS prototypes are due to non-optimized processing conditions, errors during data preparation, limitations of machine hardware, inadequate design (3D CAD model) of the prototype and sub-optimal process planning. Among all types of processing related inaccuracies, the effect caused by shrinkage is found to have major influence on the accuracy of parts produced. Materials exhibit shrinkage during thermal cycle which varies from material to material [10]. SLS process is always accompanied by shrinkage as a result of thermal and phase change effects. In practice, the shrinkage is experimentally calibrated and the same

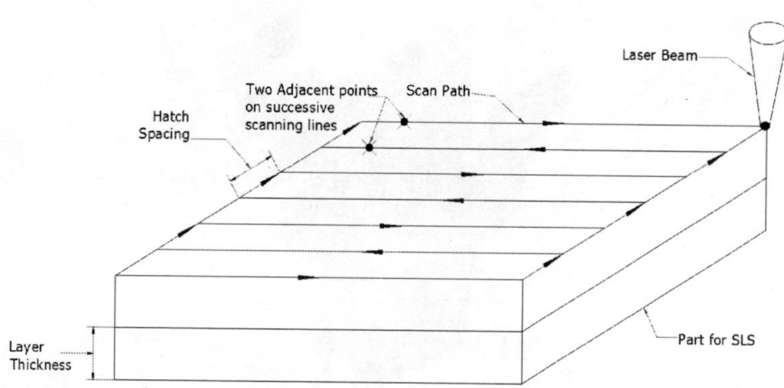

Fig. 1. Laser sintering schematics

is compensated during data preparation stage of rapid prototyping by scaling STL files. The purpose of compensation seems to be that of reducing the inaccuracies due to shrinkage and not to attempt for their elimination [11]. However, notwithstanding the complexity of shrinkage, the basic approach used today in most of the RP processes involves some form of simple shrinkage compensation. Before making the part, users are allowed to set a constant "shrinkage compensation factor" which is different for different axis to overcome the shrinkage effect [12]. The shrinkage compensation factor is found by fabricating standard test specimens and deriving a linear relationship between the nominal dimensions and fabricated part dimensions. Here when dealing with shrinkage compensation, a fundamental assumption usually made is that shrinkage is orthotropic and the shrinkage compensation scaling factors are constant along X, Y and Z axis. However, these constant scaling factors are not realistic because of the dynamic nature of the SLS process and varying layer geometry in part building. Several attempts [9–21] have been made to improve the accuracy of RP parts by either controlling or compensating the effect of shrinkage. In most of these approaches the assumption is that shrinkage is constant. Two important research issues of shrinkage being non-uniform along every direction and a method of non-uniform compensation are not explicitly attempted. Therefore in this paper a new methodology has been discussed to compensate non-uniform shrinkage in sliced layer files rather than compensating the STL files.

1. EFFECT OF DELAY TIME ON PART STRENGTH

To study the effect of delay time on part strength, the tensile test specimens were fabricated in various delay time ranges as per ASTM D638 standard. Since SLS machine doesn't have provision to fabricate parts with different delay time ranges,

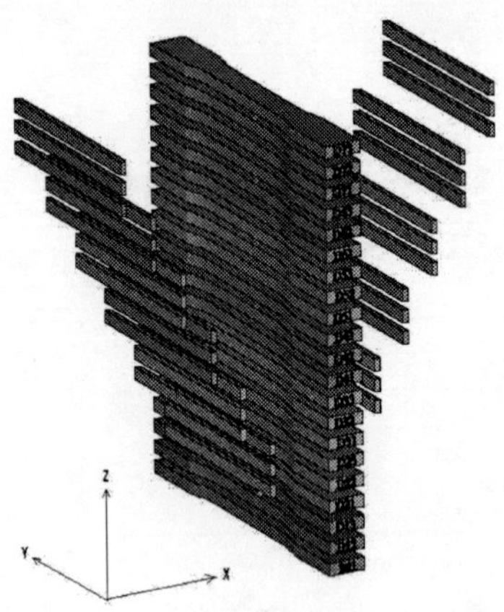

Fig. 2. Arrangement of tensile specimens on build platform for varying delay time

a novel method has been developed to design a scheme for fabrication of tensile specimens with constant delay time range. The arrangement of tensile specimens on build platform has been shown in Fig. 2. Tensile specimens were fabricated for seven different delay time ranges, ranging from $0 - 0.049$ second in the increments of 0.007 second. Fabricated specimens were tested for ultimate tensile strength on universal testing machine INSTRON 5582. The part strength has been found maximum in the delay time range $0.007 - 0.014$ second as shown in Fig. 3. This has been considered as optimum value of delay time range. An algorithm has been developed and implemented in MATLAB to find out optimum part build orientation for improved tensile strength based on delay time. Optimum build orientation is found at 60 degree orientation for tensile specimen from developed code. Further to validate the results obtained from the developed code, tensile specimens were fabricated at different part build orientations in the range of $0 - 90$ degree at an interval of 15 degree. Tensile test revealed that part strength is maximum for 60 degree orientation (Fig. 4) and same has been predicted by developed code. Therefore it is concluded that the developed algorithm is able to predict the orientation which gives maximum strength for given values of process parameters based on delay time. The developed code can handle any 3D part with various geometric shapes and multiple contours.

2. EFFECT OF PROCESS PARAMETERS ON PART STRENGTH

Since the effect of energy density on part strength has been well established, it is interesting to investigate how other parameters affecting part strength namely powder refresh rate, layer thickness, part bed temperature and hatch pattern. In

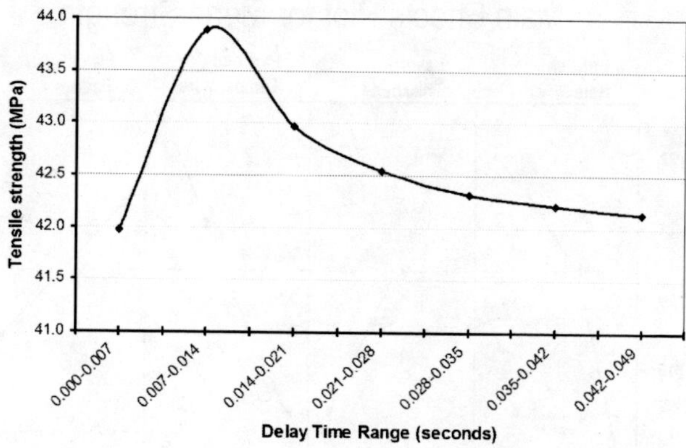

Fig. 3. Tensile strength Vs delay time range

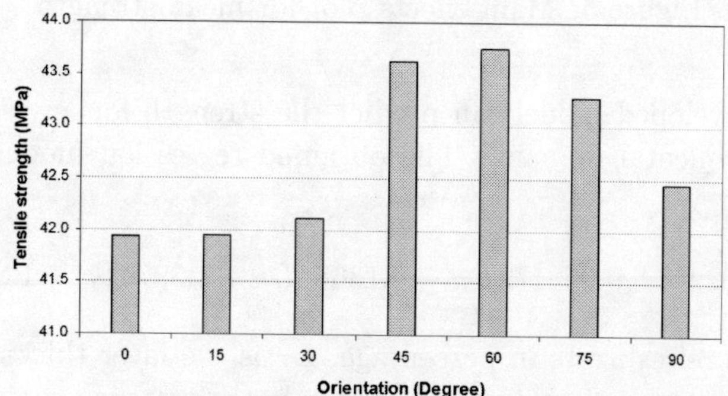

Fig. 4. Graph showing tensile strength at various orientations obtained in confirmation test

order to study the effect of selected parameters energy density has been kept constant in the present study. Investigations have been made to study the amount of variation in part strength due to variation in process parameters. Entire parameter space was explored with minimum number of experiments using Taguchi's modified L_{16} orthogonal array.

Fabricated tensile specimens were tested for ultimate tensile strength and experimental data has been analyzed by using signal to noise ratio (S/N ratio) and analysis of variance (ANOVA). Based on the results of the S/N ratio and ANOVA, optimal parameter settings for better strength have been obtained. The main effects plot for mean strength has been presented as Fig. 5. It is concluded that refresh rate effects part strength significantly in SLS.

Empirical model has been derived by linear regression using standard statistical

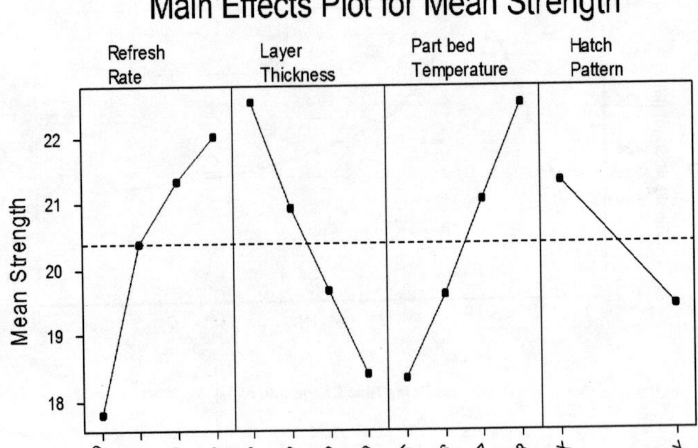

Figure 5: Main effects plot for mean strength

software. The developed model can predict the strength for any set of parameters within the experimental domain. The obtained regression model has been given below

$$S = -202.94 + 0.068172R_R - 0.13684L_T + 1.3976T_B - 1.8846H_P, \qquad (1)$$

where, R_R is the refresh rate in percentage, L_T is the layer thickness in μm, T_B is the part bed temperature in °C and H_P is the hatch pattern i.e., 1 for along X axis and 2 for along Y axis.

Regression model has been found adequate based on F-Test. Percentage contribution of the various parameters in regression model has been presented in Fig. 6. Confirmation experiment was also carried out to validate the developed empirical model. Comparison of measured strength with predicted values of strength for selected process parameters has been presented in table 1 and a good agreement between the expected and actual strength has been observed. These experiments also led to new insight of early necking observed in parts with higher refresh rates.

3. SLS OF CLAY REINFORCED POLYAMIDE

MMT is a natural inorganic material and has been used as major material to prepare PLS nanocomposites [22]. SEM micrograph of MMT (Cloisite 30B) has been presented in Fig. 7 *a* which shows layered structure of the material. In the present work, unlike the majority of previous research undertaken in the area of forming of

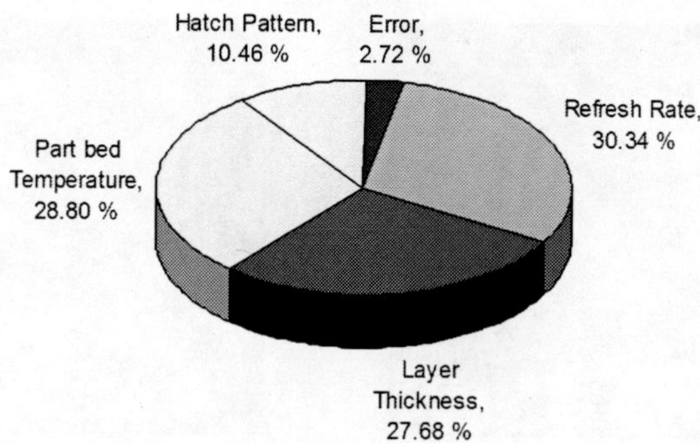

Fig. 6. Percentage contributions of process parameters on part strength

Table 1. Comparison of part strength predicted by ANOVA and confirmation experiment

S. No.	Ultimate tensile strength (MPa)		
	From developed model equation (4.10)	Expected range of values from ANOVA at confidence level of 90%	Measured value from confirmation experiment
1	27.86	26.43 – 29.29	26.93

nano-composites by melt intercalation [23], the polymer clay blended powder has directly been processed in SLS machine where no compaction or shear force was available. Before sintering, the Cloisite 30B and polyamide were mixed on a mechanical mixture [24] and mixtures of 2 and 5% clay was prepared. In Fig. 7 *b*, SEM micrograph of blend of polyamide with the clay has been shown which ensures uniform mixing, however the clay appears in agglomerated form.

Before fabricating tensile test specimens of clay/PA blend, a systematic study has been carried out by using TGA and DSC to understand the thermal behavior and possibility of sintering. TGA has been conducted on standard equipment (Perkin–Elmer TGA7) to establish thermal stability of clay. Total 30% weight loss has been observed for clay however no degradation has been observed up to 200°C. Window of sinterability, i.e, difference between melting temperature and crystallization temperature of material, is crucial while processing in SLS because sintering gets completed in a very short time and the difference in the two temperatures should be large enough [25, 26]. DSC analysis has been carried out on Perkin–Elmer DSC7 for polyamide and Cloisite 30B blend and it has been observed that the difference in two temperatures was within the prescribed limit and almost similar to virgin polyamide (Fig. 8). From these two analysis, it has been concluded that the clay/PA

Fig. 7. SEM micrograph of (*a*) Cloisite 30B clay at 3000x and (*b*) blended powder mixture of PA with 5% 30B at 500x

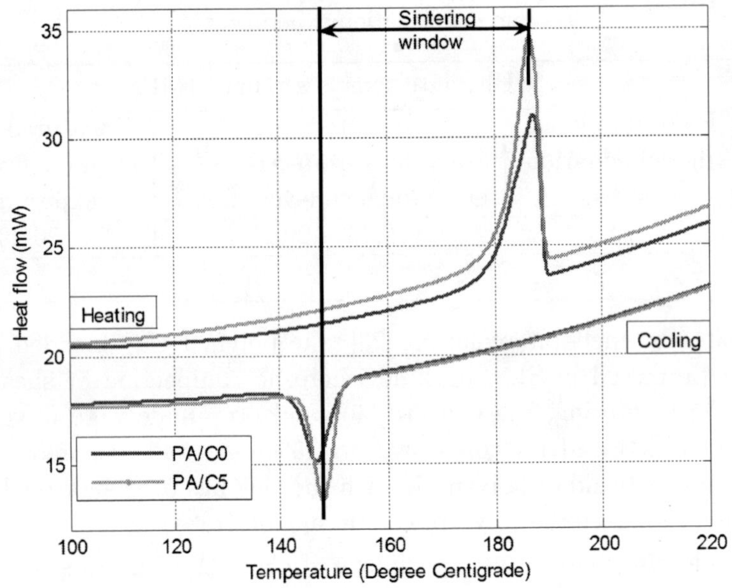

Fig. 8. DSC heating and cooling curves for PA/C0 and PA/C5

blend can be sintered by SLS process. Suitable part bed temperature has been found by conducting typical cross test as suggested by SLS machine manufacturer. Part bed temperature was found slightly higher for clay reinforced polyamide powder as compared to virgin polyamide powder.

In order to study the mechanical strength of the parts produced by SLS of clay/PA, standard tensile test specimens were fabricated. To achieve good quality parts and proper sintering, suitable process parameters were investigated. For this,

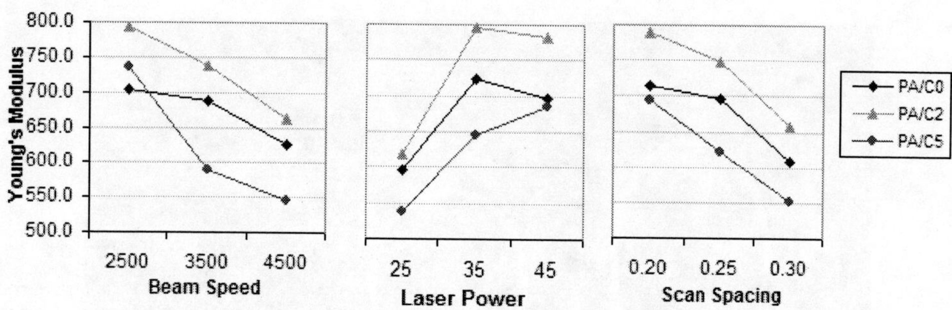

Fig. 9. Effect of process parameters on Young's modulus

preliminary trials were conducted and then entire process parameter space was explored with minimum number of experiments using Taguchi's L_9 orthogonal array. Fabricated specimens were tested on UTM and effect of various process parameters on mechanical properties of polyamide as well as clay reinforced polyamide has been obtained and presented in Fig. 9. Young's modulus of the SLS parts made from clay reinforced polyamide composite has been found more at 2% clay content however it is found lesser at 5% clay content as compared to sintered PA.

Various material characterization techniques like SEM, XRD etc. have been used to characterize the composite formed. SEM micrograph showed poor dispersion of clay in polymer matrix (Fig. 10). This forms a heterogeneous system and there has been a possibility of hindered polymerization. These may be the reasons for reduced mechanical properties of clay reinforced polyamide composite as compared to virgin PA.

To reveal nano-structural features of clay reinforced polyamide composite, XRD patterns were recorded at room temperature using PANalytical XPertPro equipment. Many samples were analyzed and in few samples intercalation or partial exfoliation has been observed. However majority of the samples did not show evidence of formation of nano-composite. Therefore, it is concluded that the possibility of formation of nanocomposites may not be denied but it is not distributed everywhere in the sample. The XRD patterns also showed evidence of reduced crystallinity which resulted into lower mechanical properties.

4. DEXEL BASED SHRINKAGE COMPENSATION FOR THE IMPROVED ACCURACY

Ragunath and Pandey [6] studied the effect of process parameters on the process and material shrinkage. They found that scan length mostly influences shrinkage in the X-direction. They also predicted that scaling factors can have a linear relation-

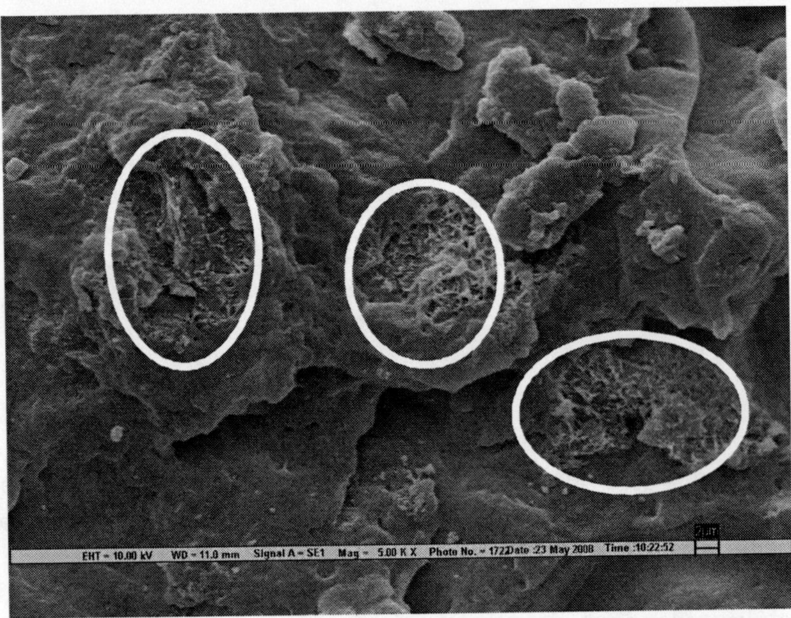

Fig. 10. SEM micrograph of PA2200/Clay composite showing heterogeneous system

ship with scan length. They derived empirical relations for percentage shrinkage in terms of scan length using Taguchi method. However they used scaling factors based on the maximum dimensions not on the individual scan lengths because of the non-availability of a dexel based compensation system. Since a complex geometry will have scan lengths varying inside the contours of a single part, a new methodology needs to be developed for compensating scan length along hatch vectors. The shrinkage model developed by Raghunath and Pandey [6] has been used for compensating shrinkage along single direction dexel space. These scaling factors (S_X, S_Y, and S_Z in %) to compensate shrinkage along X, Y, and Z directions are as given below [6]

$$S_X = 1.61191 - 0.01615L_P - 0.00964L_C,$$
$$S_Y = 0.785926 - 0.032656L_P + 0.000281B_S,$$
$$S_Z = 28.6238 + 0.000308B_S + 4.0417H_S + 0.16792T_B$$

where, L_P is the laser power in watt, B_S is the beam speed in mm/sec, H_S is the hatch spacing in mm, T_B is the part bed temperature in °C and L_C is the scan length in mm. In the present work, most of the process parameters are kept constant except scan length.

The shrinkage model in X direction of the part has been used to compensate

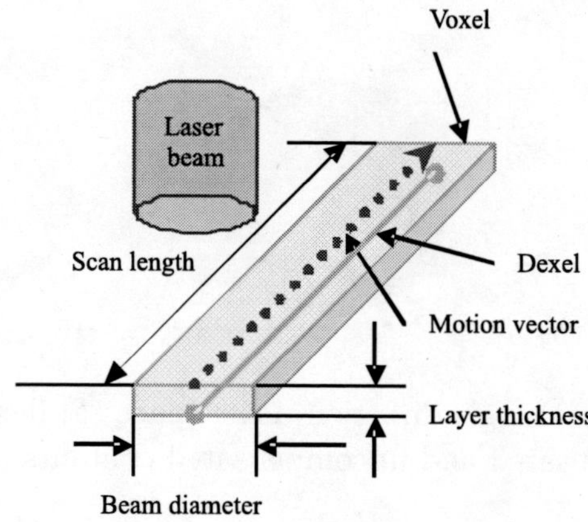

Figure 11: Dexel model of the SLS process

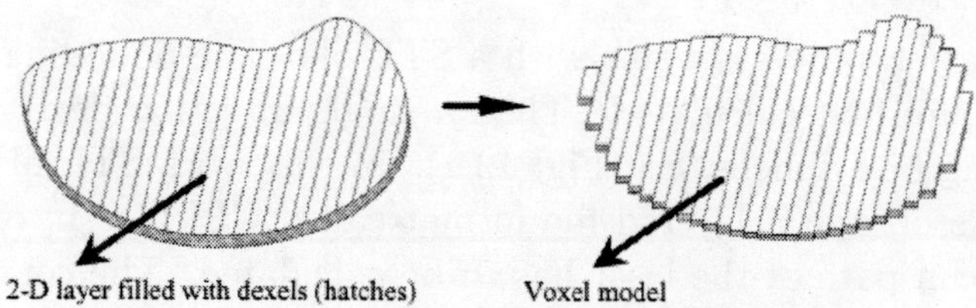

Figure 12: Dexelisation and voxelisation of a slice [30]

shrinkage along single direction dexel space. The concept of dexel was first proposed by Tim Van Hook [27] in 1986. The dexel and voxel models have been utilized in visualization of the RP process [28, 29]. The single direction dexel model of the SLS process has been given in Fig. 11. A single dexel can represent a big voxel and the scan length for the voxel corresponds to the dexel length. Fig. 12 shows the dexelisation and voxelisation for a typical 2D slice.

When the shrinkage scaling factors are constant, shrinkage compensation is relatively easy. The scaling transformation can be used to offset the vertices of the triangles of the STL file to the single scaling factor value in each X, Y and Z direction. However if the shrinkage scaling factors varies with scan length, compensation requires offsetting each scan line of the part. Therefore compensating along dexels has been preferred rather than triangles of the STL file. Generation of dexels has been easier with layered data files such as CLI file. Therefore a sliced layer file has been preferred rather than a STL file for the ease of compensation. Compensat-

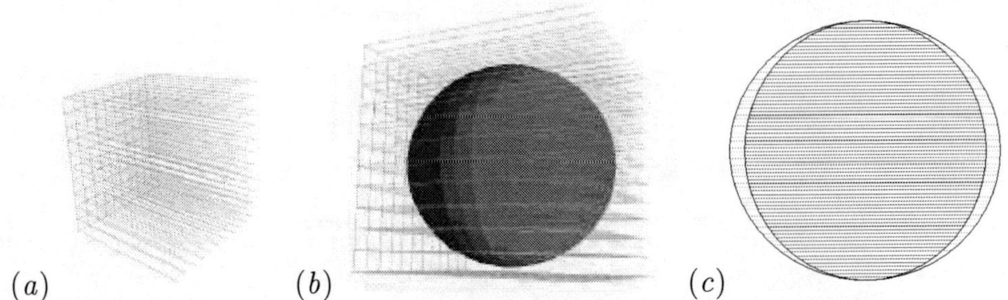

Figure 13: Dexelisation (*a*) single direction dexel space (*b*) dexelisation of a sphere (*c*) comparison of compensated and uncompensated contours of a slice

ing on sliced data has been proven successful by Tong et al. [31] for compensating machine errors for SLA and FDM processes. However due to the popularity of the STL file in RP fabrication, a slicing algorithm is integrated into the compensation system to convert STL to CLI file.

The dexelisation process begins with a STL file. The bounding box dimensions of the STL file are first calculated. Then a single direction dexel space is created for the bounding box dimensions (Fig. 13 *a*). In this work STL file was sliced into layers and was stored in a layered file format called CLI file. In order to generate dexels, entire scan path of the laser has to be calculated. The contour information was extracted and all the co-ordinate points of the contour were sorted based on whether the contour was internal or external. After reading the contour information dexels and its lengths were found using intersection of hatch lines with the contours (Fig. 13 *b*). Then compensation value for every dexel was calculated using the shrinkage model. After offsetting each dexel at its end points and preserving its center location, the contour was rebuilt with its new vertices and a new compensated CLI file was written (Fig. 13 *c*). Then this file was used for part building. Unlike the conventional compensation techniques, the compensation lengths vary non-linearly with the dexel length in this method. Since percentage shrinkage is a function of dexel length and during compensation it has been multiplied by the dexel length, the amount of compensated length varies non-linearly with dexel length.

A case study has been done to understand the effectiveness of the developed methodology in improving accuracy of the SLS parts. The design of the part was adapted from National Aerospace Standard (NAS 979) which is usually used for evaluating form tolerance capabilities of machine tools. The test part has been shown in Fig. 14 *a*. The NAS test part consists of a circular shape on a diamond shape which is on top of a square shape. The faces are identified by the numbers and are marked from F1 to F8 and edges formed out of these faces are marked from E1 to

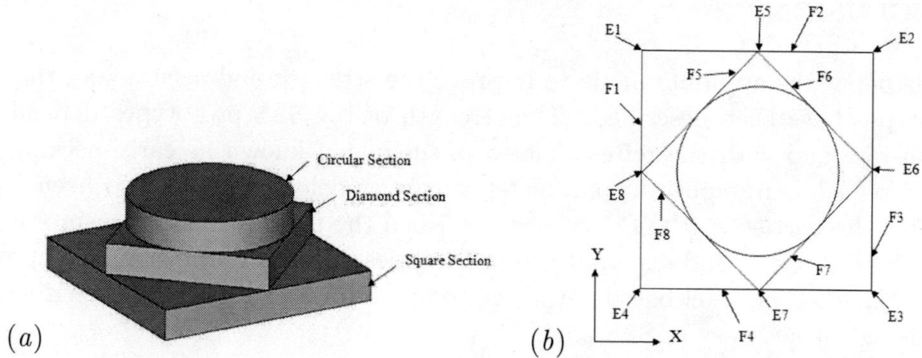

Figure 14: (a) NAS Test part for comparison of geometric tolerances; (b) Faces and edges marked on NAS Test part

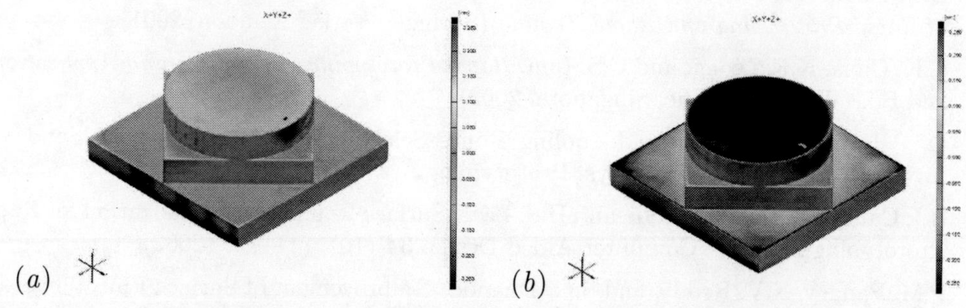

Figure 15: Deviation plots for NAS Test piece (a) Shrinkage model (b) Machine manufacturer suggested scaling factors

E8 as shown in Fig. 14 b. Two versions of the part were fired and process parameters were kept constant. For one of the parts, machine manufacturer suggested scaling factors were used and for other flexible shrinkage model has been used.

To visualize the improvement in accuracy, deviation plots were constructed using point cloud data from ATOS white light scanner for the NAS Test parts. The deviation plots along $+X$ $+Y$ $+Z$ viewing direction have been shown in Fig. 15. It can be observed that maximum deviation of shrinkage model part is within ± 0.15 mm (Fig. 15 a) while the rigid scaling part has errors more than ± 0.25 mm (Fig. 15 b). Thus it has been concluded that the new compensation approach proposed in this paper not only reduced the dimensional inaccuracy but also improved the form accuracy.

CONCLUSIONS

In this paper the attempts made to improve the strength and accuracy of the SLS prototypes have been described. The strength of the SLS prototypes depends on refresh rate and with the refresh rate a phenomenon known as early necking has been revealed. An important parameter known as delay time has also been found to affect the part strength. Paper also described the feasibility of laser sintering of nano-clay/PA blend and explored the possibilities of formation of nano-composite. A novel method of dexel based shrinkage compensation has also been described for improving the accuracy of SLS parts.

REFERENCES

1. D.T. Pham and S.S. Dimov, *Rapid Manufacturing: The technologies and applications of Rapid Prototyping and Rapid Tooling* (Springer–Verlag, London, 2001).
2. C.K. Chua, K.F. Leong, and C.S. Lim, *Rapid Prototyping: Principles and Applications*, 2nd Ed. (World Scientific, Singapore, 2003).
3. D. Pal and B. Ravi, "Rapid Tooling Route Selection and Evaluation for Sand and Investment Casting," Virt. Phys. Prototyping **2** (4), 197–207 (2007).
4. R.I. Campbell, M. Martorelli, and H.S. Lee, "Surface Roughness Visualization for Rapid Prototyping Models," Computer Aided Design **34** (10), 717–725 (2002).
5. P.M. Pandey, N.V. Reddy, and S.G. Dhande, "Improvement of Surface Finish by Staircase Machining in Fused Deposition Modeling," J. Mater. Process. Technol. **132**, 323–331 (2003).
6. N. Raghunath and P.M. Pandey, "Improving Accuracy through Shrinkage Modelling by Using Taguchi Method in Selective Laser Sintering," Int. J. Machine Tools Manufact. **47** (6), 985–995 (2007).
7. P. Radhakrishnan, P. Arivalagan, and C. Gajendran, "Effect of Pre-Contouring and Post-Contouring on the Accuracy of Parts Produced by Direct Metal Laser Sintering," Journal of the Institution of Engineers (India), Part PR: Production Engineering Division **88**, 3–6 (2007).
8. P.B. Bacchewar, S.K. Singhal, and P.M. Pandey, "Statistical Modelling and Optimization of Surface Roughness in Selective Laser Sintering Process," Proc. IMechE, Part B: J. Eng. Manufact. **221** (1), 35–52 (2007).
9. P.K. Jain, K. Senthilkumaran, P.M. Pandey, and P.V.M. Rao "Advances in Materials for Powder Based Rapid Prototyping," in *Proceedings of International conference on Recent Advances in Materials and Processing, Dec. 15–16, 2006, PSG Tech. Coimbatore, India* (2006), pp. 14.8.
10. P.K. Venuvinod and W. Ma, *Rapid Prototyping: Laser Based and Other Technologies* (Kluwer Academic Publishers, London, 2004), pp. 275–277.

11. Y. Tang, H.T. Loh, J.Y.H. Fuh, et al., "An Algorithm for Disintegrating Large and Complex Rapid Prototyping Objects in a CAD Environment," Int. J. Adv. Manuf. Technol. **25**, 895–901 (2005).
12. K. Tong, E.A. Lehtihet, and S. Joshi, "Parametric Error Modeling and Software Error Compensation for Rapid Prototyping," Rapid Prototyping J. **9** (5), 301–313 (2003).
13. P. Jacobs, "The Effects of Random Noise Shrinkage on Rapid Tooling Accuracy," Mater. Des. **21** (2), 127–136 (2000).
14. H.H. Zhu, L. Lu, and J.Y.H. Fuh, "Study on Shrinkage Behavior of Direct Laser Sintering Metallic Powder," Proc. IMechE, Part B: J. Eng. Manufact. **220** (2), 183–190 (2006).
15. X. Wang, "Calibration of Shrinkage and Beam Offset in SLS Process," Rapid Prototyping J. **5** (3), 129–133 (1999).
16. K. Manetsberger, J. Shen, and J. Muellers, "Compensation of Non-Linear Shrinkage of Polymer Materials in Selective Laser Sintering," in *Proc. Solid Free Form Fabrication symposium, University of Texas, Austin, August, 2003* (2003), pp. 346–356.
17. H.J. Yang, P.J. Hwang, and S.H. Lee, "A Study on Shrinkage Compensation of the SLS Process by Using Taguchi Method," Int. J. Machine Tools Manufact., **42** (11), 1203–1212 (2002).
18. Y. Ning, Y.S. Wong, and J.Y.H. Fuh, "Effect of Control of Hatch Length on Material Properties in the Direct Metal Laser Sintering Process," Proc. IMechE, Part B: J. Eng. Manufact. **219** (1), 15–25 (2005).
19. N. Ragunath and P.M. Pandey, "Improving Accuracy through Shrinkage Modeling by Using Taguchi Method in Selective Laser Sintering," Int. J. Machine Tools Manufact. **47** (6), 985–995 (2007).
20. Q. Dao, J.C. Frimodig, H.N. Le, et al., "Calculation of Shrinkage Compensation Factors for Rapid Prototyping (FDM 1650)," Computer Applications in Engineering Education **7** (3), 186–195 (1999).
21. W.L Wang, C.M. Cheah, J.Y.H. Fuh, and L. Lu, "Influence of Process Parameters on Stereolithography Part Shrinkage," Mater. Des. **17** (4), 205–213 (1996).
22. Y. Wang, Y. Shi, and S. Huang, "Selective Laser Sintering of Polyamide–Rectorite Composite," J. Mater.: Des. Appl. **219**, 11–15 (2005).
23. S.S. Ray and M. Okamoto, "Polymer/layered Silicate Nanocomposites: a Review from Preparation to Processing," Progr. Polym. Sci. **28** (11), 1539–1641 (2003).
24. I. Mironi-Harpaz, M. Narkis, and A. Siegmann, "Synthesis of Unsaturated-Polyster/Organo-Caly Nanocomposites: a Fundamental Approach," J. Nanostruct. Polym. Nanocompos. **1** (1), 35–43 (2005).
25. E.D. Dickens Jr., B.L. Lee, G.A. Taylor, A.J. Magistro, and H. Ng, "Sinterable Semi-Crystalline Powder and Near-Fully Dense Article Formed Therewith," US Patent No. 5,990,268 (1999).
26. H. Scholten and W. "Christoph Use of a Nylon-12 for Selective Laser Sintering," US Patent No. 6,245,281 (1998).

27. T.V. Hook, "Real Time Shaded NC Display," ACM SIGGRAPH Computer Graphics **20** (4), 15–20 (1986).
28. S.H. Choi and A.M.M. Chan, "A Virtual Prototyping System for Rapid Product Development," Computer Aided Design, **36** (5), 401–412 (2004).
29. S. Manohar, V. Chandru, and C.E. Prakash, "Voxel Based Modelling for Layered Manufacturing," IEEE Comput. Graph. Appl., 42–57 (November 1995).
30. Y. Ning, Y.S. Wong, J.Y.H. Fuh, and H.T. Loh, "An Approach to Minimize Build Errors in Direct Metal Laser Sintering," IEEE Transactions on Automation Science and Engineering **3** (1), 73–80 (2006).
31. K. Tong, E.A. Lehtihet, and S. Joshi, "Error Compensation for Fused Deposition Modeling (FDM) Machine by Correcting Slice Files," Rapid Prototyping J. **14** (1), 4–14 (2008).

STUDY OF STABLE CRACK GROWTH THROUGH AISI 4340 STEEL USING CTOD/CTOA

D.N. Jadhav[1a*] and S.K. Maiti[2]**

ABSTRACT

Two dimensional elastic-plastic finite element analysis have been carried out to characterize stable crack growth (SCG) through AISI 4340 steel, which is widely used in power plants and for which some experimental results are available in the literature. In the first part of this study, examination of suitability of crack tip opening displacement (CTOD) and crack tip opening angle (CTOA) criteria, which are considered to be related, for the characterization of mode I SCG through finite element analysis based on ABAQUS (version 6.6) software is presented. Contrary to usual expectations of a constant CTOD/CTOA governing the extension, an increasing-initially-constant-later (IICL) type of variation of CTOD/CTOA is found to be suited for the case. In the second part, usefulness of the same parametric variation is examined for mixed mode (I and II) crack growth through compact tension (CT) specimens. The details of analysis and accuracy of prediction of load variation with displacement are given. Further, variations of J integral with crack extension for mode I and mixed modes are compared.

Key words: CTOD/CTOA, stable crack growth, finite element analysis, CT specimen

INTRODUCTION

In the past numerical investigations on SCG have been conducted using various characterizing parameters such as J integral [1–2], tearing modulus [3],

---

[1]Department of Mechanical Engineering, Sardar Patel College of Engineering, Andheri — 400058, Mumbai, India

[2]Department of Mechanical Engineering, Indian Institute of Technology Bombay, Powai — 400076, Mumbai, India

[a]Currently QIP Research Scholar, Department of Mechanical Engineering, Indian Institute of Technology Bombay, Powai — 400076, Mumbai, India

[*]E-mail: *dnjadhav@iitb.ac.in*

[**]E-mail: *skmaiti@me.iitb.ac.in*, *susant02@yahoo.co.in*

CTOD/CTOA [4–7], etc. It has been found that among the candidate parameters the CTOD/CTOA is good for characterizing the SCG and its adoption is favored by the ease of implementation in finite element analysis.

Many investigators [1–7] have examined the mode I SCG through various test specimens mostly made of aluminum alloys or pipe grade steels. They have observed that CTOD/CTOA remains more or less constant throughout the SCG except for some high values at the initial stages. A comprehensive review of study of mode I SCG using CTOD criterion is given by Newman et al. [8]. Though the earlier investigators [4–8] suggest a constant CTOD/CTOA criterion for relatively thin structures, Lam et al. [9] have demonstrated that the initial high values of CTOD/CTOA play an important role in predicting SCG especially under plane strain condition. Severe underestimation of load occurs under constant CTOD/CTOA criterion. However Lam et al. have reported that use of bilinear CTOD/CTOA variation, decreasing initially and constant later, helps to predict results more close to the experimental observations.

The experimental investigations on SCG through CT specimen of AISI 4340 steel under both mode I and mixed mode (I and II) have been reported by Mourad et al. [10]. The characterization of this material based on these results has been done earlier by Krishna Kishore [11] using the CTOD/CTOA criterion. He has reported that a gradually increasing CTOD/CTOA with crack extension gives prediction of crack initiation and maximum load close to the experimental observations. The same results have been analyzed here considering an increasing-initially-constant-later (IICL) scheme of variation of CTOD/CTOA with crack extension and ABAQUS (version 6.6) software.

Based on the observations of three point bend specimen (TPB) and CT specimen geometries it has been reported in the literature by, e.g. Maiti and Mourad [13], that both mode I and mixed mode SCG in the presence of crack tip elastic-plastic deformation can be characterized by crack opening angle (COA) criterion. There is a good agreement between experimental and predicted load displacement results when the extent of crack growth is small. This issue has also been examined here for larger crack growth.

1. FINITE ELEMENT ANALYSIS

1.1. *Specimen geometry and material property*

The specimen geometry is shown in Fig. 1 *a*. The nominal sizes of CT specimen are 120 mm×120 mm×8 mm thick. The crack was made by wire-cut machining (wire diameter = 0.1 mm) giving the initial crack size a_0/W = 0.41, 0.42, 0.43, 0.44, and 0.45. The material properties employed are given in Table 1.

Table 1. Mechanical properties of AISI 4340 steel

Property	Value
Yield stress, σ_0	487 MPa
Ultimate tensile stress, σ_u	662 MPa
Young's modulus of elasticity, E	198 GPa
Poisson's ratio, ν	0.3

1.2. *Two-dimensional modeling and analysis*

A two-dimensional elastic-plastic finite element analysis is carried out to predict the variation of experimental [10] load with load-line displacement (LLD) employing the true stress-strain curve (Fig. 1 *b*) of the material and loading under displacement control. The analysis was done under plane stress condition. Though the crack tip had a finite radius, it was treated as a sharp crack following the observation of Maiti et al. [12]. The case of $a_0/W = 0.41$ was first studied. The discretization is shown in Fig. 2. There are 3581 nodes and 3978 elements of which 2922 are four noded quadrilateral elements and the rest are three noded triangles. The elements at the crack edges are mostly square of side 0.4 mm. The potential crack edges are modeled as slave and master contact surfaces, which are partially bonded. The debonding of these edges arising out of crack growth is based on the CTOD criterion. The bonding span is kept 25 mm approximately to cover the full span of experimentally observed crack extension. The part of the ligament beyond this point is constrained only in the y direction (Fig. 2). To simulate the loading condition, the loading pin hole is filled with a material with higher stiffness to avoid any severe local deformation around the load point. For loading the pin was displaced incrementally in the vertical (y) direction only.

1.3. *CTOD/CTOA criterion*

The crack extension was studied applying both constant CTOD/CTOA and proposed IICL type of variation of CTOD/CTOA (Fig. 3). The crack is allowed to extend by one element whenever the CTOD at a fixed distance (0.4 mm) behind the current crack tip reaches a critical value. The CTOA is defined as CTOD divided by the distance taken for its measurement behind the current crack tip.

1.4. *Results*

Since the analysis is done under displacement controlled condition, immediately after the release of a crack tip node, there is a drop in load from the level reached before the release. These drops are clearly seen in Fig. 4. The load corresponding to the first drop indicates the initiation load. For a comparison with the experimental

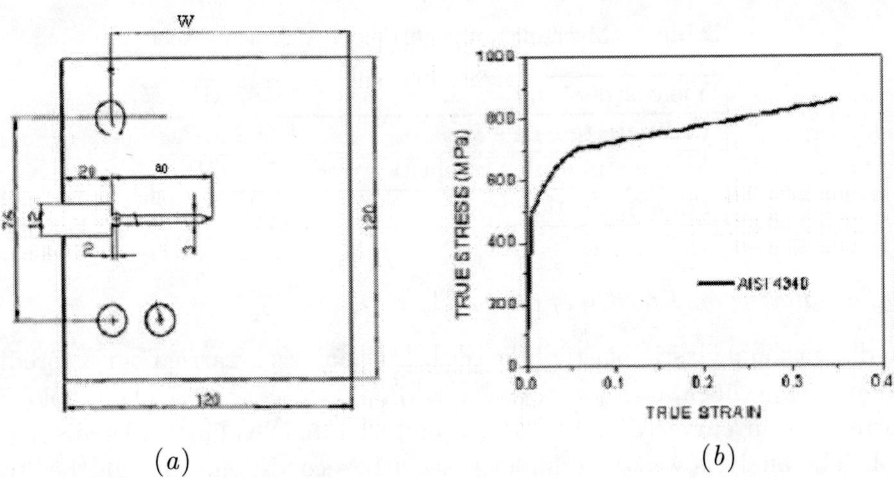

(*a*) (*b*)

Fig. 1. (*a*) CT specimen geometry. (*b*) True stress-strain curve for AISI4340 steel. All dimensions are in mm

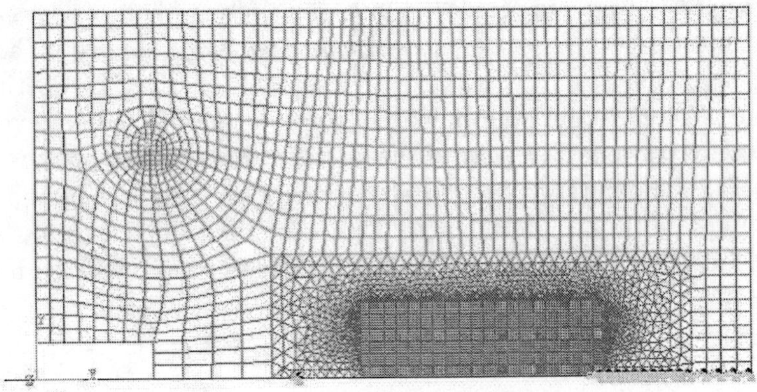

Fig. 2. Two-dimensional FE modeling for CT specimen with a_0/W=0.41

results all load-LLD plots are obtained by drawing the top envelope of a variation of the type shown in Fig. 4.

The predicted load-LLD variations are compared with the experimental results in Fig. 5. Severe crack tunnelling, 8 to 12 mm, was reported near the maximum load [10]. Because of the tunnelling actual distance between the load line and the crack tip will be more than the distance at the surface. To accommodate this, crack opening displacement at a distance 8 mm behind the load line was plotted against the load. This displacement is indicated as LLD_8. The predicted load-LLD_8 variations based

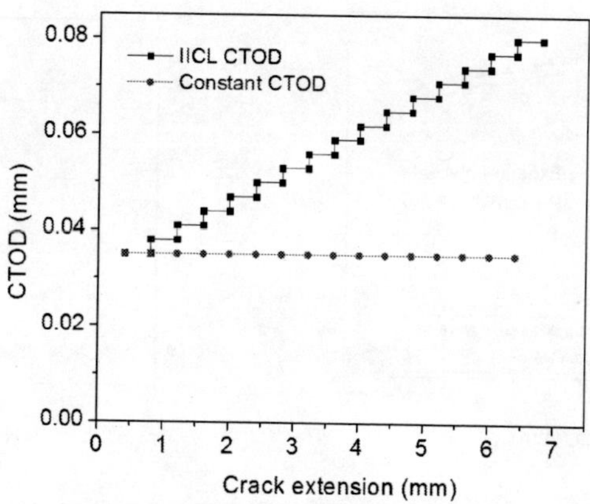

Fig. 3. Types of CTOD/CTOA variation with crack extension

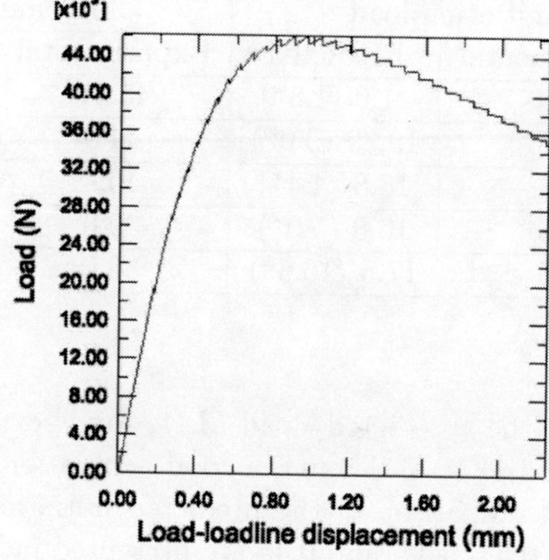

Fig. 4. A typical load-loadline displacement plot of ABAQUS output

on the same CTOD/CTOA variation predict the experimental trend better than the results based on constant CTOD/CTOA (Fig. 5). The latter shows a prominent drooping tendency and under-estimates the maximum load by 16.86% . Further, the LLD at maximum load is less by 2 mm. On the whole it appears that the 2-D FEA under plane stress condition using IICL type of CTOD/CTOA criteria appears better. For validation, further case studies with other crack lengths were carried out.

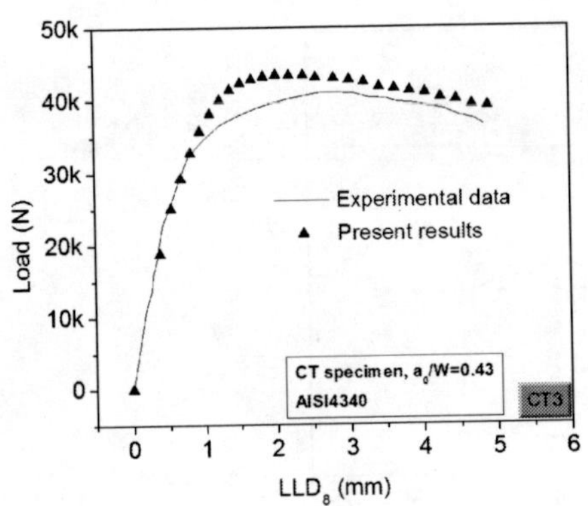

Fig. 5. Comparison of predicted load-LLD_8 and experimental results

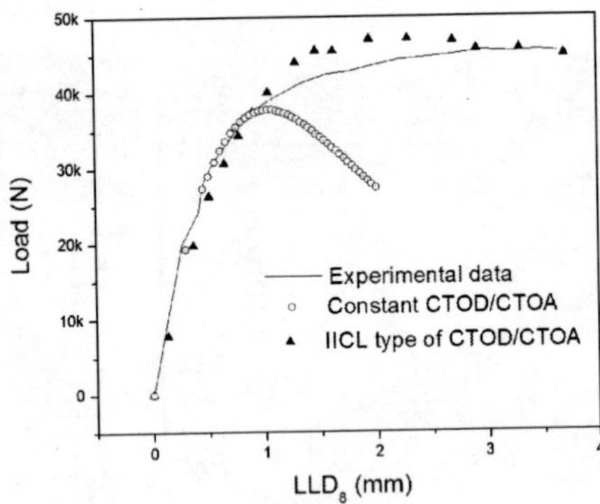

Fig. 6. Comparison of predicted load-LLD_8 and experimental results

Table 2. Comparison of experimental and predicted initiation and maximum loads

$\frac{a_0}{W}$	Initiation load		Maximum load	
	Experimental	Predicted	Experimental	Predicted
0.41	22.0	21.6 (1.8*)	45.47	47.6 (4.6*)
0.42	20.0	19.39 (3.0*)	44.0	45.5 (3.4*)
0.43	18	18.8 (4.4*)	40.4	43.4 (8.8*)
0.44	16	16.8 (5.0*)	39.0	40.3 (3.3*)
0.45	15	17.5 (16.6*)	36.0	40.0 (11.1*)

* Error %

A comparison of the predicted load-LLD_8 and experimental results for $a_0/W = 0.43$ are shown in Fig. 6. There is a good correlation between predicted and experimental data. A comparison of the predicted initiation and maximum loads for $a_0/W = 0.41$, 0.42, 0.43, 0.44, and 0.45 are presented in Table 2. The difference is less than 10% for all cases except for $a_0/W = 0.45$. The reason for increase in difference in the case of $a_0/W = 0.45$ is perhaps the reduction in distance of coarse mesh from the initial crack tip.

2. CHARACTERIZATION OF MIXED MODE SCG

The possibility of characterization of the mixed mode SCG in terms of the same IICL type of variation of CTOD/CTOA has been examined.

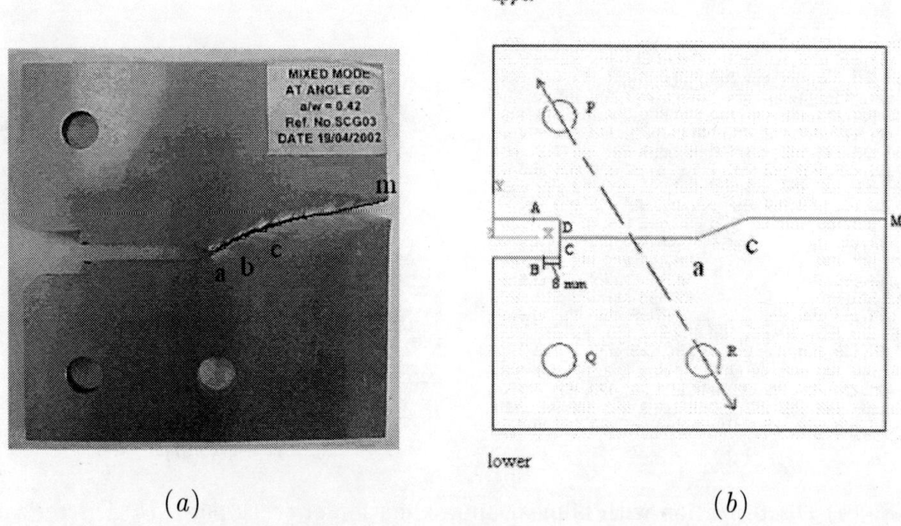

(*a*) (*b*)

Fig. 7. (*a*) Fractured CT specimen. (*b*) Schematic outline of mixed mode specimen $a_0/W = 0.42$ and 60° loading angle

2.1. *Modeling and analysis*

The discretization was done taking note of the crack path reported in [10] and approximating it by three line segments (Figs. 7 *a* and *b*). The automatic node release scheme of ABAQUS was again exploited. A typical discretization for 60° loading angle is shown in Fig. 8 *a*. There are 8322 four noded quadrilateral elements and 5550 three noded triangular elements. Ahead of the crack tip over the span a-b-c square elements of 0.4 mm side are used (Fig. 8 *b*). Elements behind the crack tip are parallelogram with two sides parallel to the crack. The 2-D elastic plastic finite element analysis was carried out using again the true stress-strain data of the material. The loading was imposed through gradual movement at the center of the upper pin and keeping other pin fixed. In order to adapt the CTOD/CTOA criterion, the resultant distance between two nodes on the opposite crack flanks at a distance of 0.4 mm behind the current crack tip was taken as CTOD. The steps followed for the analysis of a mixed mode are the same as in the case of mode I.

2.2. *Results*

The predicted load-LLD_8 variations are compared with the experimental results (Fig. 9 *a* and *b*). The predicted and experimental results are in good agreement. The other five mixed mode cases were also studied in the same manner. The initiation and maximum loads are compared with the experimental data for $a_0/W = 0.42$

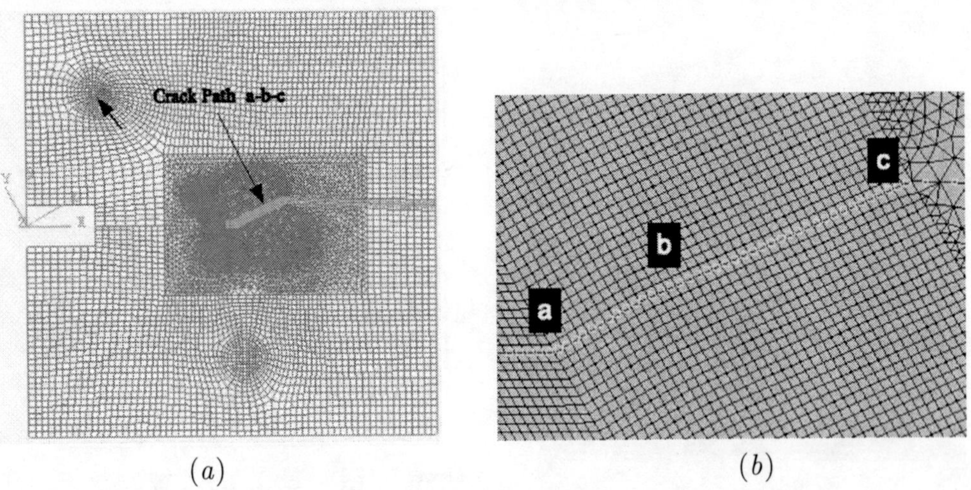

(*a*) (*b*)

Fig. 8. (*a*) Discretization with bilinear approximation of crack path. (*b*) Discretization details around the bilinear crack path

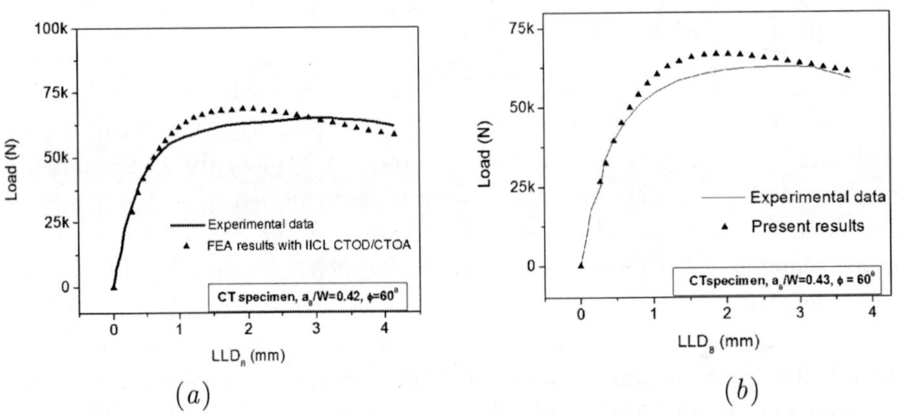

(*a*) (*b*)

Fig. 9. Comparison of predicted load-LLD_8 variations with experimental results

and 0.43 in Tables 3 and 4. The maximum difference between the theoretical and experimental loads is about 10%.

3. RESISTANCE CURVES

The J integral was computed during the mixed mode crack growth through the ABAQUS software. These curves for both mode I and mixed mode show the similar

Table 3. Results on mixed mode ($a_0/W = 0.42$)

Loading angle	Experimental load [7] (kN)		Predicted load by FEM (kN)	
	P_i	P_{max}	P_i	P_{max}
60°	30.5	65.73	29.0 (3.7*)	67.2 (2.2*)
65°	27.84	58.53	24.95 (10.3*)	63.96 (9.3*)

* Error %

Table 4. Results on mixed mode ($a_0/W = 0.43$)

Loading angle	Experimental load [7] (kN)		Predicted load by FEM (kN)	
	P_i	P_{max}	P_i	P_{max}
50°	33.83	75.43	33.06 (2.2*)	80.4 (6.6*)
60°	29.5	63.0	26.5 (10.1*)	66.8 (6.03*)
65°	27	57.53	24.7 (8.5*)	59.5 (3.4*)

* Error %

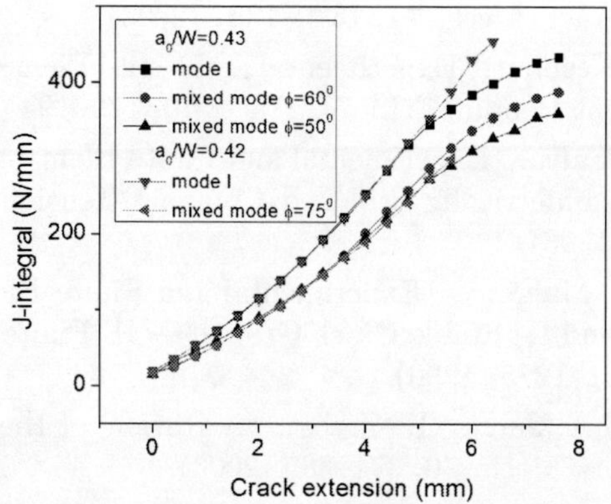

Fig. 10. Comparison of two mode I and three mixed mode J resistance curves

trend of variation (Fig. 10). This may further reinforce the characterization of SCG through AISI 4340 steel by the IICL type of CTOD/CTOA variation.

CONCLUSIONS

In case of AISI 4340 steel the initial-increasing-constant-later (IICL) type of variation of CTOD/CTOA appears to be suitable for characterizing mode I SCG. The same criterion can also be employed to study the mixed mode (I and II) SCG. With a bilinear approximation of the crack path the predicted results for mixed mode show good agreement with experimental observations. The J resistance curves in mode I and mixed modes have the similar trend of variation.

REFERENCES

1. C.F. Shih, "Relationships between the J-Integral and the Crack Opening Displacement for Stationary and Extending Cracks," J. Mech. Phys. Solids **29**, 305–326 (1981).
2. M.F. Kanninen, E.F. Rybicki, R.B. Stonesifier, et al., "Elastic-Plastic Fracture Mechanics for Two Dimensional Stable Crack Growth and Instability Problems," in *Elastic-Plastic Fracture.* Ed. by J.D. Landes, J.A. Begley, and G.A. Clarke (ASTM STP 668, Philadelphia, 1979), pp. 121–150.
3. P.C. Paris, H. Tada, Z. Zahoor, and H. Ernst, "Instability of the Tearing Mode of Elastic Plastic Crack Growth," in *Elastic-Plastic Fracture.* Ed. by J.D. Landes, J.A. Begley, and G.A. Clarke (ASTM STP 668, Philadelphia, 1979), pp. 5–36.
4. S. Mahmmod and K. Lease, "Two-Dimensional and Three-Dimensional Finite Element Analysis of Critical Crack-Tip-Opening Angle in 2024-T351 Aluminum Alloy at Four Thicknesses," Eng. Fract. Mech. **71**, 1379–1391 (2004).
5. S.K. Maiti and A.S. Kesbhat, "Experimental and Finite Element Study on Stable Crack Growth in Three Point Bending," Int. J. Fract. **60**, 179–194 (1993).
6. S.K. Maiti and P.D. Salva, "Experimental and Finite Element Study on Mode I Stable Crack Growth in Symmetrically Stiffened Compact Tension Specimen," Eng. Fract. Mech. **44**, 721–733 (1993).
7. S.K. Maiti and D.K. Mahanty, "Experimental and Finite Element Studies on Mode I and Mixed Mode (I and II) Stable Crack Growth — II. Finite Element Analysis," Eng. Fract. Mech. **37**, 1251–1275 (1990).
8. J.C. Newman Jr., M.A. James, U. Zerbst, "A Review of the CTOA/CTOD Fracture Criterion," Eng. Fract. Mech. **70**, 371–385 (2003).
9. P.S. Lam, Y. Kim, and Y.J. Chao, "The Non-Constant CTOD/CTOA in Stable Crack Extension under Plane-Strain Condition," Eng. Fract. Mech. **73**, 1070–1085 (2006).
10. A.H.I. Mourad, M.J. Alghafri, O.A. Abu Zeid, and S.K. Maiti, "Experimental Investigation on Ductile Stable Crack Growth Emanating from Wire-Cut Notch in AISI 4340 Steel," Nucl. Eng. Des. **235**, 637–647 (2005).
11. G.K. Kishore, *Characterization of Stable Crack Growth through AISI 4340 Steel and Some Related Studies* (MTech dissertation, Mechanical Engineering Department, IIT Bombay, Mumbai, 2007).
12. S.K. Maiti, S. Namdeo, and A.H.I. Mourad, "A Scheme for FEA of Mode I and Mixed Mode Stable Crack Growth and a Case Study with AISI 4340 Steel," Nucl. Eng. Des. **238**, 787–800 (2008).
13. S.K. Maiti and A.H.I. Mourad, "Criterion for Mixed Mode Stable Crack Growth-II Compact Tension Geometry," Eng. Fract. Mech. **52**, 349–359 (1995).

EFFECT OF TRANSVERSE ELASTIC GRADIENT ON THE STRESS FIELD AND STRESS INTENSITY FACTOR IN A CRACKED PLATE

R. Kommana[1], S. Wadgaonkar[1], and V. Parameswaran[1]

ABSTRACT

The problem of a through thickness crack in a transversely graded plate is addressed in this study. The asymptotic expansion of the displacement field near the tip of the crack was developed from which the first three terms in the expansion of the stress field were derived. The analysis revealed that the elastic gradient along the crack front affects neither the inverse square root nature nor the angular structure of the stress field. Subsequently an edge-cracked plate having exponential variation of elastic modulus along the thickness was prepared. The cracked plate was subjected to pure bending and the displacements on the surfaces were measured using digital image correlation. Using the derived displacement field the stress intensity factor (SIF) was determined from the experimentally measured displacements. The experimental results indicated that the SIF varies along the crack front and the extent of SIF variation is same as that of the elastic modulus.

Key words: transversely graded plate, stress intensity factor, elastic gradient, digital image correlation

INTRODUCTION

The mechanics of cracks in graded material has received extensive attention in recent years. In most of the existing studies, the elastic modulus varies in the plane of the crack and remains constant along the crack front [1–6]. Such problems are treated as plane problems and the effect of elastic gradient on the stress field and stress intensity factor is very well understood. The case of a crack oriented such that the elastic gradation is through the crack front has not received any attention so far. This situation could happen when the gradation is along the thickness of a

[1]Department of Mechanical Engineering, Indian Institute of Technology Kanpur, Kanpur — 208016, India

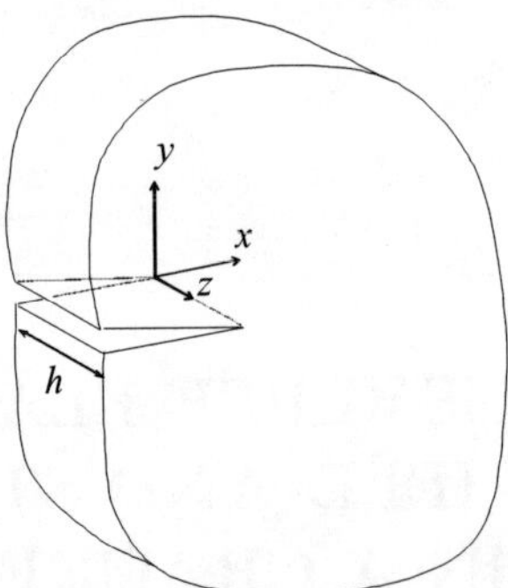

Fig. 1. Problem geometry

plate (transversely graded) having a through thickness crack. The problem cannot be considered as a plane problem and has to be treated in the three dimensional framework.

In the present study, first the asymptotic structure of the displacement and stress field near the tip of a crack in a transversely graded plate is developed by eigen function expansion [7, 8]. The first three terms in the expansion of the displacement and stress field are derived for an opening mode crack. Subsequently an edge-cracked plate having exponential variation of elastic modulus along the thickness was prepared and the crack was loaded in pure bending. The displacements near the crack tip were measured on the two surfaces of the plate using digital image correlation (DIC). The stress intensity factor (SIF) was calculated from the measured displacements using the derived displacement field. The details of the development of the displacement field and experimental measurement of the SIF are discussed in subsequent sections.

1. CRACK-TIP DISPLACEMENT AND STRESS FIELD

The geometry of the problem at hand is as shown in Fig. 1, where the straight crack front is along the z axis and the crack faces are traction free. The problem is formulated in cylindrical coordinates with the origin located at the crack tip. The elastic modulus variation is assumed as exponential (equation (1)) whereas the Poisson's ratio, ν, is assumed to be constant.

$$E(z) = E_0 e^{\alpha z}, \tag{1}$$

where E_0 is the elastic modulus at $z = 0$ and α, referred to as the non-homogeneity parameter, is a constant having dimension $(L)^{-1}$. The Navier equations of equilibrium in cylindrical coordinates can be written for this material as

$$\begin{aligned}
&\frac{\partial \vartheta}{\partial r} + (1-2\nu)\left\{\left[\nabla^2 - \frac{1}{r^2}\right]u_r - \frac{2}{r^2}\frac{\partial u_\theta}{\partial \theta} + \alpha\left[\frac{\partial u_r}{\partial z} + \frac{\partial u_z}{\partial r}\right]\right\} = 0,\\
&\frac{1}{r}\frac{\partial \vartheta}{\partial \theta} + (1-2\nu)\left\{\left[\nabla^2 - \frac{1}{r^2}\right]u_\theta + \frac{2}{r^2}\frac{\partial u_r}{\partial \theta} + \alpha\left[\frac{1}{r}\frac{\partial u_z}{\partial \theta} + \frac{\partial u_\theta}{\partial z}\right]\right\} = 0 \qquad (2)\\
&\frac{\partial \vartheta}{\partial z} + (1-2\nu)\nabla^2 u_z + 2(1-\nu)\alpha\frac{\partial u_z}{\partial z} + 2\nu\alpha\left[\frac{\partial u_r}{\partial r} + \frac{u_r}{r} + \frac{1}{r}\frac{\partial u_\theta}{\partial \theta}\right] = 0,
\end{aligned}$$

where

$$\vartheta = \frac{\partial u_r}{\partial r} + \frac{1}{r}\frac{\partial u_\theta}{\partial \theta} + \frac{u_r}{r} + \frac{\partial u_z}{\partial z}, \quad \nabla^2 = \frac{\partial^2}{\partial r^2} + \frac{1}{r}\frac{\partial}{\partial r} + \frac{1}{r^2}\frac{\partial^2}{\partial \theta^2} + \frac{\partial^2}{\partial z^2},$$

and u_r, u_θ, u_z the components of the displacements in the r, θ and z directions respectively. The displacement components are expanded as a double series as give in equation (3) [7].

$$\begin{aligned}
&u_r = \sum_{m=0}^{\infty}\sum_{n=0}^{\infty} r^{\lambda_m+n}U_n^{(m)}(\theta,z), \quad u_\theta = \sum_{m=0}^{\infty}\sum_{n=0}^{\infty} r^{\lambda_m+n}V_n^{(m)}(\theta,z),\\
&u_z = \sum_{m=0}^{\infty}\sum_{n=0}^{\infty} r^{\lambda_m+n}W_n^{(m)}(\theta,z),
\end{aligned} \qquad (3)$$

where the powers of r, λ_m ($m = 0, 1, 2, \ldots$) and the functions $U_n^{(m)}$, $V_n^{(m)}$, $W_n^{(m)}$ are yet to be determined. Substitution of equation (3) in equation (2) and collecting the terms associated with identical powers of r we can construct a set of partial differential equations as follows

$$\begin{aligned}
&2(1-\nu)[(\lambda_m+n)^2-1]U_n^{(m)} + [(\lambda_m+n)-3+4\nu]\frac{\partial V_n^{(m)}}{\partial \theta} + (1-2\nu)\frac{\partial^2 U_n^{(m)}}{\partial \theta^2}\\
&\quad = -(\lambda_m+n-1)\frac{\partial W_{n-1}^{(m)}}{\partial z}\\
&\quad - \alpha(1-2\nu)(\lambda_m+n-1)W_{n-1}^{(m)} - (1-2\nu)\frac{\partial^2 U_{n-2}^{(m)}}{\partial z^2} - \alpha(1-2\nu)\frac{\partial U_{n-2}^{(m)}}{\partial z},\\
&2(1-\nu)\frac{\partial^2 V_n^{(m)}}{\partial \theta^2} + (1-2\nu)[(\lambda_m+n)^2-1]V_n^{(m)} + [(\lambda_m+n)+3-4\nu]\frac{\partial U_n^{(m)}}{\partial \theta} \qquad (4)\\
&\quad = -(1-2\nu)\frac{\partial^2 V_{n-2}^{(m)}}{\partial z^2} - \frac{\partial^2 W_{n-1}^{(m)}}{\partial \theta\,\partial z} - \alpha(1-2\nu)\frac{\partial W_{n-1}^{(m)}}{\partial \theta} - \alpha(1-2\nu)\frac{\partial V_{n-2}^{(m)}}{\partial z},\\
&(1-2\nu)\frac{\partial^2 W_n^{(m)}}{\partial \theta^2} + (1-2\nu)(\lambda_m+n)^2 W_n^{(m)} = -(\lambda_m+n)\frac{\partial U_{n-1}^{(m)}}{\partial z} - \frac{\partial^2 V_{n-1}^{(m)}}{\partial \theta\,\partial z}\\
&\quad - 2(1-\nu)\frac{\partial^2 W_{n-2}^{(m)}}{\partial z^2} - 2\alpha(1-\nu)\frac{\partial W_{n-2}^{(m)}}{\partial z} - 2\nu\alpha(\lambda_m+n)U_{n-1}^{(m)} - 2\nu\alpha\frac{\partial V_{n-1}^{(m)}}{\partial \theta}.
\end{aligned}$$

In equation (4), the functions $U_{-1}^{(m)}, U_{-2}^{(m)}, V_{-1}^{(m)}, V_{-2}^{(m)}, W_{-1}^{(m)}, W_{-2}^{(m)}$ etc. are zero. The solution to $U_n^{(m)}, V_n^{(m)}, W_n^{(m)}$ can be obtained for each value of n. For $n = 0$ we can obtain $U_0^{(m)}, V_0^{(m)}, W_0^{(m)}$ as

$$\begin{aligned}
U_0^{(m)} &= B_1^{(m)}(z)\cos(\lambda_m+1)\theta + B_2^{(m)}(z)\sin(\lambda_m+1)\theta \\
&\quad + C_1^{(m)}(z)\cos(\lambda_m-1)\theta + C_2^{(m)}(z)\sin(\lambda_m-1)\theta, \\
V_0^{(m)} &= -B_1^{(m)}(z)\sin(\lambda_m+1)\theta + B_2^{(m)}(z)\cos(\lambda_m+1)\theta \\
&\quad + \frac{\lambda_m+3-4\nu}{\lambda_m-3+4\nu}[-C_1^{(m)}(z)\sin(\lambda_m-1)\theta + C_2^{(m)}(z)\cos(\lambda_m-1)\theta], \\
W_0^{(m)} &= A_1^{(m)}(z)\cos(\lambda_m\theta) + A_2^{(m)}(z)\sin(\lambda_m\theta).
\end{aligned} \tag{5}$$

Using the strain displacement relations and Hooke's law, the stress components are evaluated and on imposing the traction free condition on the crack face, $\sigma_{\theta\theta} = \sigma_{r\theta} = \sigma_{\theta z} = 0$, $\theta = \pm\pi$ we get the following equations

$$\begin{aligned}
&B_1^{(m)}(z)\cos(\lambda_m+1)\pi \pm B_2^{(m)}(z)\sin(\lambda_m+1)\pi \\
&\quad + \frac{(\lambda_m+1)}{(\lambda_m-3+4\nu)}[C_1^{(m)}(z)\cos(\lambda_m-1)\pi \pm C_2^{(m)}(z)\sin(\lambda_m-1)\pi] = 0, \\
&B_1^{(m)}(z)\sin(\lambda_m+1)\pi \pm B_2^{(m)}(z)\cos(\lambda_m+1)\pi \\
&\quad + \frac{(\lambda_m-1)}{(\lambda_m-3+4\nu)}[C_1^{(m)}(z)\sin(\lambda_m-1)\pi \pm C_2^{(m)}(z)\cos(\lambda_m-1)\pi] = 0, \\
&A_1^{(m)}(z)\sin(\lambda_m)\pi \pm A_2^{(m)}(z)\cos(\lambda_m)\pi = 0
\end{aligned} \tag{6}$$

For a non-trivial solution of the above set of equations, the determinant of the coefficients of functions of z must vanish and this condition is achieved for the following values of λ_m.

$$\lambda_m = \frac{m}{2}, \quad m = 0, 1, 2, \ldots \tag{7}$$

The negative values of m have been excluded in equation (7) so that the displacements are bounded as $r \to 0$. At this stage the dependence of the displacements on r is established and hence equation (3) can be written as a single series as follows

$$u_r = \sum_{n=0}^{\infty} r^{n/2} f_n(\theta, z), \quad u_\theta = \sum_{n=0}^{\infty} r^{n/2} g_n(\theta, z), \quad u_z = \sum_{n=0}^{\infty} r^{n/2} h_n(\theta, z), \tag{8}$$

where f_n, g_n, and h_n are yet to be determined. Equation (7) is substituted in equation (2) resulting in the following set of partial differential equations.

$$
\begin{aligned}
&(1-2\nu)\frac{\partial^2 f_n}{\partial\theta^2}+\Big[\frac{n}{2}-3+4\nu\Big]\frac{\partial g_n}{\partial\theta}+2(1-\nu)\Big[\frac{n^2}{4}-1\Big]f_n=-\Big(\frac{n}{2}-1\Big)\frac{\partial h_{n-2}}{\partial z}\\
&\quad-\alpha(1-2\nu)\Big(\frac{n}{2}-1\Big)h_{n-2}-(1-2\nu)\frac{\partial^2 f_{n-4}}{\partial z^2}-\alpha(1-2\nu)\frac{\partial f_{n-4}}{\partial z},\\
&2(1-\nu)\frac{\partial^2 g_n}{\partial\theta^2}+\Big[\frac{n}{2}+3-4\nu\Big]\frac{\partial f_n}{\partial\theta}+(1-2\nu)\Big[\frac{n^2}{4}-1\Big]g_n\\
&\quad=-\frac{\partial^2 h_{n-2}}{\partial\theta\,\partial z}-\alpha(1-2\nu)\frac{\partial h_{n-2}}{\partial\theta}-(1-2\nu)\frac{\partial^2 g_{n-4}}{\partial z^2}-\alpha(1-2\nu)\frac{\partial g_{n-4}}{\partial z},\\
&(1-2\nu)\Big[\frac{\partial^2 h_n}{\partial\theta^2}+\frac{n^2}{4}h_n\Big]=-\frac{n}{2}\frac{\partial f_{n-2}}{\partial z}-\frac{\partial^2 g_{n-2}}{\partial\theta\,\partial z}\\
&\quad-2(1-\nu)\frac{\partial^2 h_{n-4}}{\partial z^2}-2\alpha(1-\nu)\frac{\partial h_{n-4}}{\partial z}-n\nu\alpha f_{n-2}-2\nu\alpha\frac{\partial g_{n-2}}{\partial\theta}.
\end{aligned}
\tag{9}
$$

Equation (9) is solved for each value of n starting with $n = 0$ to determine the functions f_n, g_n, and h_n. For opening mode loading, the first four terms in the expansion of the in plane displacement components (u_r and u_θ) are as follows. Further details can be found in [8].

$$
\begin{aligned}
u_r&=B_0(z)\cos\theta+B_1(z)r^{(1/2)}\Big[\cos\frac{3\theta}{2}-(5-8\nu)\cos\frac{\theta}{2}\Big]+B_2(z)r\cos 2\theta\\
&\quad+B_3(z)r^{(3/2)}\Big[\cos\frac{5\theta}{2}+(3-8\nu)\cos\frac{\theta}{2}\Big],\\
u_\theta&=-B_0(z)\sin\theta-B_1(z)r^{(1/2)}\Big[\sin\frac{3\theta}{2}-(7-8\nu)\sin\frac{\theta}{2}\Big]-B_2(z)r\sin 2\theta\\
&\quad-B_3(z)r^{(3/2)}\Big[\sin\frac{5\theta}{2}-(9-8\nu)\sin\frac{\theta}{2}\Big].
\end{aligned}
\tag{10}
$$

It should be noticed that the solution is complete only to the extent of the dependence on r and θ at this stage and the functions $B_j(z)$ are yet to be determined. The first three terms in the expansion of the in plane stress components are as follows.

$$
\begin{aligned}
\frac{(1+\nu)}{E_0e^{\alpha z}}\sigma_{rr}&=\frac{1}{2}B_1(z)r^{-1/2}\Big[\cos\frac{3\theta}{2}-5\cos\frac{\theta}{2}\Big]+B_2(z)(1+\cos 2\theta)\\
&\quad+\frac{3}{2}B_3(z)r^{1/2}\Big[\cos\frac{5\theta}{2}+3\cos\frac{\theta}{2}\Big],\\
\frac{(1+\nu)}{E_0e^{\alpha z}}\sigma_{\theta\theta}&=-\frac{1}{2}B_1(z)r^{-1/2}\Big[\cos\frac{3\theta}{2}+3\cos\frac{\theta}{2}\Big]+B_2(z)(1-\cos 2\theta)\\
&\quad-\frac{3}{2}B_3(z)r^{1/2}\Big[\cos\frac{5\theta}{2}-5\cos\frac{\theta}{2}\Big],\\
\frac{(1+\nu)}{E_0e^{\alpha z}}\sigma_{\theta\theta}&=-\frac{1}{2}B_1(z)r^{-1/2}\Big[\cos\frac{3\theta}{2}+3\cos\frac{\theta}{2}\Big]+B_2(z)(1-\cos 2\theta)
\end{aligned}
\tag{11}
$$

$$- \frac{3}{2} B_3(z) r^{1/2} \Big[\cos \frac{5\theta}{2} - 5 \cos \frac{\theta}{2} \Big],$$

$$\frac{(1+\nu)}{E_0 e^{\alpha z}} \sigma_{r\theta} = - \frac{1}{2} B_1(z) r^{-1/2} \Big[\sin \frac{3\theta}{2} + \sin \frac{\theta}{2} \Big] - B_2(z) \sin 2\theta$$

$$- \frac{3}{2} B_3(z) r^{1/2} \Big[\sin \frac{5\theta}{2} - \sin \frac{\theta}{2} \Big].$$

One can notice that the leading term in equation (11) is the classical inverse square root term. Therefore the variation of the elastic modulus along the crack front does not affect the inverse square root singular nature of the stress field at the crack tip. Further, the angular structure of the stress field in equation (11) is same as that given by Williams for a plane crack [9]. The function $B_1(z)$, can be related to the classical mode-I stress intensity factors $k_1(z)$ as follows

$$k_1(z) = - \frac{2\sqrt{2\pi}}{1+\nu} B_1(z) E(z). \tag{12}$$

2. EXPERIMENTAL DETAILS

The transversely graded plate required for the experiment was prepared in the form of a particulate composite with glass particles embedded in an epoxy matrix through a gravity casting technique. Further details can be found in [9]. The elastic modulus of the prepared plate varied continuously from 3 GPa to 7 GPa along the thickness (z) in a near exponential manner as shown in figure 2. A single edge notched (SEN) specimen of length 160 mm, width 40 mm and thickness (h) of 12 mm was machined out of the prepared plate. A notch of 12 mm length was first machined and then the same was extended to a length of 14 mm by forcing a sharp razor blade into the notch tip. The SEN plate was loaded in bending as shown in figure 3 *a*.

The in plane displacement field on the specimen surface was measured using digital image correlation (DIC). To facilitate the DIC measurement, a random intensity pattern (see figure 3 *b*) was applied on the surfaces of the specimen using black and white aerosol paint. The photograph of the pattern was recorded before loading and at a load of 500 N using a 12 bit, 2 Mega pixel CCD camera. The in plane displacements were calculated by correlating the images using the Vic2D software (Correlated Solutions Inc., NC). The measured opening displacements on the two surfaces of the plate are shown in figure 4. The SIF was determined by fitting the derived displacement field (equation (10)) to the measured displacements. The calculated SIF is given in table 1. It can be observed that the SIF is higher on the stiffer side of the plate. The SIF on the stiffer side is 2.5 times that on the compliant side.

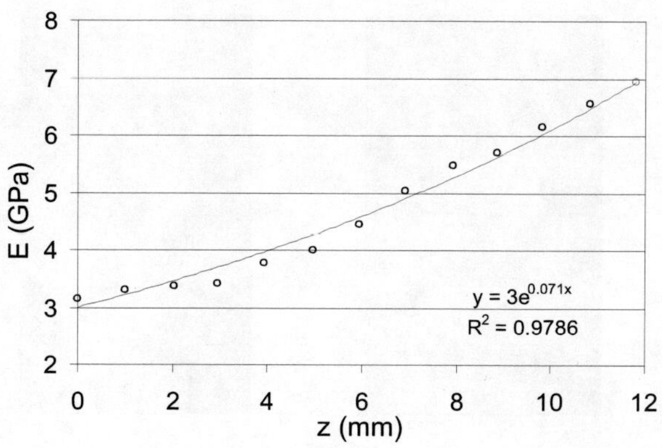

Fig. 2. Elastic modulus profile of the graded plate

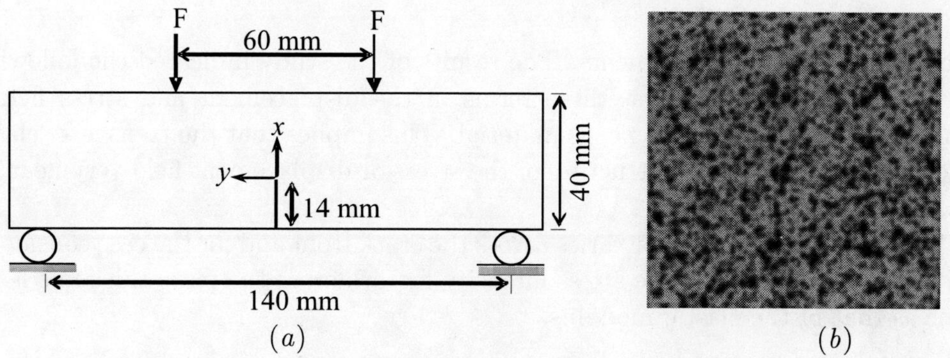

Fig. 3. (*a*) SEN plate subjected to four-point bending, (*b*) Random intensity pattern

Table 1. SIF for a graded plate

Location	E (GPa)	SIF (MPa·m$^{1/2}$)
Epoxy side ($z = 0$)	3.0	0.51
Glass bead side ($z = h$)	7.0	1.26

CONCLUSIONS

The problem of a crack in a transversely graded plate having elastic modulus varying continuously along the thickness is studied. The asymptotic structure of the displacement and stress field is brought out. A cracked graded plated was prepared and was loaded in pure bending. The displacements on the surfaces of the plate were measured using DIC. The SIF on the two surfaces of the plate was calculated

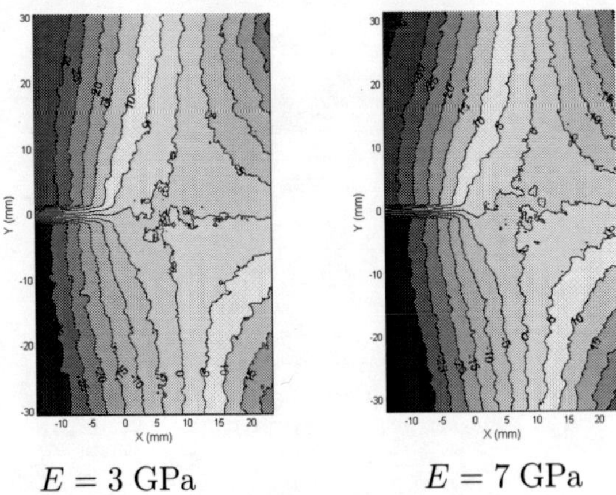

$E = 3$ GPa $E = 7$ GPa

Fig. 4. Measured displacements (μm)

from the measured displacements. The results of the study indicated the following.

The structure of the first three terms in the displacement and stress field is identical to that in a homogeneous material. This implies that the transverse elastic gradients do not affect the structure of the stress or displacement field very near the crack tip.

The stress intensity factor varies across the crack front and for the case considered the extent of variation of the stress intensity factor across the plate thickness is the same as that of the elastic modulus.

ACKNOWLEDGMENTS

The authors acknowledge the financial support from Aeronautical Research & Development Board for the DIC system used in the study.

REFERENCES

1. F. Erdogan, "Fracture Mechanics of Functionally Graded Materials," Compos. Engng **5** (7), 753–770 (1995).
2. P. Gu and R.J. Asaro, "Cracks in Functionally Graded Materials," Int. J. Solids Struct. **34** (1), 1–17 (1997).
3. P. Gu and R.J. Asaro, "Cracks Deflection in Functionally Graded Materials," Int. J. Solids Struct. **34** (24), 3085–3098 (1997).
4. C.-E. Rousseau and H.V. Tippur, "Influence of Elastic Gradient Profiles on Dynamically Loaded Functionally Graded Materials: Cracks along the Gradient," Int. J. Solids Struct. **38** (44/45), 7839–7856 (2001).

5. G. Anlas, J. Lambros, and M.H. Santare, "Dominance of Asymptotic Crack Tip Fields in Elastic Functionally Graded Materials," Int. J. Fract. **115** (3), 193–204 (2002).
6. V. Parameswaran and A. Shukla, "Asymptotic Stress Fields for Stationary Cracks along the Gradient in Functionally Graded Materials," J. Appl. Mech. **69** (3), 240–243 (2002).
7. R.J. Hartranft and G.C. Sih, "The Use of Eigen Function Expansion in the General Solution of Three Dimensional Crack Problems," J. Math. Mech. **19** (2), 123–138 (1969).
8. S.C. Wadgaonkar and V. Parameswaran, "Structure of Near Tip Stress Field and Variation of Stress Intensity Factor for a Crack in a Transversely Graded Material," J. Appl. Mech. **75** (Nov. 2008), (2008) doi:10.1115/1.2966177
9. M.L. Williams, "On the Stress Distribution at the Base of a Stationary Crack," J. Appl. Mech. **24**, 109–114 (1957).

NUMERICAL SIMULATION OF EDGE CRACK PROBLEMS USING X-FEM

I.V. Singh[1*], B.K. Mishra[1], and S. Kumar[1]

ABSTRACT

In this work, eXtended finite element method (X-FEM) has been used to obtain the solution of edge crack problems as it provides a versatile technique to model static as well as moving crack problems without re-meshing/conformal meshing. Heaviside function has been used to handle displacement discontinuity at the crack surfaces, whereas crack tip is enriched by branch functions to capture asymptotic nature of stress field. The stress intensity factors are evaluated using interaction integral approach. The effect of crack orientation has been studied for single as well as two edge cracks, whereas crack interaction for two edge cracks lying on the same face have been studied under plane stress conditions.

Key words: edge cracks, extended FEM, stress intensity factor, interaction and orientation

INTRODUCTION

Numerical simulation of the problems involving cracks is essential to quantify and predict the behavior of structures under service conditions. Thus accurate evaluation of fracture parameters such as stress intensity factors (SIF) becomes quite important for simulation based design & analysis. To simulate cracked structures, a number of methods such as finite element method (FEM), boundary element method (BEM) and finite difference method (FDM) are available. FEM has been in the forefront of numerical methods used for the simulation of fracture mechanics problems. A variety of approaches developed over the years in FEM, has helped it to establish itself as the most suited method for analyzing the asymptotic stress fields at the crack tip. However, FEM requires that the crack surface should coincide with the edge of the finite elements i.e. a conformal mesh is required besides the requirement

[1]Department of Mechanical and Industrial Engineering, Indian Institute of Technology Roorkee, Roorkee — 247667, Uttarakhand, India

[*]E-mail: *indrafme@iitr.ernet.in*, *bhanufme@iitr.ernet.in*

of special elements to handle crack tip asymptotic stresses. To overcome these difficulties, a novel approach known as extended finite element method (XFEM) has been developed to model arbitrary discontinuities without conformal mesh or re-meshing [1–4].

XFEM allows the modeling of arbitrary geometric features independently of the finite element mesh. In XFEM, the cracks are easily identified and can be handled by level a set function [5], which allows the discontinuity to be defined independent of the mesh.

Components and structures fail catastrophically ether due to pre-existing cracks or loads may develop/initiate cracks during the service life. In view of the above, analysis of cracked structures is an important issue. XFEM provides a technique where cracks in different configurations can be easily studied without the need for re-mesh or conformal mesh. In the present work, the analysis of single edge crack and effect of the presence of second edge crack on stress intensity factors (SIFs) have been performed. Once the displacement and stress field are known, the SIFs have been calculated by interaction integral approach. The current study has been performed using linear elastic fracture mechanics (LEFM) theories and analytical solutions available for two dimensional displacement fields, which are used for crack tip enrichment.

1. NUMERICAL FORMULATIONS

In this section, the governing equilibrium equations for the displacement field in elasto-statics have been discretized and implemented in context of XFEM. The domain (Ω) as shown in Fig. 1, is bounded by Γ, which is composed of Γ_c, Γ_t, and Γ_u. The strong form of equilibrium equations and boundary conditions elasto-static domain are:

$$\nabla \boldsymbol{\sigma} + \mathbf{b} = 0 \quad \text{in } \Omega, \tag{1}$$

$$\boldsymbol{\sigma}\mathbf{n} = \bar{\mathbf{t}} \quad \text{on } \Gamma_t, \tag{2}$$

$$\boldsymbol{\sigma}\mathbf{n} = 0 \quad \text{on } \Gamma_c, \tag{3}$$

$$\mathbf{u} = \bar{\mathbf{u}} \quad \text{on } \Gamma_u, \tag{4}$$

where $\boldsymbol{\sigma}$ is the stress tensor, $\mathbf{b}$ is the body force vector, $\bar{\mathbf{u}}$is the displacement vector, $\bar{\mathbf{t}}$ is the traction force and $\mathbf{n}$ is the unit normal at the crack faces i.e. Γ_c are assumed to be traction free.

The approximation for the vector function $\mathbf{u}$ with the partition of unity enrichment [6] is given by the following equation:

$$\mathbf{u}^h(\mathbf{x}) = \sum_{I=1}^{n} N_I(\mathbf{x}) \left(\sum_{\alpha=1}^{n} \psi_\alpha(\mathbf{x}) \mathbf{a}_I^\alpha \right), \tag{5}$$

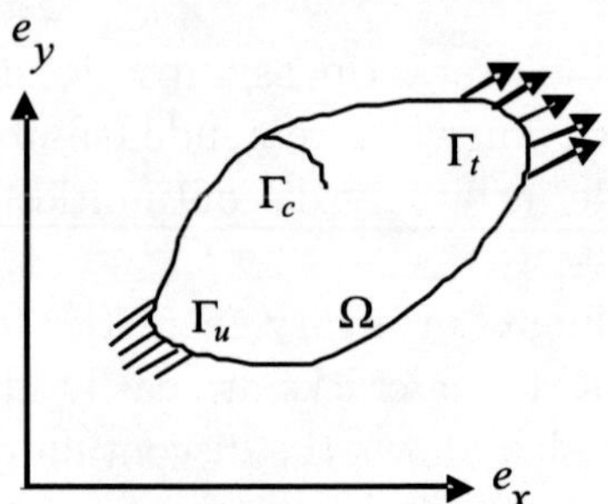

Fig. 1. Domain with a crack discontinuity

where $N_I(\mathbf{x})$ element shape function forms a partition of unity $\sum_I N_I(\mathbf{x}) = 1$, ψ_α are the enrichment functions. By using the notion of unity, the standard approximation is enriched with additional functions. In particular instance of 2-D crack modeling, the enriched displacement trial and test approximation is written as [7]:

$$\mathbf{u}^h(\mathbf{x}) = \sum_{I\in n} N_I(\mathbf{x}) \left[\mathbf{u}_I + \underbrace{H(\mathbf{x})\mathbf{a}_I}_{I\in n_r} + \underbrace{\sum_{\alpha=1}^{4} \Phi_\alpha(\mathbf{x})\mathbf{b}_I^\alpha}_{I\in n_A} \right] \tag{6}$$

where,$\mathbf{u}_I$ is the nodal displacement vector associated with the continuous part of the finite element solution, $\mathbf{a}_I$ is the nodal enriched degree of freedom associated with Heaviside (discontinuous) function, and $\mathbf{b}_I^\alpha$ is the nodal enriched degree of freedom vector associated with the elastic asymptotic crack tip functions. In Eq. (6), n is the set of all nodes in the mesh; n_r is the set of nodes whose shape function is cut by the crack face and n_A is the set of nodes whose shape function support is cut by the crack tip. For any node n_r, the support of the nodal shape function is fully cut into two disjoint pieces by the crack. Heaviside jump function $H(\mathbf{x})$, is a discontinuous function across the crack surface and is constant on each side of the crack i.e. +1 on one side and −1 on other. In this work, XFEM has been applied to the edge crack problems which are having discontinuity in primary variable i.e. $\mathbf{u}$. The crack tip enrichment (i.e. branch function enrichment) is done by adding additional degree of freedom at the tip enriched nodes using the displacement field distribution available from analytical solutions of LEFM, which is given as:

$$\Phi_\alpha(\mathbf{x}) = \left[\sqrt{r}\sin\frac{\theta}{2}, \sqrt{r}\cos\frac{\theta}{2}, \sqrt{r}\sin\theta\sin\frac{\theta}{2}, \sqrt{r}\sin\theta\cos\frac{\theta}{2}\right], \quad \alpha = 1, 2, 3, 4, \tag{7}$$

where, r and θ are the polar coordinates in local crack-tip coordinate system.

The discrete weak form for linear elasto-statics is obtained as:

$$\int_{\Omega^h} \boldsymbol{\sigma} : \delta\boldsymbol{\varepsilon}^h \, d\Omega = \int_{\Omega^h} \mathbf{b}\delta\mathbf{u}^h d\Omega + \int_{\partial\Omega_I^h} \bar{\mathbf{t}}\delta\mathbf{u}^h d\Gamma \quad \forall \delta\mathbf{u}^h \in \mathbf{U}_0^h, \tag{8}$$

where $\mathbf{u}^h \in \mathbf{U}^h$ and $\delta\mathbf{u}^h \in \mathbf{U}_0^h$ are approximate and test functions used in XFEM. By substituting the trial and test functions in above equation and using the arbitrariness of the nodal variations, a following set of discrete equations is obtained as:

$$\mathbf{Kd} = \mathbf{f}, \tag{9}$$

where d is the vector of nodal unknowns, $\mathbf{K}$ and $\mathbf{f}$ are the global stiffness matrix and external force vector respectively. The stiffness matrix and force vector are computed on element level and are assembled into global counterparts through usual finite element assembly procedure. The additional degrees of freedom arise due to the enrichment and are handled by considering fictitious nodes. The nodes lying on the crack faces are enriched by Heaviside function (H) which adds two additional degrees of freedom per enriched node, while tip nodes are enriched by branch functions (Eq. (2.7)) which adds four additional degree of freedom per tip node. The elemental contribution of $\mathbf{K}$ and $\mathbf{f}$ are as follows:

$$\mathbf{f}_i^e = \{\mathbf{f}_i^u \quad \mathbf{f}_i^a \quad \mathbf{f}_i^{b1} \quad \mathbf{f}_i^{b2} \quad \mathbf{f}_i^{b3} \quad \mathbf{f}_i^{b4}\}^T, \tag{10}$$

$$\mathbf{k}_{ij}^e = \begin{bmatrix} \mathbf{k}_{ij}^{uu} & \mathbf{k}_{ij}^{ua} & \mathbf{k}_{ij}^{ub} \\ \mathbf{k}_{ij}^{au} & \mathbf{k}_{ij}^{aa} & \mathbf{k}_{ij}^{ab} \\ \mathbf{k}_{ij}^{bu} & \mathbf{k}_{ij}^{ba} & \mathbf{k}_{ij}^{bb} \end{bmatrix} \tag{11}$$

where, sub-matrices and vectors that appear in the equation are given by

$$\mathbf{k}_{ij} = \int_{\Omega^e} (\mathbf{B}_i^r)^T \mathbf{D} \mathbf{B}_j^s \, d\Omega \quad (r,\, s = u,\, a,\, b), \tag{12}$$

$$\mathbf{f}_i^u = \int_{\partial\Omega_L^h \cap \partial\Omega^e} N_i \bar{\mathbf{t}} \, d\Gamma + \int_{\Omega^e} N_i \mathbf{b} \, d\Omega, \tag{13}$$

$$\mathbf{f}_i^a = \int_{\partial\Omega_L^h \cap \partial\Omega^e} N_i H \bar{\mathbf{t}} \, d\Gamma + \int_{\Omega^e} N_i H \mathbf{b} \, d\Omega, \tag{14}$$

$$\mathbf{f}_i^{b\alpha} = \int_{\partial\Omega_L^h \cap \partial\Omega^e} N_i \Phi_\alpha \bar{\mathbf{t}} \, d\Gamma + \int_{\Omega^e} N_i \Phi_\alpha \mathbf{b} \, d\Omega \quad (\alpha = 1,\, 2,\, 3,\, 4), \tag{15}$$

where, D is the plane stress constitutive matrix given as

$$\mathbf{D} = \frac{E}{1-\nu^2} \begin{bmatrix} 1 & \nu & 0 \\ \nu & 1 & 0 \\ 0 & 0 & \frac{1}{2}(1-\nu) \end{bmatrix}$$

and N_i is finite element shape function defined at node of the finite element. $\mathbf{B}_i^u$, $\mathbf{B}_i^a$, and $\mathbf{B}_i^b$ are the matrix of shape function derivatives given by

$$\mathbf{B}_i^u = \begin{bmatrix} N_{i,x} & 0 \\ 0 & N_{i,y} \\ N_{i,y} & N_{i,x} \end{bmatrix}, \quad \mathbf{B}_i^b = \left[\mathbf{B}_i^{b1} \quad \mathbf{B}_i^{b1} \quad \mathbf{B}_i^{b1} \quad \mathbf{B}_i^{b1}\right],$$

$$\mathbf{B}_i^a = \begin{bmatrix} (N_i H)_{,x} & 0 \\ 0 & (N_i H)_{,y} \\ (N_i H)_{,y} & (N_i H)_{,x} \end{bmatrix}, \quad \mathbf{B}_i^{b\alpha} = \begin{bmatrix} (N_i \Phi_\alpha)_{,x} & 0 \\ 0 & (N_i \Phi_\alpha)_{,y} \\ (N_i \Phi_\alpha)_{,y} & (N_i \Phi_\alpha)_{,x} \end{bmatrix} \quad (\alpha = 1, 2, 3, 4).$$

The Matlab® has been used to implement the above formulation [8].

2. RESULTS AND DISCUSSIONS

The numerical simulation has been performed for plane stress condition using the following data: size of domain = 100 mm by 200 mm, modulus of elasticity $(E) = 200$ GPa, Poisson's ratio $(\nu) = 0.3$ and far field stress $\sigma_0 = 100$ MPa. The results have been evaluated using four node quadrilateral elements. Single edge crack and double edge cracks (length 40 mm on the same edge) have been considered for the study. The domain has been discretized using a mesh size of 20×40 nodes for single edge crack and 20×50 nodes for two edge cracks. The values of stress intensity factor have been calculated using interaction integral approach [9].

2.1. *Single edge crack*

Fig. 2 shows the configuration and values of SIFs under mode-I loading for single edge crack. The effect of change in length of edge crack has been studied and it has been noticed that there is increase in SIF-I with the increase in crack length. Fig. 2 also shows that the SIF values predicted by XFEM are in close agreement with the values calculated from the analytical solutions [10]. Fig. 3 shows the results for inclined single edge crack with respect to the horizontal axis, and it has been noticed that there is a monotonous decrease in the value of SIF-I with the increase in inclination, while SIF-II first increases up to the angle of 45° then starts decreasing.

2.2. *Two edge cracks*

The change in values of SIFs for first edge crack in the presence of second crack has been analyzed. Two cracks lying on the same edge are considered for the study. Three set of results were obtained, first by changing the distance between two parallel cracks, next by changing the length of second crack, and finally by changing the orientation of second crack with respect to the horizontal axis. Figure 3 shows two parallel crack configurations. The results shown in the Figure 3 shows that there is increase in SIF-I values with the increase in the distance between two cracks, whereas SIF-II decreases with the increase of distance between them. Figure 4 shows the effect of change in length of second crack on the SIF values calculated on the first. The second crack is lying at 20 mm distance (d) from the first crack which is placed in the center of the edge. The results show that there is decrease

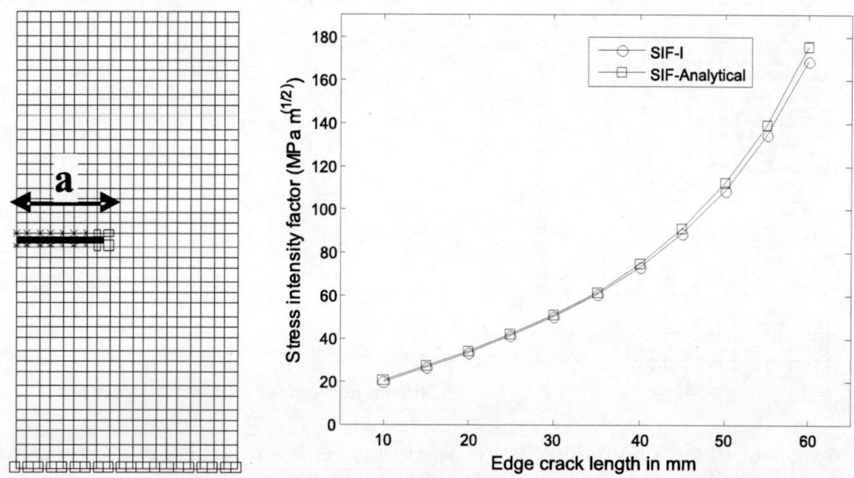

Fig. 2. Effect of the change in crack length on mode-I SIF

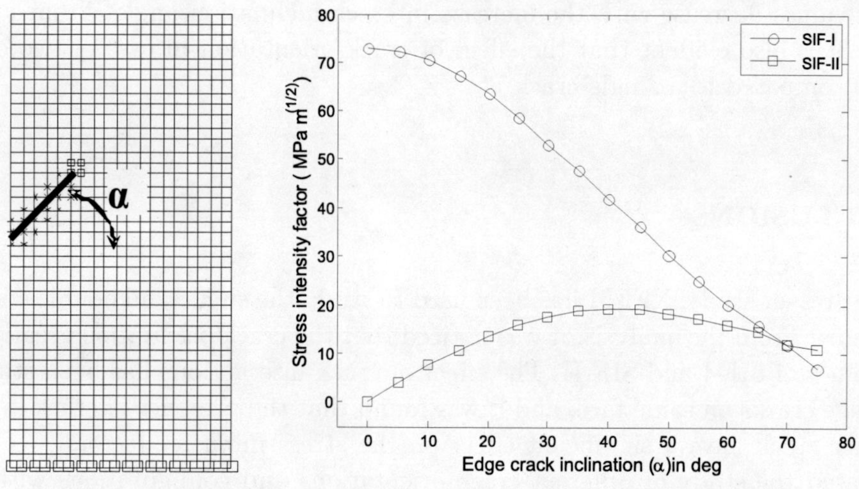

Fig. 3. Effect of the change in crack orientation (α) on SIFs

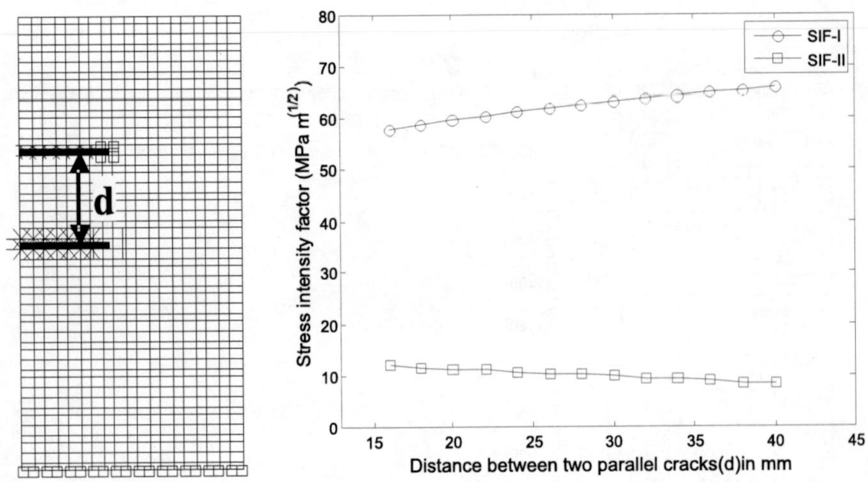

Fig. 4. Effect of the change in distance (d) between two parallel edge cracks on SIFs

in SIF-I values with the increase in the crack length, whereas SIF-II values show an increasing trend. Figure 5 shows the SIF values calculated for first crack in the presence of second inclined edge crack. The second crack has been kept at a distance of $d = 20$ mm. From the results presented in Figure 5, it has been found that there is an increase in SIF-I values with the increase in inclination angle (α), whereas SIF-II values decrease with the increase in crack inclination angle. From Figures 2 and 5, it is also evident that the effect of crack orientation in case of two cracks is less as compared to a single crack.

CONCLUSIONS

In the present study, XFEM has been used to study the single and double edge crack problems. From the analysis, it was noticed that the crack orientation greatly affects the values of SIF-I and SIF-II. The effect of crack interactions has been studied for two edge cracks on same face, and it was found that the presence of crack in vicinity of other crack have a significant effect on the stress intensity factors. XFEM has facilitated the study of different crack orientations and configurations without the need of re-meshing/conformal meshing, thereby allowing faster simulations. During analysis, it was also noticed that X-FEM requires no change in the mesh for the change in crack geometry, and there is no need to align the element boundaries along the crack faces.

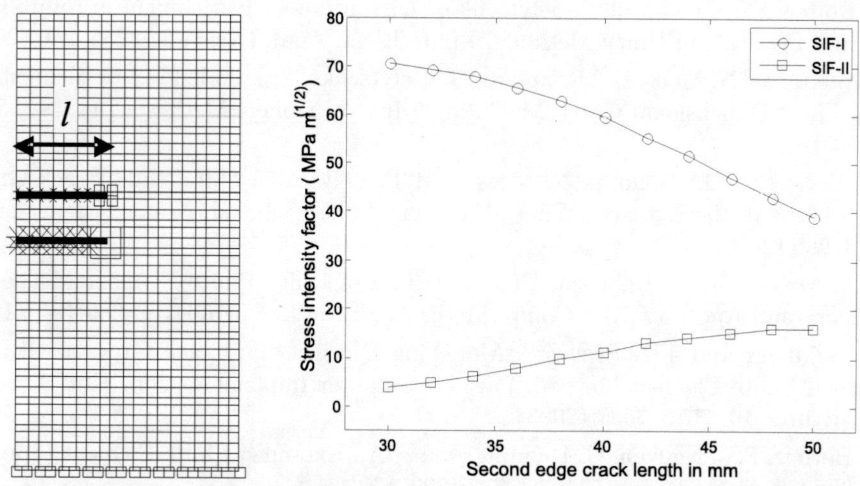

Fig. 5. Effect of the change in the length of second crack on SIFs

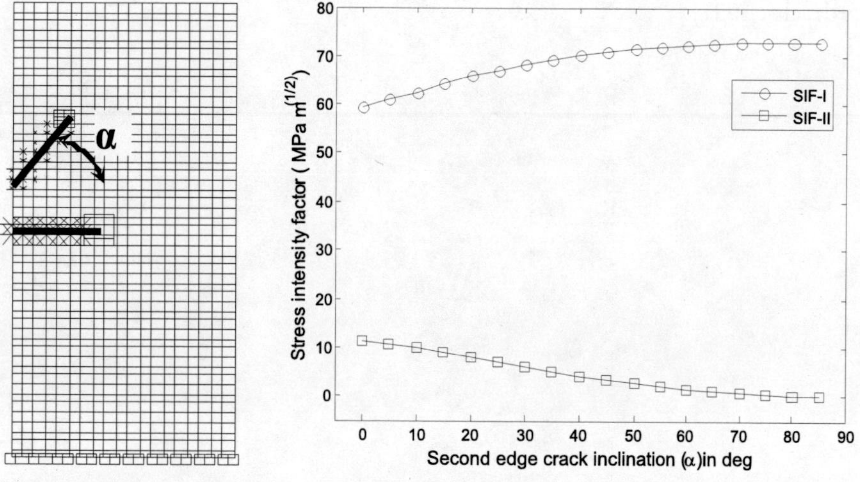

Fig. 6. Effect of the change in the orientation of second crack on SIFs

REFERENCES

1. T. Belytschko and T. Black, "Elastic Crack Growth in Finite Elements with Minimal Remeshing," Int. J. Numer. Methods Engng **45**, 601–620 (1999).

2. N. Moes, J. Dolbow, and T. Belytschko, "A Finite Element Method for Crack Growth without Remeshing," Int. J. Numer. Meth. Engng **46**, 131–150 (1999).

3. J. Dolbow, N. Moes, and T. Belytschko, "Discontinuous Enrichment in Finite Elements with a Partition of Unity Method," Finite Elem. Anal. Desisn **36**, 235–260 (2000).
4. N. Sukumar, N. Moes, B. Moran, and T. Belytschko, "Extended Finite Element Method for Three-Dimensional Crack Modelling," Int. J. Numer. Meth. Engng **48**, 1549–1570 (2000).
5. M. Stolarska, D.L. Chopp, N. Moes, and T. Belyschko, "Modelling Crack Growth by Level Sets in the Extended Finite Element Method," Int. J. Numer. Meth. Engng **51**, 943–960 (2000).
6. J.M. Melenk and I. Babuska, "The Partition of Unity Finite Element Method: Basic Theory and Applications," Comp. Meth. Appl. Mech. Engng **139**, 289–314 (1996).
7. N. Sukumar and J.H. Prevost, "Modelling Quasi-Static Crack Growth with the Extended Finite Element Method. Part I: Computer Implementation," Int. J. Solids and Structures **40**, 7513–7537 (2003).
8. S. Bordas, P.V. Nguyen, C. Dunant, et al., "An Extended Finite Element Library," Int. J. Numer. Meth. Engng **71**, 703–732 (2007).
9. B.N. Rao and S. Rahman, "An Interaction Integral Method for Analysis of Cracks in Orthotropic Functionally Graded Materials," Comput. Mech. **32**, 40–51 (2003).
10. D. Broek, *Elementary Engineering Fracture Mechanics*, 4th rev. ed. (Kluwer Academic, Dordrecht, 1991).

NUMERICAL MODELLING OF FAILURE IN MATERIALS

M.D. Ghouse[1*], R.P. Nair[1**], C.L. Rao[1***], and B.N. Rao[1****]

ABSTRACT

Failure of materials can be modelled using continuum mechanics as well as discrete particle mechanics. The paper will present simulation of failure in a brittle material (concrete) using a unit cell based approach, where the material is modelled as a two phase material having aggregates and cement mortar. Smeared crack model is used to model the failure of mortar matrix. Uniaxial tensile test is simulated numerically for different aspect ratios and different shapes of aggregates and the results are presented in the paper. The discrete particle modelling of a material is done using a discrete element method. The discrete element method was used to model the deformations in a one dimensional as well as a two-dimensional deformation of steel. The preliminary results using a discrete element method are promising and the extensions of the method to simulate disintegration in material are indicated in the paper.

Key words: unit cell approach, smeared crack model, discrete element method, failure in materials, numerical simulation

INTRODUCTION

Failure in materials is an important phenomenon that will need to be modelled, in order to capture the failure in structures. Traditionally, failure is treated as a limiting phenomenon of a deformational process and hence numerical methods that are used to simulate deformation in a continuum are used to predict the failure in materials. In recent times, attempts are also being made to simulate failure (indicted

[1]Indian Institute of Technology Madras, Chennai — 600036, Tamil Nadu, India
[*]E-mail: *gousewiz902@gmail.com*
[**]E-mail: *rajeshpnair@gmail.com*
[***]E-mail: *lakshman@iitm.ac.in*
[****]E-mail: *bnrao@iitm.ac.in*

by separation of particles) independently, using special numerical technique. In this paper, some attempts to simulate failure in a brittle material (concrete) using the continuum approach, as well as simulations of failure in a ductile material (steel) using discrete element approach will be introduced and some general conclusions will be drawn regarding the relative merits of both the processes.

1. NUMERICAL MODELING OF CONCRETE FAILURE USING A UNIT CELL APPROACH

1.1. *Introduction*

Concrete is a composite consisting of cement, fine aggregates, coarse aggregates and water. Generally, aggregates are mixed with cement paste matrix to form concrete. Concrete has a highly heterogeneous microstructure. When microstructure of concrete is observed, broadly it can be visualized to be consisting of two phase's *viz.*, hardened cement paste matrix and coarse aggregates. To determine the macroscopic behavior of heterogeneous concrete is an important problem in engineering applications. The current phenomenological models for concrete deformation, which assume concrete as a homogeneous material, cannot correlate the heterogeneous structure of the material with the mechanical behavior of the material. For obtaining a deeper understanding of the physical processes involved in the mechanical response, models considering the heterogeneous nature of concrete are required.

Wittmann [1] has identified three structural levels for concrete modeling. These are microscopic, mesoscopic, and macroscopic levels. At the microscopic level, the models of material science based on real mechanisms are used to describe qualitatively and quantitatively the complex physical, chemical and mechanical behavior of hardened cement paste. Characteristic edge length of this scale is of the order 10^{-1} mm, whereas the characteristic length scales at the meso and macro scales are of the order of 10^2 mm and 10^3 mm and above respectively. The mesoscopic level is introduced to account for the composite nature of the material and its inhomogeneities. Modeling of concrete failure process, at mesoscale aims at studying the role of concrete heterogeneity on the failure processes taking place in concrete.

Simulation of the material response in a two-dimensional continuum using lattice models is being attempted by several researchers in the recent past. Vervuurt et al. [2] used lattice models with beam elements to numerically simulate wedge splitting test of concrete structure. Slowik et al. [3] studied the fracture processes of concrete using truss elements. Compression, direct tension and wedge splitting tests were simulated numerically using lattice models. Wittmann et al. [1] used two dimensional plane stress elements to compute the effective elastic modulus and the effective diffusion coefficient. The same approach was later extended to predict fracture in direct tension specimens. In these investigations, the fictitious crack approach was used to model matrix phase and linear elastic property for the aggregates. The load-displacement response was reasonably predicted with these models, and cracking patterns qualitatively matched experimental data. More recently, Wang et al. [4] carried this investigation further using triangular elements for mesh generation. A nonlinear finite element method suitable for mesoscopic study of concrete was developed. The proposed model was applied to study the behavior of concrete under uniaxial tension. These authors observed that the crack initiated at the aggregate-matrix interface, and the descending part of stress-strain curve is not an invariant property and depends on gauge length. In this paper, a continuum model is used for numerically simulating compression behavior of concrete using a unit cell approach.

In the current work, representative volume element of concrete is idealized as a unit cell consisting of one aggregate surrounded by hardened cement paste matrix. This unit cell approach will help us to study the material response and its sensitivity to parameters like: a) effect of change in properties of constituent material on overall structural behavior b) initiation of crack c) propagation of crack in matrix phase. The unit cell analysis has been adopted earlier to study material behavior of other material as well. Pettermann and Suresh [5] have established the unit cell approach to predict the material properties of piezoelectric composites, consisting of brittle hard fibers embedded in a ductile matrix. The complete set of overall moduli is predicted for various fiber arrangements and fiber cross section. These results are compared to semi analytical bounds from the literature. The dependence on the fiber volume fraction is also predicted very well. The current work aims at simulating the nonlinear mechanical response of a unit cell of concrete consisting of a single aggregate embedded in a cement matrix in a two dimensional domain. The sensitivity of the mechanical response is studied with respect to various parameters and is compared with experimental results available from the literature.

1.2. *Mechanical modeling of unit cell*

In this paper representative volume element of concrete is idealized as a unit cell consisting of single aggregate surrounded by hardened cement paste matrix. Aggregate is idealized to have round shape. The idealized unit cell used in numerical simulation is shown in the Fig. 1.

1.2.1. Modelling aggregate. Coarse aggregate represent around 40–50% of the concrete volume. Gravels constitute the majority of coarse aggregate used in concrete followed by the crushed stones. The shape of the aggregate particles depends on the aggregate type. In general, gravel aggregates have rounded shape while

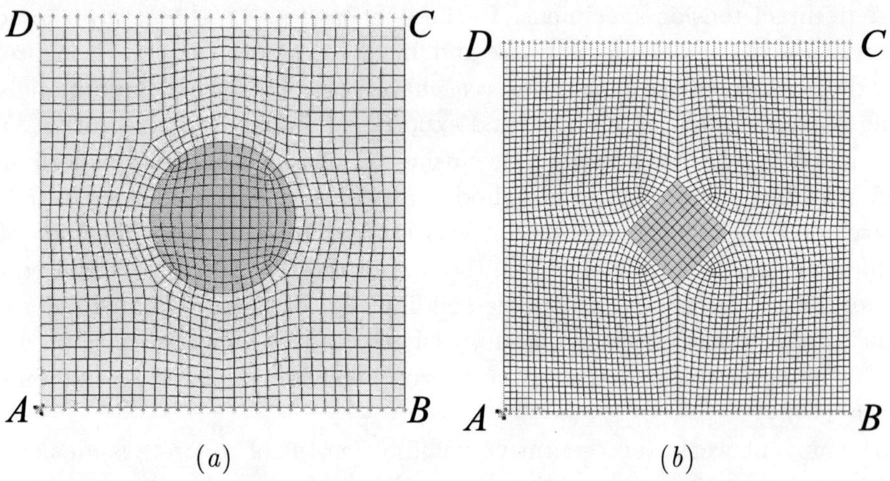

Fig. 1. (*a*) Unit cell with round aggregate, (*b*) Unit cell with angular aggregate

crushed stone aggregates have an angular shape. In this work aggregate is modeled as round and angular. Aggregate is modeled as a linear-isotropic material. It behaves linearly throughout the analysis. A linear stress-strain relation for aggregate is given by Eq. (1)

$$\begin{Bmatrix} \sigma_x \\ \sigma_y \\ \tau_{xy} \end{Bmatrix} = \frac{E}{1-\upsilon^2} \begin{pmatrix} 1 & \upsilon & 0 \\ \upsilon & 1 & 0 \\ 0 & 0 & \frac{1-\upsilon}{2} \end{pmatrix} \begin{Bmatrix} \varepsilon_x \\ \varepsilon_y \\ \gamma_{xy} \end{Bmatrix}. \tag{1}$$

1.2.2. Modelling mortar matrix. Mortar matrix shows softening behavior after reaching the tensile strength which can be captured from stress separation curves. This softening behavior is due to the formation of fracture process zone. The material in this zone softens progressively due to microcracking. This softening behavior is localized in a fictitious crack, where stress transfer normal to the crack can still take place. The stress transfer capacity as function of the crack width is given by the strain-softening diagram shown in Fig. 2. The area under stress-softening diagram is fracture energy (G_f) of mortar matrix. Fracture energy of matrix is generally determined experimentally using a notched specimen loaded in flexure. The value for G_f is obtained by computing the area under the load-deflection curve and dividing it by the net cross-section of the specimen above the notch. In this work G_f is calculated form CEB-FIP model code 1990 which recommends the use of Eq. (2)

$$G_g = \alpha_f \left(\frac{f_{\rm cm}}{f_{\rm cmo}} \right)^{0.7}, \tag{2}$$

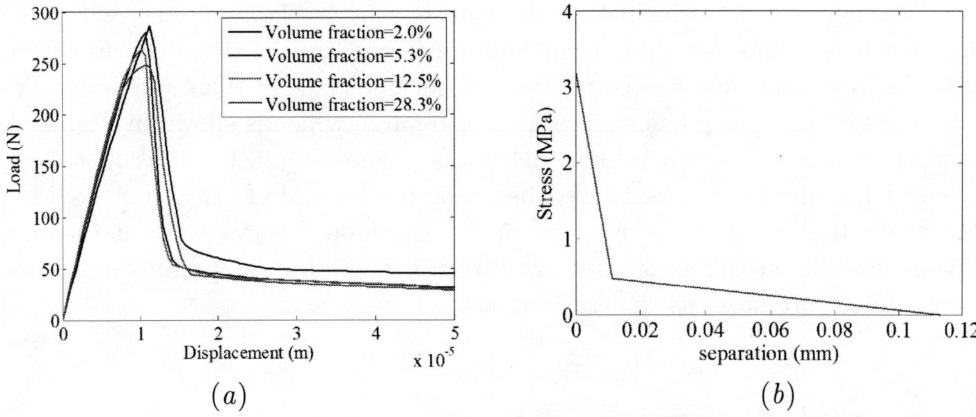

Fig. 2. (*a*) Stress-Separation curve, (*b*) Load-displacement curves obtained for round aggregates

where α_f is a coefficient, which depends on the maximum aggregate size whose value is taken as 0.02, f_{cmo} is equal to 10 MPa and f_{cm} is the compressive strength of matrix. Stress-separation relation is calculated from CEB-FIP model code 1990 which recommends a bilinear stress-crack opening relationship given in Eq. (3)

$$\sigma_{\text{ct}} = \begin{cases} f_{\text{ctm}}\Big(1 - 0.85\dfrac{w}{w_1}\Big) & \text{for } 0.15 f_{\text{ctm}} \leq f_{\text{ctm}}, \\ \dfrac{0.15 f_{\text{ctm}}}{w_c - w_1}(w_c - w) & \text{for } 0 \leq 0.15 f_{\text{ctm}}, \end{cases} \quad (3)$$
$$w_1 = \frac{2G_f}{f_{\text{ctm}}} - 0.15 w_c, \quad w_c = \beta_f \frac{G_f}{f_{\text{ctm}}},$$

where f_{ctm} is direct tensile strength in MPa, σ_{ct} is tensile stress in MPa, w_1 is crack opening, w_c is crack opening for $\sigma_{\text{ct}} = 0$ and β_f is coefficient depends on maximum aggregate size whose value is taken as 8.

In this work a brittle crack model is used. This model uses a smeared crack model to represent the discontinuous brittle behavior in concrete. It does not track individual "macro" cracks: instead, constitutive calculations are performed independently at each material point of the finite element model. The presence of cracks enters into these calculations by the way in which the cracks affect the stress and material stiffness associated with the material point. A simple Rankine criterion is used to detect crack initiation. This criterion states that a crack forms when the maximum principal tensile stress exceeds the tensile strength of the brittle material. As soon as the Rankine criterion for crack formation has been met, we assume that a first crack has formed. The crack surface is taken to be normal to the direction of the maximum tensile principal stress. Subsequent cracks may form with crack surface normals in the direction of maximum principal tensile stress that is orthogonal to the directions of any existing crack surface normals at the same point. Post failure behavior is specified as stress-separation curve which is shown in Fig. 2. An important feature of the cracking model is that, whereas crack initiation is based on Mode I fracture only, postcracked behavior includes Mode II as well as Mode I. The Mode II shear behavior is based on the common observation that the shear behavior depends on the amount of crack opening. More specifically, the cracked shear modulus is reduced as the crack opens.

1.3. *Results and discussions*

Uniaxial tensile loading is applied on the unit cell and is numerically simulated using finite element method. While following a unit cell approach, special emphasis is given for boundary conditions. Pettermann and Suresh [5], Kouznetsova et al. [6] have described about the periodic boundary condition to be applied on a unit cell. According to Kouznetsova et al. [6], for uniaxial loading simple boundary conditions are sufficient. Fig. 1 shows discretised unit cell with applied simple boundary conditions, where node A is fixed, node D is constrained in x-direction and edge AB is constrained in y-direction. Displacement controlled load is applied along the edge CD. The material properties used for the analysis are given in the Table 1. Load-displacement curves obtained for round aggregates in uniaxial tension are shown in Fig. 2 *b*. Simulations are conducted for different aggregate volume fraction. Initiation of failure is observed at the interface of aggregate and mortar matrix. It is observed from the simulation that the volume fraction of aggregate affects the strength of concrete. As the volume fraction of aggregate increases, the strength of the concrete is found to decrease. Similar trends are observed for simulations in compression. The decrease in strength that is predicted for the composite using the current model is in close agreement with the observations that are observed experimentally in concrete. Simulations were conducted for a unit cell with angular aggregate oriented 90° with horizontal axis. Load-displacement curves are as shown in Fig. 3 *b*. Load-displacement curves show similar pattern as obtained from unit cell with round aggregates but the strength of the unit cell with angular aggregate is higher than the unit cell with round aggregate. This may be because of better interlocking provided by angular aggregates with mortar matrix than the round aggregates. Area under load-displacement curve for angular aggregate is more than round aggregates which indicate fracture energy for angular aggregate is more than the round aggregate which shows brittleness of concrete is decreased.

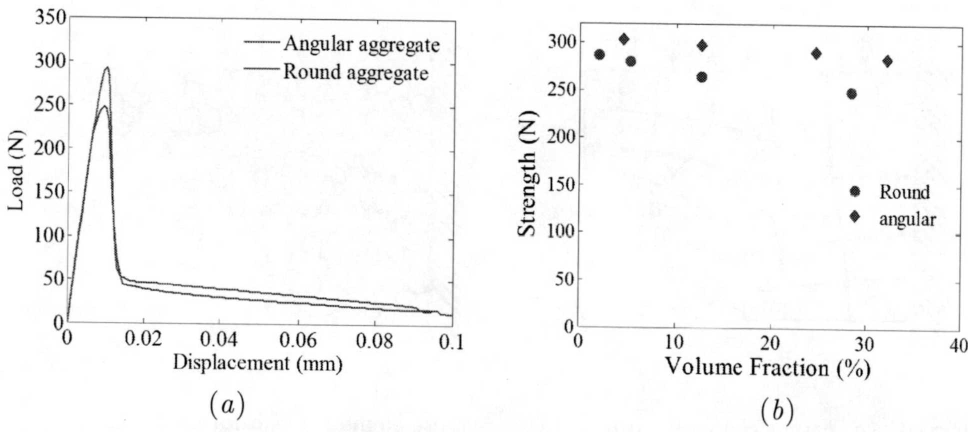

Fig. 3. (*a*) Variation of strength with aggregate volume fraction for round and angular aggregate, (*b*) Load-displacement curve for round and angular aggregate of same aspect ratio

Table 1. Mechanical properties of constituent materials

	Mortar matrix	Aggregate
Young's modulus (GP)	28.7	86.7
Poisson's ratio	0.27	0.3
Fracture energy (Nm/m^2)	44.85	—

2. DISCRETE ELEMENT METHOD FOR BALLISTIC IMPACT PROBLEMS

2.1. *Introduction*

Discrete Element Method (DEM) is a powerful numerical technique used to model solid and particulate media. It is similar to Finite Element Method (FEM) except that the elements can break and can separate away easily. Each discrete elements is a rigid element and is having deformability at the contact points only. The discrete elements are assumed to be connected between each other by normal and shear springs. The deformability of the springs is taken as the deformability of the contact points. The stiffness at the contact points is determined by taking a unit cell of the numerically modeled geometry and applying the uni-axial tension. Two basic laws are used in DEM which is (a) Force displacement law to find the force on each element from the displacement at the contact and (b) Newton's second law of motion to find displacement of discrete elements resulting from the forces acting on it. Typical comparison of finite element mesh verses discrete element mesh is shown

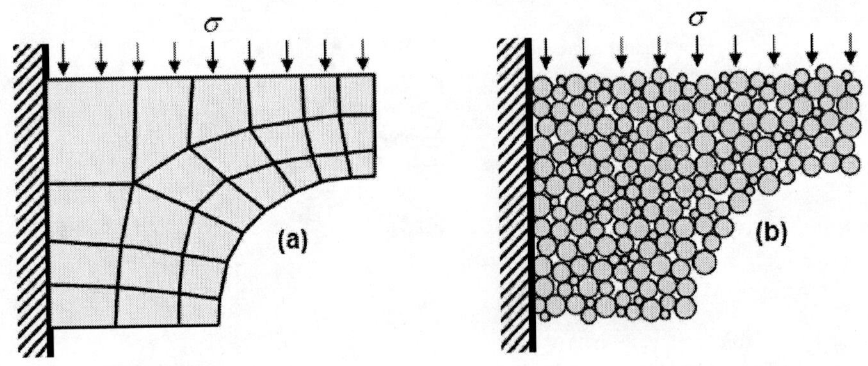

Fig. 4. (*a*) Finite element mesh and (*b*) Discrete element mesh for cantilever beam subjected to uniform distributed load

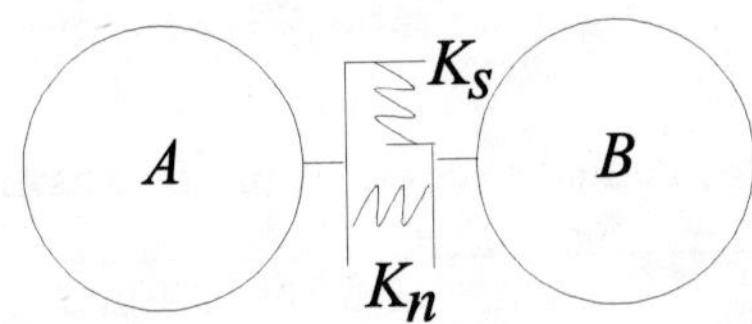

Fig. 5. Discrete element interaction model for bonded contact

in Fig. 4. Discrete element interaction model for bonded contact is shown in Fig. 5.

2.2. *Literature review on DEM*

Cundall et.al [7] introduced discrete element method for modeling granular elements. The DEM is based on use of an explicit numerical scheme in which the interaction of particles is monitored contact by contact and the motion of the particles modeled particle by particle. Cundall compared the DEM results by comparing the force vector plots with corresponding plots obtained from photo elastic analysis. Federico et.al [8] used 2-D circular discrete elements for modeling the solid and particulate materials including concrete beams subject to impact loading etc. Contact failure is based on criteria that if the tensile force in a certain contact exceeds a limit value, the contact brakes and can no longer support tensile forces. Determination of Inter element normal and shear stiffnesses is calculated from a DEM model of a unit cell of material. The details of the stiffness matrix determination are well explained in his Ph.D thesis [9]. Polygonal elements are introduced by Camborde et.al [10] in order to describe zero porosity numerical medium. In his paper, micro concrete behavior is studied using the polygonal elements and the results are matching well with experimental results.

Basic steps in DEM. The basic steps in DEM are as given below.

a) Modeling the geometry of material as an assembly of discrete element particles;

b) Finding the local parameters including inter element stiffness (K_n and K_s);

c) Defining the boundary conditions and loading for the modeled geometry;

d) Defining the criteria for the separation of discrete element particles;

e) Setting up a contact detection technique for each discrete element to find its neighboring elements after separation of discrete element particles took place.

2.3. *Objective and scope of DEM on ballistic impact study*

Ballistic impact is the study which deals with projectile hitting the target and observing its effects in terms of deformation and fragmentation of the target. The need to model ballistic impact condition is for several applications of impact such as nuclear and military applications, where we encounter high velocity projectiles hitting a target. Numerical modeling of the process helps to reduce the expense of doing repeated experiments and also to optimize the geometry and material properties of target and penetrator. The importance of ballistic impact study is to know the failure pattern of target and to find the depth of penetration associated with the impact of a penetrator on a thick target.

The objective is to develop a new 2-D polygonal discrete element to model ballistic impact simulation for finding depth of penetration of projectile in the case of thick target and residual velocity in the case of think target. For knowing the physics of penetration 1-D discrete element method was implemented on the ballistic impact study and validation with experimental results is done before exploring the 2-D Discrete elements application. The scope of present study is to explore the feasibility of DEM as a substitute for costly experimental setups in defense organizations like DRDO, BARC etc.

DEM using non-circular elements. Discrete elements are basically used for non-continuum bodies like sand, rock and concrete etc. It can also be used to model continuous body if the deformation compatibility is satisfied at the element boundaries. In most of the literatures for DEM, discrete elements are circular discs in two dimensional domain and spheres in three dimensional domains due to ease of defining the contact points. But the disadvantage of circular elements are (a) voids will be generated excessively which cause poor numerical modeling of geometry and (b) excessive element rotation can occur in numerical simulation. In order to avoid these disadvantages in our present work, we use bar elements in 1-D domain and polygonal elements in 2-D domain to model zero porosity numerical medium.

Basic formulations for DEM are as shown from Eqn. (4)–(8). Eqn. (4)–(8) is the force displacement law and Eqn. (6)–(8) is law's of motion.

$$\vec{F}_n = k_n \vec{n}, \tag{4}$$

$$\vec{F}_s = k_s \vec{s}_h \tag{5}$$

where $\vec{F}_n$ is the normal force between two discrete elements, k_n is the normal spring stiffness and $\vec{n}$ is the normal displacement, $\vec{F}_s$ is the shear force between two discrete elements, k_s is the shear spring stiffness and $\vec{s}_h$ is the shear displacement between two discrete elements.

Equation of motion is applied to each discrete element. Explicit integration of the Newton's equation is done by a time-centered scheme as shown from equation (6) to (9).

$$\ddot{\vec{x}}^{(t)} = \frac{\sum \vec{F}_i^{(t)}}{m_D E}, \tag{6}$$

$$\dot{\vec{x}}^{(t+\Delta t)} = \dot{\vec{x}}^{(t-\Delta t)} + 2\ddot{\vec{x}}^{(t)} \Delta t, \tag{7}$$

$$\vec{x}^{(t+2\Delta t)} = \vec{x}^{(t-\Delta t)} + 2\dot{\vec{x}}^{(t+\Delta t)} \Delta t, \tag{8}$$

where $\sum \vec{F}_i^{(t)}$ is the sum of all contact force acting on ith discrete element and m_{DE} is the mass of a discrete element, $\ddot{\vec{x}} = \begin{Bmatrix} a_x \\ a_y \end{Bmatrix}$ is the acceleration, $\dot{\vec{x}} = \begin{Bmatrix} v_x \\ v_y \end{Bmatrix}$ is the velocity, and $\vec{x} = \begin{Bmatrix} d_x \\ d_y \end{Bmatrix}$ is the displacement of a discrete element.

2.4. *Problems solved in 1-D domain using DEM*

DEM method based on equation of motion is used to solve different problems in one dimensional domain. Initially cantilever bar, made up of steel, is subject to cyclic velocity. Its behavior is observed numerically using DEM and results are compared with analytical solution. After comparing the results, bullet penetration on thick target is studied in one dimensional domain. In the 1-D DEM model the elements are allowed to deform only in the direction of velocity of impact. The portion of target which is in line with penetrator alone is modeled in 1-D domain, since that portion alone is subjected to high compression. It is modeled in 1-D DEM as a rod impacting another stationary rod as shown in Fig. 6 using one dimensional discrete

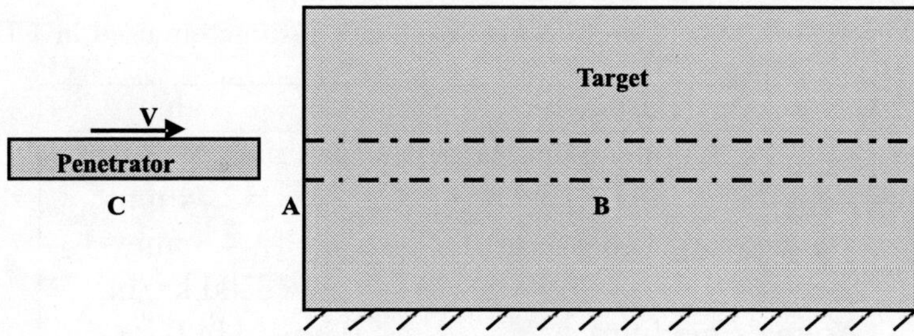

Fig. 6. Schematic diagram of penetrator and target arrangement for ballistic impact

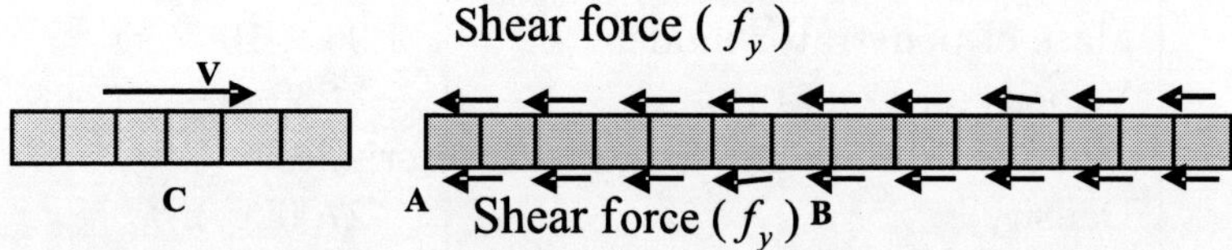

Fig. 7. One dimensional discrete element model for impact analysis incorporating shear forces on the boundaries of discrete element of target

bar elements. The depth of penetration is possible to obtain using DEM. The depth of penetration of penetrator is calculated from the total distance travelled by the first discrete element of penetrator at the end of DEM simulation. The effect of boundary is taken care by introducing shear force in the target elements which is shown in Fig. 7. The shear force on the target element is given by

$$f_y = \frac{\pi}{\sqrt{3}} dh\sigma_{\text{ot}} \tag{9}$$

where d and h are the diameter of penetrator and height of target discrete element respectively, σ_{ot} is the flow stress [10] of the target element which is given in Table 2. Fig. 8 shows the comparison of 1-D DEM results with analytical, experimental and FDM results. Fig. 9 is the variation of depth of penetration for different impact velocities of bullet. As expected when the impact velocity increases the depth of penetration also increases with respect to time.

2.5. *Two dimensional discrete element simulations*

From 1-D discrete element simulations, the physics of penetration is able to obtain like depth of penetration, contact force etc. Need to study the 2-D discrete element simulations is to simulate the multiple fracture propagation . More accurate result to experiment is possible to obtain in 2-D domain. In 1-D domain only normal

Table 2. Penetrator, target, and discrete elements parameters used in 1-D modeling

Parameter	Value
Copper Penetrator characteristics	
Length	12.74 mm
Diameter of Penetrator	4.8 mm
Density	8500 kg/m^3
Young's modulus	114 GPa
Height of a discrete element	6.37×10^{-2} mm
Number of discrete element used	200
Flow stress	443 MPa
Mass of penetrator model	1.96×10^{-3} kg
Velocity of Penetrator	230 m/s
Aluminium Target characteristics	
Density	2700kg/m^3
Young's modulus	70 GPa
Thickness of target	25 mm
Diameter of target	50 mm
Height of a discrete element	1.25×10^{-1} mm
Number of discrete element used	200
Flow stress	776 MPa
Yield Stress	20 MPa
Mass of target model	1.22×10^{-3} kg
Discrete element model parameters	
Time step for simulation	2.744×10^{-9} s
Total number of time steps	700

stiffness (K_N) exists, where in 2-D domain both normal stiffness (K_N) and shear stiffness (K_s) has to be taken care in the discrete element analysis. Mostly the 2-D discrete elements are circular disc due to simplicity in defining the normal at the contact points. But as stated in 2.3 the non porous medium can be modeled much better by non circular elements such as polygonal elements. So in the present study for 2-D simulation, polygonal elements are chosen for the DE analysis. As an initial step rhombus elements are used to model a cantilever subjected to sinusoidal load. The modeled cantilever beam consists of unit cells of rhombus elements. Similar to circular discrete elements, the inter element stiffness for the rhombus elements is calculated by applying uni-axial tension on the unit cell. The inter element connection for the unit cell, which is made up of four rhombus discrete elements, is as shown in Fig. 10. Here, the DEM results are obtained in terms of displacements as well as forces for each discrete elements of the cantilever beam.

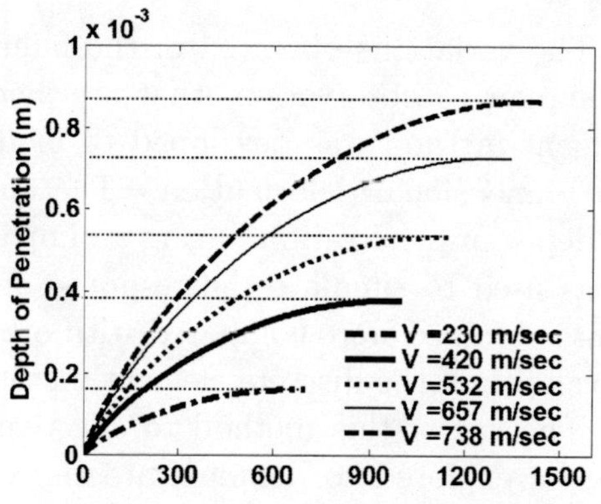

Fig. 8. Comparison of 1-D Analytical, DEM, and FDM results with experimental results

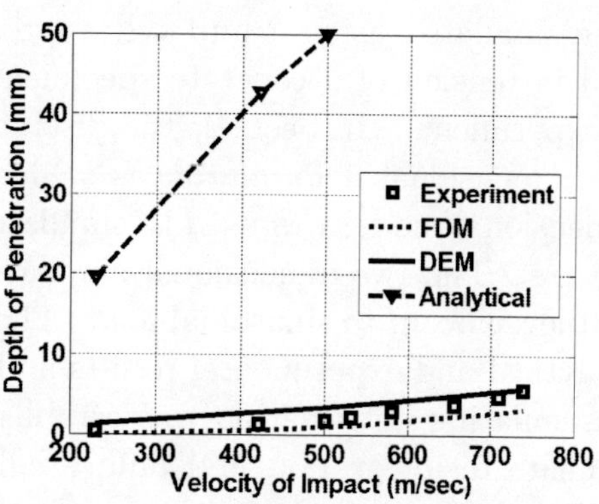

Fig. 9. Depth of penetration calculated using 1-D DEM for different impact velocities

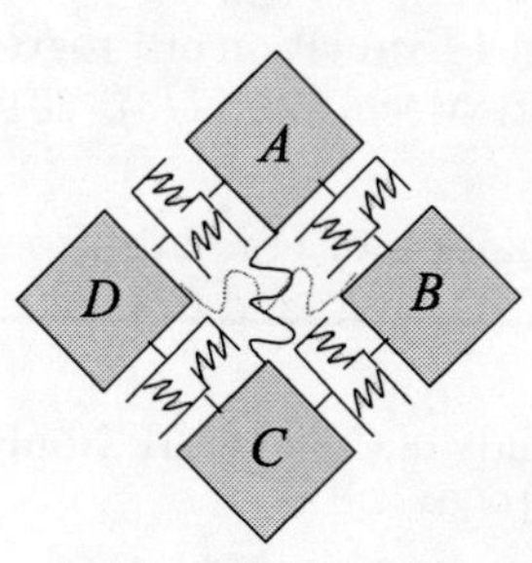

Fig. 10. Unit Cell using rhombus elements

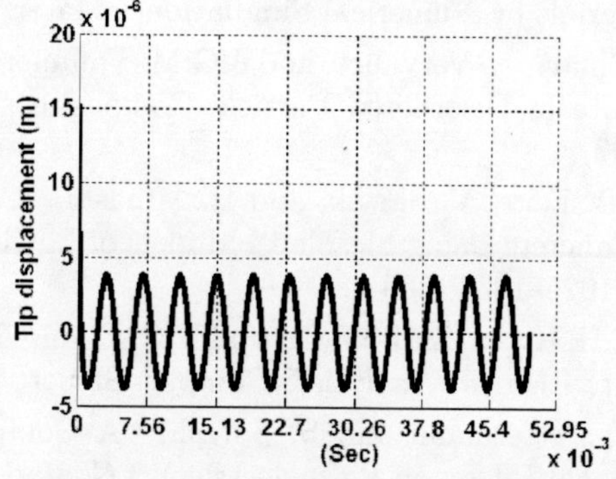

Fig. 11. Variation of displacement at the free end of cantilever beam using DEM

For the cantilever beam subjected to sinusoidal load, the analytical as well as DEM results using rhombus elements show similar trends for the tip displacement. The variation of vertical displacement at the free end of the cantilever beam calculated using DEM formulations in MATLAB 7.0 platform is as shown in Fig. 11. It is observed that tip displacement is stabilized to constant amplitude with progress in time since there was no damping applied between discrete elements.

2.6. *Conclusion*

In this work, two simple examples of simulation of deformation and failure processes using numerical techniques, is presented. In Sect. 1, the failure processes in concrete were simulated using a unit cell approach. The variations observed of the failure load in tension of a concrete specimen, are similar to observations that are made in experiments. In Sect. 2, the discrete element method was developed to model one dimensional deformation as well as two dimensional deformation. The one-dimensional model was used to simulate the depth of penetration during an impact process. The two-dimensional simulation was used to simulate the response of a cantilever beam to sinusoidal load. The results obtained were compared with other analytical and experimental results and the validity of the discrete element method in simulating deformation was established. The use of this method to effectively simulate disintegration and failure will now be explored to demonstrate the full utility of this method.

REFERENCES

1. H. Sadouki and F.H. Wittmann, "On the Analysis of the Failure Process in Composite Materials by Numerical Simulation," Mater. Sci. Engng **104**, 9–20 (1988).
2. B. Chiaia, A. Vervuurt, and J.G.M. Vanmier, "Lattice Model Evaluation of Progressive Failure in Disordered Particle Composites," Engng Fract. Mech. **57** (2/3), 301–318 (1997).
3. J.P.B. Leite, V. Slowik, and H. Mihashi, "Computer Simulation of Fracture Processes of Concrete Using Mesolevel Models of Lattice Structures," Cement and Concrete Res. **34**, 1025–1033 (2004).
4. A.K.H. Kwan, Z.M. Wang, and H.C. Chan, "Mesoscopic Study of Concrete II: Nonlinear Finite Element Analysis," Comput. Struct. **70**, 545–556 (1999).
5. H.E. Pettermann and S. Suresh, "A Comprehensive Unit Cell Model: A Study of Coupled Effects in Piezoelectric 1-3 Composites," Int. J. Solids Struct. **37**, 5447–5464 (2000).
6. V. Kouznetsova, W.A.M. Brekelman, and F.P.T. Baaijens, "An Approach to Micro-Macro Modeling of Heterogeneous Materials," Computat. Mech. **27**, 37–48 (2001).
7. P.A. Cundall and D.L. Strack, "A Discrete Numerical Model for Granular Assemblies," Geotech. **29** (1), 47–65 (1979).
8. F.A. Tavarez and M.E. Plesha, "Discrete Element Method for Modeling Solid and Particulate Materials," Int. J. Numer. Meth. Engng **70**, 379–404 (2007).
9. F.A. Tavarez, "Discrete Element Method for Modeling Solid and Particulate Materials," Doctoral Thesis (University of Wisconsin-Madison, 2005).
10. F. Camborde, C. Mariotti, and F.V. Donze, "Numerical Study of Rock and Concrete Behaviour by Discrete Element Modeling," Comput. Geotech. **27**, 225–247 (2000).

LIGHT WEIGHT, HIGH STRENGTH MATERIAL AND COMPONENT DEVELOPMENT AT NPL, USING NOVEL PROCESSING TECHNIQUES

A.K. Gupta[1]

ABSTRACT

The ever increasing demand for light materials possessing high strength, high modulus with low density have catalyzed the development of number of high-strength aluminum (Al) & magnesium (Mg) alloys and a variety of composite materials. These are potential materials for aerospace and automobile applications. The enhancement of properties in these materials is primarily being met by the development of newer alloys and other hi-tech materials in conjunction with the use of novel primary and secondary processing techniques. These techniques help to engineer the microstructure of the material for achieving enhanced properties.

At the National Physical Laboratory, the Metals & Alloys Group, has been engaged in the process and technology development of high performance strategic structural materials. The thrust is on developing materials, which are light in weight, possess high strength and modulus and thus find application as potential materials in aerospace, automobile and other general engineering industries. The emphasis of research is on the primary and secondary processing of different non-ferrous metals, alloys, composites and other hi-tech materials.

These materials are being synthesized using variety of techniques, like, Liquid Metallurgy, Powder Metallurgy and Spray Atomization & Deposition technique, which has been developed in-house and are being secondary processed using hot extrusion and cold/warm forging. Number of components have been developed for aerospace and automobile applications in recent years, which include, motor body component (reduced scale) for Milan missile, aerospace grade Al- alloy rivet wires for HAL, Al-Li light weight inserts for ISRO, aerospace grade MMC circular tubes as compressive strut members, Mg-alloy square tubes for VSSC, oval shaped tube, as skid landing gear, for Advanced Light Helicopter for HAL, etc.

Apart from these, R&D efforts are also underway to develop bulk nanostructured metallic materials employing ECAP technique, nanocomposites using high energy powder ball

[1]Division of Engineering Materials, National Physical Laboratory, Dr. K.S. Krishnan Road, New Delhi — 110012, India

milling at cryogenic temperatures followed by high pressure/rapid sintering consolidation and functionally gradient MMCs using centrifugal casting, Recently work is being initiated on the development of Al & Mg sheets employing twin roll casting technique. This paper would summarize the development efforts being carried out at NPL in developing these materials and components.

INTRODUCTION

High strength, high modulus and low density are the prime material property requirements for most of the components for aerospace, automobile and also for some general engineering industrial applications. This has led to the development of number of high-strength Al & Mg-alloys and a variety of composite materials, including metal matrix composites. These property requirements are met by development and engineering these newer materials with sustained innovations in process technologies. The ability to design and manufacture components, with optimum weight reduction, effectively requires a understanding of material development using primary and secondary processing and structure-property correlations.

The requirement of light-weight and high-strength materials in the automobile industry has been driven by stringent standards worldwide for improved fuel economy and low exhaust emissions. Al and Mg are two most commonly used light engineering alloys and have basically similar mechanical properties and can be used for similar applications. The advantage of Mg in comparison to Al is that it allows a weight saving of the order of 25% which makes them interesting materials for automobile applications.

Over the last few decades, the light weight alloys have undergone rapid advancements. The discovery of precipitation hardening has led to higher strength levels achievable in Al-alloys. Constant efforts are being made ever since to develop components using Al-alloys, such as, Al-Cu, Al-Mg, Al-Li, etc., and other composites, especially for aerospace and automobile applications

Mg-alloys are currently finding increasing usage in low-weight structural applications for automobile and aerospace industries, due to their low density, high specific strength and modulus, good vibration damping and excellent machinability, leading to potential weight savings in order to meet the increasing demands for lower emissions and more economic use of fuel. However, several problems associated with this material are mainly its poor strength, low room temperature ductility and poor corrosion resistance, etc. The problems of strength and corrosion, associated with these alloys, have more or less been addressed with the recent advances in alloy chemistry of these alloys. However, owing to its low ductility (due to its hcp structure) the problem of deformation-forming still remains major bottle-neck for its usage in automobile and other applications. Therefore, there has been a considerable interest in automobile industries worldwide to develop Mg-alloys that possess improved room temperature ductility.

PROCESS & TECHNOLOGY DEVELOPMENT AT NPL

The Metals & Alloys Group at NPL has been engaged in the process and technology development of high performance, strategic materials & components mainly for aerospace and automobile applications. The emphasis of research is on the primary & secondary processing of different non-ferrous metals, alloys, composites and other hi-tech materials, such as, Al-based alloys & composites including Al-Si and different grades of Mg-alloys, etc. The main two areas of development at NPL are (i) Materials development and (ii) Component development.

The group has set-up primary processing & modern metal forming facilities including a well equipped characterization laboratory. The primary material processing techniques include stir-casting, centrifugal casting, powder metallurgy including high energy cryogenic ball milling and in-house developed spray-forming unit. These processes are used to synthesize a host of Hi-tech materials for strategic applications, like, hypereutectic Al-Si alloy, Mg-alloys, MMCs, Nanocomposites, Functionally Gradient Materials, etc.

These materials are then secondary processed using hot extrusion, cold forging (employing 500 ton vertical hydraulic multi-purpose press) and other forming techniques to develop components for different applications. The group has been working on the design/development of toolings for extrusion and optimizing the hot extrusion parameters of various MMCs with different matrix and reinforcement materials, which include, Al-alloy (2124/6061) reinforced with SiCp, chopped carbon fibre, ZrO_2 flyash, etc.

At NPL, the thrust is presently on developing Al and Mg alloys with refined microstructure and enhanced properties using both, novel primary processing and secondary processing techniques. Significant amount of work has been carried out on the development of a variety of Al and Mg-alloys by primary processing using spray-forming technique. The extrusion processing of different Al and Mg-based alloys has also been carried out to develop components in form of rods, tubes and sections. Further, the light weight & high strength structural tubular components of Al-based metal matrix composites (MMCs), were also developed using hot-extrusion technique.

SYNTHESIS OF Mg & Al ALLOYS USING SPRAY-FORMING TECHNIQUE

Spray-forming, which is a rapid solidification process, leads to increased solid solubility, sharp reduction in grain size and micro-segregation effects resulting in refined and equiaxed microstructure with substantial microstructural homogeneity in the spray-formed deposit. The inert conditions required for spray-forming minimizes oxidation and other deleterious reactions especially for reactive materials like Mg-alloys. All of these lead to enhancement of mechanical properties of the spray-formed alloys. Thus, the attractive combination of microstructure and mechanical properties, achievable through rapid solidification, have prompted the use of spray-forming process for synthesis of novel alloys with improved properties.

Mg-ALLOYS

Different Mg-alloys, such as, AZ31, AZ91C, EZ33, WZ155 and WZ103 have been successfully spray-formed at NPL with improved microstructure & properties. These alloys have been spray-formed by depositing the atomized melt droplets on a rotating substrate, using argon as the atomizing gas and optimizing the processing parameters.

The spray-formed deposits were characterized using optical microscope, scanning electron microscope/energy dispersive spectrometer and X-ray diffraction. The optical micrograph of the cast mother AZ31 alloy, used for spray-forming, is shown in Fig. 1 *a*. This microstructure shows large grains of magnesium ($\sim$ 200–800 μm) surrounded by significantly large worm-shaped $Mg_{17}Al_{12}$ intermetallic aggregate ($\sim$ 50–250 μm), as identified by EDS analysis, in the intergranular region. A typical spray-formed deposit of Mg-alloys AZ31 is shown in Fig 1 *b*.

Fig. 2 shows optical micrographs of spray-formed deposit from different regions of the spary formed deposit. Fig. 2 *a* shows the optical micrograph of spray-formed alloy from the interior of spray-formed deposit (region *a*, which shows fine equiaxed grains of about 20–50 μm with $Mg_{17}Al_{12}$ intermetallic particles of size 6–9 μm uniformly distributed mainly along the grain boundaries.

It is apparent from Fig. 1 *a* and Fig. 2 *a* that the microstructure of both, the cast and spray-formed alloy, consists of $Mg_{17}Al_{12}$ intermetallic compound, distributed in a magnesium matrix, but the microstructure is greatly refined on spray-forming. The presence of $Mg_{17}Al_{12}$ intermetallic phase in the cast and spray-formed alloys has also been confirmed by X-ray diffraction studies (Fig. 3). The microstructure is found to be uniform and equiaxed in the entire central core of the deposit (Fig. 2 *a–c*).

The porosity was observed to be about 2.5–3.5% in the central core of the spray-formed deposit but increases near the peripheral region (7–8%) as shown in

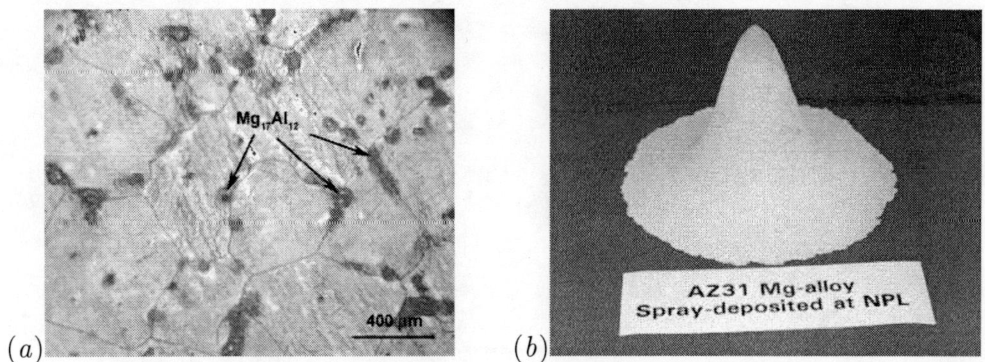

Fig. 1. AZ31 Mg-alloys (*a*) microstructure of as-cast alloy, (*b*) bell-shaped spray-formed deposit

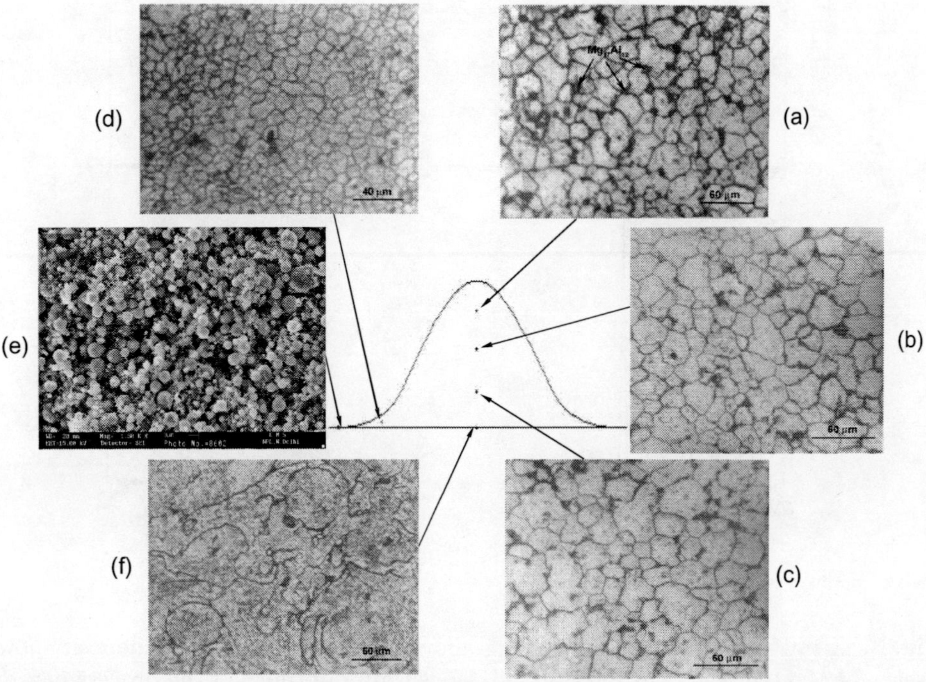

Fig. 2. Optical micrographs of spray-formed deposit from (*a*) central region *a*, (*b*) central region *b*, (*c*) central region *c*, (*d*) peripheral region *d*, (*e*) extreme peripheral region *e* (unetched), (*f*) bottom exterior region *f*

Fig. 2 *d* The porosity becomes very high in the extreme peripheral region ($\sim 25\%$) of the spray-formed deposit, where the microstructure exhibited loosely bound pre-

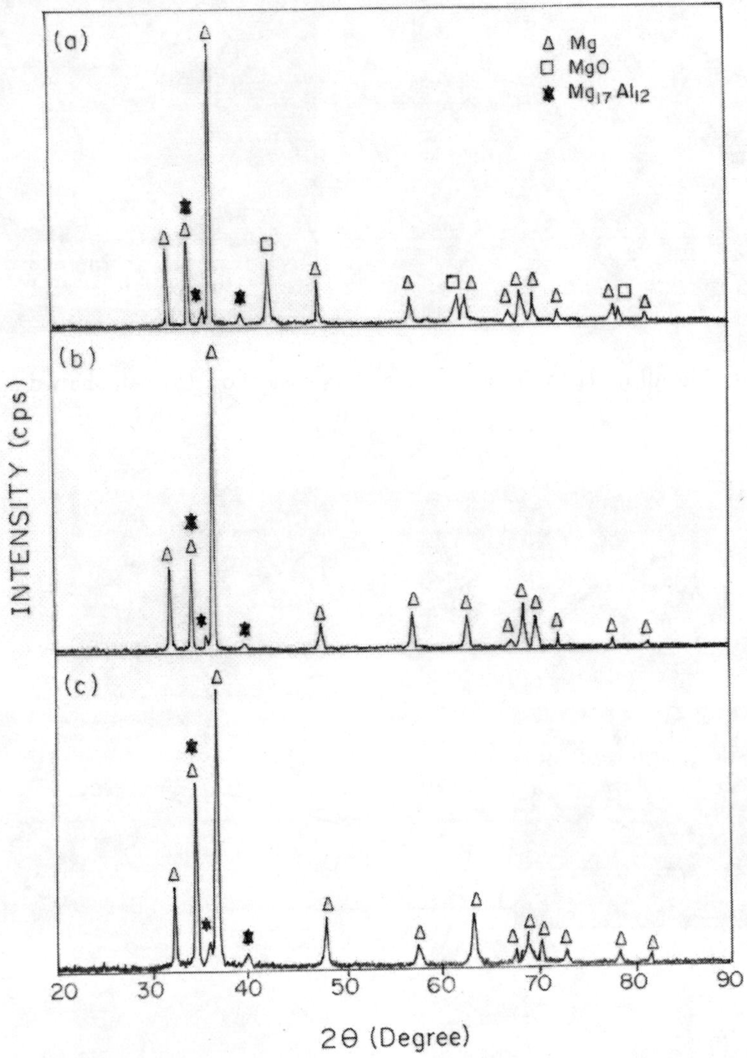

Fig. 3. X-ray diffraction patterns of Mg-alloy AZ31 (*a*) spray-formed deposit showing presence of MgO–phase, (*b*) spray-formed deposit after optimizing the process parameters showing absence of MgO–phase, (*c*) cast mother alloy

solidified particles of 1–4 μm (Fig. 2 *d*).

Fig. 3 shows X-ray diffraction patterns of spray-formed and cast AZ31 alloy, indicating the phases present. Initial spray-forming experiments indicated a presence of MgO phase in the spray-formed deposits as is evident from X-ray diffraction patterns shown in Fig. 3 *a*. It is well known that magnesium is very prone to oxidation especially at high temperatures, therefore, special precautions had to be taken in order to avoid oxidation in the spray-formed deposit. The absence of the MgO peaks in X-ray diffraction pattern (Fig. 3 *b*) clearly indicates that the precautions taken were effective in avoiding oxidation in spray-formed Mg alloys. The X-ray diffraction patterns in Fig. 3 also suggest the presence of intermetallic compound $Mg_{17}Al_{12}$ phase in both, the cast as well as spray-formed alloys.

The mechanical properties of Mg-alloys AZ31 spray-formed deposits exhibited higher strength, ductility and microhardness as compared to their counterpart cast master alloy. Typically for AZ31 alloys, while the cast alloy posses UTS, elongation and elastic modulus of 178 MPa, 7.8% and 43 GPa, respectively, the spray-formed alloy exhibited these properties as 228 MPa, 13% and 44 GPa, respectively. These mechanical properties of the spray-formed alloys are expected to increase significantly after extrusion due to reduction in porosity.

Al-ALLOYS

Al-Si is one of the important alloys used currently for different wear-based automobile applications, such as, engine block/head, pistons, gear boxes, etc. due to its high wear resistance, low CTE & good mechanical properties. Presently this alloy is made by LM based routes, which have some limitations such as, (*a*) leads to large & coarse primary Si particles, (*b*) non-uniform microstructure, (*c*) synthesis of hypereutectic Al-Si alloys with fine grain size not possible. Also their ductility is limited which is caused by the cast microstructure composed of plate-like Si aggregates embedded in Al-matrix. All these problems can be overcome by spray-forming which breaks down the plate-like Si structure and a fine microstructure is evolved having a particulate-type morphology, thus improving its properties. Spray-forming being a rapid solidification process, also increases the solid solubility limit in these Al-Si alloys, which is one of the important parameter used to engineer the wear/thermal properties of these alloys. Hypereutectic Al-Si, which can be synthesized using spray-forming employing rapid solidification have increased applications not only in automobile industry for wear-based applications but more recently as thermal management materials for electronic packaging.

Both eutectic and hypereutectic Al-Si alloys have been spray-formed at NPL by optimizing the process parameters. Fig. 4 *a* and *b* show the microstructures of as-cast and spray-formed hypereutectic Al-30Si, respectively. The microstructure of spray formed alloy exhibited a fine globular particulate-type morphology (1–3 μm) of Si distributed uniformly in the ductile Al-matrix in contrast to much larger acicular primary phase with plate-like microstructure (200–600 μm) observed in the as-cast alloys. This fine microstructure observed in spray-formed alloys is responsible for its improved properties.

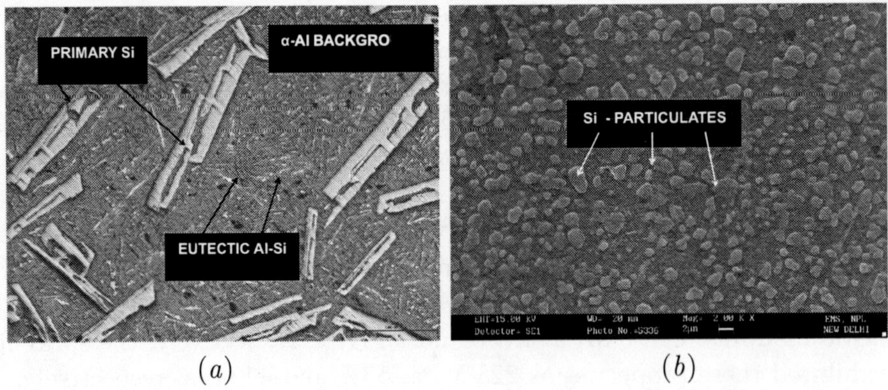

(*a*) (*b*)

Fig. 4. (*a*) As-cast hypereutectic Al-30Si, (*b*) Spray formed hypereutectic Al-30Si

Mg-BASED EXTRUDED COMPONENTS

The tubular components are mostly fabricated employing extrusion process to produce tubes in desired shapes & sizes followed by other processing methods such bending or hydroforming, etc to produce components in useable shapes. However, these processing methods require high ductility in order to form them at room temperature. In contrast to Al-alloys, conventional Mg-alloys possess limited room temperature ductility.

Although an incremental ductility enhancement in Mg-alloys has been achieved by several methods, however, efforts are underway worldwide to achieve substantial increase in ductility in Mg alloys, employing several methods. At NPL, certain RE-based Mg-alloys with refined microstructure have been developed using hot extrusion under optimized extrusion parameters. The Mg-alloys were extruded in the form of rods of different shapes, such as, square, rectangular and circular cross-section, using different process parameters.

EXTRUDED Mg-ALLOY RODS

Several Mg-alloys, such as, AM30 (Mg-3Al-0.5Mn), AM50 (Mg-5Al-0.5Mn), Mg-RE, etc and pure Mg were extruded in the form of rods using different process parameters. The Mg-RE extruded alloys indicated an increase in grain size with increasing extrusion temperature and the microstructure and mechanical properties were found to be optimum at the temperature of 400°C.

In order to study the effect of ER on the microstructure, Mg-RE alloys were extruded in the form of circular rods at the optimized temperature of 400°C and at three different ERs (extrusion ratios) of 9:1, 25:1 and 36:1 and the results are shown

(a) G.S.= $38 \pm 17\ \mu$m (b) G.S.= $59 \pm 25\ \mu$m (c) G.S.= $78 \pm 33\ \mu$m

Fig. 5. Microstructures of Mg-RE alloy extruded at (a) ER=9:1, (b) ER=25:1, and (c) ER=36:1

in Fig. 5. This figure suggests that the grain size increases with increasing ER which may be due to the fact that increasing extrusion ratio results in increased effective billet temperature which leads to coarser grain size upon dynamic recrystallization.

Under these optimized process conditions of temperature and ER, different Mg-alloys were extruded and the comparison of their tensile strength is presented in Fig. 6. This figure suggests that the alloy Mg-Al-RE shows the highest strength.

The fractured tensile specimens were examined under SEM and the tensile fractographs of pure Mg and Mg-RE alloy are shown as Fig. 7 *a* and *b*, respectively. In the case of pure Mg (Fig 7 *a*), coarse grained fractured surface is observed, which is indicative of poor ductility in this case. The Mg-RE alloy (Fig. 7 *b*), shows the fractured surface to be fine grained with small sized dimples, which is indicative of ductile fracture. The primary reason for such a contrast in fracture behavior has been attributed to an improvement in texture in extruded Mg-RE alloys, which is currently being investigated.

EFFECT OF DIE SHAPE ON EXTRUDED MG-ALLOY TUBES

Mg-alloy AM 30 was extruded at an optimized temperature of 400°C in the form of circular tubes with 3 mm wall thickness and 30 mm OD, using porthole and conical dies. Figure 8 *a*–*d* shows optical micrographs of tube, extruded using the porthole die and conical dies, in both, transverse and longitudinal directions. The tubes extruded using porthole die, exhibit equiaxed grain structure with an average grain size $\sim$ 16 μm, although a few elongated grains are also observed in longitudinal direction. However, in the case of tube extruded using conical die, the average grain size is coarser ($\sim$ 30 μm). The finer grains in the case of porthole die compared to conical die product are probably due to higher amount of plastic deformation

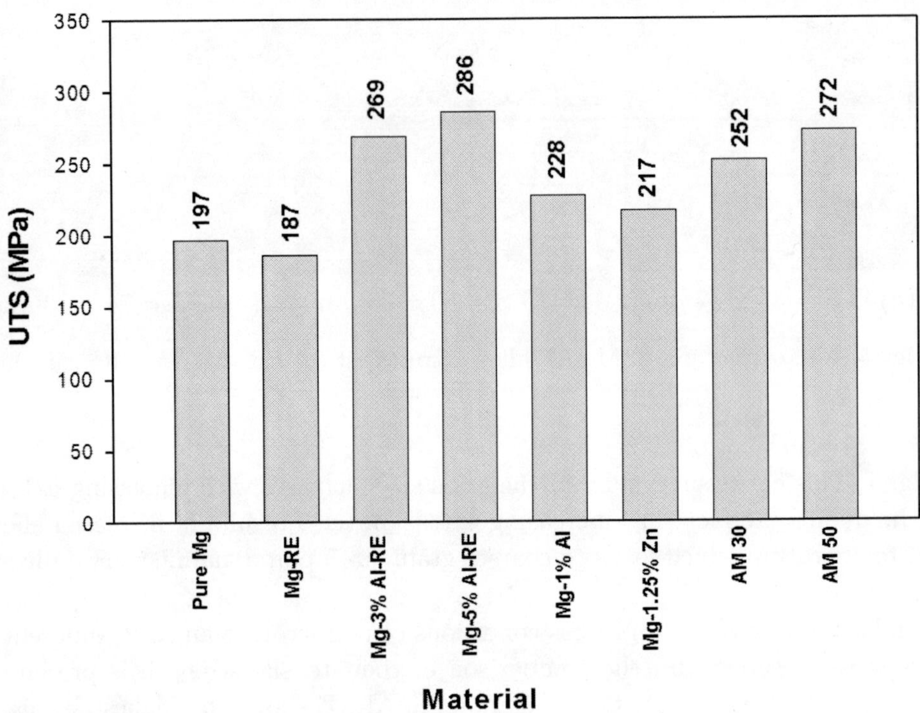

Fig. 6. Tensile strength of different Mg alloys extruded under optimized process conditions

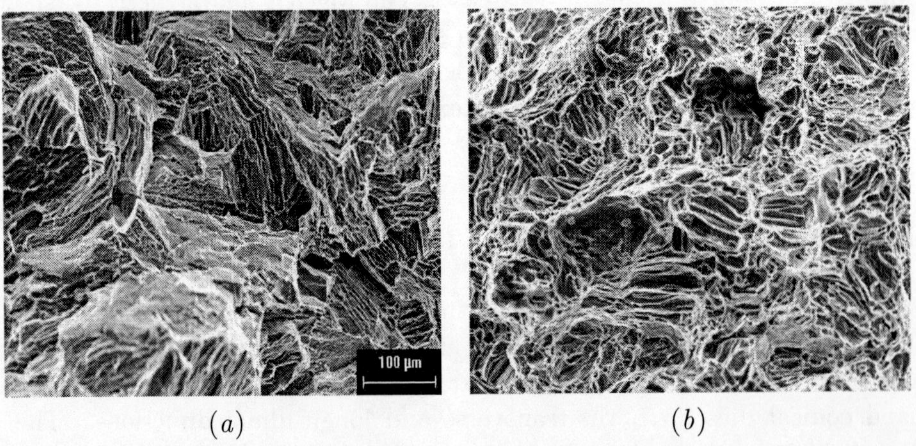

(*a*) (*b*)

Fig. 7. SEM micrographs of fractured surfaces of (*a*) pure Mg showing limited ductility and (*b*) Mg-RE showing high ductility

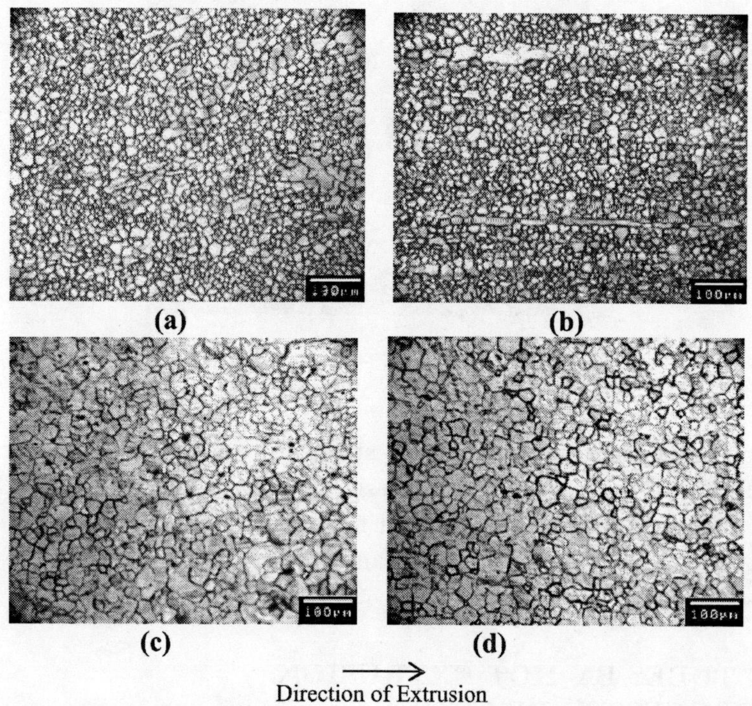

Fig. 8. Microstructure of extruded tube - using porthole and conical die, (*a*) porthole-transverse section, (*b*) porthole-longitudinal section, (*c*) conical-transverse section and (*d*) conical-longitudinal

imparted during the extrusion using porthole die.

EXTRUDED MAGNESIUM SQUARE TUBE

Square tubes of Mg-alloy ZK30 were developed employing hot-extrusion technique for obtaining defect-free tubes with improved properties. These tubes (Fig. 9), having dimensions of 30 mm OD, 3 mm wall thickness, and 500 mm length, exhibited improved mechanical properties of UTS — 280 MPa, Yield Strength — 205 MPa and Elongation of 10%.

Al-BASED MMCS AND COMPONENTS

Al-based MMCs have been finding wide applications in automotive & aerospace applications on account of their light weight, high specific strength & stiffness and amenable for production by a number of processing techniques. At NPL, different

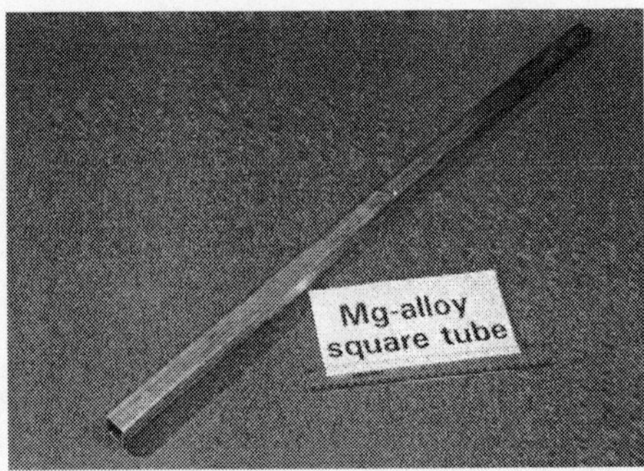

Fig. 9. ZK30 square tube developed at NPL

primary processing techniques, such as, stir-casting technique, powder metallurgy, spray-forming have been used to synthesize a variety of MMCs, which have been secondary processed using hot-extrusion to produce different components. Several MMC components have been developed and a typical component of MMC rod/tube will be described.

Al-MMC TUBES BY HOT EXTRUSION OF PM PROCESSED BILLETS

Al-alloy/SiCp MMCs were synthesized using the powder metallurgy (PM) technique, which involved blending of SiC particulates (particle sizes: 1.9, 3.0, and 14 μm) with 2124 Al alloy powder (average particle size 94 μm) in varying volume fractions (10, 15, and 20%). These blended powders were degassed and subsequently CIPed followed by hot pressing under vacuum. The hot pressed billets were further hot extruded in the form of circular rods on a 500-ton vertical hydraulic press under optimized process conditions. Fig. 10 *a* shows photograph of these circular rods with different ERs of 9:1, 16:1, and 36:1. Typical microstructure of the MMC rods is shown in Fig. 10 *b*, which represents uniform distribution of SiCp reinforcement in the matrix material.

The above development work was further extended to the fabrication of MMC tubes for possible application as drive shaft in automobile. The Al-alloy/SiCp MMC tubes have been shown in Figure 11 and their mechanical properties are shown in Table 1.

DEVELOPMENT OF Al-Li/SiCp MMCs

Al5Li/SiCp MMCs were developed with an aim to explore the possibility for their application as novel material with low density and high stiffness, suitable for aerospace

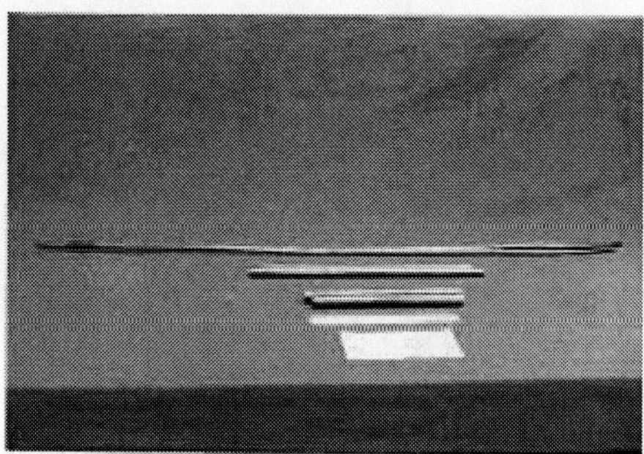

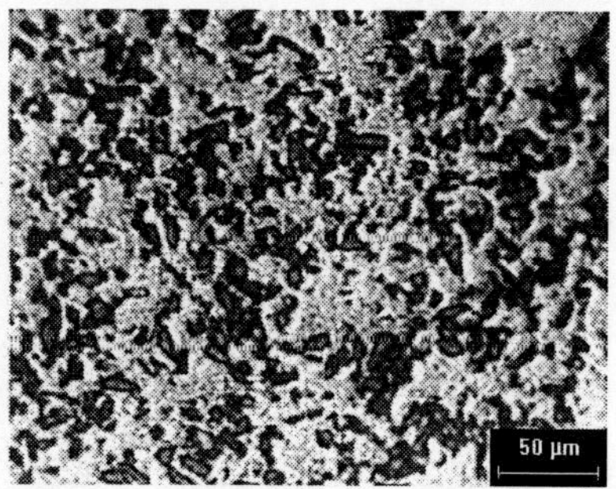

Fig. 10. (*a*) Al-SiCp MMC rods at ERs of 9:1, 16:1, and 36:1, and (*b*) optical micrograph showing uniform distribution of SiC particulates in metallic matrix

Fig. 11. 2124 Al/SiCp MMC circular tubes

Table 1. Mechanical properties of MMC tubes 2124 Al+15% SiCp (1.9 μm)

Property	Experimental Value
YS (MPa)	380
UTS (MPa)	480
Elongation (%)	4.0
E (GPa)	95
Buckling properties (L/D ratio: 2.5)	
Compressive YS (MPa)	261
Ultimate buckling stress (MPa)	453

and automobile applications. Although, Li when alloyed with Al increases the elastic modulus and yield strength considerably along with decreasing the density, however,

Table 2. Mechanical properties Al-5Li/SiCp MMC rods

Material	UTS (MPa)	Elong. (%)	E (GPa)	ρ (g/cm^3)
Al-5Li	410	1.3	82	2.34
Al-5Li/13% SiC	405	0.5	107	2.46
Al-5Li/17% SiC	420	0.4	115	2.50

Li has limited solid solubility of 2.5% in Al. In this work, Al-5%Li powder was synthesized by rapid solidification technique to increase solid solubility and Al5Li/SiCp MMCs were synthesized using powder metallurgy technique with varying amount of ceramic reinforcement SiCp. Finally, the MMC billets were hot-extruded on a 500-ton hydraulic press at a temperature of 400°C and ER of 36:1. Mechanical properties of the extruded composites are shown in Table 2. It may be observed that the elastic modulus was enhanced substantially to 115 GPa and the density of 2.5 g/cm^3 was obtained, although at the expense of reduction in ductility.

SYNTHESIS OF Al-30Si NANOCOMPOSITES

Al-30Si nanocomposites have been synthesized using high energy mechanical ball milling followed by high temperature consolidation of these nano-powders at high pressures to obtain nanocomposites with superior mechanical and thermal properties.

Al and Si elemental powders (both $\sim$ 100 μm) were ball milled in stoichiometric proportions for extended time periods at high speeds under controlled Argon gas atmosphere in a high-energy ball mill. The resulting powders were characterized using XRD, SEM and TEM. It was found that with increasing milling time the crystallite sizes of both Al and Si decrease sharply initially and finally become nearly constant at high milling periods. However, the rms lattice strains in these powders increased with increasing milling time. The DSC results indicate that melting point of Al decreases with decrease in its crystallite size.

These nanopowders were cold compacted and the green nanocomposites were then consolidated using rapid sintering at high pressure (4–5 GPa) employing a belt-type high pressure assembly on a 1000-ton hydraulic press. The high pressure sintering temperature ($\sim$ 500°C) with rapid heating rate (300–400°C/min) were employed so that the grain growth in these nanocomposites was avoided, which was confirmed by XRD analysis. Results indicate that the Vickers microhardness of nanocomposite increases more than three folds as compared to similar microcomposite consolidated at identical conditions. Also the consolidated nanocomposite show much higher compressive strength as compared to their microcomposite counterpart.

FUNCTIONALLY GRADIENT MMCs

Functionally Gradient Materials (FGM) are a new class of engineering materials in which the composition and/or the microstructure varies in one specific direction. Such a variation leads to corresponding change in the material's properties whose gradient can be tailored depending on the processing conditions to suit a particular application.

In centrifugal casting technique, the MMC feedstock melt is introduced in a rotating mould which is subsequently cooled, resulting in a compositional gradient in the cylindrical cast MMC product produced mainly from the difference in the centrifugal force produced by the difference in density between the molten metal and ceramic particles. A centrifugal casting unit, consisting of a meting furnace and rotating mould enclosed in a holding furnace, with automatic melt delivery system, has been installed at NPL. This work is underway to synthesize Functionally Gradient MMC cylindrical tubular products with a gradient of the ceramic reinforcement along its wall thickness.

BULK NANOSTRUCTURED METALLIC MATERIALS EMPLOYING ECAP TECHNIQUE

Equal Channel Angular Pressing (ECAP) is a relatively new technique among several processes available for the synthesis of bulk nanomaterials. The importance of this technique lies in imposition of very large strains without the introduction of any concomitant changes in the cross-sectional dimensions of a billet as compared to commercial industrial metal-forming processes (rolling, extrusion, etc), which involve considerable reduction in area for the similar strain values. ECAP is carried out by pressing a well-lubricated billet in multiple passes through a die having two channels of equal cross section, intersecting usually at an angle of 90° to 120°. Severe Plastic Deformation is incorporated by simple shear in a thin layer at the crossing plane of the two channels resulting in the refinement of microstructure drastically. At NPL, some work has been carried out using this process for Al and Mg-alloys by designing a few die/punch assemblies with different included angles. Substantial reduction in grain size has been achieved in Al & Mg-alloys employing several passes. Further the improvement in strength and ductility, by refining the microstructure in these alloys using this technique, is currently under progress.

Al & Mg ALLOY SHEETS EMPLOYING TWIN ROLL CASTING TECHNIQUE

Twin Roll Casting (TRC) is the state-of-the-art method for producing sheets of Al,

Mg other metallic alloys directly from melt. This process enables metallic sheets to be produced in thicknesses of less than 5–6 mm, thus superseding the conventional process, in which alloy melt is subjected to direct chill casting in the form of thick slabs. These slabs are homogenized at a higher temperature and subjected to repeated rolling and reheating till they are reduced to the desirable thickness. This conventional process of producing metallic sheets is very time consuming and costly. However, TRC process is an one-step process that can produce metallic sheets at a reduced cost coupled with higher productivity. Moreover, this process leads to better microstructure and mechanical properties of the sheet product. This is one of the frontier areas where NPL has recently initiated its R&D.

CONCLUSIONS

The R & D on light weight & high strength materials is increasingly being pursued worldwide due to high fuel costs & global environmental concerns to reduce vehicular emissions. Mg & Al alloys and their composites are the main materials on which substantial technology development work is being carried out at NPL. The thrust is on the technology development of materials, processes and components by sustained innovations in primary and secondary processing of these light alloys to enhance their material properties leading to better performance of the product.

ACKNOWLEDGEMENTS

I would like to thank the Director, National Physical Laboratory, New Delhi for his permission to publish this work. I would also like to gratefully acknowledge the project grant and fruitful collaboration with General Motors, India. My sincere thanks to all the members of my group, especially, Mr. R.C. Anandani, Dr. Rajeev Chopra, Dr. Ajay Dhar, Mr. Rajiv Sikand, and Mr. Vipin Jain.

EXPERIMENTAL EVALUATION OF FLOW STRESS AND FRACTURE PROPERTIES OF SA516 (GR.70) STEEL AT HIGH STRAIN RATES FOR CONTAINER APPLICATIONS FOR TRANSPORTATION OF RADIOACTIVE MATERIALS

V.M. Chavan[1a*], S.K. Maiti[2], B.K. Dutta[3], S. Sharma[1], and R.G. Agrawal[1]**

ABSTRACT

The regulatory standards concerned with the transportation of radioactive materials require that the container design have to maintain structural integrity under hypothetical accident conditions such as 9m drop on unyielding target. This implies that influence of strain rate on the material behavior has to be considered in design and safety analysis of containers. In this paper results on dynamic flow stress and fracture toughness are presented for SA516 (Gr70) steel at elevated strain rates. A Split-Hopkinson-Pressure-Bar (SHPB) setup was used to evaluate stress-strain characteristics under compression at normal and elevated temperatures using cylindrical specimens. For fracture under dynamic loading, drop weight tests at sub-normal temperatures were considered. From the test results it is observed that the flow stress and fracture toughness of the alloy is sensitive to the rate of loading and test temperature.

Key words: SA516 steel, high strain rate, Spilt Hopkinson pressure bar, dynamic fracture toughness, drop weight test, shipping cask

---

[1]Division of Refuelling Technology, Bhabha Atomic Research Centre, Trombay — 400085, Mumbai, India

[2]Department of Mechanical Engineering, Indian Institute of Technology Bombay, Powai — 400076, Mumbai, India

[3]Division of Reactor Safety, Bhabha Atomic Research Centre, Trombay — 400085, Mumbai, India

[a]Currently Research Scholar at Mechanical Engineering Department, Indian Institute of Technology Bombay, Powai — 400076, Mumbai, India

[*]E-mail: *vivekmc@barc.gov.in*

[**]E-mail: *skmaiti@me.iitb.ac.in*, *susant02@yahoo.co.in*

INTRODUCTION

Use of thick walled monolithic ferritic steel (SA516) constructions for radioactive materials shipping casks is on the rise in recent years. To prevent release of radioactive material to environment during its transportation, it is necessary to ensure that the container material does not lead to any unstable crack growth at temperatures of $-40°$C under high amplitude dynamic loading arising out of its free fall from a 9 m height followed by a fire engulfing for 30 minutes. Safety assessment of the container design under these extreme loading conditions is often based on a combined approach involving material testing and numerical simulations. In terms of material testing this means that the flow stress and fracture properties of the material must be evaluated through testing at high strain rates.

SA516 carbon steel is a ferritic-pearlitic steel with manganese (0.85–1.2% by weight) as an alloying element. The addition of Mn helps to lower the ductile brittle transition temperature (DBTT) [1, 2]. Silicon (0.15–0.4% by weight) is added to suppress the cementite formation. Carbon and nitrogen are the interstitial atoms in the alloy; these are responsible for dynamic strain aging (DSA) phenomena at medium temperatures. Serrated stress-strain curves observed in tensile tests are due to the presence of DSA. The alloy has been extensively studied for obtaining mechanical properties under quasi-static loading conditions at normal and elevated temperatures [3–6]. Study of effect of radiation embitterment as well as synergetic effect of the interstitial impurity atoms coupled with radiation induced defects on the fracture toughness, ductility and shifting the DSA to higher temperature has been made [4, 5]. Mechanical properties of the alloy are strongly influenced by microstructural parameters, especially the pearlite content as well as size, morphology and distribution in the ferritic matrix. In a large SA516 carbon steel cask the microstructure is often not homogeneous over the wall thickness. Fracture toughness of the alloy is reported to be significantly affected by loading rate and temperature. At higher rates of loading, there is a remarkable reduction in fracture toughness and the temperature at which the fracture resistance becomes minimum shifts to higher level [3].

In this paper results on dynamic flow stress and fracture toughness are presented for SA516 (Gr70) steel at elevated strain rates. A Split-Hopkinson-Pressure-Bar (SHPB) setup was used to evaluate stress-strain characteristics under compression using cylindrical specimens at strain rates 1000 s^{-1} and 3000 s^{-1} at temperatures ranging from 30°C to 400°C. Another objective of the present work has been to develop a suitable technique for experimental evaluation of fracture properties of the steel subjected to impact loading. For fracture under dynamic loading, drop weight tests at sub-normal temperatures were considered. These tests were carried out using pre-notched three-point-bend (TPB) specimens of size similar to the stan-

Table 1. Chemical composition of SA516 (Gr.70) steel plate (wt %)

C	Mn	Si	P	S	Ni	Cr	Cu
0.24	1.14	0.2	0.016	0.022	47 (ppm)	30 (ppm)	180 (ppm)

dard Charpy V-notched (CVN) specimens (10 mm×10 mm×55 mm long). A 2 mm strain-gauge was bonded close to the crack tip to record the strain history. The striker force history was also recorded in each test through high-speed data acquisition system of the test setup developed. The dynamic stress intensity factor, hence the dynamic fracture toughness, corresponding to the crack initiation has been determined through the strain gauge data and the crack-tip field solution.

1. TEST SETUP

All the tests were carried out on specimens machined from ASTM SA516 (Gr. 70) carbon steel plate (85 mm thick). Chemical analysis of the steel shows the composition of the material as follows.

Cylindrical samples of 5 mm diameter and 5 mm length were used for the compression tests. Twenty numbers of high strain rate tests and five tests at unit strain rate have been done using the cylindrical specimens. Molybdenum disulphide was used as a lubricant between specimen and anvils of the setup for high temperature tests and commercial grease was used for the normal temperature tests to avoid friction between specimen and anvils during the compression tests.

Initially, compression tests at a strain rate of 1 s^{-1} were performed using servo hydraulic testing machine over a range of temperature form 30°C to 400°C. Elevated temperatures were attained using an arrangement of heating furnace mounted on machine and a temperature controller. Ceramic bars were used to compress the sample. The temperature is measured using a K-type of thermocouple and a multimeter. A computer interfaced with the testing machine has been employed for test control and data acquisition. Force and displacement are measured during the test. A true strain exceeding 40% have been obtained during the test.

High strain rate testing was done using SHPB setup (Fig. 1). The main components of this setup consist of a gas gun, a striker bar, an incident bar, a transmission bar and a high speed strain gauge data acquisition-cum-analysis system. The striker bar is inserted towards the breach chamber end in the barrel of the gas gun. The sample to be tested is sandwiched between the incident and transmission bar. Once the test conditions are ready, the specially designed breach valve is activated mechanically, causing the expanding gas pressure to accelerate the striker bar. Thus the striker bar is propelled at high velocity towards free end of the gun barrel. Imme-

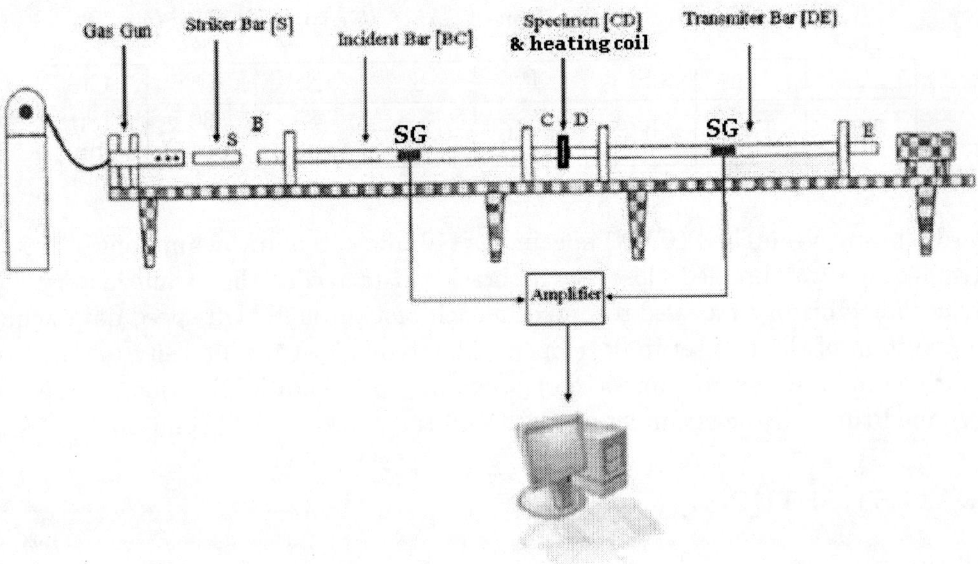

Fig. 1. Schematic of SHPB setup used for compression tests at elevated strain rates

diately in front of the free end of gun barrel, the striker bar collides with the incident bar. Upon impact, an elastic compression wave propagates down the incident bar toward the sample. On reaching the sample due to repeated wave propagation in it, the sample deforms plastically. A part of the wave goes through to the transmission bar (transmitted pulse) and the remaining part is reflected back into the incident bar (reflected pulse), each of which is picked up by the strain gauges mounted on the corresponding bars. Further details of high strain rate testing using SHPB can be found elsewhere [7, 8].

For fracture toughness evaluation, the TPB specimens were tested in the L-S orientation, i.e. the length of the specimen is along the longitudinal direction of the plate; the notch and crack plane is in the thickness direction and the width of the specimen is along the transverse direction of the plate. The notch is created by wire cut EDM using 0.1 mm diameter wire. The strain gauge is bonded to the specimen surface at a distance of 2.5 mm from the crack tip. The gauge axis is oriented 90° to the crack propagation direction. The strain gauges used are resistance type gauge with 120Ω resistances. The size of backing foil of the strain gauge is 1 mm×1 mm. The small gauges are used for their accuracy of strain measurement near the crack tip where the spatial gradients of the strain field are substantial [9, 10]. Similar gauge is pasted on the end face of each specimen. This gauge is used to help in temperature compensation. Thermocouples were employed to measure the temperature of the

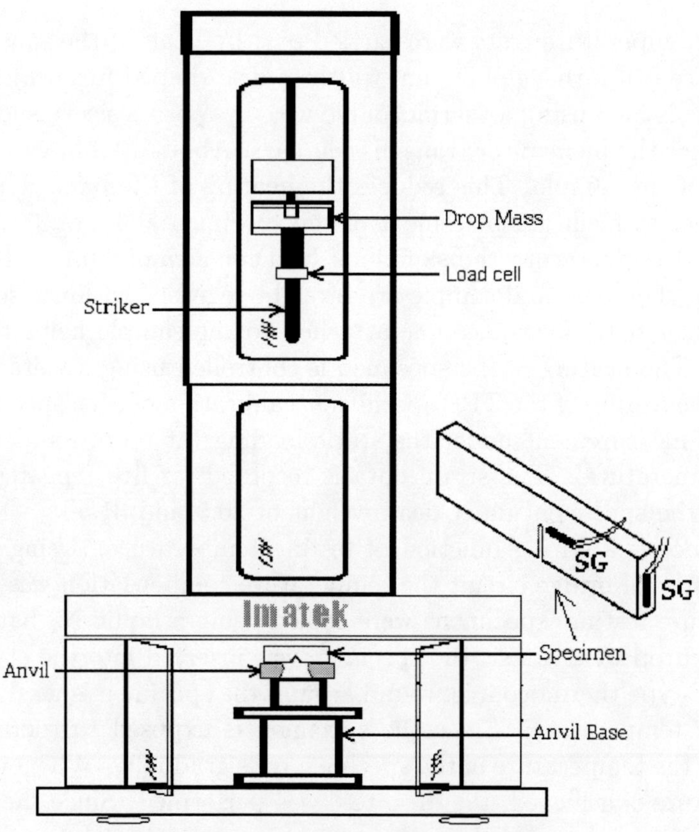

Fig. 2. Schematic of drop weight machine used for fracture tests at elevated strain rates

specimen. All drop weight tests were done on a drop weight testing machine (Fig. 2, Make: M/s Imatek, UK). The machine allows test of the TPB specimens under different test conditions. The drop mass can be changed. The drop velocity can also be varied by automatic adjustment of the drop height. The striker weight is instrumented with a load cell capable of measuring up to 75 kN. A signal conditioner along with data acquisition system capable of sampling speed up to 10 million samples per second was a part of the recording system. During testing, additional data acquisition channel was used for measurement of the strain.

2. TEST PROCEDURE

Dynamic tests using SHPB set-up at a strain rate of 1000 s^{-1} and 3000 s^{-1} were performed over a range of temperature from 30°C to 400°C. Peak strains of about 20% and 30% were obtained for tests at strain rates of 1000 s^{-1} and 3000 s^{-1} respectively.

The elevated temperature tests were carried out, by heating the sample through a small resistance coil furnace of 30 mm width (a disc shaped furnace). During heating the sample is held using a thermocouple wire by a brass sleeve such that it does not touch either the incident or transmission bars. About 10 mm gap is maintained on each side of the sample. This reduces the heating of the bars. The furnace can slide over a bar to facilitate placement of the specimen at its center. After reaching the desired temperature the sample is held for a minimum of 3 to 4 minutes before testing. The bars and sample are brought together within a second prior to firing the striker bar. Thermocouple attached to the sample helps to measure its temperature. The heating of the specimen is controlled using a voltage-regulator.

Prior to the testing of the TPB specimens, calibration of each specimen is carried out by strain measurement under the static loading (at room temperature as well as at low temperature). The strain data is recorded for five repeated loading and unloading of the specimen under dead weight of 20.5 and 31.5 kg. The individual gauge resistance is a strong function of testing temperature. Using an additional gauge and a special bridge circuit the temperature compensation was achieved. For low temperature testing, specimens were cooled using a liquid N_2 bath at $-196°$C. Just before a drop weight test, the specimen was inserted into the cryogenic bottle for cooling. T-type thermocouple wound around the specimen is used for measuring the specimen temperature. Since the specimen is exposed to room temperature environment, the temperature of the specimen rises gradually. Whenever the desired test temperature is achieved, the drop test was performed. Since the test duration is very small; it was assumed that the specimen remained at the same temperature during the test.

3. TEST RESULTS

The comparisons of test results obtained at two high strain rates (1000 s^{-1} and 3000 s^{-1}) over a range 30°C to 400°C are shown in Fig. 3. The oscillatory nature of stress-strain data is due to the stress wave effects. The true stress vs. plastic strain curves obtained through curve-fitting are shown in Fig. 4. To facilitate a comparison, the true stress vs. plastic strain curves at unit strain rate (1 s^{-1}) were also evaluated over the same range of temperatures (Fig. 5 *a*). Fig. 5 *b* shows the variation of the flow stress as a function of test temperature and strain rate. The flow stress is considered to be average of yield (true stress at 0.2% strain) and maximum stresses obtained in the test. At normal temperatures the flow stress rises with the strain rates. At higher temperatures, 300°C–400°C, the reverse, i.e. the softening behavior, is observed. That is the flow stress drops with an increase in temperature.

Fig. 6 *a* shows the three striker force records for TPB specimens tested with drop

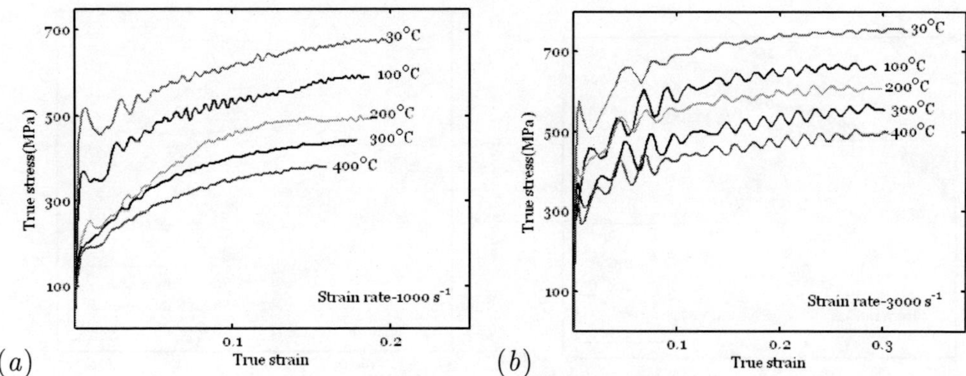

Fig. 3. True stress vs. true strain properties over a range of temperatures at two strain rates. (*a*) 1000 s^{-1} and (*b*) 3000 s^{-1}

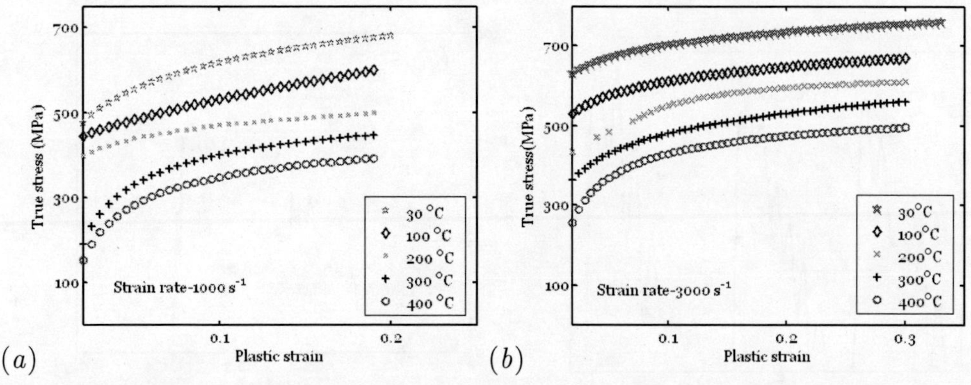

Fig. 4. True stress vs. plastic strain properties over a range of temperatures at two strain rates. (*a*) 1000 s^{-1} and (*b*) 3000 s^{-1}

weight velocity 1.5 m/s at −45°C. Since this temperature is within the transition temperature for this material, the test data falls in between the lower and upper shelf regime of the toughness vs. temperature characteristics of the material. Therefore, limited amount of ductile deformation and crack extension before instability occurs at about 0.31 ms. The strain history obtained through the strain gauge mounted on one of the specimens is shown in Fig. 6 *b*. The measured strain data can be correlated to the dynamic stress intensity factor (SIF) through the solution for elastic crack-tip strain field. The dynamic fracture toughness corresponding to the SIF at the instant of crack initiation has been determined as follows. Under quasi-

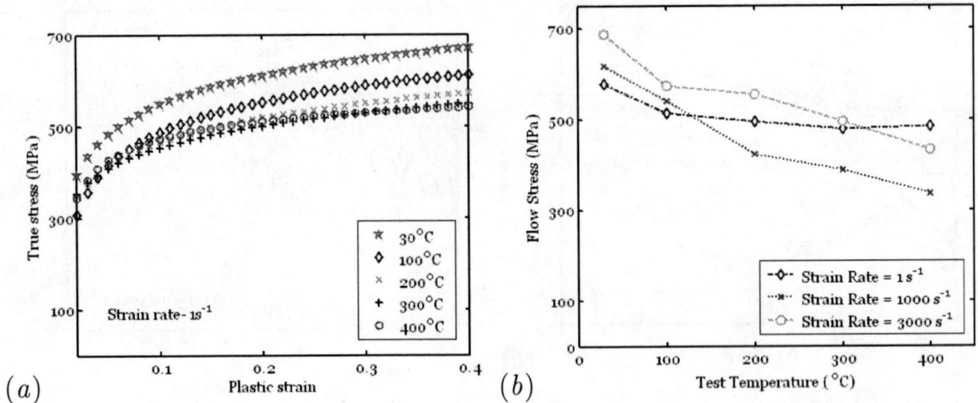

Fig. 5. (a) True stress vs. plastic strain properties over a range of temperatures at unit strain rate, (b) Flow stress vs. temperature characteristics at high strain rates

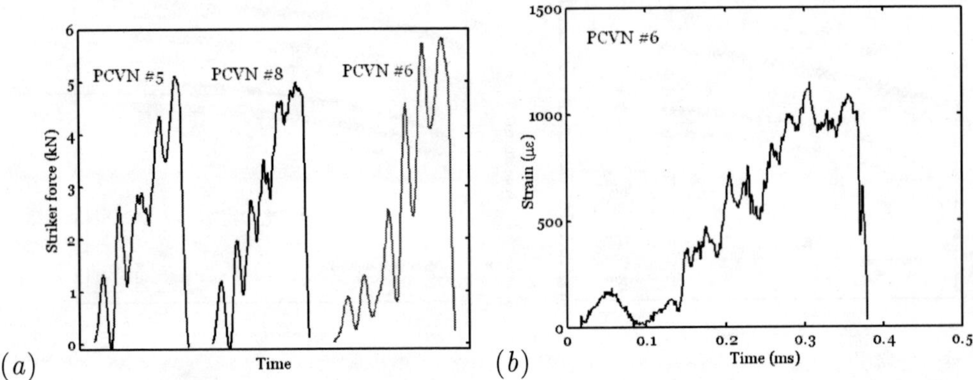

Fig. 6. (a) Striker force histories in drop weight tests for TPB specimens, (b) Strain history on a TPB specimen

static loading, the SIF

$$K_I = Y\Big(\frac{a}{W}, S, B\Big) F_{\text{static}} = Y\Big(\frac{a}{W}, S, B\Big) f(\varepsilon_{\text{static}})$$

and under impact loading the corresponding SIF

$$K_I(t) = f(\varepsilon_{\text{dyn}}(t)) Y\Big(\frac{a}{W}, S, B\Big) = F_{\text{dyn}}(t) Y\Big(\frac{a}{W}, S, B\Big).$$

The time variation of stress intensity factor for the TPB specimen is shown in Fig. 7. In spite of the oscillatory nature of the striker force, the rise of SIF can be assumed monotonic till the peak value at the onset of crack extension $t \approx 310$ μs.

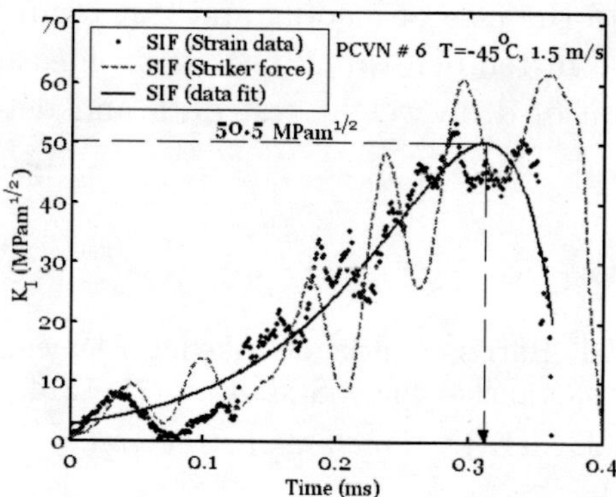

Fig. 7. Stress intensity factor (SIF) history for a TPB specimen

After that there is an abrupt variation of the SIF due to the unstable crack extension. A comparison between the SIFs obtained through the longitudinal strain and striker force recordings (Fig. 7) indicates that improved results are obtainable through the strain measurements. This is because the oscillation amplitudes are lower in the second case.

SUMMARY

Since it is necessary to ensure that the container carrying radioactive material does not lead to any unstable crack growth under high amplitude dynamic loading at temperatures of $-40°C$ arising out of a free fall from 9 m height, its design must be based on material properties evaluated under dynamic loading conditions. This means that the flow stress and fracture properties of the container material shall be evaluated through testing at high strain rates. In this paper dynamic flow stress and fracture toughness of SA516 (Gr70) steel plate have been evaluated under high strain rates over a range of temperatures (30°C to 400°C). A Split-Hopkinson-Pressure-Bar (SHPB) setup was used to evaluate stress-strain characteristics under compression using cylindrical specimens at strain rates $1000\ s^{-1}$ and $3000\ s^{-1}$ at normal and elevated temperatures.

For dynamic fracture studies drop weight tests at sub-normal temperatures were considered. The striker force history and strain history from strain gauge mounted near crack-tip were recorded in each test through high-speed data acquisition system. The inertial effects play an important role during impact tests of the TPB specimens. Local field measurement possible through appropriate strain gauging can help in the measurement of the DSIF, hence the dynamic fracture toughness of the material.

From the test results it is observed that the flow stress and fracture toughness of the steel is sensitive to the rate of loading and test temperature. The flow stress reduces with an increase in temperature. The strain measurement scheme can be applied for interpretation of drop weight test data and determination of dynamic facture toughness.

REFERENCES

1. J.A. Rinebolt and W.J. Harris, "Effect of Alloying Elements on Notch Toughness of Pearlitic Steels," Transactions of the A.S.M. **43**, 1175–1214 (1951).
2. A.S. Tetelman and A.J. McEvily, *Fracture of structural materials* (John Wiley & Sons Inc., New York, 1967).
3. J.H. Yoon, B.S. Lee, Y.J. Oh, and J.H. Hong, "Effects of Loading Rate and Temperature on J-R Fracture Resistance of an SA516-Gr.70 Steel for Nuclear Piping," Int. J. Press. Vess. Piping **76**, 663–670 (1999).
4. A.A. Marengo and J.E.P. Ipina, "Pressure Vessel Steel Fracture Toughness in the Regime from Room Temperature to 400°C," Nuclear Eng. Des. **167**, 215–222 (1996).
5. C.S. Seok and K.L. Murty, "Effect of Dynamic Strain Aging on Mechanical and Fracture Properties of A516Gr70 Steel," Int. J. Press. Vess. Piping **76**, 945–953 (1999).
6. K.L Murty and C.S. Seok, "Fracture in Ferritic Reactor Steel – Dynamic Strain Aging and Cyclic Loading," JOM **53** (7), 23–26 (2001).
7. P.S. Follansbee, "The Hopkinson Bar," in *Mechanical Testing, ASM Handbook*, Vol. 8 (ASM International, Materials Park, OH, 1985), pp. 198–203.
8. S. Nemat-Nasser, "Introduction to High Strain Rate Testing," in *Mechanical Testing and Evaluation, ASM Handbook*, Vol. 8 (ASM International, Materials Park, OH, 2000), pp. 939–959.
9. J.W. Dally and R.J. Sanford, "Strain Gauge Methods for Measuring the Opening Mode Stress-Intensity Factor K_I," Expt. Mech. **27**, 381–388 (1988).
10. H.J. MacGillivray and D.F. Cannon, "The Development of Standard Methods for Determining the Dynamic Fracture Toughness of Metallic Materials," in *Rapid Load Fracture Testing ASTM STP 1130*. Ed. by R. Chona and W.R. Corwin (Amer. Soc. Test. Matl., Philadelphia, 1992), pp. 161–179.

THE NUMERICAL MODELLING OF COUPLED ELASTOVISCOPLASTIC FINITE DEFORMATION AND DAMAGE PROCESSES USING THE SPLITTING METHOD

V.N. Kukudzhanov[1*] and V.S. Synyuk[2]**

ABSTRACT

Integration of the constitutive equations of ductile fracture models is analyzed in this paper. The decomposition method is applied to the Gurson's and Kukudzhanov's models. The analysis of validity of this method is done. It was shown that Kukudzhanov's model describes a large variety of materials since it involves residual stress and viscosity.

INTRODUCTION

Finite strain problems, such as metal forming, where plastic strains are by two to three orders of magnitude larger than elastic strains, involve both the problems: choosing of the appropriate material model and integration scheme. In this paper we consider a material model for void-containing ductile solids that has been developed by Gurson [3], and micromechanical model of fracture that was suggested by Kukudzhanov [4]. A brief description of these models is given in Section 1. The distinctive feature of the constitutive equations of the plasticity theory is that in addition to differential relations, these equations involve also a finite relation that constrains the stress tensor invariants. Owing to this, various mathematical methods can be used. In this paper we use the decomposition method, proposed by Kukudzhanov [5]. This method is based on decomposition of the constitutive relations with respect to physical processes. A brief description of the method is given

[1]Ishlinsky Institute for Problems in Mechanics of the Russian Academy of Sciences, Moscow, Russia

[2]The E.O. Paton Electric Welding Institute, Kiev, Ukraine

[*]E-mail: *kukudz@ipmnet.ru*

[**]E-mail: *svs-gr235@yandex.ru*

in Section 2. We also use the backward Euler method in those cases where it is more appropriate. A detailed analysis of this method for the integration of a class of pressure-dependent plasticity laws is presented in [1]. The paper closes with a numerical example of a uni-axial tension test on a square element to illustrate the basic model features. All the variables of the dimension of MPa that used throughout, we suppose to be measured in terms of the initial yield stress σ_{y0}. In such case we have a dimensionless form of equations.

1. ELASTOVISCOPLASTIC CONSTITUTIVE RELATIONS

The kinds of elastoplastic constitutive models considered in the paper are described in this section.

1.1. *General formulation of the elastoviscoplastic constitutive relations*

We assume the strain-rate decomposition

$$\dot{\varepsilon} = \dot{\varepsilon}^{\mathrm{el}} + \dot{\varepsilon}^{\mathrm{pl}}, \tag{1.1}$$

where $d\varepsilon$ is the differential change of the total strain, $d\varepsilon^{\mathrm{el}}$ is the differential of the elastic strain and $d\varepsilon^{\mathrm{pl}}$ is the differential of the inelastic (plastic) strain. For the case of linear elasticity, we write

$$\dot{\sigma} = \mathbf{D} : \dot{\varepsilon}^{\mathrm{el}}, \tag{1.2}$$

where $\mathbf{D}$ is the fourth-order elasticity tensor. In the following, we assume linear isotropic elasticity, so that

$$D_{ijkl} = 2G\, \delta_{ik}\delta_{jl} + \left(K - \tfrac{2}{3}G\right) \delta_{ij}\delta_{kl},$$

where G and K are the elastic shear and bulk moduli respectively, and δ_{ij} is the Kronecker delta.

The yield function involves the first and second invariants of the stress tensor and is given by

$$\Phi(p, q, H) = 0, \tag{1.3}$$

where $p = -\frac{1}{3}\boldsymbol{\sigma} : \mathbf{I}$ is the hydrostatic stress, $q = \left(\frac{3}{2}\mathbf{s} : \mathbf{s}\right)^{1/2}$ is the equivalent stress and H_i, $i = 1, 2, \ldots, n$, is a set of scalar state variables, $\mathbf{I}$ is the second order identity tensor and $\mathbf{s}$ is the stress deviator. The function Φ is defined such that when $\Phi < 0$, the response is purely elastic. The flow rule is written as

$$\dot{\varepsilon}^{\mathrm{pl}} = \dot{\lambda}\frac{d\Phi}{d\boldsymbol{\sigma}}, \tag{1.4}$$

where λ is a positive scalar. Using (1.3), the flow rule becomes

$$\dot{\boldsymbol{\varepsilon}}^{\mathrm{pl}} = \dot{\lambda}\left(-\frac{1}{3}\frac{\partial\Phi}{\partial p}\mathbf{I} + \frac{\partial\Phi}{\partial q}\mathbf{n}\right), \tag{1.5}$$

where $\mathbf{n} = \frac{3}{2q}\mathbf{s}$. According to [1] $\mathbf{n}$ can be determined as

$$\mathbf{n} = \frac{3}{2q^{\mathrm{el}}}\mathbf{s}^{\mathrm{el}}.$$

Using this notation, the stress tensor can be written as

$$\boldsymbol{\sigma} = -p\mathbf{I} + \frac{2}{3}q\mathbf{n}. \tag{1.6}$$

The plasticity model is completed by describing the evolution of the state variables with continuing plastic straining

$$\dot{H}_i = h_i(\dot{\boldsymbol{\varepsilon}}, \boldsymbol{\sigma}, H_j). \tag{1.7}$$

1.2. *Gurson's plasticity model*

In this section we briefly discuss the material model developed by Gurson for void-containing ductile solids. Based on a rigid-plastic upper-bound solution for spherically symmetric deformations around a single spherical void, Gurson [3] proposed the yield condition

$$\Phi = \left(\frac{q}{\sigma_y}\right)^2 + 2q_1 f \cosh\left(-\frac{3}{2}\frac{q_2 p}{\sigma_Y}\right), - (1 + q_3 f^2) \tag{1.8}$$

where σ_y is the equivalent tensile flow stress representing the actual microscopic stress state in the matrix material, and f is the current void volume fraction. The parameters q_1, q_2 and q_3 were introduced by Tvergaard [6] in an attempt to make the predictions of Gurson's equations agree with numerical studies of the materials containing periodically distributed circular cylindrical voids. The yield surface given by Equation (1.8) becomes that of von Mises (which is assumed in Gurson's model to hold for the matrix material) when $f = 0$, but whenever the void volume fraction is non-zero, there is an effect of hydrostatic stress on the plastic flow.

In Gurson's model the microscopic equivalent plastic strain $\bar{\varepsilon}_m^{\mathrm{pl}}$, to vary is assumed according to the equivalent plastic work expression

$$(1 - f)\sigma_y \, d\bar{\varepsilon}_m^{\mathrm{pl}} = \boldsymbol{\sigma} : d\boldsymbol{\varepsilon}^{\mathrm{pl}}. \tag{1.9}$$

The change in void volume fraction during an increment of deformation is partly due to the growth of existing voids and partly due to the nucleation of new voids by cracking or interfacial decohesion of inclusions or precipitated particles. Accordingly we write

$$df = df_{nuc} + df_{gr}. \tag{1.10}$$

The matrix is assumed to satisfy the plastic incompressibility condition but, because of the existence of voids, the macroscopic response does not. The change of the void volume fraction due to growth is related to the change of the total volume as

$$df_{gr} = (1 - f)d\boldsymbol{\varepsilon}^{\mathrm{pl}} : \mathbf{I}. \tag{1.11}$$

In this model, voids nucleation is considered to be controlled by plastic strain

$$df_{nuc} = A\, d\bar{\varepsilon}_m^{\mathrm{pl}}. \tag{1.12}$$

As suggested by Chu and Needleman [2] the parameter A is chosen so that the nucleation strain follows a normal distribution with mean value ε_N and standard deviation s_N:

$$A = A(\bar{\varepsilon}_m^{\mathrm{pl}}) = \frac{f_N}{s_N\sqrt{2\pi}} \exp\left(-\frac{1}{2}\left(\frac{\bar{\varepsilon}_m^{\mathrm{pl}} - \varepsilon_N}{s_N}\right)^2\right), \tag{1.13}$$

where f_N is the volume fraction of void nucleating particles.

1.3. *Micromechanical model of fracture*

In this section we briefly discuss the micromechanical model of ductile fracture, as suggested by Kukudzhanov [4]. The model is based on the phenomenological theory of dislocations and micro-defect development. According to this model, the plastic deformation and fracture is a single process caused by the motion of dislocations and also by generation and development of microdefects. The deformation process is assumed to consist of two stages. In the first stage, the material is subjected to plastic deformation due to the motion of dislocations in crystal grains of the material. Part of these dislocations that have accumulated at the grain boundaries, leads to residual microstresses (it is assumed that the deviatoric components of the residual stress tensor are proportional to the tensor of accumulated flux of dislocations). On the macrolevel, it leads to an elastoviscoplastic flow with simultaneous hardening of the material. In this model the stress tensor consists of active and residual parts

$$\boldsymbol{\sigma} = \boldsymbol{\sigma}^a + \mathbf{s}^r. \tag{1.14}$$

Active stress is determined by using the associated flow rule

$$\dot{\boldsymbol{\varepsilon}}^{\rm pl} = \dot{\lambda}\frac{d\Phi}{d\boldsymbol{\sigma}^a} \tag{1.15}$$

and the von Mises type flow condition for elastoviscoplastic material with kinematic and isotropic hardening

$$\Phi = q^a - \sigma_y = 0, \tag{1.16}$$
$$\sigma_y = \left(1 + 3\mu_1\bar{\varepsilon}^{\rm pl}\right)^N + \psi(\tau_1\dot{\bar{\varepsilon}}^{\rm pl}),$$

where $q^a = \sqrt{\frac{3}{2}\mathbf{s}^a : \mathbf{s}^a}$ is the equivalent active stress; $\bar{\varepsilon}^{\rm pl} = \sqrt{\frac{2}{3}\mathbf{e}^{\rm pl} : \mathbf{e}^{\rm pl}}$ is the equivalent plastic strain; function Φ determines the influence of strain rate on the yield stress, $\psi(0) = 0$. For residual stress we have the well-known law of kinematic hardening

$$\mathbf{s}^r = 2\alpha\mathbf{e}^{\rm pl}. \tag{1.17}$$

The second stage begins after the intensity of the dislocation flux accumulated at the boundaries of grains attains a critical level, beyond which the process of annihilation of dislocations begins (it is assumed that the flux of annihilation of dislocations is proportional to the amount of dislocations accumulated at the inclusions). This process is accompanied by disclinations of grains with the formation of voids between them. On the macrolevel, this process leads to relaxation of the internal stresses and softening of the material. The appearance of voids changes the behavior of the material essentially. In this case the material consists of the matrix and the voids, i.e., it is a two-phase material. At this stage, the flow condition for active stress is

$$\Phi = \left(\frac{q^a}{\sigma_y}\right) + 2q_1 f\cosh\left(\frac{3}{2}\frac{q_2 p^a}{\sigma_y}\right) - (1 + q_3 f^2) = 0. \tag{1.18}$$

Voids nucleates once the equivalent residual stress q^r achieves the critical value q^r_c. The volume of nucleated voids /0 is considered to be a small disturbance and it is determined from (1.18) when $\sigma_y = 0.99q^a$, $q^a = q^a_t$, $p^a = p^a_t$. At this stage of deformation one also has to use Equations (1.9) and (1.11) for state variables. The relaxation of residual stress is described by equation

$$\dot{\mathbf{s}}^r + 2\mu_2\frac{Q\left(q^r - q^r_f\right)}{\tau_2 q^r}\mathbf{s}^r = 2\alpha\dot{\mathbf{e}}^{\rm pl}, \tag{1.19}$$

where q^r_f is the final equivalent residual stress in the first stage; function Q, that characterizes the relaxation, is supposed to be a linear function of its argument, τ_2 is relaxation time of residual stresses.

2. DECOMPOSITION METHOD

Decomposition method for the integration of equations governing the behavior of elastoviscoplastic media was proposed by Kukudzhanov [5]. The idea of this method is based on the fact that an additive operator can be replaced by a multiplicative operator on a small time interval.

It was shown that for the classical elastoplastic medium independent of the time scale, the decomposition leads to an algebraic powerlaw equation for the correction coefficient of the elastic solution. For elastoviscoplastic media, the decomposition leads to a differential equation for the correction coefficients, which in this case are functions of time. The general scheme of decomposition looks as follows. We decompose the equation

$$\dot{\boldsymbol{\sigma}} = \mathbf{D} : (\dot{\varepsilon} - \dot{\varepsilon}^{\mathrm{pl}}). \tag{2.1}$$

The predictor is taken at $\dot{\varepsilon}^{\mathrm{pl}} = 0$, which corresponds to the elastic material. In this case, for each step Δt, one should solve the elastic problem

$$\dot{\boldsymbol{\sigma}} = \mathbf{D} : \dot{\varepsilon} \tag{2.2}$$

subjected to the initial conditions obtained at the previous step for the complete elastoplastic problem.

The corrector is taken at $\dot{\varepsilon} = 0$ in equation (2.1). Using equation (1.4), equations (2.1) and (1.7) becomes

$$\dot{\boldsymbol{\sigma}} = -\dot{\lambda}\mathbf{D}\frac{\partial\Phi}{\partial\boldsymbol{\sigma}}, \quad \dot{H}_i = h_i(-\dot{\lambda}\frac{\partial\Phi}{\partial\boldsymbol{\sigma}}, \boldsymbol{\sigma}, H_j). \tag{2.3}$$

The second relation in (2.3) provides the evolutionary equations for the internal parameters that characterize the structure of the material, hardening, damage, porosity and other properties.

For an elastoviscoplastic medium, the relaxation is completed at the moment $t = t_*$ the steady-state plasticity condition (1.3) has been satisfied.

For the classical (equilibrium) elastoplastic medium, the properties of which are independent of the change in the time scale, one can eliminate the time t from equations (2.3) and proceed to the variable λ, to obtain

$$d\boldsymbol{\sigma} = -d\lambda\mathbf{D}\frac{\partial\Phi}{\partial\boldsymbol{\sigma}}, \quad dH_i = \tilde{h}_i(-d\lambda\frac{\partial\Phi}{\partial\boldsymbol{\sigma}}, \boldsymbol{\sigma}, H_j). \tag{2.4}$$

Solving equations (2.4) subjected to the initial conditions resulting from the solution of the elastic problem $\boldsymbol{\sigma}(\lambda_0) = \boldsymbol{\sigma}^{\mathbf{el}}$, $H_i(\lambda^n) = H_i^n$, we find the variables $\boldsymbol{\sigma}^{n+1}$ and H_i^{n+1} as functions of λ, $\boldsymbol{\sigma}^{\mathbf{el}}$, H_i^{el}

$$\boldsymbol{\sigma} = \boldsymbol{\sigma}(\lambda, \boldsymbol{\sigma}^{\mathbf{el}}, H_i^{\mathrm{el}}), \quad H_i = H_i(\lambda, \boldsymbol{\sigma}^{\mathbf{el}}, H_j) \quad \text{for} \quad \lambda = \lambda^{n+1}. \tag{2.5}$$

Substituting these relations into the plasticity condition (1.3), we obtain the equation

$$\Phi(p(\lambda), q(\lambda), H_i(\lambda), \lambda) = 0. \tag{2.6}$$

Solving this equation and substituting the resulting λ into (2.5) we obtain the final solution of the problem.

For the elastoviscoplastic medium, the plasticity condition is of differential type. In this case, equation (1.3) for λ leads to the differential equation

$$\Phi(p(\lambda), q(\lambda), H_i(\lambda), \lambda, \dot{\lambda}) = 0. \tag{2.7}$$

This decomposition scheme is stable, if the predictor scheme for the solution of the elastic problem is stable and there exist solutions of equations (2.6) and (2.7).

3. NUMERICAL INTEGRATION OF THE CONSTITUTIVE RELATIONS

3.1. *Integration of the constitutive relations of GTN model by the decomposition method*

During the constitutive calculations, where stresses and state variables are updated, the total strain e is known. The aim of this integration is to find stresses and state variables at the end of the time increment $t + \Delta t$.

In this section we apply the decomposition method for the integration of the constitutive equations of Gurson's plasticity model. For simplicity we assume that there in no nucleation of voids, i.e. $df_{nuc} = 0$, and that the yield stress is constant, $\sigma_y = \sigma_{y0} = \text{const}$.

At the predictor step one has to solve the elastic problem. Taking into account (1.6), equation (2.2) can be written as

$$\dot{p}^{\text{el}} = -K\dot{\varepsilon} : \mathbf{I}, \quad \dot{q}^{\text{el}} = 2G\dot{\varepsilon} : \mathbf{n}. \tag{3.1}$$

We integrate these equations taking the solutions at the previous step t as the initial conditions. The result can be written as

$$p^{\text{el}} = p_t + \Delta p^{\text{el}}, \quad q^{\text{el}} = q_t + \Delta q^{\text{el}}, \tag{3.2}$$

where $\Delta p^{\text{el}} = -K\Delta\varepsilon : \mathbf{I}$, $\Delta q^{\text{el}} = 2G\Delta\varepsilon : \mathbf{n}$.

At the correction step, the first equation in can be written as

$$\frac{d\sigma}{d\lambda} = -K\frac{\partial\Phi}{\partial p}\mathbf{I} - 2G\frac{\partial\Phi}{\partial q}\mathbf{n}. \tag{3.3}$$

Projecting this equation onto $\mathbf{I}$ and $\mathbf{n}$, and using equation (1.6), we find

$$\frac{dp}{d\lambda} = -K\frac{\partial\Phi}{\partial p}, \tag{3.4}$$

$$\frac{dq}{d\lambda} = -6Gq \tag{3.5}$$

Since $df = df_{gr}$, equations (1.11) and (1.5) give

$$\frac{df}{d\lambda} = -(1-f)\frac{d\Phi}{dp} \tag{3.6}$$

Initial conditions for equations (3.5) — (3.6) are taken from the solution of the elastic problem at the predictor step

$$p(\lambda_0) = p_0 = p^{\mathrm{el}}, q(\lambda_0) = q_0 = q^{\mathrm{el}}, f(\lambda_0) = f_0 = f_t. \tag{3.7}$$

Equations (3.4) and (3.6) gives

$$dp = \frac{K}{(1-f)}df. \tag{3.8}$$

Integrating this equation with initial conditions (3.7), we find

$$f = 1 - (1-f_0)l^{-\frac{p-p_0}{K}}. \tag{3.9}$$

Eliminating $d\lambda$ from equations (3.4) and (3.5), we find

$$\frac{dp}{dq} = \frac{K}{6G}\frac{1}{q}\frac{\partial\Phi(p,q)}{\partial p} \quad \text{for} \quad \Phi(p,q) = \varphi(p)\psi(q). \tag{3.10}$$

The problem of integrating the elastoplastic equations reduces to the solution of one differential equation and one non-linear algebraic equation

$$\frac{dp}{dq} = \alpha\frac{1}{q}\psi(p), \quad p(q_0) = p_0, \tag{3.11}$$

$$\Phi(p,q) = q^2 + \phi(p) = 0, \tag{3.12}$$

where $\alpha = \frac{K}{6G}$, $\psi(p) = \frac{\partial\Phi}{\partial p}$, $\phi(p) = 2q_1 f(p)\cosh\left(\frac{3}{2}q_2 p\right) - (1 + q_3 f^2(p))$.

We use the backward Euler method for the integration of equation (3.11). Writing this equation in the incremental form, we obtain a non-linear algebraic equation. Thus, the problem reduces to the solution of the set of two non-linear equations

$$F(\Delta p, \Delta q) = \Delta p - \alpha\frac{\Delta q}{q}\psi(p) = 0, \tag{3.13}$$

$$\Phi(\Delta p, \Delta q) = q^2 + \phi(p) = 0, \tag{3.14}$$

where $\Delta p = p - p_0$ and $\Delta q = q - q_0$ are unknown. These equations are solved by means of Newton's method.

When p and q are known, the stress tensor is calculated using (1.6). The computational advantages of the decomposition method, as compared with standard methods, are achieved if at the correction step some equations can be integrated analytically.

3.2. *Integration of the constitutive relations of the micromechanical model*

We use the decomposition method to integrate the constitutive equations of plasticity by Mises with kinematical hardening.

Equations (1.15) and (1.16) give

$$\dot{\varepsilon}^{\mathrm{pl}} = \dot{\lambda}\frac{3}{2q^\alpha}\mathbf{s}^\alpha = \dot{\Lambda}\mathbf{s}^\alpha, \quad \dot{\Lambda} = \frac{3}{2q^\alpha}\dot{\lambda}. \tag{3.15}$$

Equation (2.4) for stresses relaxation can be written as

$$d\mathbf{s}^\alpha = -2G\mathbf{s}^\alpha d\Lambda. \tag{3.16}$$

Integrating this equation with initial condition $\Lambda = \Lambda_0$, $\mathbf{s}^\alpha(\Lambda_0) = \mathbf{s}^{\mathrm{el}}$, we find

$$\mathbf{s}^\alpha = \mathbf{s}^{\mathrm{el}}X, q^\alpha = q^{\mathrm{el}}X, \tag{3.17}$$

where $X = l^{-2G(\Lambda - \Lambda_0)}$ is the correction coefficient of the elastic solution.

Using equations (3.15) and (3.17), and taking into account that $\dot{\varepsilon}^{\mathrm{pl}} = \dot{\mathbf{e}}^{\mathrm{pl}}$ one can find

$$\dot{\bar{\varepsilon}}^{\mathrm{pl}} = \sqrt{\frac{2}{3}\dot{\mathbf{e}}^{\mathrm{pl}} : \dot{\mathbf{e}}^{\mathrm{pl}}} = \frac{2}{3}\dot{\Lambda}q^{\mathrm{el}}l^{-2G(\Lambda-\Lambda_0)} = -\frac{1}{3G}q^{\mathrm{el}}\dot{X}. \tag{3.18}$$

Integrating this equation we obtain

$$\bar{\varepsilon}^{\mathrm{pl}} = \bar{\varepsilon}^{\mathrm{pl}}_t - \frac{1}{3G}q^{\mathrm{el}}(X-1), \tag{3.19}$$

where $\bar{\varepsilon}^{\mathrm{pl}}_t$ is the equivalent plastic strain at the previous time step.

Substituting the values $\bar{\varepsilon}^{\mathrm{pl}}$ and $\dot{\bar{\varepsilon}}^{\mathrm{pl}}$ in the flow condition (1.16), we have

$$\Phi = q^{\mathrm{el}}X - \left[1 + 3\mu_1\left(\bar{\varepsilon}^{\mathrm{pl}}_t - \frac{1}{3G}q^{\mathrm{el}}(X-1)\right)\right]^N - \psi\left(-\frac{1}{3G}q^{\mathrm{el}}\tau_1\dot{X}\right) = 0, \tag{3.20}$$

where $\psi(x) = \alpha\tanh(x)$.

We use the backward Euler method to solve this differential equation. Approximating $\dot{X}$ with $\dot{X}=(X-1)/\Delta t$, we obtain a non-linear equation for X which is solved by means of Newton's method. Using equation (1.17), we find the equivalent residual stress

$$q^r = 3\alpha\bar{\varepsilon}^{\text{pl}}. \tag{3.21}$$

In the first stage of deformation, the stress tensor can be written as

$$\boldsymbol{\sigma} = -p^\alpha \mathbf{I} + \frac{2}{3}(q^\alpha + q^r)\mathbf{n}, \tag{3.22}$$

where $p^\alpha = -K(\varepsilon : \mathbf{I})$ is the elastic pressure.

To integrate the constitutive relations at the second stage of deformation, after nucleation of voids, we use the numerical method, proposed by Aravas [1]. In this paper we apply this method for the rate-dependent plasticity.

Flow rule (1.15) in the incremental form at the end of the increment is written

$$\Delta\varepsilon^{\text{pl}} = \frac{1}{3}\Delta\varepsilon_p \mathbf{I} + \Delta\varepsilon_q \mathbf{n}, \tag{3.23}$$

where

$$\Delta\varepsilon_p = -\Delta\lambda\frac{\partial\Phi}{\partial p^\alpha}, \quad \Delta\varepsilon_q = -\Delta\lambda\frac{\partial\Phi}{\partial q^\alpha}. \tag{3.24}$$

Elimination of $\Delta\lambda$ gives

$$\Delta\varepsilon_p\frac{\partial\Phi}{\partial q^\alpha} + \Delta\varepsilon_q\frac{\partial\Phi}{\partial p^\alpha} = 0. \tag{3.25}$$

Using incremental form of equations (1.1), (1.5) and (3.24) we can rewrite equation (1.2) as

$$\boldsymbol{\sigma}^\alpha = \boldsymbol{\sigma}^{\text{el}} - K\Delta\varepsilon_p\mathbf{I} - 2G\Delta\varepsilon_q\mathbf{n}, \tag{3.26}$$

where $\boldsymbol{\sigma}^{\text{el}} = \boldsymbol{\sigma}_t + \Delta\boldsymbol{\sigma}^{\text{el}}$.

Projecting the elasticity equation (3.26) onto $\mathbf{I}$ and $\mathbf{n}$, and using equation (1.6), we find

$$p^\alpha = p^{\text{el}} + K\Delta\varepsilon_p, \tag{3.27}$$

$$q^\alpha = q^{\text{el}} - 3G\Delta\varepsilon_q, \tag{3.28}$$

where $p^{\text{el}} = p_t + \Delta p^{\text{el}}$, $q^{\text{el}} = q_t + \Delta q^{\text{el}}$.

Equations for the state variables can be written as follows.

Taking into account that $f = f_t + \Delta f$, we can write the expression for porosity

$$f = \frac{f_t + \Delta\varepsilon_p}{1 + \Delta\varepsilon_p}. \tag{3.29}$$

For an equivalent plastic strain in the matrix material, we can write

$$\Delta\bar{\varepsilon}_m^{\rm pl} = \frac{-\Delta\varepsilon_p p^\alpha + \Delta\varepsilon_q q^\alpha}{(1-f)\sigma_y(\bar{\varepsilon}_m^{\rm pl})}, \tag{3.30}$$

$$\sigma_y = (1 + 3\mu_1\bar{\varepsilon}_m^{\rm pl})^N + \alpha\tanh(\Delta\bar{\varepsilon}_m^{\rm pl}\tau_1/\Delta t), \tag{3.31}$$

where $\bar{\varepsilon}_m^{\rm pl} = (\bar{\varepsilon}_m^{\rm pl})_t + \Delta\bar{\varepsilon}_m^{\rm pl}$.

Equations (3.30) and (3.31) are solved with given values of $\Delta\varepsilon_p$ and $\Delta\varepsilon_q$.

The problem of integrating the elastoviscoplastic equations reduces to the solution of the set of equations: (1.18), (3.25), (3.27) – (3.31). These equations are solved using Newton's method. We choose Aep and Aeq as the primary unknowns, treating equations (1.18) and (3.25) as the basic equations.

When $\Delta\varepsilon_p$ and $\Delta\varepsilon_q$ are known, the active stress tensor is calculated using (3.26).

Projecting equation (1.19) onto $\mathbf{n}$ and taking into account that $\mathbf{n}$ is collinear with s^r/q^r and $\dot{\mathbf{e}}^{\rm pl} = \dot{\varepsilon}_q\mathbf{n}$, we find

$$\dot{q}^r + \frac{2\mu_2}{\tau_2}Q\left(q^r - q_f^r\right) = 3\alpha\dot{\varepsilon_q}. \tag{3.32}$$

We use the backward Euler scheme to integrate this equation. Taking into account that Q is supposed to be a linear function of its argument, we can write

$$\Delta q^r + 2\mu_2\frac{\Delta t}{\tau_2}\left(q^r - q_f^r\right) = 3\alpha\Delta\varepsilon_q, \tag{3.33}$$

where we suppose that $2\mu_2 = G$. Solving this equation, we find

$$q^r = q_f^r + \left(q_0^r - q_f^r\right)/(1+y), \tag{3.34}$$

where $q_0^r = q_t^r + 3\alpha\Delta\varepsilon_q$, $= G\frac{\Delta t}{\tau_2}$.

Residual stress tensor is written in the form

$$\mathbf{s}^r = \frac{2}{3}q^r\mathbf{n}. \tag{3.35}$$

Finally, the solution of the problem is determined by equations (1.14), (3.26) and (3.30).

Linearization moduli. In an implicit finite element code, the equilibrium equations are written at the end of the increment, resulting in a set of non-linear equations for the nodal unknowns. If a full Newton scheme is used to solve these non-linear equations, one needs to calculate the so-called "linearization moduli" that define the variation of stress at $t + \Delta t$ caused by a variation of the total strain at $t + \Delta t$, and they are needed to define the stiffness matrix (Jacobian) for the Newton loop used to solve the overall discretized equilibrium equations. The procedure of calculation of the linearization moduli has been described by Aravas [1].

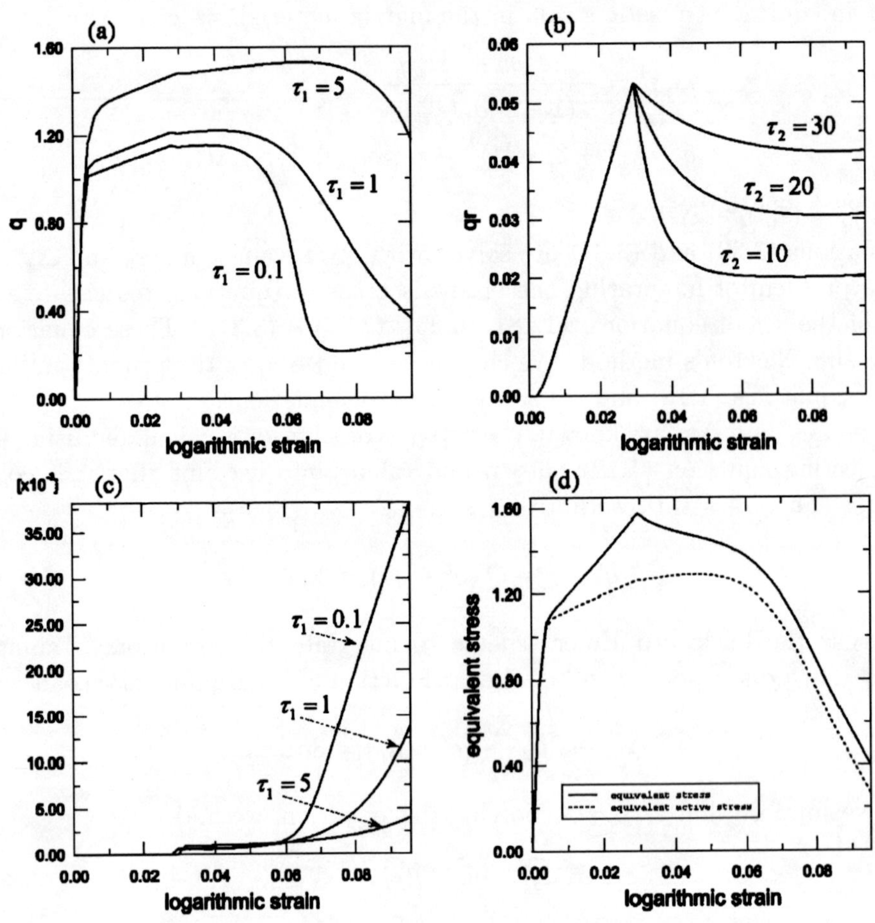

Fig. 1

4. NUMERICAL EXAMPLE

The results of the finite element analysis of a micromechanical model are represented in Figure 1. The elastic-plastic properties of the matrix material common for all calculations are specified by $\overset{\circ}{A} = 500$, $\nu = 0.3$, $\sigma_{y_0} = 1$, $\mu_1 = 5$, $N = 0.4$, $a = 1$, $q_1 = 1.5$, $q_2 = 0.2$, $q_3 = 2.25$, $q_f^r = 0.01$; for (a), (b) and (c): $a = 1$, $q_c^r = 0.05$; for (a) and (b): $r_2 = 30$; for (c): $n = 5$; for (d): $a = 5$, $q_c^r = 0.3$, $\tau_1 = 1$, $\tau_2 = 10$.

Figure 1 *a* shows that larger strain rates (parameter τ_1 is responsible for the strain rate) leads to larger resistance of the material. Figure 1 *b* shows that smaller stain rate leads to larger voids growth, i.e. material damage occurs sooner in smaller strain rates. Figure 1 *c* shows the influence of the parameter τ_2 on the rate of the residual stress relaxation. And finally, Figure 1 *d* illustrates a correlation between the active and full equivalent stress. Depending on its parameters, the micromechanical model describes a large variety of material properties.

CONCLUSIONS

Application of the decomposition method to the Gurson's and Kukudzhanov's models has been demonstrated in this paper. It was shown that the computational advantages of decomposition method (as compared with standard methods) are achieved if at the correction step, some equations can be integrated analytically. Otherwise, the backward Euler scheme is more appropriate. Constitutive relations of the micromechanical model, based on the phenomenological theory of dislocations, were integrated using decomposition and backward Euler methods. It was shown that this model describes a large variety of materials as it involves kinematic hardening with residual stress relaxation after voids nucleation and viscosity.

ACKNOWLEDGMENTS

The research was financially supported by the Russian Foundation for Basic Research (under grants Nos. 08-01-91302-IND_a and 06-01-00523-a).

REFERENCES

1. N. Aravas, "On the Numerical Integration of a Class of Pressure-Dependent Plasticity Models," Intern. J. Numer. Meth. Engng **24** (7), 1395–1416 (1987).
2. C.C. Chu and A. Needleman, "Void nucleation effects in biaxially stretched sheets," J. Engng Mater. Technol. **102**, 249–256 (1980).
3. A.L. Gurson, "Continuum theory of ductile rupture by void nucleation and growth: Part I - Yield criteria and flow rules for porous ductile materials," J. Engng Mater. Technol. **99**, 2–15 (1977).
4. V.N. Kukudzhanov, "Micromechanical model of fracture of an inelastic material and its application to the investigation of strain localization," Mech. Solids **34** (5), 58-69 (1999).
5. V.N. Kukudzhanov, "Decomposition method for elastoplastic equations," Mech. Solids **39** (1), 73–80 (2004).
6. V. Tvergaard, "Influence of voids on shear band instabilities under plane strain condition," Int. J. Fracture **17** (4), 389–407 (1981).

A HIGHER-ORDER LINEAR LAMINATED COMPOSITE SHELL THEORY

T. Kant[1*] and P. Desai[1]**

ABSTRACT

A class of shell theories, which are derived from the three-dimensional elasticity equations by expanding the displacement vector in Taylor's series in the thickness coordinate, is first reviewed. A theory for general shell deformation is then developed based on a higher-order kinematic model for anisotropic layered materials including sandwiches. The theory accounts for the effects of transverse shear deformation, transverse normal stress and transverse normal strain with an implicit nonlinear distribution of the tangential displacement component through the shell thickness.

Key words: higher-order shear-normal deformation theory, thick shells, numerical integration technique, anisotropic layered material.

INTRODUCTION

Structural shell forms-a body bounded by two curved surfaces, are extremely important structural elements in applications such as nuclear reactors, pressure vessels, spacecrafts, missiles, etc. The development of the theoretical model (the theory) to describe its behavior is an area of continuing interest. The so-called *thin shell theory*, originally formulated by Love in 1888, is firmly established (Love, 1888, 1944). It is used extensively for analytical solutions and numerical analysis and further is continuously being applied to new problems to generate much needed design data. However, its use in many practical applications involving complex geometries and loadings, cut-outs, branches, intersections, contact problem involving shells, laminated shells, etc. is not at all effective because of the implicit simplistic assumptions

[1]Department of Civil Engineering, Indian Institute of Technology Bombay, Powai — 400076, Mumbai, India

[*]E-mail: *tkant@civil.iitb.ac.in*

[**]E-mail: *payaldesai79@gmail.com*

in the theory. This calls for the development of more refined higher-order shell theories.

Here, we are concerned with the derivation of a particular higher-order general shell theory. The behavior of the shell, in any theory, is governed by the behavior of an appropriate reference surface. This necessitates the transformation of the three dimensional (3D) elasticity equations into an approximate system of two dimensional (2D) shell equations. This transformation is an essential feature of any plate or shell theory. The situation, especially in shells, is further complicated by the coupled nature of the membrane and bending behavior. These coupled deformations, in the form of stretching and curvature change of the reference surface, are required in predicting the strains/stresses that exist throughout the shell space.

Review articles [1,2] discuss on some of the aspects of the mechanics of deformable solid media including nontraditional theories of shells and plates. Use of Reissner mixed variational theorem for 2D modelling of flat and curved, multilayer structures is devoted in [3]. A historical review of zig-zag theories that have been developed for the analysis of multilayered structures (plates and shells) is given in [4]. The accuracy of shell theories, viz., Flugge, Sanders, Love and Donnell with respect to 3D elasticity solution, for cross-ply laminated circular cylindrical shells under static loads is examined in [5]. Analytical solutions of a general cross-ply doubly curved panel of rectangular planform for the static response under transverse load and free vibration are found in [6]. The effects of the laminated material response, transverse shear, and transverse normal strains are included. A complete (particular as well as complementary) Fourier solution to the boundary-value problem of static response under transverse load of a general cross-ply thick doubly curved panel of rectangular planform is presented in [7].

A meshless method for modelling of cross-ply laminated elastic shells by a higher-order theory and multiquadrics is given in [8]. A review on refined shell theories for composites, in the last two decades, is being attempted here as follows. A higher-order displacement model for the behavior of symmetric and unsymmetric laminated composite and sandwich cylindrical shells is presented in [9]. A C^0 finite element formulation for flexure-membrane coupling behavior of symmetric and asymmetric laminated cylindrical shells based on a higher-order displacement model is developed in [10]. This theory incorporates a realistic nonlinear variation of displacements through the shell thickness, and eliminates the use of shear correction coefficient/s. The discrete element chosen is a nine-noded quadrilateral with nine degrees of freedom per node. A C^0 finite element space discretization procedure for a general fibre reinforced composite cylindrical shell theory based on a higher-order displacement model is employed in [11]. A refined global approximation theory of thin multilayered anisotropic shells is developed in [12]. Solutions for unsymmetrical loading on a simply supported orthotropic circular cylindrical shell using Flugge shell theory are contained in [13]. The bending behavior of unidirectional sandwich panels ("wide" or "narrow" beams) with a compressible "soft" core and piezoelectric active face sheets is investigated through a higher-order theory in [14]. After a review of the shear deformation plate and shell theories, a consistent third-order theory for composite shells is presented in [15]. The thermal stresses in cylindrical shells subjected to local heating by using the equations which include the effects of shear deformation is described in [16]. The problem of interlaminar stresses in a cross-ply laminated cylindrical shell under radial pressure is studied in [17]. A computationally efficient, layerwise shear-deformation theory, for improving the accuracy of stress and strain predictions in the analysis of laminated shells with arbitrary thickness, is presented in [18]. Results from an analytical investigation of the behavior of composite circular cylinders of axially variable thickness subjected to internal and external surface loading are given in [19].

1. A NEW SHELL THEORY FORMULATION

The displacement expansions for a general shell are written concisely as,

$$\begin{aligned} U_1 &= u_1 + zu_1' + \frac{1}{2}z^2u_1'' + \frac{1}{6}z^3u_1''', \\ U_2 &= u_2 + zu_2' + \frac{1}{2}z^2u_2'' + \frac{1}{6}z^3u_2''', \\ U_3 &= u_3 + zu_3' + \frac{1}{2}z^2u_3'' + \frac{1}{6}z^3u_3''', \end{aligned} \tag{1}$$

The general shell equations, so derived [20], are particularized for an axisymmetric shell, more specifically for a cylinder, for which the longitudinal (U_1) and the transverse (U_3) displacement components define the displacement field and Eq. (1) reduces to,

$$\begin{aligned} U_1 &= u_1 + zu_1' + \frac{1}{2}z^2u_1'' + \frac{1}{6}z^3u_1''', \\ U_3 &= u_3 + zu_3' + \frac{1}{2}z^2u_3'' + \frac{1}{6}z^3u_3'''. \end{aligned}$$

Cylindrical shell coordinates are shown in Fig. 1. The boundary-value problem is formulated in terms of 16 differential equations called the *intrinsic equations* involving only 16 unknowns-called the *intrinsic dependent variables.* The other variables are related to the intrinsic variables by algebraic relations. One of the two coordinates describing the problem is selected as a preferred coordinate in the basic method of analysis. In the present case, due to symmetric loads, the problem is described in terms of φ — the meridional coordinate only and therefore the choice of the preferred coordinate is a straight-forward one. Intrinsic equations are derived

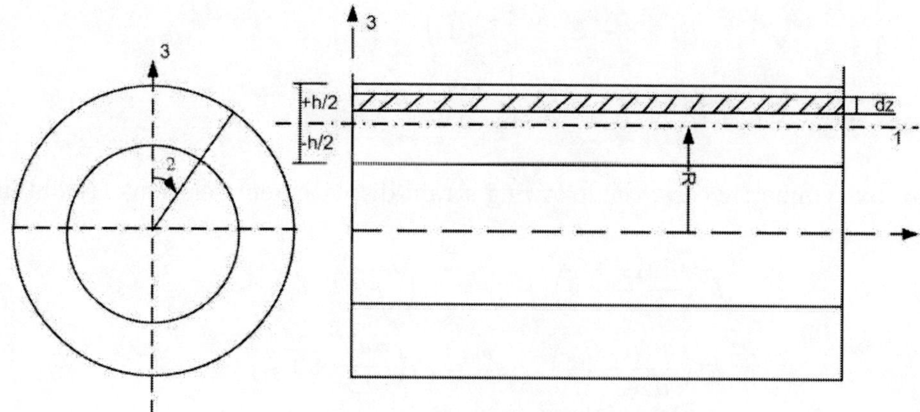

Fig. 1. Coordinates for cylindrical shell

consisting of a system of 16 first-order ordinary differential equations with respect to the meridional coordinate, containing the 16 intrinsic variables which naturally appear in the boundary conditions on the edge $\varphi = a$ constant. For the present problem, $\mathbf{y}$, defining a vector of intrinsic variables consists of,

$$\mathbf{y} = (u_1, u_1', u_1'', u_1''', u_3, u_3', u_3'', u_3''', N_{11}, M_{11}, P_{11}, M_{11}^*, Q_{11}, S_{11}, T_{11}, S_{11}^*)^t.$$

The governing equations are arranged in such a way that the derivatives of the intrinsic variables with respect to φ (axial coordinate) for a given value of φ are computed in a straight-forward manner, when the physical and geometrical parameters of the shell and the intrinsic variables themselves are known at that value of φ. Intrinsic equations and the associated auxiliary relations are derived in a systematic manner for symmetrically loaded orthotropic/layered shells of revolution. These are numerically integrated by a specially developed procedure called the segmentation method [21, 22].

2. STRAIN–DISPLACEMENT RELATIONS — SYMMETRIC LOADS

General three dimensional strain-displacement relations are written as follows,

$$\varepsilon_1 = \frac{1}{1+k_1 z}\Big(\varepsilon_1^\circ + z\varepsilon_1' + \frac{1}{2}z^2\varepsilon_1'' + \frac{1}{6}z^3\varepsilon_1'''\Big), \quad \varepsilon_3 = u_3' + zu_3'' + \frac{1}{2}z^2 u_3''',$$

$$\varepsilon_2 = \frac{1}{1+k_2 z}\Big(\varepsilon_2^\circ + z\varepsilon_2' + \frac{1}{2}z^2\varepsilon_2'' + \frac{1}{6}z^3\varepsilon_2'''\Big),$$

$$\gamma_{12} = \frac{1}{1+k_1 z}\Big(\beta_1^\circ + z\beta_1' + \frac{1}{2}z^2\beta_1'' + \frac{1}{6}z^3\beta_1'''\Big) + \frac{1}{1+k_2 z}\Big(\beta_2^\circ + z\beta_2' + \frac{1}{2}z^2\beta_2'' + \frac{1}{6}z^3\beta_2'''\Big),$$

$$\gamma_{13} = \frac{1}{1+k_1 z}\Big(\mu_1^\circ + z\mu_1' + \frac{1}{2}z^2\mu_1'' + \frac{1}{6}z^3\mu_1'''\Big),$$
$$\gamma_{23} = \frac{1}{1+k_2 z}\Big(\mu_2^\circ + z\mu_2' + \frac{1}{2}z^2\mu_2'' + \frac{1}{6}z^3\mu_2'''\Big).$$

For axisymmetric case, the following strain-displacement relations are obtained:

$$\varepsilon_1^\circ = k_1\Big(\frac{du_1}{d\varphi} + u_3\Big), \quad \varepsilon_1' = k_1\Big(\frac{du_1'}{d\varphi} + u_3'\Big),$$
$$\varepsilon_1'' = k_1\Big(\frac{du_1''}{d\varphi} + u_3''\Big), \quad \varepsilon_1''' = k_1\Big(\frac{du_1'''}{d\varphi} + u_3'''\Big),$$
$$\varepsilon_2^\circ = k_2(u_1 \cot\varphi + u_3), \quad \varepsilon_2' = k_2(u_1' \cot\varphi + u_3'),$$
$$\varepsilon_2'' = k_2(u_1'' \cot\varphi + u_3''), \quad \varepsilon_2''' = k_2(u_1''' \cot\varphi + u_3'''),$$
$$\mu_1^\circ = k_1\Big(\frac{du_3}{d\varphi} - u_1\Big) + u_1', \quad \mu_1' = k_1\frac{du_3'}{d\varphi} + u_1'' - k_1 u_1',$$
$$\mu_1'' = k_1\Big(\frac{du_3''}{d\varphi} + u_1''\Big) + u_1''', \quad \mu_1''' = k_1\Big(\frac{du_3'''}{d\varphi} - u_1'''\Big),$$
$$\beta_1^\circ = 0, \quad \beta_1' = 0, \quad \beta_1'' = 0, \quad \beta_1''' = 0,$$
$$\beta_2^\circ = 0, \quad \beta_2' = 0, \quad \beta_2'' = 0, \quad \beta_2''' = 0,$$
$$\mu_2^\circ = 0, \quad \mu_2' = 0, \quad \mu_2'' = 0, \quad \mu_2''' = 0.$$

3. STRESS RESULTANTS

Following stress-displacement relations are derived for a laminated shell:

$$\begin{aligned}
N_{11} = {} & \Big(K_{11} + \frac{C_{11}}{R}\Big)\frac{du_1}{d\varphi} + \Big(C_{11} + \frac{D_{11}}{R}\Big)\frac{du_1'}{d\varphi} + \Big(\frac{1}{2}D_{11} + \frac{1}{2}\frac{F_{11}}{R}\Big)\frac{du_1''}{d\varphi} \\
& + \Big(\frac{1}{6}F_{11} + \frac{1}{6}\frac{H_{11}}{R}\Big)\frac{du_1'''}{d\varphi} + \{K_{12}k_2\}u_3 + \Big(\frac{1}{R}C_{12} + K_{13} + \frac{F_{13}}{R}\Big)u_3' \\
& + \Big(\frac{1}{2R}D_{12} + C_{13} + \frac{D_{13}}{R}\Big)u_3'' + \Big(\frac{1}{6R}F_{12} + \frac{D_{13}}{2} + \frac{F_{13}}{2R}\Big)u_3''', \\
M_{11} = {} & \Big(C_{11} + \frac{D_{11}}{R}\Big)\frac{du_1}{d\varphi} + \Big(D_{11} + \frac{F_{11}}{R}\Big)\frac{du_1'}{d\varphi} + \Big(\frac{1}{2}F_{11} + \frac{1}{2}\frac{H_{11}}{R}\Big)\frac{du_1''}{d\varphi} \\
& + \Big(\frac{1}{6}H_{11} + \frac{1}{6}\frac{I_{11}}{R}\Big)\frac{du_1'''}{d\varphi} + \Big\{\frac{C_{12}}{R}\Big\}u_3 + \Big(\frac{1}{R}D_{12} + C_{13} + \frac{D_{13}}{R}\Big)u_3' \\
& + \Big(\frac{1}{2R}F_{12} + D_{13} + \frac{F_{13}}{R}\Big)u_3'' + \Big(\frac{1}{6R}H_{12} + \frac{F_{13}}{2} + \frac{H_{13}}{2R}\Big)u_3''',
\end{aligned}$$

$$
\begin{aligned}
P_{11} = \frac{1}{2}\Big[&\Big(D_{11} + \frac{F_{11}}{R}\Big)\frac{du_1}{d\varphi} + \Big(F_{11} + \frac{H_{11}}{R}\Big)\frac{du_1'}{d\varphi} + \Big(\frac{1}{2}H_{11} + \frac{1}{2}\frac{I_{11}}{R}\Big)\frac{du_1''}{d\varphi}\\
&+ \Big(\frac{1}{6}I_{11} + \frac{1}{6}\frac{J_{11}}{R}\Big)\frac{du_1'''}{d\varphi} + \Big\{\frac{D_{12}}{R}\Big\}u_3 + \Big(\frac{1}{R}F_{12} + D_{13} + \frac{F_{13}}{R}\Big)u_3'\\
&+ \Big(\frac{1}{2R}H_{12} + F_{13} + \frac{H_{13}}{R}\Big)u_3'' + \Big(\frac{1}{6R}I_{12} + \frac{H_{13}}{2} + \frac{I_{13}}{2R}\Big)u_3'''\Big],\\
M_{11}^* = \frac{1}{6}\Big[&\Big(F_{11} + \frac{H_{11}}{R}\Big)\frac{du_1}{d\varphi} + \Big(H_{11} + \frac{I_{11}}{R}\Big)\frac{du_1'}{d\varphi} + \Big(\frac{1}{2}I_{11} + \frac{1}{2}\frac{J_{11}}{R}\Big)\frac{du_1''}{d\varphi}\\
&+ \Big(\frac{1}{6}J_{11} + \frac{1}{6}\frac{R_{11}}{R}\Big)\frac{du_1'''}{d\varphi} + \Big\{\frac{F_{12}}{R}\Big\}u_3 + \Big(\frac{1}{R}H_{12} + F_{13} + \frac{H_{13}}{R}\Big)u_3'\\
&+ \Big(\frac{1}{2R}I_{12} + H_{13} + \frac{I_{13}}{R}\Big)u_3'' + \Big(\frac{1}{6R}J_{12} + \frac{I_{13}}{2} + \frac{J_{13}}{2R}\Big)u_3'''\Big],\\
Q_{11} = &\Big(K_{15} + \frac{C_{15}}{R}\Big)\frac{du_3}{d\varphi} + \Big(C_{15} + \frac{D_{15}}{R}\Big)\frac{du_3'}{d\varphi} + \Big(\frac{1}{2}D_{15} + \frac{1}{2}\frac{F_{15}}{R}\Big)\frac{du_3''}{d\varphi}\\
&+ \Big(\frac{1}{6}F_{15} + \frac{1}{6}\frac{H_{15}}{R}\Big)\frac{du_3'''}{d\varphi} + \Big(K_{15} + \frac{C_{15}}{R}\Big)u_1'\\
&+ \Big(C_{15} + \frac{D_{15}}{R}\Big)u_1'' + \Big(\frac{1}{2}D_{15} + \frac{F_{15}}{2R}\Big)u_1''',\\
S_{11} = &\Big(C_{15} + \frac{D_{15}}{R}\Big)\frac{du_3}{d\varphi} + \Big(D_{15} + \frac{F_{15}}{R}\Big)\frac{du_3'}{d\varphi} + \Big(\frac{1}{2}F_{15} + \frac{1}{2}\frac{H_{15}}{R}\Big)\frac{du_3''}{d\varphi}\\
&+ \Big(\frac{1}{6}H_{15} + \frac{1}{6}\frac{I_{15}}{R}\Big)\frac{du_3'''}{d\varphi} + \Big(C_{15} + \frac{D_{15}}{R}\Big)u_1'\\
&+ \Big(D_{15} + \frac{F_{15}}{R}\Big)u_1'' + \Big(\frac{1}{2}F_{15} + \frac{H_{15}}{2R}\Big)u_1'''\\
T_{11} = \frac{1}{2}\Big[&\Big(D_{15} + \frac{F_{15}}{R}\Big)\frac{du_3}{d\varphi} + \Big(F_{15} + \frac{H_{15}}{R}\Big)\frac{du_3'}{d\varphi} + \Big(\frac{1}{2}H_{15} + \frac{1}{2}\frac{I_{15}}{R}\Big)\frac{du_3''}{d\varphi}\\
&+ \Big(\frac{1}{6}I_{15} + \frac{1}{6}\frac{J_{15}}{R}\Big)\frac{du_3'''}{d\varphi} + \Big(D_{15} + \frac{F_{15}}{R}\Big)u_1'\\
&+ \Big(F_{15} + \frac{H_{15}}{R}\Big)u_1'' + \Big(\frac{1}{2}H_{15} + \frac{I_{15}}{2R}\Big)u_1''',\\
S_{11}^* = \frac{1}{6}&\Big(F_{15} + \frac{H_{15}}{R}\Big)\frac{du_3}{d\varphi} + \Big(H_{15} + \frac{I_{15}}{R}\Big)\frac{du_3'}{d\varphi} + \Big(\frac{1}{2}I_{15} + \frac{1}{2}\frac{J_{15}}{R}\Big)\frac{du_3''}{d\varphi}\\
&+ \Big(\frac{1}{6}J_{15} + \frac{1}{6}\frac{R_{15}}{R}\Big)\frac{du_3'''}{d\varphi} + \Big(F_{15} + \frac{H_{15}}{R}\Big)u_1'\\
&+ \Big(H_{15} + \frac{I_{15}}{R}\Big)u_1'' + \Big(\frac{1}{2}I_{15} + \frac{J_{15}}{2R}\Big)u_1''',
\end{aligned}
$$

$$A = \left(K_{31} + \frac{C_{31}}{R}\right)\frac{du_1}{d\varphi} + \left(C_{31} + \frac{D_{31}}{R}\right)\frac{du_1'}{d\varphi} + \left(\frac{1}{2}D_{31} + \frac{1}{2}\frac{F_{31}}{R}\right)\frac{du_1''}{d\varphi}$$
$$+ \left(\frac{1}{6}F_{31} + \frac{1}{6}\frac{H_{31}}{R}\right)\frac{du_1'''}{d\varphi} + \left\{\frac{K_{32}}{R}\right\}u_3 + \left(\frac{1}{R}C_{32} + K_{33} + \frac{C_{33}}{R}\right)u_3'$$
$$+ \left(\frac{1}{2R}D_{32} + C_{33} + \frac{D_{33}}{R}\right)u_3'' + \left(\frac{1}{6R}F_{32} + \frac{D_{33}}{2} + \frac{F_{33}}{2R}\right)u_3''',$$
$$B = \left(C_{31} + \frac{D_{31}}{R}\right)\frac{du_1}{d\varphi} + \left(D_{31} + \frac{F_{31}}{R}\right)\frac{du_1'}{d\varphi} + \left(\frac{1}{2}F_{31} + \frac{1}{2}\frac{H_{31}}{R}\right)\frac{du_1''}{d\varphi}$$
$$+ \left(\frac{1}{6}H_{31} + \frac{1}{6}\frac{I_{31}}{R}\right)\frac{du_1'''}{d\varphi} + \left\{\frac{C_{32}}{R}\right\}u_3 + \left(\frac{1}{R}F_{32} + D_{33} + \frac{F_{33}}{R}\right)u_3'$$
$$+ \left(\frac{1}{2R}F_{32} + D_{33} + \frac{F_{33}}{R}\right)u_3'' + \left(\frac{1}{6R}H_{32} + \frac{F_{33}}{2} + \frac{H_{33}}{2R}\right)u_3''',$$
$$N_{22} = (K_{21})\frac{du_1}{d\varphi} + (C_{21})\frac{du_1'}{d\varphi} + \left(\frac{1}{2}D_{21}\right)\frac{du_1''}{d\varphi} + \left(\frac{1}{6}F_{21}\right)\frac{du_1'''}{d\varphi}$$
$$+ \left\{\frac{K_{22}}{R} - \frac{C_{22}}{R}\right\}u_3 + \left(\frac{1}{R}C_{22} - \frac{D_{22}}{R^2} + K_{23}\right)u_3'$$
$$+ \left(\frac{1}{2R}D_{22} - \frac{F_{22}}{2R^2} + C_{23}\right)u_3'' + \left(\frac{1}{6R}F_{22} - \frac{H_{22}}{6R^2} + \frac{D_{23}}{2}\right)u_3''',$$
$$M_{22} = (C_{21})\frac{du_1}{d\varphi} + (D_{21})\frac{du_1'}{d\varphi} + \left(\frac{1}{2}F_{21}\right)\frac{du_1''}{d\varphi} + \left(\frac{1}{6}H_{21}\right)\frac{du_1'''}{d\varphi}$$
$$+ \left\{\frac{C_{22}}{R} - \frac{D_{22}}{R^2}\right\}u_3 + \left(\frac{1}{R}D_{22} - \frac{F_{22}}{R^2} + C_{23}\right)u_3'$$
$$+ \left(\frac{1}{2R}F_{22} - \frac{H_{22}}{2R^2} + D_{23}\right)u_3'' + \left(\frac{1}{6R}H_{22} - \frac{I_{22}}{6R^2} + \frac{F_{23}}{2}\right)u_3''',$$
$$P_{22} = \frac{1}{2}(D_{21})\frac{du_1}{d\varphi} + (F_{21})\frac{du_1'}{d\varphi} + \left(\frac{1}{2}H_{21}\right)\frac{du_1''}{d\varphi} + \left(\frac{1}{6}I_{21}\right)\frac{du_1'''}{d\varphi}$$
$$+ \left\{\frac{D_{22}}{R} - \frac{F_{22}}{R^2}\right\}u_3 + \left(\frac{1}{R}F_{22} - \frac{H_{22}}{R^2} + D_{23}\right)u_3'$$
$$+ \left(\frac{1}{2R}H_{22} - \frac{I_{22}}{2R^2} + F_{23}\right)u_3'' + \left(\frac{1}{6R}I_{22} - \frac{J_{22}}{6R^2} + \frac{H_{23}}{2}\right)u_3''',$$
$$M_{22}^* = \frac{1}{2}(F_{21})\frac{du_1}{d\varphi} + (H_{21})\frac{du_1'}{d\varphi} + \left(\frac{1}{2}I_{21}\right)\frac{du_1''}{d\varphi} + \left(\frac{1}{6}J_{21}\right)\frac{du_1'''}{d\varphi}$$
$$+ \left\{\frac{F_{22}}{R} - \frac{H_{22}}{R^2}\right\}u_3 + \left(\frac{1}{R}H_{22} - \frac{I_{22}}{R^2} + F_{23}\right)u_3'$$
$$+ \left(\frac{1}{2R}I_{22} - \frac{J_{22}}{2R^2} + H_{23}\right)u_3'' + \left(\frac{1}{6R}J_{22} - \frac{R_{22}}{6R^2} + \frac{I_{23}}{2}\right)u_3'''.$$

For deriving the equilibrium equations, the principle of minimum potential energy is chosen due to its simplicity and also because its application gives simultaneously the natural boundary conditions that are to be used with the theory.

Under axisymmetric loading, $\tau_{12} = 0$ and $\tau_{23} = 0$. Further, all shell variables are independent of θ. Making these simplifications, the following eight equations are obtained:

$$\frac{dN_{11}}{d\varphi} + \frac{k_2}{k_1}\cot\varphi(N_{11} - N_{22}) + Q_{11} + \frac{1}{k_1}(p_1 + p_1^-) = 0,$$

$$\frac{dQ_{11}}{d\varphi} + \frac{k_2}{k_1}\cot\varphi Q_{11} - N_{11} - \frac{k_2}{k_1}N_{22} + \frac{1}{k_1}(p_3 + p_3^-) = 0,$$

$$\frac{dM_{11}}{d\varphi} + \frac{k_2}{k_1}\cot\varphi(M_{11} - M_{22}) - \frac{Q_{11}}{k_1} + S_{11} + \frac{1}{k_1}(m_1^-) = 0,$$

$$\frac{dS_{11}}{d\varphi} + \frac{k_2}{k_1}\cot\varphi S_{11} - M_{11} - \frac{k_2}{k_1}M_{22} - \frac{1}{k_2}A + \frac{1}{k_1}m_3^- = 0,$$

$$\frac{dP_{11}}{d\varphi} + \frac{k_2}{k_1}\cot\varphi(P_{11} - P_{22}) - \frac{S_{11}}{k_1} - T_{11} + \frac{1}{k_1}(n_1^-) = 0,$$

$$\frac{dT_{11}}{d\varphi} + \frac{k_2}{k_1}\cot\varphi T_{11} - P_{11} - \frac{k_2}{k_1}P_{22} - \frac{1}{k_1}B + \frac{1}{k_1}n_3^- = 0,$$

$$\frac{dM_{11}^*}{d\varphi} + \frac{k_2}{k_1}\cot\varphi(M_{11}^* - M_{22}^*) + S_{11}^* - \frac{1}{k_1}T_{11} + \frac{1}{k_1}(o_1^-) = 0,$$

$$\frac{dS_{11}^*}{d\varphi} + \frac{k_2}{k_1}\cot\varphi S_{11}^* - M_{11}^* - \frac{k_2}{k_1}M_{11}^* - \frac{1}{k_1}C + \frac{1}{k_1}(o_3^-) = 0.$$

Following boundary conditions on an edge of constant φ:

$$\begin{aligned}
&u_1 = \bar{u}_1, \quad u_3 = \bar{u}_3, \quad u_1' = \bar{u}_1', \quad u_3' = \bar{u}_3',\\
&u_1'' = \bar{u}_1'', \quad u_3'' = \bar{u}_3'', \quad u_1''' = \bar{u}_1''', \quad u_3''' = \bar{u}_3''',\\
&N_{11} = \bar{N}_{11}, \quad Q_{11} = \bar{Q}_{11}, \quad M_{11} = \bar{M}_{11}, \quad S_{11} = \bar{S}_{11},\\
&P_{11} = \bar{P}_{11}, \quad T_{11} = \bar{T}_{11}, \quad M_{11}^* = \bar{M}_{11}^*, \quad S_{11}^* = \bar{S}_{11}^*.
\end{aligned}$$

4. AN EXAMPLE

A diaphragm supported (0°/core/0°) three layered cylinder is analyzed and the shell solution is compared with elasticity solution. The cylinder is subjected to an external pressure of 1000 kg/cm^2. The geometrical and material properties are $R = 0.5$ m and $l = 2$ m. Again, In view of symmetry about mid length, only half shell is analyzed. Following material properties [23] are taken for the (0°/core/0°) sandwich cylinder:

- *Face material* properties are

$$E_r = 6.894\times10^6, \quad E_\theta = 172.36\times10^6, \quad E_z = 6.894\times10^6,$$

$$G_{zr} = 1.378\times10^6, \quad \nu_{\theta r} = 0.25, \quad \nu_{zr} = 0.25, \quad \nu_{\theta z} = 0.25;$$

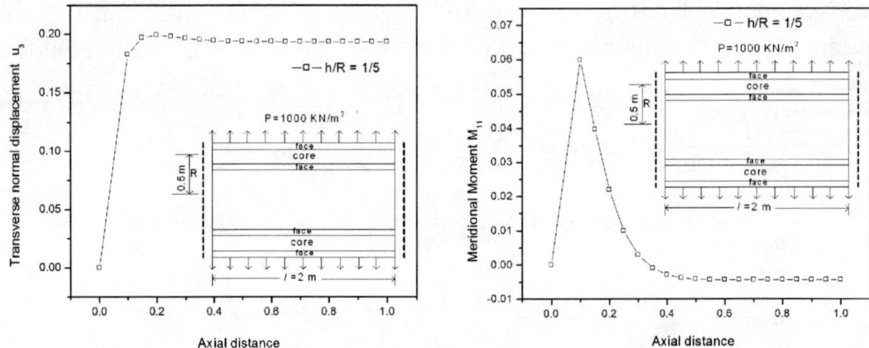

Fig. 2. Distribution of transverse normal displacement and meridional moment

- *Core material* properties are

$$E_r = 3.44 \times 10^6, \quad E_\theta = 0.275 \times 10^6, \quad E_z = 0.275 \times 10^6,$$
$$G_{zr} = 0.413 \times 10^6, \quad \nu_{\theta r} = 0.0199, \quad \nu_{zr} = 0.0199, \quad \nu_{\theta z} = 0.25.$$

Number of segments and number of subdivisions within each segment that are found suitable for the numerical study are summarized below for any future reference. For $h/R = 1/5$, number of segments based on convergence criterion are chosen as 400, number of subdivision within each segment is taken as 5 and length of each segment turns out to be as 0.0025 m.

Following non-dimensional parameters are used for presentation of results.

$$\bar{u}_3 = \frac{Eh}{p_\circ R^2} u_3, \quad \bar{M}_{11} = \frac{4}{p_\circ Rh} M_{11}, \quad \bar{Q}_{11} = \frac{4}{p_\circ \sqrt{Rh}} Q_{11}, \quad \bar{N}_{11} = \frac{1}{\nu p_\circ R} N_{11}.$$

Figs. 2–3 show the distribution of transverse normal displacement, meridional moment, transverse shear and circumferential force in thick sandwich shell. Fig. 4 shows the circumferential moment in thick sandwich shell. Following points are observed from the results: (i) Maximum values of M11, M22 and Q11 are directly proportional to h, because stiffness and hence the bending action increases with increase in h, (ii) Maximum value of N11 increases with decrease in h due to increased flexibility. N22 attains the same maximum value away from the edge, the distance being more in the case of thicker shell due to increased edge disturbance. The same holds good for u3 too, (iii) The edge effect is seen to travel more in the case of larger h confirming a known fact.

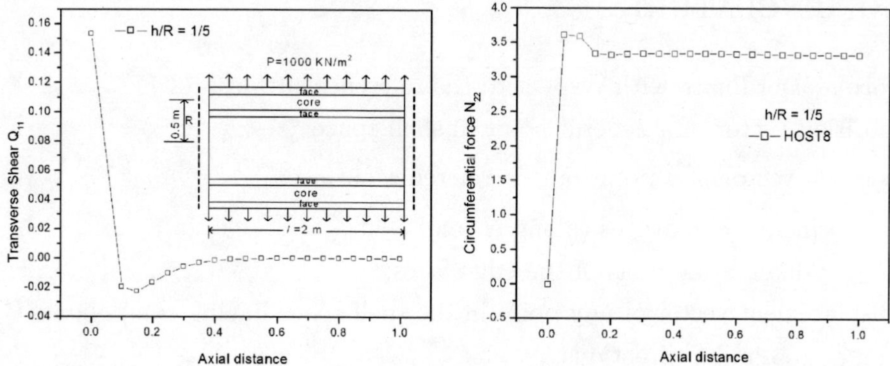

Fig. 3. Distribution of transverse shear and circumferential force

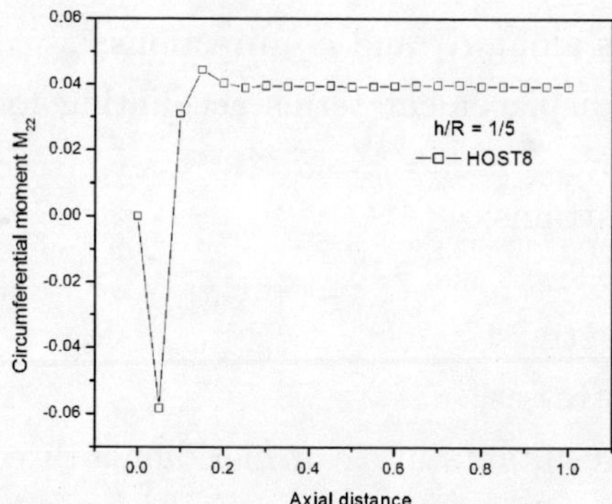

Fig. 4. Circumferential moment in thick sandwich shell

5. CONCLUDING REMARKS

A higher-order shell theory is presented for an axisymmetric finite length diaphragm supported cylinder. The complete derivation is new. The formulation incorporates the effects transverse shear deformation with warping of the transverse cross-section, transverse normal strain and transverse normal stress. The displacement model is consistent and represents refined modes of both bending and membrane deformations. Besides thick shells, this theory should find immense application for laminated shells where the distribution of the tangential displacement components through the thickness is nonlinear.

6. NOMENCLATURE

z — normal coordinate with respect to the reference surface;

$\underset{\sim}{R}$ — position vector of a general point in shell space;

$\underset{\sim}{r}$ — position vector of a point on the reference surface;

k_1, k_2 — principal curvatures (along α_1 and α_2);

dS_1, dS_2 — differential areas of the edge faces;

$\underset{\sim}{U}$ — displacement vector of any point in the shell space having components U_1, U_2, U_3 along α_1, α_2 and z directions;

u_i (u_1, u_2, u_3) — linear displacements of a point of the reference surface along α_1, α_2 and z directions;

u_1', u_2' — rotations of the normal to the reference surface in α_1 and α_2 directions;

u_1'', u_2'' — curvature terms along α_1 and α_2 directions;

u_3'', u_3''' — higher-order displacement terms accounting for the transverse normal stress and strain;

ε_i (ε_1, ε_2, ε_3) — normal strains;

γ_{ij} — shearing strains;

σ_i (σ_1, σ_2, σ_3) — normal stresses;

τ_{12}, τ_{13}, τ_{23} — shearing stresses;

N_{11}, M_{11}, etc. — stress-resultants in α_1 or meridional direction;

N_{22}, M_{22}, etc. — stress-resultants in α_2 or circumferential direction;

Q_{11}, S_{11}, etc. — transverse stress–resultants in z direction on a transverse plane with α_1 — a constant;

Q_{22}, S_{22}, etc. — transverse stress–resultants in z direction on a transverse plane with α_2 — a constant;

$\bar{u}_i$, $\bar{u}_i'$, $\bar{u}_i''$, $\bar{u}_i'''$ — specified boundary displacement values;

$\bar{N}_{11}$, $\bar{M}_{11}$, etc. — specified boundary stress resultant values;

h — thickness of a homogeneous isotropic/orthotropic shell;

E_{ij} — elasticity coefficients defining the stress-strain matrix;

K_{ij}, C_{ij}, D_{ij}, F_{ij}, H_{ij}, I_{ij}, J_{ij}, etc. — coefficients used in defining the stress-resultants for layered-system

α_1 or φ — meridional coordinate of a shell of revolution;

α_2 or θ — circumferential coordinate of a shell of revolution;

p^+, p^-, m^+, m^-, n^+, n^-, o^+, o^- — load vectors on the positive and negative external z surfaces and having components along α_1, α_2 and z directions;

REFERENCES

1. S.A. Ambartsumian, "Contributions to the Theory of Anisotropic Layered Shells," ASME Appl. Mech. Rev. **15** (4), 245–249 (1962).
2. S.A. Ambartsumian, "Nontraditional Theories of Shells and Plates," ASME Appl. Mech. Rev. **55** (5), 35 (2002).
3. E. Carrera, "Developments, Ideas, and Evaluations Based upon Reissner's Mixed Variational Theorem in the Modeling of Multilayered Plates and Shells," ASME Appl. Mech. Rev. **54** (4), 301 (2001).
4. E. Carrera, "Historical Review of Zig-Zag Theories for Multilayered Plates and Shells," ASME Appl. Mech, Rev. **56** (3), 287 (2003).
5. K. Chandrashekhara and D.V.T.G. Pavan Kumar, "Assessment of Shell Theories for the Static Analysis of Cross-Ply Laminated Circular Cylindrical Shells," Thin-Walled Struct. **22**, 291–318 (1995).
6. R.A. Chaudhuri and H.R.H. Kabir, "Static and Dynamic Fourier Analysis of Finite Cross-Ply Doubly Curved Panels Using Classical Shallow Shell Theories," Compos. Struct. **28**, 73–91 (1994).
7. R.A. Chaudhuri and H.R.H. Kabir, "Fourier Solution to Higher Order Theory Based Laminated Shell Boundary-Value Problem," AIAA J. **33** (9), 1681–1688 (1995).
8. A. Ferreira, C. Roque, and R. Jorge, "Modelling Cross-Ply Laminated Elastic Shells by a Higher-Order Theory and Multiquadrics," Comput. Struct. **84**, 1288–1299 (2006).
9. T. Kant and M. Menon, "Higher-Order Theories for Composite and Sandwich Cylindrical Shells With C^0 Finite Elements," Comput. Struct. **33** (5), 1191–1204 (1989).
10. T. Kant and M. Menon, "Estimation of Interlaminar Stresses in Fibre Reinforced Composite Cylindrical Shells," Comput. Struct. **38** (2), 131– 147 (1991).
11. T. Kant and M. Menon, "A Finite Element-Difference Computational Model for Stress Analysis of Layered Composite Cylindrical Shells," Fin. Elem. Anal. Des. **14**, 55–71 (1993).
12. G. Kulikov, "Refined Global Approximation Theory of Multilayered Plates and Shells, ASCE J. Engng Mech. **127** (2), 119–125 (2001).
13. D. Paliwal and D.K. Gupta, "Orthotropic Pressure Vessels Subject to Local Loads," Int. J. Press. Vessels Piping **76**, 387–392 (1999).
14. O. Rabinovitch, J.R. Vinson, and Y. Frostig, "High-Order Analysis of Unidirectional Sandwich Panels with Piezolaminated Face Sheets and Soft Core," AIAA J. **41** (1), 110–118 (2003).
15. J.N. Reddy and R.A. Arciniega, "Shear Deformation Plate and Shell Theories: From Stavsky to Present," Mech. Adv. Mater. Struct. **11**, 535–582 (2004).
16. K. Shirakawa, "Thermal Stresses in Cylindrical Shells Based on Improved Theory," Nucl. Engng Des. **68**, 53–59 (1981).
17. X. Wang, W. Cai, and Z.Y. Yu, "An Analytic Method for Interlaminar Stress in a Laminated Cylindrical Shell," Mech. Adv. Mater. Struct. **9** (2), 119–131 (2002).

18. Z. Xu, C. Aditi, and H.S. Kim, "An Efficient Layerwise Shear-Deformation Theory and Finite Element Implementation," J. Reinforced Plast. Compos. **23** (2), 131 (2004).
19. A.M. Zenkour and S.A. Jeddah, "Stresses in Cross-Ply Laminated Circular Cylinders of Axially Variable Thickness," Acta Mech. **187**, 85–102 (2006).
20. P. Desai, "Stress Analysis of Finite Length Cylinders-Some Studies," PhD Thesis (Indian Institute of Technology Bombay, Powai, Mumbai 400076, 2008).
21. T. Kant and S. Patil, "Numerical Analysis of Pressure Vessels Using Various Shell Theories," in *Tech. Rep. Research Report IIT-B/CE 79-1* (Indian Institute of Technology Bombay, India, Civil Engineering Department, 1979).
22. T. Kant and C.K. Ramesh, "Numerical Integration of Linear Boundary Value Problems in Solid Mechanics by Segmentation Method," Int. J. Num. Meth. Engng **17**, 1233-1256 (1981).
23. N.J. Pagano, "Exact Solutions for Composite Laminates in Cylindrical Bending," J. Compos. Mat. **3**, 398–411 (1969).

FRACTURE CHARACTERISTICS OF FIBROUS CONCRETE COMPOSITES

K.G.C. Babu[1]

ABSTRACT

Fracture toughness in recent years is recognized and accepted to be one of the most important material properties. There are several efforts that explain the fracture characteristics of metals reasonably well by taking into account the characteristics at the microscopic levels as well as to accommodate the different geometries. While these have been successful to an acceptable degree in metals, the scenario in composites like concrete has not reached the same degree of success. The primary difficulty is in understanding and accommodating the macro conglomerate characteristics of these materials. The different sizes of aggregates and the corresponding effects of intrusions and defect locations can not be defined clearly. While the complex aggregate distribution in concrete composites is difficult to represent mathematically, the problem is complicated further through the introduction of fibers in the system. This paper brings to fore some of these aspects and suggests a broad outline to address the same.

Key words: concrete, fracture, defects, intrusions, aggregate, fiber

INTRODUCTION

In recent years, a lot of emphasis is being placed on, not only the design, development and understanding of the behavior of structures in service, but also in critically appraising the failure mechanisms to have a better focus on the safety aspects. In this context fracture mechanics is considered to be a better and a more reliable tool in understanding the behavior of materials and structures. The concepts of ductility as can be seen from plasticity, strain hardening, strain softening and post cracking or post yield behavior in materials and structures are well entrenched in design, even if only through empirical relationships.

[1]Department of Ocean Engineering, Indian Institute of Technology Madras, Chennai — 600036, Tamil Nadu, India. E-mail: *kgbabu18@yahoo.com*

Efforts to understand fracture the fracture characteristics of metals taking into account the characteristics at the microscopic and atomic levels as well as to accommodate the different geometries have had a reasonable progress [1]. However, the scenario in composites like concrete has not reached the same degree of success. The primary difficulty is in understanding and accommodating the macro defects and conglomerate characteristics of these materials in to the mathematical formulations. The different sizes of aggregates and the corresponding effects of intrusions and defect locations are not well defined even in well designed concretes [2]. While these complexities of the aggregate distribution in a concrete composite is certainly difficult to represent mathematically, the problem is further complicated through the introduction of fibers in the system.

One way is to look and mathematically address the intrusions (both aggregate and fibers if any) and defect system of such a complex material through a statistical distribution of the constituents and defects. Needless to say that the efficacy of these solutions should be ascertained through proper experimental verifications that establish the constants needed to define the behavior. The other method is to simplify the whole system as a pseudo homogeneous material without recognizing these factors, and present simplistic solutions. One important difficulty even in normal concrete is that the actual system is mostly dependent on the total volumes of each size fraction and the distribution of the different aggregates. In view of this, it is proposed to look into these parameters more closely and come up with theoretical formulations that are amenable to solutions with the unknowns being taken care through a few well defined empirical constants that are established through specific experimental investigations required.

1. FRACTURE MECHANICS

Fracture mechanics is the study of the response and failure of structures as a consequence of crack initiation and propagation. The earliest theories on the strength of materials were primarily based on the equilibrium of forces of attraction and repulsion between the atoms, primarily for metals. Later theories were based on an understanding of the lattice structure and the defects in the systems. Griffith is one of the earliest to explain the strength of solids through the inherent flaws which result in stress concentrations leading to stresses that are several folds higher than the applied stress at the crack tip in the 1920s. Griffith formulated the concept that a crack in a component will propagate if the total energy of the system is lowered with crack propagation (valid for brittle materials only). In the 1940s, Irwin extended the theory for ductile materials. This energy needed to create the new crack extensions, over the years, has developed into the science of fracture mechanics.

As an example take the case of a small slit cut in a plate under uniform tension

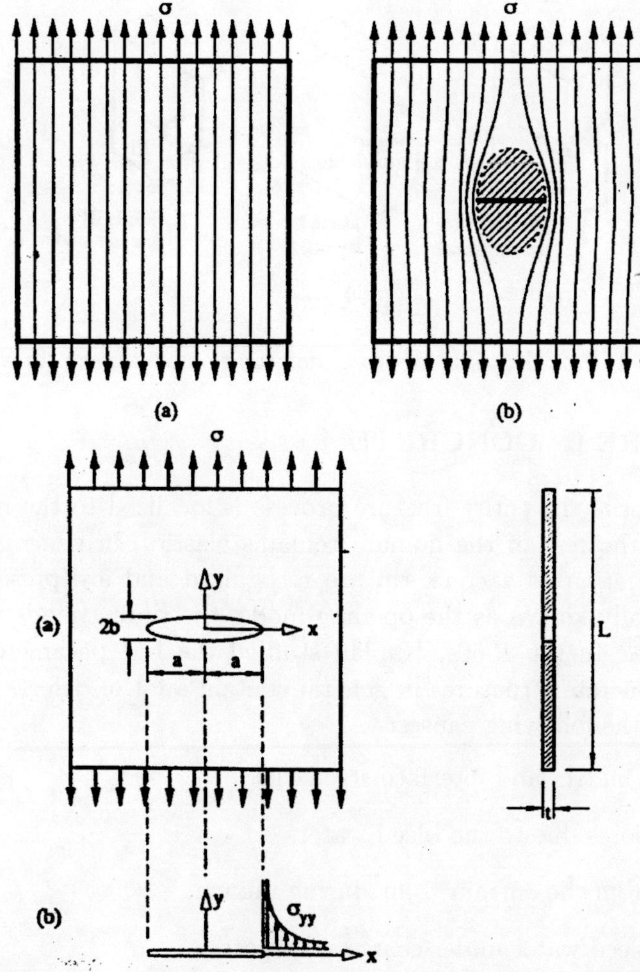

Fig. 1. Principal stress trajectories in a thin sheet and stress concentration effects

(Fig. 1). It is easy to see that the stress flow lines get together near the crack tip with stress relaxation immediately above and below the crack. Theoretically the stress tends to infinity at the tip of the crack. A complete theoretical formulation as suggested by Griffith is not in the scope of the present discussion. Fracture toughness is a measure of the energy required to break and can be obtained by simply calculating the area under the stress deformation relationship. In metals the standard Charpy and Izod tests are employed to measure the fracture toughness directly through the energy absorbed in the breaking process.

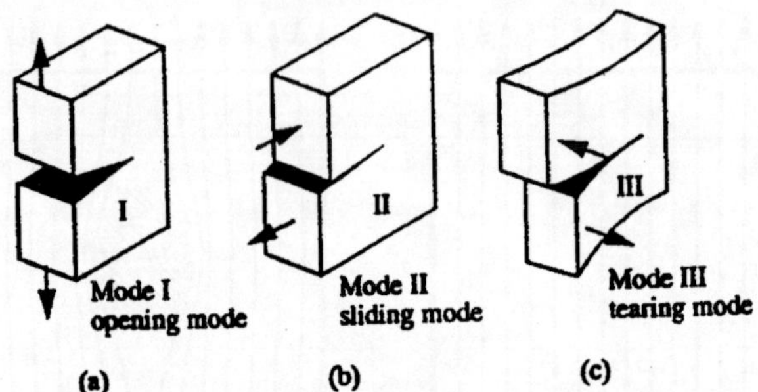

Fig. 2. Possible modes of deformation at a crack tip

2. FRACTURE IN CONCRETE

In a brittle material the entire fracture process is localized to the regions near the crack tip while the rest of the domain remains elastic. In general three possible modes of deformation at a crack tip are recognized and are presented in Fig. 2. These are generally known as the opening mode, the sliding mode and the tearing modes of failure. In the 1960s, Kaplan studied the FM parameters of concrete. Concrete and concrete structures in general contain a lot of defects which could be due to many of the following causes:

- aggregate, matrix and interface strengths,
- capillary pores due to the bleed water,
- air voids from the entrained air during mixing,
- lenses of bleed water under coarse aggregates,
- shrinkage cracks (plastic, drying, thermal etc.),
- structural cracks due to external loading etc.

These micro cracks coalesce during the fracture process leading to the failure ultimately. Also, in quasi brittle and non-homogeneous materials like concretes the following parameters need to be better understood [3] (Fig. 3):

a) aggregate interlock,

b) fracture surface tortuosity, and

c) resultant sliding friction at the fracture surfaces.

Apart from these the specimen size and geometry were also seen to be affecting the fracture characteristics [4].

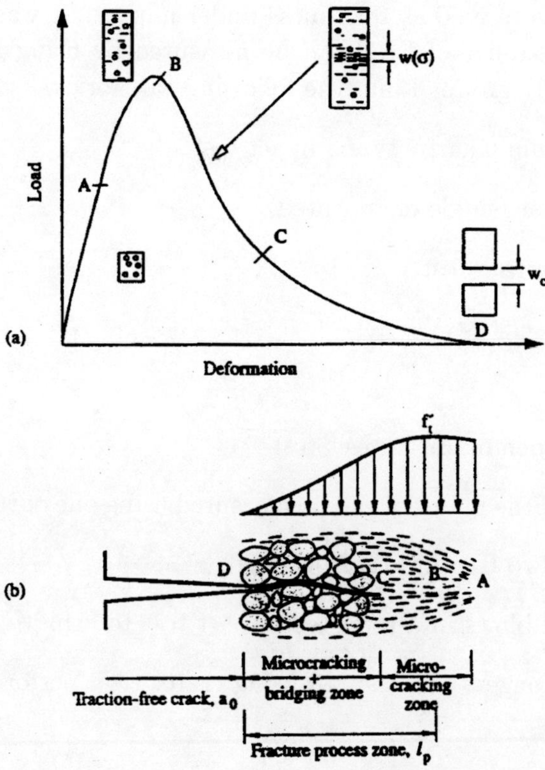

Fig. 3. Load deformation response and fracture process zone in concrete

3. TEST METHODS FOR FRC TOUGHNESS CHARACTERIZATION

There are several test methods employed by the different research workers in the field of concrete fracture mechanics. It is not possible to even remotely discuss and compare the merits and demerits of the various systems adopted. However, broadly it can be seen that there are two schools of thought — one that concentrated on obtaining the fracture energy absorbed by a specimen through essentially a single impact (sometimes even repeated impacts), and two those who try to measure a stress displacement type of measurement (like load–deflection, load–CMOD) and assessing more clearly the ductility in the system rather than just the total energy absorbed in fracture. A lot of work related to this has been in the area of fiber reinforced concretes, which imparts ductility to an otherwise brittle material [5].

Many investigators have shown that addition of fibers greatly increases the energy absorption and cracking resistance of concrete under impact. This energy absorption of FRC is termed as toughness under impact. It was reported by many that the impact resistance of FRC can be measured by using number of different test methods, broadly grouped into the following categories:

- weight pendulum Charpy type impact test,
- drop weight test (single or repeated),
- constant strain rate test,
- projectile impact test,
- explosive test,
- instrumented pendulum impact test.

The resistance of the material can be measured using one of the following criteria,

- energy needed to fracture the specimen,
- the number of blows in a repeated impact test to achieve a specified distress,
- the size of damage — (measured using crater size, perforation or scab or the size etc).

The measured performance can be used to compare different material compositions or to design a structural system that should withstand certain kinds of impact loads. The results from these tests should be interpreted very carefully because they depend on a number of factors like:

- specimen geometry,
- loading configuration,
- loading rate,
- test system compliance,
- prescribed failure criteria, etc.

Alternatively, tests are designed to understand the crack formation and growth (primarily dictated by a predetermined failure path through notches provided in the specimen), which will show the ductility of the system more clearly. A typical figure showing this stress strain characteristics of a high strength fiber reinforced concrete composite in compression is presented in Fig. 4, for a general understanding of the ductility characteristics of a system. The three point and four point bending test

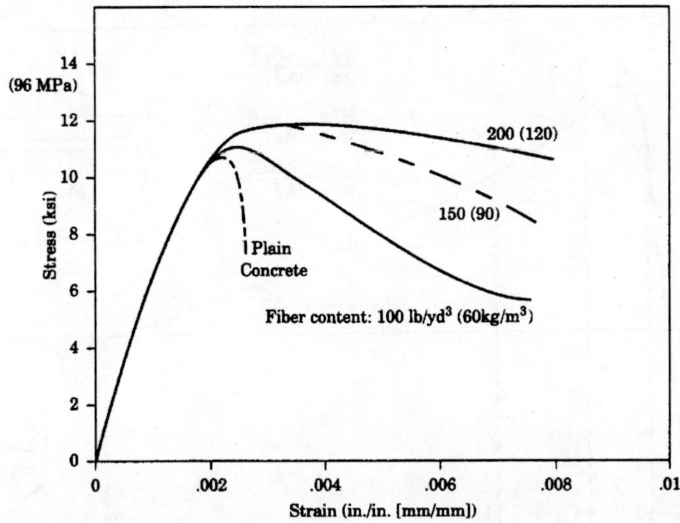

Fig. 4. Stress strain behavior of high strength FRC concretes in compression

setup of a notched beam is generally adopted for obtaining the load–deformation or load–CMOD characteristics of an FRC concrete member. These types of test setups are essentially developed to understand the ductility as well as the residual strength characteristics of fibrous composites. In many of these cases a fracture toughness index (I_t) is determined as defined by the following ratio (Fig. 5)

$$I_t = \frac{\text{(Area under the load deflection curve until the load reaches zero for a fiber composite)}}{\text{(Area under the load deflection curve until the load reaches zero for plain matrix)}}.$$

4. ASTM METHOD

The ASTM C 1018 standard method [6] is based on determining the amount of energy required first to deflect and crack the FRC beam loaded at its third points and then to selected multiples of the first-crack deflection to assess the *toughness indices* (Fig. 5). Toughness indexes I_5, I_{10}, I_{20}, I_{30}, etc., are then calculated by taking the ratios of the energy absorbed to a certain multiple of first-crack deflection and the energy consumed up to the occurrence of first crack. Expressed in general terms:

$$I_N = \frac{\text{(Energy absorbed up to a certain multiple of first crack deflection)}}{\text{(Energy absorbed up to the first crack)}}.$$

The subscripts N in these indexes are based on the elasto–plastic analogy such that, for a perfectly elasto–plastic material, the index I_N would have a value equal

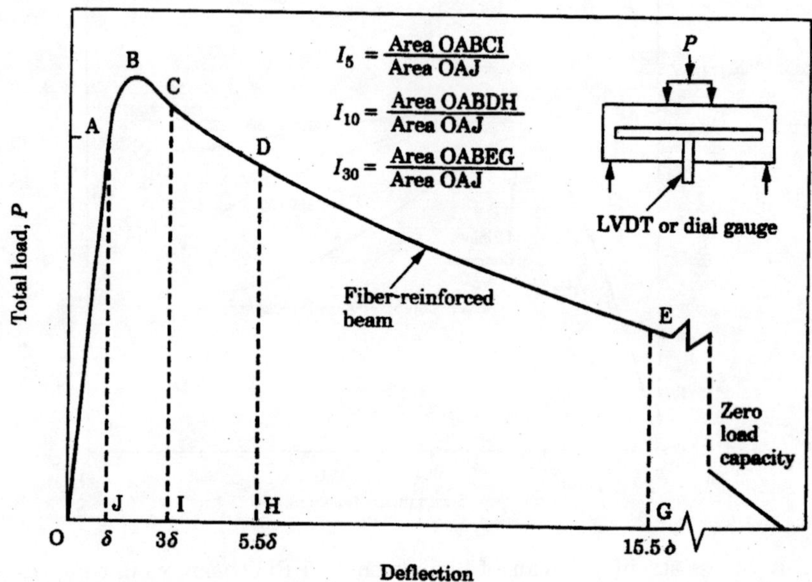

Fig. 5. Toughness indices and residual strength factors

to N. The scheme, thus, compares a given FRC with a conceptual material that behaves in an ideally elasto–plastic manner. Implicitly, the scheme also assumes that plain concrete is ideally brittle and, hence, the various toughness indexes in its case assume a constant value of 1. The strength remaining in the material is characterized by the residual strength factors (R) derived from the toughness indexes (Fig. 5). Expressed in general terms $R_{M,N}$, the residual strength factor between indexes I_M and I_N ($N > M$) is expressed as

$$R_{M,N} = C(I_N - IM),$$

where constant $C = 100/(N - M)$ chosen such that for an ideally elasto-plastic material the residual strength factors assume a value equal to the stress at which the elastic-to-plastic transition takes place. The plain concrete, with its ideally brittle response, therefore, has residual strength factors equal to zero. Both toughness indexes and the residual strength factors provide information on the shape of the load-deflection plot and are presumably independent of the specimen size and other testing variables.

5. JSCE METHOD

In this technique, the area under the load-versus-deflection plot up to a load point deflection of span/150 is obtained [7]. From this measure of flexural toughness, a flexural toughness factor (FT) is calculated. Note that the flexural toughness factor (FT) has the units of stress such that its value indicates, in a way, the post-matrix cracking residual strength of the material when loaded to an arbitrary deflection of span/150. Clearly, the flexural toughness factor is dependent on specimen geometry and other resting variables. The chosen deflection of span/150 for its calculation is purely arbitrary and not based on serviceability considerations.

The development of existing fracture mechanics standards and the introduction of new ones should ensure sufficient flexibility to satisfy the needs of research workers and practicing engineers. At the present time, the existing standards are probably of greater interest to the research worker. This situation is perfectly acceptable since testing for comparison purposes between the different matrices, fibers, influence of fiber volume and length, etc. is carried out in research laboratories. Nevertheless, taking a longer-term view of standards, it is important to ensure that standards are capable of being used by practicing engineers involved in the assessment of existing structures. Structural assessment is often accompanied by taking cores from existing structures and hence some thought should be given to the possible use of cylindrical test specimen geometries.

6. PRESENT PROPOSAL

In a nutshell the above recognizes the need for a fresh look at the fracture toughness of cementitious composites, both from the perspective of more homogeneous materials with microscopic defects like metals and also the defects originating from macro conglomerate characteristics of a cementitious composite.

The primary difficulty is in understanding and accommodating the macro conglomerate characteristics of these materials. The different sizes of aggregates and the corresponding effects of intrusions and defect locations can not be clearly defined even in well designed concretes. While the complex aggregate distribution in a concrete composite is certainly difficult to represent mathematically, the problem is complicated further through the introduction of fibers in the system [8] (Fig. 6).

One method is to look and mathematically address the intrusions and defect system of such a complex material through statistical methods. But the efficacy of these solutions can not be ascertained without a proper experimental verification. The other method is to simplify the whole system as a pseudo homogeneous material without recognizing these factors, and present simplistic solutions. One important difficulty is that the actual system is mostly dependent on the total volumes of each

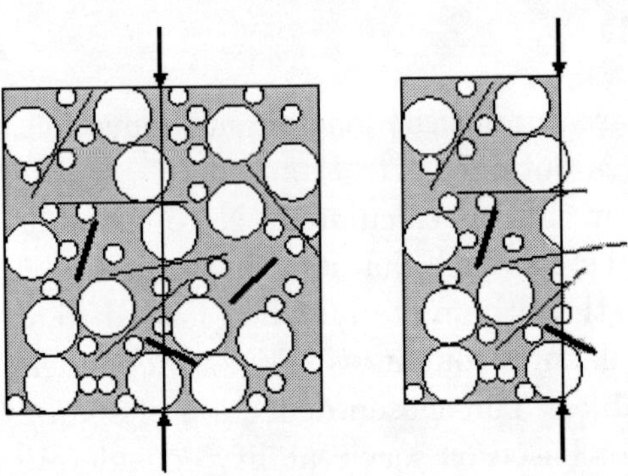

Fig. 6. Effects of aggregate and fiber intrusions on fracture

size fraction and the distribution of the different aggregates. In view of this, it is proposed to look into these parameters more closely and come up with theoretical solutions based on a few well defined parameters through experimental investigations for setting up the empirical constants required.

REFERENCES

1. J.P.B. Leite, V. Slowik, and H. Mihashi, "Computer Simulation of Fracture Processes of Concrete Using Mesolevel Models of Lattice Structures," Cement Concrete Res. **34** (6), 1025–1033 (2004).
2. B. Cotterell, "The Past, Present, and Future of Fracture Mechanics," Engng Fract. Mech. **69** (5), 533–553 (2002).
3. B.L. Karihaloo, A. Carpinteri, and M. Elices, "Fracture Mechanics of Cement Mortar and Plain Concrete," Adv. Cement Based Mater. **1** (2), 92–105 (1993).
4. F.E. Amparanoa, Y. Xib, and Y. Sook Rohb, "Experimental Study on the Effect of Aggregate Content on Fracture Behavior of Concrete," Engng Fract. Mech. **67** (1), 65–84 (2000).
5. P.N. Balaguru and S.P. Shah, *Fiber Reinforced Cement Composites* (McGraw-Hill, New York, 1992).
6. ASTM C 1018, *Standard Test Method for Flexural Toughness and First Crack Strength of Fiber Reinforced Concrete*, ASTM Standards, 4.02 (1990).
7. JSCE-SF4, *Method of Test for Flexural Strength and Flexural Toughness of Fiber Reinforced Concrete*, JSCE (1984).
8. J. Llorca and M. Elices, *Influence of specimen geometry and size on fracture of fiber-reinforced ceramic-matrix composites*, Engng Fract. Mech. **44** (3), 341–358 (1993).
9. Z.P. Bažant, *Concrete Fracture Models: Testing and Practice*, Engng Fract. Mech. **69** (2), 165–205 (2002).

TRANSIENT ANALYSIS OF FRP COMPOSITE STRUCTURES USING HIGHER-ORDER FLAT FACET ELEMENTS ON PARALLEL COMPUTERS

M. Shah[1], R. Khare[2], and T. Kant[3*]

ABSTRACT

In this investigation, finite element formulation has been developed for transient analysis of FRP composite plates. Under transient loading, damages such as delamination, matrix cracking etc. occur in laminated composites and accurate stress analysis is essential from design point of view. The higher-order shear deformation theory has been used in current study for more realistic evaluation of interlaminar shear stresses. The solution of equation of motion is obtained by using Newmark Beta Method. The transient response of present model is validated with available literature for uniformly distributed loading using eight noded element. The transient analysis is compute intensive and use of higher-order theories demand extra computational requirements. In the present work, parallel algorithms have been developed and implemented on cluster of workstations. The elemental stiffness matrix generation and assembly are carried out sequentially. For solution of linear equations ($Ax = B$) many methods are available such as conjugate gradient method and multi-frontal methods, which are effective in parallelization. Some researchers have developed software using these methods and made available in the form of library on workstation as well as on parallel clusters, using MPI and OpenMP standards. In present work finite element algorithm is integrated with developed parallel library: Watson Sparse Matrix Package (WSMP). The parallel algorithms are implemented on distributed and shared memory clusters and typical test cases are solved. A study on the time performance has been carried out and the effectiveness of parallel algorithms to handle large size transient analysis problems is demonstrated.

Key words: laminated composites, higher order theories, transient analysis, Newmark beta method, WSMP, shear stresses

---

[1]Centre for Development of Advanced Computing (C-DAC), India

[2]Shri G.S. Institute of Technology and Science, Indore, Madhya Pradesh — 452003 India

[3]Department of Civil Engineering, Indian Institute of Technology Bombay, Powai — 400076, Mumbai, India

[*]E-mail: *tkant@civil.iitb.ac.in*

INTRODUCTION

Structural elements made of fiber-reinforced composite material are used extensively in aerospace, automobile, civil, marine and other related weight sensitive engineering applications, primarily due to their high strength-to-weight and stiffness-to-weight ratios and also due to their an isotropic material properties which can be tailored through variation of the fiber orientation and stacking sequence. As fiber-reinforced laminates have been playing an important role in high performance structures for the last three decades, the need to have accurate understanding of their structural behavior under design environment is apparent. The increasing application of fiber-reinforced composites in space vehicles (e.g. high performance air craft) has generated considerable interest in the analysis of laminated plates. Most structures, whether they are used in land, sea or air, are subjected to dynamic loads during their operation. Therefore, there exists a need for assessing the transient response of laminated plates. Advanced multilayer composite structures include laminated plates and shells, which are composed of plies, or laminas of fiber reinforced material oriented in an optimum manner so as to achieve the maximum strength–to-weight ratios. Engineers have a wide scope in composite structural design because of the variety of constituent materials, which can be employed, and the numerous options in fiber orientations and the laminas 'arrangements. Initiation and growth of this failure mode is due to interlaminar stresses acting through the thickness. Further, in plane lamina stresses are essential to ensure the strength requirements. Thus, theories, which can predict the complete behaviors, become necessary for better understanding of the complex failure mechanism and strength of multiplayer composite structures. Thus, higher-order flat facet element [1] is used in present FE formulations. Generally the problem of transient analysis using FEM is data intensive, involving finer finite element meshes. The total number of variables in the model can go up to few millions. This is particularly true for 3-D complex geometries. In order to have reasonable turn-around time to solve FEM models, high-performance computing platforms can be effectively used. While traditional vector supercomputer architectures have continued to improve in performance, the growth in the performance of microprocessors has proceeded at a far more rapid rate. The price-performance ratio of vector supercomputers lags far behind than that of today's microprocessor machines. However, individual microprocessors are not able to solve large problem, which leads to cluster super computing. In present work finite element algorithm is integrated with developed parallel library: Watson Sparse Matrix Package (WSMP) [4, 5] to solve equation system $AX = B$.

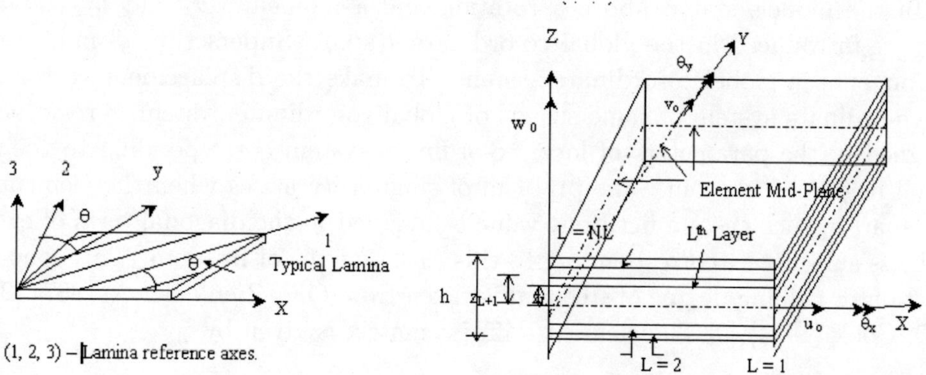

Fig. 1. Element laminate geometry with positive set of lamina/laminate reference axes, displacement components and fibre orientation

1. FORMULATION

1.1. *Finite element analysis*

A flat facet element of a composite laminated plate or shell consisting of laminae with isotropic/orthotropic material properties oriented arbitrarily in space is considered and shown in Fig. 1.

The components of displacements [1] are assumed to be as follows:

$$\begin{aligned} u &= u_0 + z^2 u_0^* + z\theta_y + z^3\theta_y^*, \\ v &= v_0 + z^2 v_0^* - z\theta_x - z^3\theta_x^*, \\ w &= w_0. \end{aligned} \tag{1}$$

The thickness normal deformation effect is not considered here as the linear and quadratic terms in the expansion of transverse displacement w are excluded. The continuum displacement vector at the mid-plane can thus be defined as

$$\mathbf{d} = \{u_0, v_0, w_0, \theta_x, \theta_y, u_0^*, v_0^*, \theta_x^*, \theta_y^*\}^t. \tag{2}$$

In the above relations, the terms u, v and w are the displacements of a generic point (x, y, z) in the laminate domain in the x, y, and z directions, respectively and t is the time. The parameters u_0, v_0 are the in-plane displacements and w_0 is the transverse displacement of a point (x, y) on the laminate middle plane. The functions θ_x, θ_y are the rotations of the normal to the laminate middle plane about x- and y-axes, respectively. The parameters u_0^*, v_0^*, θ_x^* and θ_y^* are the higher-order terms in the Taylor's series expansion. Eight noded isoparametric element is used.

In the models stated above a rotation and a moment (Θ_z^g and M_z^g) about Z axis are introduced in the global co-ordinate system. Superscript 'g' indicates the components in global co-ordinate system. To make the displacement vector in local co-ordinate system of same size as of global co-ordinate system, zero values are assigned in the parameters of local co-ordinate system corresponding to the additional degree of freedom. The problem of singularity arises when the elements are coplanar, in such cases a fictitious value is assigned to the diagonal term of element stiffness sub-matrix corresponding to this (last) degree of freedom. This concept of rotation or drilling degree of freedom is taken from O.C. Zienkiewicz (1987) [2] and R.D. Cook (1994) [3] The element stiffness matrix is given by

$$\mathbf{K}_{ij}^e = \int_{-1}^{+1}\int_{-1}^{+1} \mathbf{B}_i^T \mathbf{D} \mathbf{B}_j |\mathbf{J}|\, d\xi\, d\eta, \tag{3}$$

where, $|\mathbf{J}|$ is the determinant of standard jacobian matrix, $\mathbf{B}$ is the strain displacement matrix and $\mathbf{D}$ is the constitutive matrix. The element mass matrix is given by

$$\mathbf{M} = \int_{\text{Area}} \mathbf{N}^T \bar{\mathbf{m}} \mathbf{N}\, d(\text{Area}), \tag{4}$$

where $\hat{\mathbf{K}}$ is the effective stiffness matrix and $\hat{\mathbf{F}}(t + \Delta t)$ is the effective force vector. The parameter is defined as where, shape function matrix, $\mathbf{N} = [N_1, N_2, \ldots, N_{NN}]$ and

$$\mathbf{m} = \begin{bmatrix} I_1 & 0 & 0 & 0 & 0 & 0 & 0 & 0 & 0 \\ 0 & I_1 & 0 & 0 & 0 & 0 & 0 & 0 & 0 \\ 0 & 0 & I_1 & 0 & 0 & 0 & 0 & 0 & 0 \\ 0 & 0 & 0 & I_2 & 0 & 0 & 0 & 0 & 0 \\ 0 & 0 & 0 & 0 & I_2 & 0 & 0 & 0 & 0 \\ 0 & 0 & 0 & 0 & 0 & I_3 & 0 & 0 & 0 \\ 0 & 0 & 0 & 0 & 0 & 0 & I_3 & 0 & 0 \\ 0 & 0 & 0 & 0 & 0 & 0 & 0 & I_4 & 0 \\ 0 & 0 & 0 & 0 & 0 & 0 & 0 & 0 & I_4 \end{bmatrix} \tag{5}$$

The parameters I_1, I_2 and (I_3, I_4) are normal inertia, rotary inertia and higher-order inertia terms respectively. They are

$$(I_1, I_2, I_3, I_4) = \sum_{L=1}^{n} \int_{h_l}^{h_{L+1}} (1, z^2, z^4, z^6)\rho^L\, dz, \tag{6}$$

where ρ^L is the material density of the Lth layer.

1.2. *Transient dynamic analysis*

By neglecting damping, the governing equation can be written as

$$\mathbf{M}\ddot{\mathbf{d}} + \mathbf{K}\mathbf{d} = \mathbf{q}(t). \tag{7}$$

At time $t + \Delta t$, Eq. (7) is written as

$$\mathbf{M}\ddot{\mathbf{d}}_{t+\Delta t} + \mathbf{K}\mathbf{d}_{t+\Delta t} = \mathbf{q}(t + \Delta t), \tag{8}$$

where $\mathbf{M}$ and $\mathbf{K}$ are the mass and stiffness matrices and $\mathbf{q}$, $\mathbf{d}$, $\ddot{\mathbf{d}}$ are the force, displacement and acceleration vectors respectively. Newmark-β method [2] is employed to obtain the solution to this equation. Accordingly, the velocity and acceleration vectors at $t + \Delta t$ are written as

$$\dot{\mathbf{d}}_{(t+\Delta t)} = \dot{\mathbf{d}}_{(t)}\Delta t\{(1-\gamma)\ddot{\mathbf{d}}_{(t)} + \gamma\ddot{\mathbf{d}}_{(t+\Delta t)}, \tag{9}$$

$$\ddot{\mathbf{d}}_{(t+\Delta t)} = \frac{1}{\beta \Delta t^2}(\mathbf{d}_{(t+\Delta t)} - \mathbf{d}_{(t)}) - \frac{1}{\beta \Delta t}\dot{\mathbf{d}}_{(t)} - \Big(\frac{1}{2\beta} - 1\Big)\ddot{\mathbf{d}}_{(t)}. \tag{10}$$

The parameters β, γ are constants whose values depend on the finite difference scheme used in the calculations. Here, we have used the constant average acceleration method, which is implicit and unconditionally stable. For this method β is 0.25 and γ is 0.5. Substituting Eq. (11) into Eq. (9), we obtain

$$\hat{\mathbf{K}}\mathbf{d}_{(t+\Delta t)} = \hat{\mathbf{F}}(t + \Delta t), \tag{11}$$

where $\hat{\mathbf{K}}$ is the effective stiffness matrix and $\hat{\mathbf{F}}(t + \Delta t)$ is the effective force vector. The parameters are defined as

$$\hat{\mathbf{K}} = \frac{1}{\beta(\Delta t)^2}\mathbf{M} + \mathbf{K}, \tag{12}$$

$$\hat{\mathbf{F}}(t + \Delta t) = \mathbf{h}(t) + \mathbf{q}(t + \Delta t), \tag{13}$$

where $\mathbf{h}(t)$ is given as

$$\mathbf{h}(t) = \mathbf{M}\Big(\Big[\frac{1}{\beta \Delta t^2}\Big]\mathbf{d}_{(t)} + \Big[\frac{1}{\beta \Delta t}\Big]\dot{\mathbf{d}}_{(t)} + \Big[\frac{(1-2\beta)}{2\beta}\Big]\ddot{\mathbf{d}}_{(t)}\Big). \tag{14}$$

The unknown in Eq. (14) are the displacement vector $\mathbf{d}_{(t+\Delta t)}$ and the force vector $\mathbf{q}(t + \Delta t)$, since the displacement, velocity and acceleration at time "t" are known at every point inside the plate. The Hamilton variational principle [2] is used to derive the laminate finite element equations of motion.

2. PARALLEL IMPLEMENTATION

2.1. *PARAM — Machine architecture and environment*

The developed software is ported on PARAM padma, which is parallel distributed-memory MIMD architecture, developed by Centre for Development of Advanced Computing (C-DAC). System is composed of 18 p550 nodes and one p570 node. Each p550 computer node consisted of 8 Power5 'SMP on-a-chip' CPUs at 1.49 GHz. One p570 computer node consists of 16 Power5 'SMP on-a-chip' CPUs at 1.9 GHz. The present studies are done on Gigabit net (10/100MB) interconnection network, which is a low-latency, high-bandwidth point-to-point link. The main parallel programming model supported on PARAM is message passing. In the present studies the Message Passing Interface (MPI) and pthreads library is used. MPI is a communication library for both parallel computers and workstation networks. With MPI the portability of parallel programming with message passing interface has been ensured. It has been developed as a standard for message passing and related operations during 1994 [6]. In the present work Fortran 90 (xlf) compiler is used, on IBM p550 and p570, with optimization flags. The computational structural mechanics group of C-DAC has successfully carried out collaborative R and D projects with Indian Institute of Science, Bangalore and ICAD, Moscow [12, 13].

2.2. *Parallel algorithm*

The parallel finite element program is built using parallel linear simulations equation solver WSMP. A finite element technique adopted here, in general, has three major computational stages. First involves formation of element level matrices, vectors and assembling to global matrixes. Second involves solution of a system of linear simultaneous equations. And finally calculation of stress. Out of above three stages second stage is computational intensive and considered for parallelization in this work. For the current work cluster of symmetric multi-processors (SMPs) are targeted, as this seems to be the architecture of choice for future large machines. Explicit message passing Interface is used for communication between node (single SMP node) and Pthreads is used for communication within shared memory

2.3. *Data distribution*

Data distribution and work distribution is very important in parallel algorithm. As host processor performs analysis and distribution of data, memory required by host processor is more than other processes. In case of large problem it may cause memory and workload imbalance. Also performance of linear simultaneous equation solver is dependent on number of non-zero on each processor. It is observed that with equal

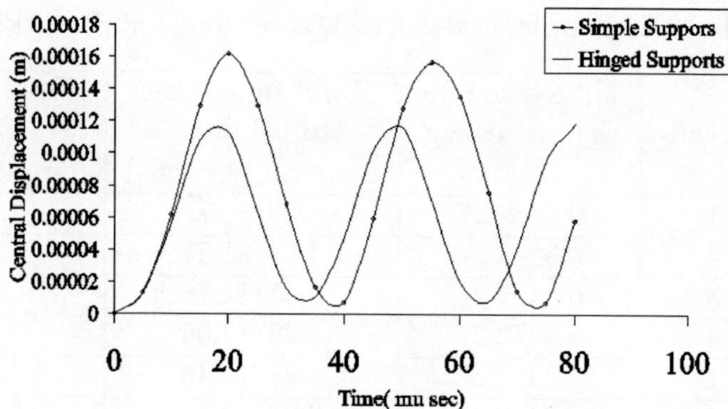

Fig. 2. Transient Response for two boundary conditions

number of non-zero on each process gives better performance for solver. In present study we are using data distribution as non-zero in first row of stiffness matrix on host processes and equal number of remaining non-zero on remaining processes so as to get both advantages of solving large size problem with good performance.

2.4. *Solver library interface*

Many researchers are working on development of parallel linear simultaneous equation solver. Some of them has developed solver in the form of library and made available on workstation as well as parallel clusters using MPI and OpenMP/Pthreads. Some researcher has done comparative study of performance of available solver for different problems. In the present study we have used Watson Sparse matrix Package, WSMP solver [4, 5]. WSMP is high performance, robust and easy to use software package for solving large sparse system of linear equation. It is available in parallel and serial form with symmetric and un-symmetric equation solving capacity. For communication between two processes it uses MPI library and on SMP nodes it can also be used with Pthreads.

3. RESULTS AND DISCUSSIONS

3.1. *Numerical validation*

The following problem has been considered to validate the numerical results. The central displacements as shown in Fig. 2 for both the test cases (Simply and Hinged supported boundary conditions) matches exactly with the results obtained by C. Meimaris and J.D. Day [8].

Table 1. Timing obtained of Case1/Case2 for parallel algorithm

Processes	Threads	Time taken for pre-processing data	Total time taken to factorize equation (seconds)	Total time taken for single iteration (seconds)
2	1	0.34/2.27	7.32/34.77	1.70/5.23
	2	0.34/2.27	4.84/21.21	1.53/5.12
	3	0.34/2.27	3.45/14.78	1.55/5.13
	4	0.34/2.27	3.42/14.56	1.56/5.17
4	1	0.34/2.27	4.28/19.43	1.61/6.71
	2	0.34/2.27	3.27/13.82	1.85/6.80
8	1	0.34/2.27	3.61/15.46	2.35/6.89
	2	0.34/2.27	3.55/15.10	2.33/6.88

A two layered antisymmetric crossply [0/90] square plate of length 250 mm and thickness 50 mm has been considered. A plate is loaded by a uniform pressure step load at time= 0 of magnitude 100 kPa. The material properties are as follows,

$$E_1 = 25 \text{ GPa},\ E_2 = E_3 = 1 \text{ GPa},\ G_{12} = G_{13} = 0.5 \text{ GPa},\ G_{23} = 0.2 \text{ GPa}$$
$$\nu_{12} = \nu_{13} = \nu_{23} = 0.25,\ \rho = 1.0 \text{ kg·m}^{-3},\ a = b = 25 \text{ cm},\ h = 5 \text{ cm}.$$

The time step size was taken as 5 μseconds, and the study was done for 100 μseconds.

3.2. *Performance of parallel algorithm*

In this section, we present test case for evaluation and benchmarking of developed algorithm. The above test case is used with 60 × 60 element mesh, resulting 11041 nodes, 3600 elements (for 100 time steps) and Case2 for 100 × 100 element mesh, resulting 30401 nodes, 10000 elements (for 100 time steps) subjected to simply supported boundary condition using eight noded isoparametric element. Total number of degree of freedom used for Case1 and Case2 are 110410 and 304010 respectively.

It is observed from above Table 1. that, with increasing the number of processors the factorization time is reducing while time for single iteration is near about constant. Also with increasing number of thread per process, factorization time is reducing and time for single iteration is near about constant. With increasing the problem size, the more scalability can be observed. The combined approach of distributed memory and shared memory parallelization is found beneficial. Inter process communication poses limitations for getting scalability on parallel machines. It is important for structural designers to carry out simulation studies for large FE meshes in reasonable amount of time. The implementation of parallel code, as observed from results above, can fulfil this objective.

CONCLUSIONS

A program, for finite element transient analysis of composites structures has been developed using higher order shear deformation theories and implemented on PARAM padma (high performance parallel cluster of SMPs). The present parallel algorithm uses library solver based on multi frontal solvers. A test case with finer grids is used to test performance on parallel computers. The applicability and scalability of the algorithms is established here.

REFERENCES

1. T. Kant and R.K. Khare, "A Higher-Order Facet Quadrilateral Composite Shell Element," Int. J. Num. Meth. Engng **40**, 4477–4499 (1997).
2. O.C. Zienkiewicz and R.L. Taylor, *The Finite Element Method*, 4th ed., Vol 2: *Solid and Fluid Mechanics,Dynamic and Non-Linearity* (McGraw-Hill, New Delhi, 1987).
3. R.D. Cook, "Four Nodded 'Flat' Shell Element: Drilling Degrees of Freedom, Membrance-Bending Coupling, Warped Geometry, and Behaviour," Comput. Struct. **50**, 549–555 (1994).
4. A. Gupta, *WSMP: Watson Sparse Matrix Package Part I — Direct Solution of Symmetric Sparse Systems*, IBM Research Report (IBM T.J. Watson Research Center, New York, 2000).
5. A. Gupta, *WSMP: Watson Sparse Matrix Package Part II — Direct Solution of Symmetric Sparse Systems*, IBM Research Report (IBM T.J. Watson Research Center, New York, 2000).
6. W. Groupp, E. Lusk, and A. Skjellus, *Using MPI Portable Parallel Programming with Message Passing Interface*, (The MIT Press, Cambridge, Massachusetts, London, 1994).
7. R. Chandra and L. Dagum, *Parallel Programming in OpenMP* (Morgan Kaufma Publishers, SanFrancisco, 2000).
8. C. Meimaris and J.D. Day, "Dynamic Response of Laminated Anisotropic Plates," Comput. Struct. **55**, 269–278 (1995).
9. T. Kant and M.P. Menon, "Higher-Order Theories for Composite and Sandwich Cylindrical Shells with C^0 Finite Elements," Comput. Struct. **33**, 1191–1204 (1989).
10. T. Kant and B.S. Manjunatha, "An Unsymmetric FRC Laminate C^0 Finite Element Model with Twelve Degrees of Freedom per Node," Int. J. Engng Comput. **5** (4), 300–308 (1988).
11. M.S. Shah, A.P. Kumbhar, and B.B. Mahanta, "Parallel Finite Element Stress Analysis of FRP Composites under Low Velocity Impact using Higher-Order Flat Facet

Elements," in *Conceptual Approach To Structural Design. Proc. of 3rd Specialty Conf., Singapore* (2005), pp. 175–181.

12. V.L. Yakushev and M.S. Shah, "Simulation of Non-linear Stability Analysis in Thin Walled Structures on Parallel Computers," Int. J. Comput. Appl. Techn. **24** (4), 218–225 (2005).
13. M.S. Shah and V.L. Yakushev, "Parallel Non-linear Stability Analysis in Thin-Walled Structures," in *Proceedings. Indo-Russian workshop on Problems in Nonlinear Mechanics of Solids with Large Deformation. November 22–24, 2006, IIT Delhi* (IIT Delhi, New Delhi, 2006), pp. 175–181.

THERMOELASTIC DAMPING IN METAL–MATRIX COMPOSITE MATERIAL

B.K. Mishra[1] and I.V. Singh[1*]

ABSTRACT

In the present work, a 2-D finite element model of a particulate composite material is studied. The element size of the order of size of the particulate reinforcement is used. A large domain size is taken so that it contains a large number of particles so that averaging over the volume of the domain becomes meaningful. Corresponding to a given volume fraction, a uniform but random distribution of elements with the thermoelastic properties of the reinforcement material is created. Material properties of the matrix material are assigned to the remaining elements. The domain is subjected to a time harmonic uniform state of stress. Stress and strain distribution in the entire domain is obtained. Next, the one way coupled heat conduction equation is solved to obtain the temperature distribution in the domain due to thermoelastic effect. The temperature distribution so obtained is used to find energy dissipation following an entropic approach. The ratio of the energy dissipated to the elastic strain energy stored gives the volume averaged specific damping capacity of the composite. The volume averaged specific damping capacity of an Al–SiC composite has been found for different volume fraction of the reinforcement.

Key words: metal–matrix composites, thermoelastic damping

INTRODUCTION

The damping of an engineering structure is important for vibration control and fatigue endurance. High damping is required to control the amplitude of the resonant vibration response. Energy dissipation in a structure may occur due to inherent material damping or by external means such as friction at the joints. In the conventional materials the inherent material damping is usually small. The composite

[1]Department of Mechanical and Industrial Engineering, Indian Institute of Technology Roorkee, Roorkee — 247667, Uttarakhand, India

[*]E-mail: *indrafme@iitr.ernet.in*, *bhanufme@iitr.ernet.in*

materials offer attractive opportunities for creating materials with high inherent material damping. This coupled with the superior strength to weight ratio promises light and flexible structures with high material damping resulting in enhanced dynamic performance.

Inherent material damping of a material depends on many energy dissipation mechanisms. Thermoelastic damping, first postulated by Zener [1] is one of them. Thermoelastic damping is induced when irreversible heat flow takes place in a cyclic manner within an elastic material. When a non-homogenous time dependent stress field is generated in an elastic material, a non-homogenous temperature increase takes place due to thermoelastic effect. An irreversible heat flow takes place inside the elastic body from the relatively hotter part to the cooler parts. Since the stress field and hence the temperature rise is cyclic, an alternating flow of heat takes place between different parts of the body resulting in energy dissipation.

Thermoelastic damping is normally very small in homogenous materials. Composite materials exhibit higher damping than the conventional materials. Moreover, composites offer an opportunity to design materials with superior properties than the individual components. This is particularly so for the case of inherent material damping. Due to the importance of developing materials with better damping properties without compromising their stiffness properties, thermoelastic damping in metal-matrix composite is being investigated in the present work.

1. THEORETICAL FORMULATION

In the present work, the specific damping capacity of a metal matrix composite is obtained using FEM. A representative volume of a metal matrix composite is generated and discretized such that it contains large number of elements of the reinforcement particles as well as that of the matrix material. A uniform but random distribution of particles corresponding to the volume fraction of the reinforcement material is created using a random number generation scheme. The domain is subjected to an in-plane hydrostatic state of stress varying harmonically with time at a given excitation frequency. It is assumed that the time varying stress and strain fields in the elastic domain can be obtained by superimposing the harmonic term on the static solution. Thus the elastic analysis is quasi-static. The stress field thus obtained is substituted in the one-way coupled heat conduction equation for the same representative volume of the composite. For the heat conduction problem, adiabatic boundary condition is assumed. The solution of the heat conduction equation gives the temperature field which is not in phase with the applied stress field. Due to this phase difference, energy is lost during a complete cycle. The energy dissipation in a cycle is obtained through the entropic approach suggested by Kinra and Milligan [2]. The volume averaged specific damping capacity is obtained by dividing the energy dissipated in a cycle by the maximum stored elastic energy in the elastic body.

The governing equilibrium equations for the displacement field for the elastic analysis have been discretized using FEM. The domain (Ω) is bounded by Γ, which is composed of Γ_t and Γ_u. The equilibrium equations and boundary conditions for the elastic domain are:

$$\nabla \cdot \boldsymbol{\sigma} + \mathbf{b} = \mathbf{0} \quad \text{in } \Omega, \tag{1}$$

$$\boldsymbol{\sigma} \cdot \mathbf{n} = \bar{\mathbf{t}}, \tag{2}$$

$$\mathbf{u} = \bar{\mathbf{u}}, \tag{3}$$

where $\boldsymbol{\sigma}$ is the stress tensor, $\mathbf{b}$ is the body force vector, $\bar{\mathbf{u}}$ is the prescribed displacement vector, $\bar{\mathbf{t}}$ is the traction force and $\mathbf{n}$ is the unit normal to the boundary Γ_t.

Following the usual process of FEM formulation, the following matrix equation is obtained.

$$\mathbf{Kd} = \mathbf{f}, \tag{4}$$

where $\mathbf{d}$ is the vector of nodal unknowns, $\mathbf{K}$ and $\mathbf{f}$ are the global stiffness matrix and external force vector respectively. The external load vector $\mathbf{f}$ may be expressed as $\mathbf{f} = \mathbf{f}_0 \cdot e^{i\omega t}$ and following the quasi-static assumption, the solution may be assumed to be $\mathbf{d} = \mathbf{d}_0 \cdot e^{i\omega t}$. Thus, equation (2.4) is converted to an equivalent static problem governed by

$$\mathbf{Kd}_0 = \mathbf{f}_0. \tag{5}$$

Solving equation (2.5), displacement, strain and stress fields are obtained. It should be noted that inertia dependent terms are not included in equation (2.1). Also the solution for the stress and strain field is obtained assuming isothermal conditions. These assumptions are justified in the light of very small effect of inertia considerations [3] and very small increase in temperature due to thermoelastic effect [2, 4]. The strain field obtained through the analysis of the elastic problem, is used in the linear one-way coupled heat conduction equation

$$T_{,ii} - \frac{C}{k}\frac{\partial T}{\partial t} = \frac{E\alpha T_0}{k(1-2\nu)}\frac{\partial \epsilon_{kk}}{\partial t}, \tag{6}$$

where T is the increase in the temperature of the solid due to thermoelastic effect, ϵ_{kk} is the sum of normal strains, T_0 is the reference temperature of the unstressed elastic body, C, k, α are specific heat per unit volume, thermal conductivity and coefficient of thermal expansion of the elastic body. Since ϵ_{kk} is time harmonic with frequency ω, the steady state temperature excursion of the body will also be time harmonic at the same frequency but may have a phase with the elastic strain components. Thus, assuming $\epsilon_{kk} = \epsilon_{kk0} \cdot e^{i\omega t}$ and $T = \theta \cdot e^{i\omega t}$, equation (2.6) reduces to

$$\theta_{,ii} - i\omega \frac{C}{k}\theta = i\omega \frac{E\alpha T_0}{k(1-2\nu)}\epsilon_{kk0}, \tag{7}$$

ϵ_{kk0} is readily obtained from the quasi-static solution of the elasticity problem as explained earlier. In the next step finite element formulation of the heat conduction problem is carried out following the usual procedure and a system equation is obtained as

$$\mathbf{K}_T\theta = \mathbf{f}_T. \tag{8}$$

An adiabatic boundary condition is assumed at the boundaries and the temperature field is obtained. Since, the $\mathbf{K}_T$ and the $\mathbf{f}_T$ matrices are complex, the amplitude of the temperature excursion turns out to be complex indicating that temperature field is not in phase with the elastic strain field. This is the reason behind thermoelastic damping. The energy dissipated in a cycle is obtained using the entropic approach [2, 4]. The rate of entropy production due to irreversible heat transfer is given by

$$\frac{ds_p}{dt} = \frac{k}{(T_0^2)(T_{,i} \cdot T_{,i})}. \tag{9}$$

The entropy production in one cycle per unit volume is obtained as

$$\Delta s = \int_0^{2\pi/\omega} \frac{ds_p}{dt}\,dt = \frac{\pi k}{\omega T_0}(\theta_{,i}\bar{\theta}_{,i}), \tag{10}$$

where $\bar{\theta}_{,i}$ is the complex conjugate of $\theta_{,i}$. The total energy dissipation in the domain may be obtained by

$$\Delta W = T_0 \int_V \Delta S\, dV. \tag{11}$$

The maximum amount of stored elastic energy W can be found from the elastic stress and strain fields. The volume averaged specific damping capacity of the material ψ is defined as

$$\psi = \frac{\Delta W}{W}. \tag{12}$$

2. RESULTS AND DISCUSSIONS

The numerical simulation is carried out on a representative volume of the size 2 mm×2 mm using a 20 × 20 mesh of uniform size. The element size is of the order of 100 μm. The material properties are assigned to the elements using a uniformly distributed random numbers in the range $(0, 1)$ corresponding to each element. If the random number corresponding to a particular element is less than

Table 1. Thermoelastic properties of Aluminum and SiC [3]

Properties	Material	
	Aluminum	SiC
Modulus of Elasticity, E (GPa)	70	460
Poissons' Ratio, ν	0.33	0.33
Density, ρ (kg/m^3)	2.70×10^3	3.26×10^3
Thermal conductivity, k (J/(s·m·K))	222.0	90.0
Specific Heat per unit Volume, C (J/(m^3·K))	2.43×10^6	4.33×10^6
Coefficient of Thermal Expansion, α (K^{-1})	23.6×10^{-6}	4.30×10^{-6}

the volume fraction V_f of the composite then material properties of the reinforcement is assigned to that element. Otherwise, the properties of the matrix material are assigned to the element. In the numerical simulation, Aluminum matrix with SiC particle reinforcement is considered. Volume fraction of the composite has been varied from 0.05 to 0.35. The circular frequency of excitation ω has been varied from 1 rad/s to 10^4 rad/s. The thermo-mechanical properties of the two materials are given in Table 1.

The variation of volume averaged specific damping capacity ψ with excitation frequency ω for different volume fraction V_f of the composite is shown in Fig. 1. It is evident that, the specific damping capacity at first increases with an increase in frequency. After attaining a peak, it again starts decreasing. This is consistent with the results obtained by other researchers [2, 3, 4]. The reason for this is attributed to the heat conduction mechanism in a thermoelastic solid. At very low frequency, the rate of temperature change is very slow and essentially it results in almost isothermal conditions and hence the heat conduction between different parts of the body is not much. On the contrary, at very high frequencies, heat does not have much time to get conducted and again a low value of thermoelastic damping is observed. For the particle size considered, the thermoelastic damping peak is obtained at around 100 rad/s which agree well with works of Bishop and Kinra [4]. The strong dependence of specific damping capacity near the thermoelastic damping peak frequency on the volume fraction of the composite is worth noting. As the volume fraction of the reinforcement is increased, the specific damping capacity increases. Introduction of particulates in the matrix increases the heterogeneity of the elastic as well as thermal field resulting in greater thermo-mechanical mismatch resulting in higher damping. It is concluded that like other properties of a composite, the thermoelastic damping capacity can also be tailored to suit the requirements by proper selection of matrix and reinforcement material and volume fraction of the composite.

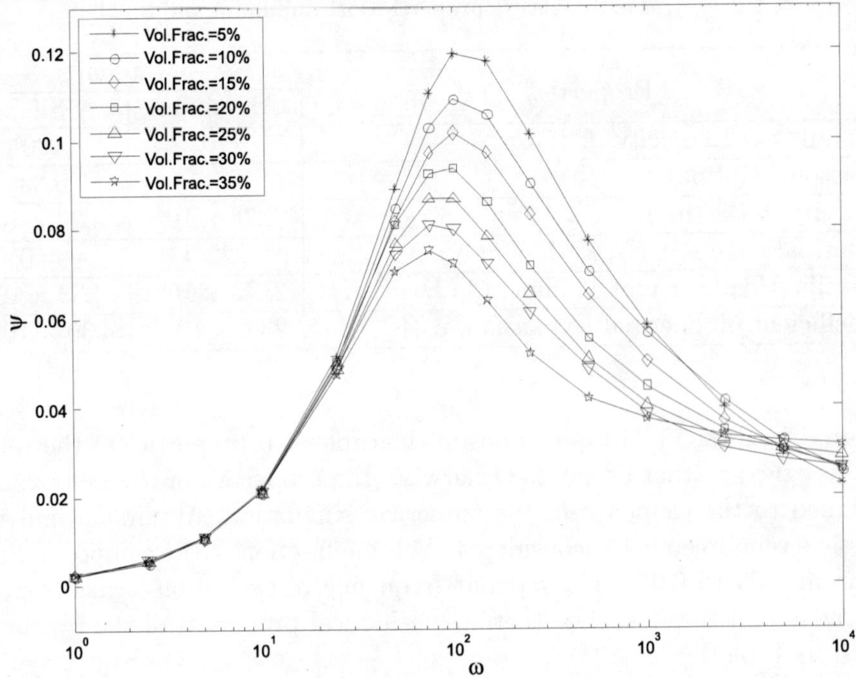

Fig. 1. Variation of volume averaged specific damping capacity ψ with excitation frequency ω for different volume fraction V_f

CONCLUSIONS

In the present study, thermoelastic damping of an Aluminum/SiC composite has been evaluated using FEM. The volume averaged damping specific damping capacity is found to depend on the excitation frequency and the volume fraction of the composite. Thermoelastic damping peak is obtained at around 100 rad/s. The volume averaged damping specific damping capacity increases with an increase in the volume fraction. It is concluded that metal-matrix composite materials with high damping capacity can be developed by proper selection of matrix and reinforcement material and volume fraction of the composite.

REFERENCES

1. C. Zener, "Internal Friction in Solids I. Theory of Internal Friction in Reeds," Phys. Rev. **52** (3), 230–235 (1938).
2. V.K. Kinra and K.B. Milligan, A Second Law Analysis of Thermoelastic Damping," J. Appl. Mech. **61** (1), 71–76 (1994)

3. S.K. Srivastava, B.K. Mishra, and S.C. Jain, "Dynamic Effects in Elastothermodynamic Damping of Metal–Matrix Composites with Spherical Reinforcements," ASME Trans. J. Vibr. Acoust. **121** (4), 476–481 (1999)
4. J.E. Bishop and V.K. Kinra, "Analysis of Elastothermodynamic Damping in Particle-Reinforced Metal–Matrix Composite Materials," Metallurg. Mater. Trans. A **26** (11), 2773 2783 (1995)

NONLINEAR DYNAMICS AND CHAOS CONTROL OF VEHICLE SUSPENSION SYSTEM

R.D. Naik[1] and P.M. Singru[1]

ABSTRACT

This paper deals with dynamical behavior of a nonlinear suspension system. Nonlinear theories are used to find the condition of chaotic vibration in case of quarter car model excited by road profile. We investigated the strange chaotic attractor and its unstable periodic orbits for the system. Results show that there exists a chaos in the system for some parameter range. Finally we have proposed experimental method to investigate chaos in the vehicle suspension system and different methods to control the chaos in

Key words: nonlinear suspension, chaos, quatercar

INTRODUCTION

Recently attention is mainly paid to the research on intelligent vehicle suspension system with hysteretic damping (such as electro-rheological and magneto-rheological fluid damping). Multivalued and non-smooth hysteresis will lead to many complicated behaviors such as bifurcation and chaos [1]. Chaotic is a term assigned to a class of motions in deterministic physical and mathematical systems whose time history has a sensitive dependence on initial conditions. The investigation on chaotic motion in dynamical systems has attracted much attention for many years.

Shaohua et al. [1] investigated a possible chaotic motion in a nonlinear vehicle suspension system which is subjected to multi-frequency excitation from road surface. The Melnikov function is used to derive the critical condition for the chaotic motion, and then it is investigated that the effects of parameters in nonlinear damping on the chaotic field. The path from quasi-periodic to chaotic motion is found via Poincaré map and Lyapunov exponents. Litak et al. [3] studied the Melnikov criterion to examine a global homoclinic bifurcation and transition to chaos in the case

[1]Mechanical Engineering Group, BITS Pilani Goa Campus, Goa — 403726, India

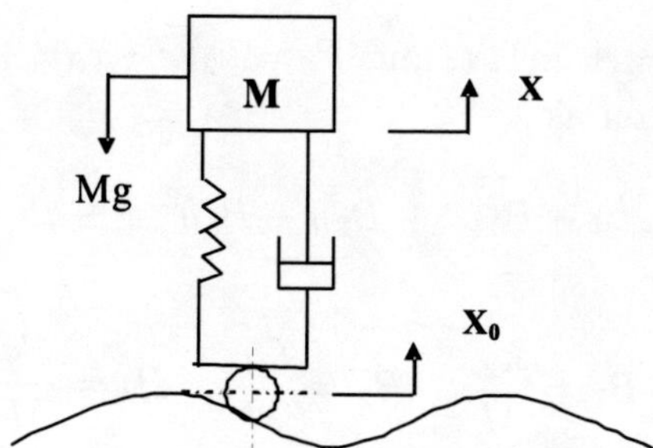

Fig. 1. The quarter-car model with nonlinear damping and stiffness

of a quarter car model excited by a road surface profile consisting of harmonic and noisy components. By analyzing the potential an analytic expression is found for the homoclinic orbit. The road profile excitations including harmonic and random characteristics as well as the damping are treated as perturbations of a Hamiltonian system. The critical Melnikov amplitude of the road surface profile is found, above which the system can vibrate chaotically. This transition is analyzed for different levels of noise and illustrated by numerical simulations. Litak et al. [4] investigated Melnikov criterion to examine a global homoclinic bifurcation and transition to chaos in the case of a quarter car model excited kinematically by the road surface profile. By analyzing the potential an analytic expression is found for the homoclinic orbit. By introducing a harmonic excitation term and damping as perturbations, the critical Melnikov amplitude of the rough surface profile is found, above which the system can vibrate chaotically.

In this paper condition for chaotic motion in hysteretic nonlinear suspension system is investigated. The chaos in the system is determined by using Poincaré map, bifurcation diagram and Lyapunov exponents. The existence of chaos in the system is verified via phase plane plot and time history. Synchronization and control of chaos in nonlinear vehicle suspension system has not been studied before. Finally we have proposed some control schemes to suppress the chaotic motion in the suspension system.

1. THE QUARTER-CAR MODEL

The one degree of freedom quarter-car model with hysteretic nonlinear damping is studied as shown in Fig. 1. The equation of motion of the system is given by

$$M\ddot{x} + k_1(x - x_0) + mg + F_z = 0, \tag{1}$$

where M is the mass of the body, k is the suspension stiffness; F_z is hysteretic nonlinear damping force, x_0 the road excitation and x the body's vertical displacement.

Now let $y = x - x_0$, $x_0 = A\sin\Omega' t$, and $F_z = k_2(x - x_0)^3 + C_1(\dot{x} - \dot{x}_0) + C_2(\dot{x} - \dot{x}_0)^3$. Equation 1 can be written as

$$\ddot{y} + \omega^2 y + B_1 y^3 + B_2\dot{y} + B_3\dot{y}^3 = -gF\sin\Omega' t, \tag{2}$$

where

$$\omega^2 = \frac{k_1}{M}, \quad B_1 = \frac{k_2}{M}, \quad B_2 = \frac{C_1}{M}, \quad B_3 = \frac{C_2}{M}, \quad F = A\Omega'^2.$$

The parameters are defined as $M = 240$kg, $k_1 = 160.000$N/m, $k_2 = -300.000$N/m^3, $C_1 = 250$ N·s/m, $C_2 = -25$ N·s^3/m^3 The corresponding dimensionless equation of motion can be written for a scaled time variable $\tau = \omega t$ as

$$\ddot{y} + y + ky^3 + \alpha\dot{y} + \beta\dot{y}^3 = -g' + A\Omega^2\sin(\Omega t), \tag{3}$$

where

$$k = \frac{B_1}{\omega^2} = \frac{k_2}{k_1}, \quad \alpha = \frac{B_2}{\omega} = \frac{C_1}{\sqrt{k_1 M}}, \quad \beta = B_3\omega = C_2\sqrt{\frac{k_1}{M}}, \quad \text{and} \quad \Omega = \frac{\Omega'}{\omega}.$$

Equation (2) in state space form can be written as

$$\begin{aligned} \dot{z}_1 &= z_2, \\ \dot{z}_2 &= -z_1 - kz_1^3 - \alpha z_2 - \beta z_2^3 - g' + A\Omega^2\sin(\Omega t). \end{aligned} \tag{4}$$

2. NUMERICAL SIMULATION

Equation (3) is integrated numerically using fourth order Runge-Kutta method. The Poincaré map is plotted. Poincaré maps are generated by sampling the system stroboscopically with a period $T = 2\pi/\omega$. The periodic and chaotic motions can be distinguished from the bifurcation diagram, while the quasi-periodic and chaotic motion may be confused. However, they can be found by using Lyapunov exponents. Lyapunov exponents may be used to measure the sensitive dependence of the initial conditions. The phase portraits for different values of road excitation amplitude are shown in Fig. 2–4 and the corresponding Lyapunov exponents are shown in the Table 1.

3. EXPERIMENTAL SETUP

Experimental setup will be developed for nonlinear suspension system to investigate chaos in the vehicle suspension system. A diagram showing the major components of an experimental setup is shown in Fig. 8. The source of vibration is an external

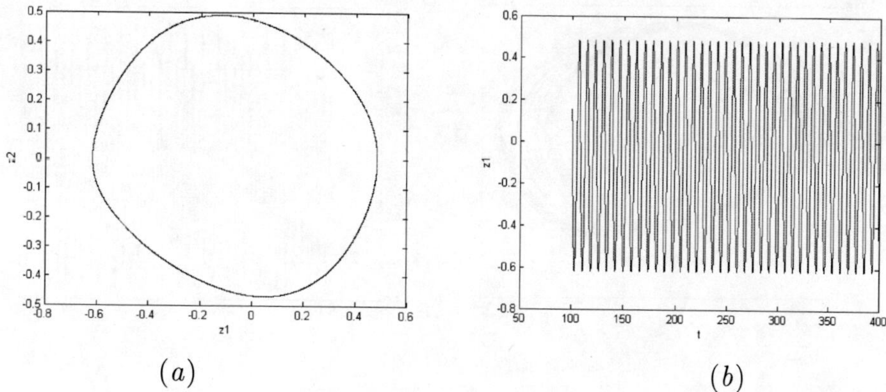

Fig. 2. Phase portraits and time histories when $\Omega = 0.8$ and $A = 0.31$

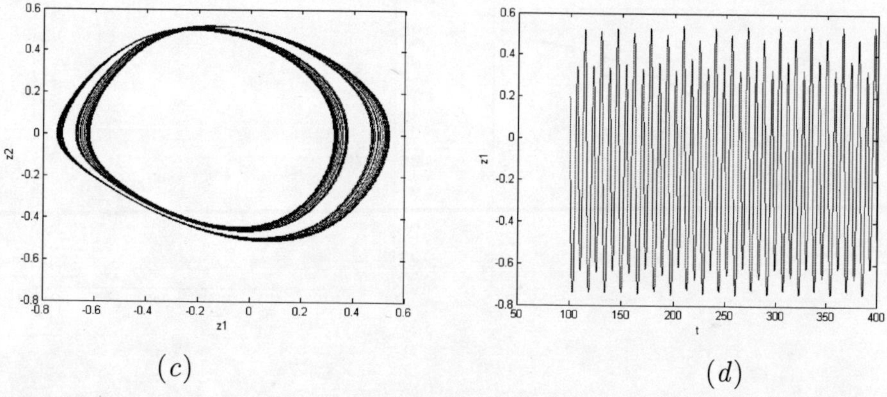

Fig. 3. Phase portraits and time histories when $\Omega = 0.8$ and $A = 0.407$

Table 1

No.	A	Lyapunov exponents	Nature
(a)	0.31	-0.1625	Regular
(b)	0.41	0.0354	Chaotic

energy source. The major elements include transducers to convert physical variables into electronic voltages, a data acquisition and storage system, graphical display such as oscilloscope and data analysis computer. The main objective of the experimental setup will be to develop poincaré maps and to determine criteria for chaos in the

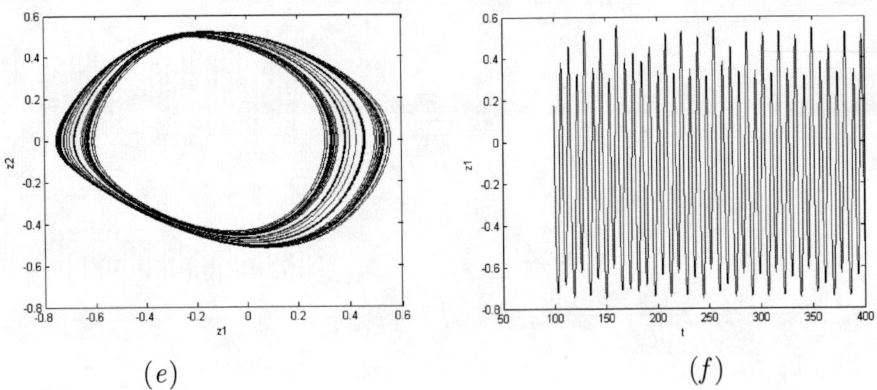

Fig. 4. Phase portraits and time histories when $\Omega = 0.8$ and $A = 0.41$

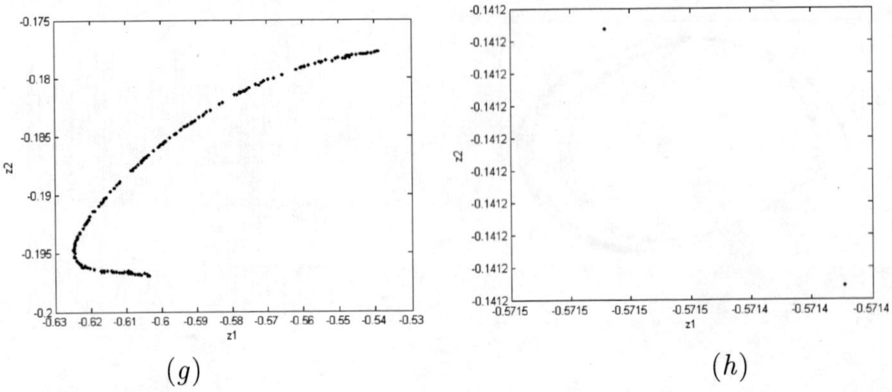

Fig. 5. Poincaré map for $\Omega = 0.8$ and (g) $A = 0.41$, (h) $A = 0.31$

system.

4. PROPOSED METHODOLOGY FOR CONTROL OF CHAOS

Nonlinear dynamic behavior has been discussed in the previous section. It is found that the system exhibits both regular and chaotic motions. Unpredictable behavior due to chaotic motion is undesirable phenomena in the nonlinear system. In order to improve the performance of the system and to avoid chaotic phenomena, we need to convert a chaotic motion to a periodic motion. In this section, we focus on the control of chaotic motion. For this purpose, state feedback control, delayed feedback control and control of chaos by using synchronization have been proposed.

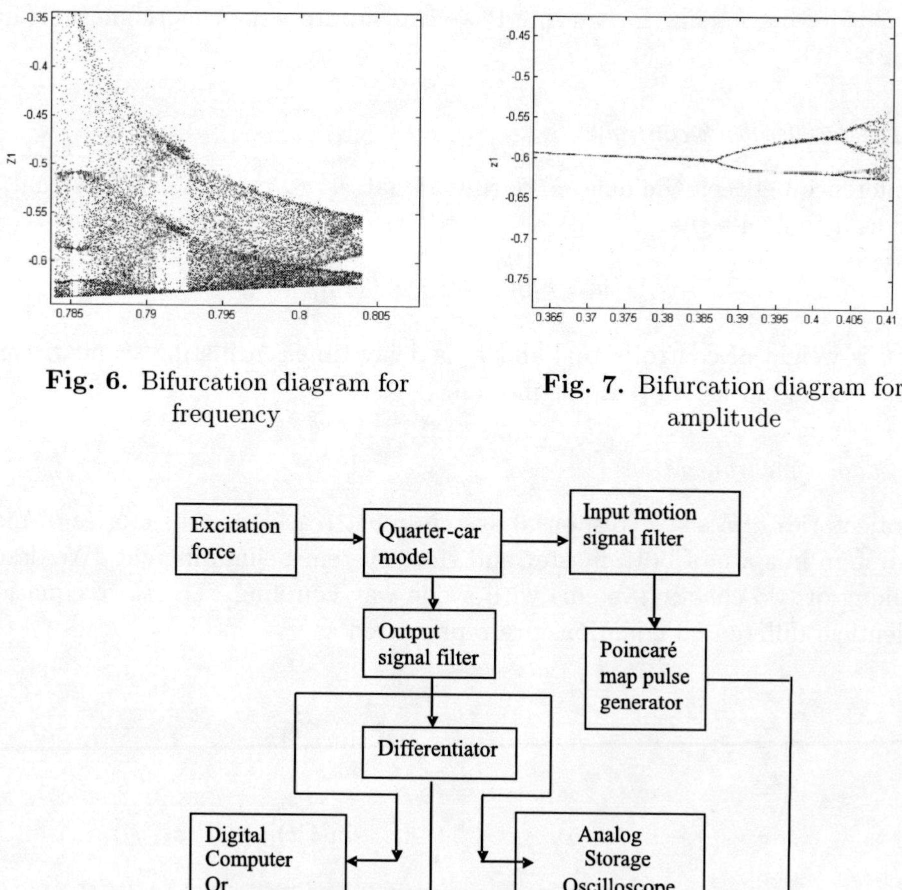

Fig. 6. Bifurcation diagram for frequency

Fig. 7. Bifurcation diagram for amplitude

Fig. 8. Experimental set up to measure the poincaré map of the chaotic system

4.1. *State feedback control*

Consider n-dimensional dynamical system $\dot{x} = f(x,t)$, where $x(t) \in R^n$ is the state vector, and $f = (f_1, \cdots, f_i, f_n)$, where f_i is linear or nonlinear function and f includes at least one nonlinear function. Let $f_k(x,t)$ is the key nonlinear function that leds to chaotic motion in the system. Then only one term of state feedback of an available system variable x_m is added to the equation that includes f_k as follows

$$\dot{x} = f_k(x,t) + Px_m, \quad k,\, m \in \{1,\, 2,\, \ldots,\, n\}, \tag{5}$$

where P is feedback gain. By varying P we can suppress the chaotic motions in the system.

4.2. *Delayed feedback control*

The difference between the delayed output signal $\beta(\tau - \tau_d)$ and output signal $\beta(\tau)$ is used as a control signal

$$u = k[\beta(\tau - \tau_d) - \beta(\tau)], \tag{6}$$

where k is weight of control signal and τ_d is delay time. Adjusting k and τ_d we can convert chaotic motion to periodic motion.

4.3. *Chaos synchronization*

The trajectories of a slave (response) system must track the trajectories of master drive system in spite of both master and slave system being different. We describe the linking of two chaotic systems with a one way coupling. The state equation of two identical differential equations are represented as

$$\begin{aligned} \dot{z}_1 &= z_2, \\ \dot{z}_2 &= -z_1 - kz_1^3 - \alpha z_2 - \beta z_2^3 + A\Omega^2 \sin(\Omega t), \end{aligned} \tag{7}$$

$$\begin{aligned} \dot{y}_1 &= y_2, \\ \dot{y}_2 &= -y_1 - ky_1^3 - \alpha y_2 - \beta y_2^3 + A\Omega^2 \sin(\Omega t) + \in F(z_2, y_2), \end{aligned} \tag{8}$$

where $\in$ is coupling parameter, $F(z_2, y_2)$ is the coupling function. We define an error function

$$E(t) = |z_1 - y_1| + |z_2 - y_2| + |\dot{z}_1 - \dot{y}_1| + |\dot{z}_2 - \dot{y}_2|,$$

when the value of $E(t)$ is less than 10^{-6}, the synchronization of these two identical systems are achieved.

5. NUMERICAL RESULTS

It may be seen from Fig. 3 and Fig. 4 that the phase trajectories are complicated and time history curve is not periodic. From bifurcation diagram Fig. 6 it is clear that period doubling route to chaos occurs in a small range $\Omega \in [0.784, 0.804]$. Fig. 7 shows that the chaos occurs in the system for $\Omega = 0.8$ when $A = 0.407$ to $A = 0.41$. The dominant Lyapunov exponents for $\Omega = 0.8$ and $A = 0.31$ is -0.1625 showing regular dynamic behavior and for $A = 0.41$ it is 0.0354 showing a chaotic behavior.

CONCLUSION

This study addresses dynamical behavior and control of chaotic dynamics of a suspension system with hysteretic nonlinearity. The dynamic behavior is observed using different nonlinear tools. It is found that for certain parameters there exists a chaos in the system. In this paper computational methods and controlling of chaos have been proposed to study the dynamical behaviors of the nonlinear suspension system. In future we have to investigate the chaos experimentally.

REFERENCES

1. L. Shaohua, Y. Shaopu, and G. Wenwu, "Investigation on Chaotic Motion in Hysteretic Non-Linear Suspension System with Multi-Frequency Excitations," Mechanics Research Communications **31** (2), 229–236 (2004).
2. G. Litak, M. Borowiec, M. Friswell, and W. Przystupa, "Chaotic Response of a Quarter-Car Model Forced by a Road Profile with a Stochastic Component," Chaos. Solit. Fractals (2009) (in press).
3. G. Litak, M. Borowiec, M. Friswell, and K. Szabelski, "Chaotic Vibration of a Quarter-Car Model Excited by the Road Surface Profile," Communicat. Nonlin. Sci. Numer. Simulat. **13**, 1373–1383 (2008).
4. G. Litak, M. Borowiec, M. Ali, and L.M. Saha, "Pulsive Feedback Control of a Quarter Car Model Forced by a Road Profile," Chaos. Solit. Fractals **33**, 1672–1676 (2006).
5. R. Wang, J. Deng, and Z. Jing, "Chaos Control in Duffing System," Chaos. Solit. Fractals **27**, 249-257 (2006).
6. H. Chen, "Chaos and Chaos Synchronization of a Symmetric Gyro with Linear-Plus-Cubic Damping," J. Sound Vibr. **255**, 719–740 (2002).
7. Y. Chu, J. Zhang, X. Li, et al., "Chaos and Chaos Synchronization for a Non-Autonomous Rotational Machine Systems," Nonlin. Anal.: Real World Appl. **9**, 1378–1393 (2008).
8. C. Chaohong, X. Zhenyuan, and X. Wenbu, "Converting Chaos into Periodic Motion by State Feedback Control," Automatica **38**, 1927–1933 (2002).
9. F.C. Moon, *Chaotic and Fractal Dynamics: An introduction for Applied Scientist and Engineers* (Wiley-Interscience, New York, 1992).

INDEX